W9-AED-852

# SPACEFLIGHT IN THE ERA OF AERO-SPACE PLANES

Russell J. Hannigan
with Foreword by
David C. Webb

**KRIEGER PUBLISHING COMPANY**

**MALABAR, FLORIDA**
**1994**

CARL A. RUDISILL LIBRARY
LENOIR-RHYNE COLLEGE

TL
795.5
.H36
1994
May 1996

Cover painting "Exploring the Space Frontier" by Robert T. McCall, commissioned and provided courtesy of Honeywell Space Systems Group.

Original Edition 1994

Printed and Published by
**KRIEGER PUBLISHING COMPANY**
**KRIEGER DRIVE**
**MALABAR, FLORIDA 32950**

Copyright © 1994 by Krieger Publishing Company

All rights reserved. No part of this book may be reproduced in any form or by any means, electronic or mechanical, including information storage and retrieval systems without permission in writing from the publisher.
*No liability is assumed with respect to the use of the information contained herein.*
Printed in the United States of America.

FROM A DECLARATION OF PRINCIPLES JOINTLY ADOPTED BY A COMMITTEE OF THE AMERICAN BAR ASSOCIATION AND A COMMITTEE OF PUBLISHERS:

This publication is designed to provide accurate and authoritative information in regard to the subject matter covered. It is sold with the understanding that the publisher is not engaged in rendering legal, accounting, or other professional service. If legal advice or other expert assistance is required, the services of a competent professional person should be sought.

**Library of Congress Cataloging-In-Publication Data**

Hannigan, Russell J. (Russell James), 1963–
Spaceflight in the era of aero-space planes/Russell J. Hannigan
—Original ed.
p. cm.
ISBN 0-89464-046-1 (alk. paper)
1. Reusable space vehicles. 2. Launch vehicles (Astronautics)
I. Title. II. Title: Aero-space planes.
TL795.5.H36 1994
629.47—dc20 92-37687
CIP

10 9 8 7 6 5 4 3 2

# Contents

# Foreword

There were some important works in the sixties and seventies detailing our early efforts to develop a winged machine that would use air-breathing engines to fly to space as a single stage to orbit vehicle. Here is another, written with authority and feeling by aerospace engineer Russell Hannigan, who makes the subject come alive both for those who have come to it recently and those who already have professional expertise in the field.

He first brings us up to date (early 1993) by reviewing the technological innovations and developments that have occurred since the sixties which bring this daunting concept tantalizingly closer to reality. He then analyzes the efforts that different nations have made and are making, to design, build, and test air-breathing hypersonic aero-space planes. Interestingly, he does not restrict himself to winged, air-breathing engined vehicles under the category of aero-space planes, but includes the newly developed, single-stage rocket engined vehicles such as the Delta Clipper X-1 (recently tested by McDonnell Douglas) in this classification. The Delta Clipper may have neither wings nor air-breathing engines, but its lifting body planform is designed to provide a cross-range capability of more than 1,500 miles—greater than that of the winged Space Shuttle.

His analysis of the U.S., German, Japanese, and other efforts to develop air-breathing hypersonic space launch vehicles is masterful and makes fascinating reading. But, he does not stop there. Of equal and perhaps even greater import is his discussion of the economic implications that will stem from the introduction of such vehicles into the world's space launch fleets. His thesis here should be required reading for all those who believe, as some put it, that "going to space is expensive, so it always will be expensive"—that is, the demand for space launches is inelastic.

Russell Hannigan plainly demonstrates the fallacy of this argument, showing that it is expensive only because we continue to build launch vehicles which cannot be tested (in the manner in which aircraft are tested), can only be launched once, and which we cannot get back. He demonstrates that once you change that paradigm by introducing vehicles which are fully reusable, which can be tested many times before they are put into service, and once in service can be operated like airplanes with multiflight, rapid turnaround capabilities, you radically alter the situation. With regularly scheduled launches, and similar reliability and maintenance requirements to those found in airline operations, there will be a drastic reduction in launch costs. This, in turn, will result in lower payload costs: once it is possible to launch replacements immediately and relatively inexpensively, it will be neither necessary nor economic to "over-engineer" payloads in response to the enormous costs and lengthy delays that now occur in placing them in orbit.

In this time of mounting pessimism around the world regarding the future of our space programs, it gives one great hope to have someone of the younger generation with all the requisite technical and political knowledge to finally provide an answer to the many Cassandras of today who continually cry that space transportation is inevitably expensive and must always remain so. We need this kind of reinvigorating critical, yet constructive analysis, rather than just accepting the current thinking as either the first or the last word on how we are going to go to space.

DAVID C. WEBB, Ph.D.

# Preface

I am a product of the Shuttle era. Like many people, I was led to believe that the Shuttle would enable cost-effective, routine space activities—a new era when ordinary men and women could soar into space almost as easily as flying across the Atlantic. At university in England, I remember getting up at 4 a.m. on a cold January morning in 1984 to listen to President Reagan announce America's intention to build a permanently manned space station. So much was promised: the Shuttle, space stations, human solar system exploration, the industrialization of space . . . the future.

By 1985 reality was beginning to dawn. Unabashed, I got my first job in the space industry with the hope that many of the problems being encountered were simply growing pains. Shortly thereafter, January 28, 1986, came to pass, and nothing seems the same since.

In 1992 *eight* Shuttle flights were successfully achieved—one below 1985's all-time high of *nine*—at an average cost in excess of $500 million per flight. The scaled-down Space Station Freedom is still 4 years or so from first element launch, there are no commercial space factories or made in space products, and the Moon seems farther away in the night sky than it ever has. As for Europe, their first venture into crewed spaceflight has been put on indefinite hold until sometime next century. We seem to be going nowhere very quickly, while spending a fortune in the process. As much as I am committed to space, if this is the best we can do—or want to do—then perhaps we shouldn't bother at all.

Even with the current reality, I believe our future in space remains as promising as it ever has. I want nothing more than to see an expanding space infrastructure, permanent occupation of the Moon, and sojourns to Mars. There is a real and continuing need for global communication services, constellations of Earth observation and weather satellites, and scientific and astronomy missions. Eventually, space has the potential to become just another domain for human utilization and industrialization—perhaps allowing the cleaning up of the three terrestrial domains, land, sea, and air, so ravaged over the last century.

However, if we continue to treat space as something sacred, or succumb to the belief that space is expensive *just because it is expensive*, then we're never going to get anywhere. Indeed, we may simply price ourselves out of the space business now that the political foundations on which most space programs were built have all but collapsed. But there really isn't any inherent reason why space activities must always be so expensive, ponderous in their execution, and subject to costly delays. Yet, an examination of our first 35 years in space clearly suggests the contrary. To me, the reason is obvious: space activities are expensive today because we cannot get to and from space when we want, as frequently and regularly as our needs dictate, without fear of our payloads being dumped into the ocean, and above all, at an affordable price.

Transportation is the common denominator of human exploration and trade. But, we'll never realize our potential in space as long as Shuttles cost in excess of $500 million per flight and are incapable of making more than eight, frequently delayed flights per year. Medium capacity expendable rockets, typically costing $100 million per launch, are even more restrictive as they must succeed (or fail catastrophically) and are incapable of returning payloads to Earth. Further, today's space launchers are the only form of transportation requiring a brand new vehicle to be built for every single mission. The profound consequence of which is that it is impossible to perform shakedown testing of the vehicle before it carries an expensive payload—including humans—into orbit.

Aero-space planes—vehicles operated like aircraft that regularly fly to and from orbit—are capable of bringing space transportation more in line with terrestrial transportation. I didn't write this book because I am fascinated with these high-tech machines—even though they are indeed fascinating. At the end of the day, it really doesn't matter if aero-space planes use scramjets or have wings or land vertically, as long as they provide an affordable, user-friendly means of transporting people and payloads to and from space. My interest has always been in what aero-space planes can *do*, rather than what they *are*.

There is a clear need for some equivalent of a "railroad to space." The industrial revolution would never have occurred without affordable, reliable, available, two-way transportation—period. Of course, the railroads themselves did not create the industries that fuelled the industrial revolution. Yet, the industrial revolution would never have occurred without user-friendly transportation. I don't know if aero-space planes will unleash a "space industrial revolution"—our limited space activities to date provide little indication either way. What seems very clear is that there will never be a space industrial revolution without aero-space planes. Short of a revolution, aero-space planes can be justified purely on today's space activities, if only to save money.

Throughout this book I have endeavored to simplify complex technical issues as far as possible while maintaining the essential meaning—there is only one mathematical equation and I've tried to avoid using graphical representations except where absolutely necessary. As a result, the book should be seen as a discussion of what I consider to be the prevalent issues which, I hope, will be of interest to technical, political, and economic analysts and observers alike. This is important because, I believe, one of the most basic problems with space today is that many decisions are made on a misguided appreciation of the issues involved and their ramifications. In this sense, my book tries to provide a strategic perspective of space, exploring the integration between many of the key technical, political, and economic issues driving space activities today.

Despite the length, I feel that I have only skimmed the surface of most subjects, occasionally delving deeper into specific areas of interest. Also, because of the magnitude of the research task, emphasis is placed primarily on U.S. and European space activities, although I've tried to maintain a healthy appreciation of other national space programs. The book is far from definitive, and I certainly would not claim to have addressed every possible scenario, particularly in the areas of the aero-space plane development and possible mission options. Rather, I've attempted to say what might be possible, hopefully providing a starting point from which further discussion can evolve. Aero-space planes remain immature and highly misunderstood systems, making it practically impossible to engage in a fully balanced discussion.

Another important aspect is that this book was written during a period of great upheaval in space programs (and also in the political landscape; as the final draft was nearing completion, the Soviet Union decided to become a Commonwealth of Independent States). However, the underlying messages remain the same. Although Hermes, the Columbus Free-Flyer and other major programs may have been radically scaled back or even cancelled by the time this book reaches print, the rationale used for so many years to justify such programs hasn't changed and continues to underpin most space activities today. Whatever the status of the current major space programs, they nevertheless make excellent case studies.

My objective has never been to lambaste the space industry, but to rationally assess the fundamental problems, and then discuss what I consider are realistic solutions and their implications. I am sensitive to the fact that my conclusions may challenge the conventional "wisdom." Therefore, in putting this book together I have tried to stick to the facts and concentrate on the basics, avoiding, if possible, getting swept into second-order issues. This is particularly the case with commercial space activities where, despite the rhetoric of many commercial space entrepreneurs, I believe we will never see any true growth beyond communications satellites without drastic reductions in launch costs—by as much as a factor of 100—coupled with equally drastic improvement in launch rates.

The development of humanity through the ages has been characterized by our ability to push back the limits of seemingly insurmountable barriers. If as individuals, nations, or as a species we truly intend to explore, exploit, and industrialize the space frontier, then building a "railroad, highway, or airliner" to space is unequivocally the highest possible priority. Cost-effective, user-friendly space transportation is not just another project fighting for funds—*it is a mandatory prerequisite*. I believe this is slowly beginning to be realized in pockets of the space community, but it is amazing how something so obvious can take so long to be understood.

As we commemorate the 500th anniversary of Columbus's voyage to the New World, it is important to remember that the opportunity he created would never have occurred without the technology, heritage, and expertise needed to construct the fully reusable, testable, reliable, available, and affordable ships he first set sail in. The Shuttle and expendable rockets have nothing in common with the *Niña, Pinta*, and *Santa Maria*. I firmly believe that we will only be in a position to truly discover the "New World" of space once we have the functional equivalent of Columbus's ships. This means true space transportation systems; this means *aero-space planes*.

# Acknowledgments

I was warned that authors almost invariably underestimate the magnitude of the task needed to write a book. I am no exception. Fortunately, I have been able to draw upon the resources of many people, without whom this book would have never been completed. In particular, I'd like to thank David C. Webb, a man of the highest standing and whose constant encouragement and support carried me through the frustrations and torment of writing this book. I am indebted to his vision, his strength, and his friendship.

Also, I'd like to thank Courtney Stadd, Bill Haynes, John Cassanto, Ann Deering, Alan Bond, Jerel Whittingham, John Sved, Anita Gale, Mike Dillon, Roy Gibson, John Logsdon, Heinz Pfeffer, Paul Czysz, and Jess Sponable among others for reading and commenting on the various chapters I sent them. I would like to express my gratitude to those individuals listed below for finding time in their busy schedules to see me as well as check my text. However, responsibility for the content of the book remains solely my own.

Thanks go to CREST for supporting the initial phase of my research and to the many people who provided pictures and illustrations.

I would like to express a special thanks to all the people at Krieger Publishing Company for helping make my first publication all the more enjoyable. To Robert Krieger for his offer to take on the task of printing my words, and to Mary Roberts for her total patience, understanding, and quality.

Above all I would like to dedicate this book to my wife, Allison, who sacrificed so much during the course of writing. I simply would never have finished this book without her skills at proofreading and copyediting and, in many cases, for rewriting large parts of the text into English. Indeed, she rivalled me in her effort when we routinely put in 16 to 17 hour days during the last few weeks of writing—a time when the very thought of having a free weekend seemed a distant luxury. I owe her much, and count myself extremely fortunate to have found a partner and friend with whom to share a commitment to spaceflight.

Finally, greatest thanks of all go to my mother, Mollie, for instilling in me the belief that *anything* really is possible if you want it enough.

* * * * * * *

During the course of researching this book, I had the distinct pleasure of meeting and interviewing a number of individuals for whom I have great admiration. It was a very refreshing experience which, more than anything else, has given me hope for the future of space.

**Bruce Abell** Former Senior Research Fellow, Hudson Institute, Washington, D.C.*

**Robert Barthelemy** Former Program Manager, National Aero-Space Plane, Joint Program Office, Wright-Patterson AFB, Ohio.*

**Lee Beach** Former Director, National Aero-Space Plane Office, NASA Langley Research Center, Virginia.*

**Alan Bond** President, Reaction Engines Ltd., England, and inventor of the HOTOL engine, the RB-545.*

**John Cassanto** President, Instrumentation Technology Associates, Pennsylvania, and successful pioneer of commercial microgravity operations.

**Paul Czysz** The Oliver L. Parks Professor of Saint Louis University Parks College and former NASP Principal Scientist at McDonnell Douglas.*

**Ann Deering** Former Assistant Vice President, Johnson & Higgins, New York.

**Paul Donguy** Deputy General Manager, Hyperspace, France.*

**Anita Gale, James Kirkpatrick, Jeffrey Morrow, Rand Simberg** Rockwell International Corporation, Space Transportation Systems Division, California.

**William Gaubatz** Director, Special Projects and SSRT Program Manager, McDonnell Douglas Space Systems Company, California.

**Roy Gibson** Former Director General of ESA and the British National Space Centre.

**James Greene** Republican Subcommittee Staff Director, Subcommittee on Technology and Competitiveness and Committee on Science, Space, and Technology.

**Jacques Hauvette** Director, Hyperspace, France.*

**William Haynes** Freelance Space Systems Analyst in California, and former Skylab crew systems manager.

**Dave Helas, David Salt** Formerly of British Aerospace (Space Systems) Ltd., England.

**Horst Hertrich** Responsible for Sänger activities within the German Federal Ministry for Research and Technology (BMFT).

**Keith Hindley** President, Technology Detail, England, and expert on technology activities in the Commonwealth of Independent States.*

**Maxwell Hunter** President/CEO of Space Guild and originator of the Space Ship Experimental concept that led to the Single Stage To Orbit program.

**Heribert Kuczera** German Hypersonic Technology Program Manager, Deutsche Aerospace/MBB Space Communications and Propulsion Systems Division, Germany.*

**Timothy Kyger** Legislative Assistant for Dana Rohrabacher, U.S. House of Representatives.*

**Lt. Col. Pat Ladner** Former Director, Test Operations of the Strategic Defense Initiative Office at the Pentagon, Washington, D.C., and former manager of the Single-Stage Rocket Technology Program.

**Bill Lawrence** Corporate Manager, Advanced Systems and R&D, General Dynamics Corporation, Virginia.

**Congressman Tom Lewis** Republican Member of Congress for Florida, U.S. House of Representatives.*

**John Logsdon** Director, Space Policy Institute, The George Washington University, Washington, D.C.*

**Gleb Lozino-Losinsky** General Director-Chief Designer, Research/Production Firm "Molnija," Moscow.

**Congressman Dave McCurdy** Democratic Member of Congress for Oklahoma, U.S. House of Representatives.*

**Frank Miles** Writer and Television Producer, former Independent Television News Science Correspondent.*

**Robert Parkinson** Executive, Future Space Infrastructure Department of British Aerospace (Space Systems) Ltd., England, and co-originator of HOTOL.*

**Sir Geoffrey Pattie** Member of Parliament in the United Kingdom, and former Minister for Trade and Industry.*

**Heinz Pfeffer** Head, Future Launcher Office, European Space Agency, Paris.*

**John Pike** Associate Director for Space Policy, Federation of American Scientists, Washington, D.C.

**Mark Pross** Former International Relations Specialist, U.S. General Accounting Office, Washington, D.C.

**Maj. Gen. Milnor Roberts** President, The High Frontier, Virginia.*

**Congressman Dana Rohrabacher** Republican Member of Congress for California, U.S. House of Representatives.*

**Johannes Spies** Deutsche Aerospace/MBB-ERNO, Germany, and originator of the LART concept.

**Major Jess Sponable** Test Operations of the Strategic Defense Initiative Office at the Pentagon, Washington, D.C., and former program manager of the Single-Stage Rocket Technology Program.

**Courtney Stadd** Former Senior Director-Commercial Space, U.S. National Space Council, Washington, D.C.

**Lt. Gen. Thomas Stafford** Vice President, Stafford, Burke and Hecker, Inc., Chairman of the Synthesis Group, and former Gemini, Apollo, and Apollo-Soyuz astronaut.*

**John Sved** Deustche Aerospace/MBB-ERNO, Germany.

**George van Reeth** Former ESA Director for Administration, currently President of the International Space University.*

**Richard Webb** Senior Estimating Specialist, Economics Analysis, Contracts and Estimating, General Dynamics Space Systems Division, San Diego.

**Michael Weeks** Former Deputy Director, National Aero-Space Plane Office, NASA, Washington, D.C.*

**Robert Williams** Originator of NASP (Copper Canyon) and former NASP Program Manager.*

**Tatsuo Yamanaka** Director, Space Technology Research Group, National Aerospace Laboratory, Japan.

**German Zagainov** Director, Central Aerohydrodynamics Institute (TsAGI), Moscow.*

* Interviews recorded on tape.

# Introduction

# Access Is the Key to Space

*Near the horizon, a gleaming silver tower bathed in floodlights, stood the last of the Saturn V's, for almost twenty years a national monument and a place of pilgrimage. Not far away, looming against the sky like a man-made mountain, was the incredible bulk of the Vertical Assembly Building, still the largest single structure on Earth.*

*But these things now belonged to the past, and he was flying towards the future. As they banked, Dr. Floyd could see below him a maze of buildings, then a great airstrip, then a broad, dead-straight scar across the flat Florida landscape—the multiple rails of a giant launching track. At its end, surrounded by vehicles and gantries, a spaceplane lay gleaming in a pool of light, being prepared for its leap to the stars.*

—Arthur C. Clarke, *2001: A Space Odyssey* [1]

This is a book about spaceflight as it could be in the era of "true" space transportation systems—in the era of *aero-space planes*. Aero-space planes are the space transportation equivalent of the clipper ships, the canals, the railroads, the highways, and the airliners. They are fully reusable vehicles that can climb out of the Earth's deep gravitational well and fly into Earth orbit, deliver or recover a payload, then return to the ground where they are turned around like aircraft, loaded with new cargo, and readied for another mission weeks or days later.

Fundamentally, aero-space planes are about making space easier and less expensive to access, enabling space programs to be conducted more quickly and at lower cost than today. Aero-space planes are also about enabling new opportunities for the exploration, exploitation, and perhaps, the future industrialization of space. In essence, aero-space planes are about building a "railroad to space"; they are about an era when accessing space becomes truly routine; and they are about bringing space down to Earth.

The emergence of aero-space planes and their impacts on spaceflight are treated in the four major sections of this book. Section 1 *Space Today* provides an overview of the current status of major space programs around the world. Observers of space activities in the late 1980s and early 1990s are all too aware that programs like Space Station Freedom, Hermes, and Columbus have and continue to suffer cost overruns, are constantly being scaled back, and find their schedules stretched by years. Many have asked whether such programs provide value for money and even if they are achievable at all. These issues are discussed in Chapter 1. It is conjectured that the fundamental reason why space programs are so expensive, restrictive, and take so long, is simply because of the current generation of launch systems. Chapter 2 explains the intimate relationship between current space transportation systems and the space missions they launch, with emphasis placed on space station-type programs where regular support launches are required. Chapter 3 then provides a discussion of why partially reusable and expendable launch systems are expensive to use, fly infrequently, fail periodically, and are subject to delays. Understanding the critical issues related to the space transportation infrastructure is key to finding constructive solutions.

Section 2 *Aero-Space Planes in Perspective* provides insight into the world of these true space transportation systems. Aero-space planes hold the potential of providing a radical improvement in accessing space, drastically reducing launch costs while simultaneously increasing launch frequency, reliability, and safety. Such capabilities, as noted in Chapter 4, are inherent characteristics of all transportation systems, whether boats, automobiles, airliners, or aero-space planes. Specifically, these vehicles are fully reusable, capable of undergoing incremental or shakedown testing, and can safely abort a mission and return the vehicle and the user's expensive payload to Earth. Throughout the discussion, the word *potential* stands out. Until the last few years the general consensus has been that such vehicles are not technically feasible, as noted in Chapter 5. Recent developments, primarily in the areas of materials and propulsion systems, have begun to suggest, however, that the first generation of aero-space planes might be—or very soon might be—a realistic proposition. In Chapters 6 to 10, a number of the more notable aero-space plane programs, including the U.S. National Aero-Space Plane program, the British HOTOL and joint British/Russian Interim-HOTOL, the German Sänger concept, and the U.S. Delta Clipper vehicles, are discussed in order to give an insight into the nature of such efforts, the factors shaping their startup and design, and the prospects for their future development.

Section 3 *Implementing Aero-Space Planes* attempts to discuss the range of issues that must be addressed before the first aero-space plane fleet begins to supplant current launch systems. As an introduction, Chapter 11 provides an overview of the lessons of the Space Shuttle from political, technical, and economic perspectives. Specifically, the decision rationale that led to the selection of the Shuttle configuration that flies today is provided, and reasons are outlined as to why the Shuttle failed to meet—and never could have met—its hoped-for operational and economic objectives. Chapter 12 discusses the problems of funding high-technology programs, particularly those that are both very risky and expensive. The difference in the cost between developing the enabling critical technologies and actually constructing vehicles is described. Chapter 13 explores the relative value of and problems with international cooperation in the development of aero-space planes. Chapter 14 includes a broad discussion of many of the legal, insurance, and regulatory issues that come into play with the development of a vehicle that is both an aircraft and spacecraft simultaneously. Environmental factors are also discussed, together with the issues associated with certifying aero-space planes for launch. Chapter 15 discusses the fundamental issue of tolerating failures. Aero-space planes will, like all transportation systems, suffer failures that can lead to an interruption in space access if an alternative launch means is not available. Potential backup launch solutions needed to minimize such interruptions are discussed. Finally, Chapter 16 explores many of the problems associated with determining the capability of an aero-space plane configuration, and the relative merits of various development and operational strategies, including a discussion of private versus government involvement.

Section 4 *New Directions in Spaceflight* essentially returns to the discussion presented in Section 1. How might today's space activities change given the availability of the first aero-space plane fleet, one that is capable of flying 50 missions per year to and from space, at about $10 million per dedicated launch with a payload of around 7–10 tonnes? Answering this question specifically helps keep the examples used within the realm of what is known, thereby avoiding "leap-of-faith" arguments. Comparing the current type of missions launched on expendable rockets and the Shuttle with what could be achieved when launched on aero-space planes is instructive in conveying the importance of cost-effective, user-friendly transportation. Chapter 17, therefore, provides a detailed explanation as to why a drastic reduction in launch costs, coupled with a significant increase in launch opportunities, could significantly reduce a typical space mission cost and schedule. Chapter 18 then discusses the possible impacts of aero-space planes on a range of unmanned space missions, including technology development, science and astronomy, Earth observation, communication satellites, and military operations. Chapter 19 moves on to explore some possible impacts of aero-space planes on human spaceflight. It is conjectured that the cost of building and maintaining space stations and lunar and Mars outposts would fall significantly with the availability of a transportation system that could launch regularly, reliably, and at low cost. More profoundly, aero-space planes may be fundamental to the technical and operational feasibility of such human spaceflight activities. Finally, Chapter 20 attempts to explain why the much-hyped commercialization of space, specifically the manufacturing of products in space, cannot even be considered a *potential* until the advent of aero-space planes. Once aero-space planes are a demonstrated success, the opportunities for exploiting and industrializing space are likely to become clearer. One scenario is described, showing the importance of commercial space to facilitating space activities that are rational, justifiable, and allow self-sustaining growth.

The premise under which this book was written is simple: space is the fourth domain for human activity after the land, sea, and air. Like those domains confined to Earth, space is a resource and can be utilized for scientific, strategic, and commercial applications where the benefits outweigh the cost. In this sense, for the space resource to be exploited to the fullest, the methods of accessing it should be as unconstraining as possible. Specifically, transportation to and from space should be regular, frequent, reliable, safe, and above all, as inexpensive as possible. If technology is not a boundary, getting to and from space should ultimately be as inexpensive and easy as flying on a Boeing 747 or using a freight train. Should the myriad of technical problems be resolved, aero-space planes have the potential to make space just another domain for widespread human activity.

*Access is the key to space.*

1. Clarke, A. C., *2001: A Space Odyssey*, (Arrow Books Limited, London, 1988 and published by Hutchinson, London), pp. 39–40. (Reprinted with permission.)

# SECTION 1

# SPACE TODAY

*As though it were the signal for a concerted demonstration on behalf of ships of the sail, the passing of 1850 marked the beginning of a prolonged outburst of popular enthusiasm which was strikingly reflected in the press. Nothing like it had happened before and, such is the complexity of modern society that in all probability, nothing like it will ever be seen again. News space was far more valuable in 1851 than today, yet for two years and more the great dailies snatched with avidity at every scrap of clipper ship information.*

—Carl Cutler, *The American Clipper Ship*
(Halcyon House, New York, 1930)

*Extreme skeptics argued that railroads were too crude to insure regular service, that the sparks thrown off by belching engines would set fire to building and fields, and that speeds of 20 or 30 miles an hour could be "fatal to wagons, road and loading, as well as human life." More sober critics questioned the ability of railroads to provide low cost transportation, especially for heavy freight.*

—Robert William Fogel, *Railroads and American Economic Growth*
(John Hopkins Press, 1964)

*Transportation is a key element in the social and economic organization of an industrially developed society. As the construction of railroads in America influenced the location of cities, so the development of streetcar lines helped to shape them. Both forms of transportation contributed to employment opportunities and economic growth, indeed to the course of the industrial revolution.*

—Mark Rose, *Interstate: Express Highway Politics*
(The University of Tennessee Press, 1979)

*The greatest tribute that can be paid to commercial aviation today is the fact that it is taken so much for granted . . . Aviation is important because of the ways it influences everyday living. It affects everyday life because it affects social, economic, and political surroundings. It affects modern man because it affects the way all people live and work and because it has influenced the nature of relationships with other nations. It enables people to travel with greater ease and to cover the globe with unique and beneficial contributions of their cultural systems.*

—Robert Kane, Allan Vose, *Air Transportation*
(Kendall/Hunt Publishing Co., 1979)

# Chapter 1

# What Price Space?

## The Triumph of Spaceflight

In August 1989, *Voyager 2* flashed past Triton, the largest moon of the planet Neptune, and glimpsed a world unlike any other visited before. One which was, until this point, just a minuscule smudge of light in the best Earth-bound telescope. Voyager 2 showed Triton to be an extraordinary place, resplendent with hues of blue and pink, with geysers ejecting ice crystals 50 kilometers into the thin upper atmosphere. Eleven years after launch and 4.5 billion kilometers from Earth, a distance so great that even light takes over 4 hours to traverse, Voyager 2 brought its last act to a grand finale. This was a fitting climax to what is regarded by many as the greatest exploration mission in history: the "Grand Tour" of the solar system.

We live in an extraordinary period of human history. A thousand years from now, historians will write about one of the most significant moments in history—when mankind left the cradle of Earth for the first time. Superlatives cannot adequately describe the successes like those of the Voyager missions. Indeed, who would have thought back in 1903 when the Wright Brothers first demonstrated the ability to fly in heavier-than-air machines, that a mere 66 years later the gigantic Saturn 5 rocket (Figure 1.1), weighing some 3,000 tonnes, would send American astronauts to the surface of the Moon?

An overview of the world's space activities shows a series of triumphs in the last 35 years. Of course, this success has not been limited to the United States alone. In a string of firsts, the then Soviet Union launched the first artificial satellite, Sputnik 1; the first man, Yuri Gagarin; the first woman, Valentina Tereshkova; the first picture of the back side of the moon, Luna 3; the first pictures of the surface of another planet, Venera 9 on Venus; and the first permanently crewed space station, Mir. The nations of Europe have combined their individual expertise to create a space program today that can challenge the preeminence of the United States and C.I.S. in many areas. The European Ariane series of launchers dominates the global commercial launch industry, boosting more than 50% of all commercial satellites. Although Japan has not achieved the kind of spectacular feats typifying the space programs of other nations, it has set in place a long-range, step-by-step strategy that may lead to the spreading of the "Japanese Miracle" into space. China has also been a steady player in the space business and has been the only other country allowed by the United States and Europe to launch some foreign commercial satellites. Other countries such as India and Israel have launched their own indigenously built satellites on their own rockets.

A ring of approximately 100 satellites orbits 35,000 kilometers above the Earth once every 24 hours, appearing to be stationary in the sky. These geosynchronously orbiting spacecraft serve a variety of practical applications and

**Figure 1.1** Apollo 11 launches into history (*NASA*).

needs. National and international telecommunications satellites are capable of receiving telephone, television, and other signals, and then retransmitting them over a wide area or to specialized locations on the surface of the Earth. The Early Bird satellite enabled instantaneous transatlantic television to be broadcast for the first time in 1965, and brought into our common culture a new phrase, "Live by satellite." Today, communications satellites, such as those operated by the 120-nation organization Intelsat, carry more than half the total of international telephone traffic, and the high volume of international trade we see today is a direct consequence of the ability of these satellites to provide affordable, reliable and available communications services.

In addition, and perhaps more profoundly, these satellites have made it possible for people all over the world to *see and hear* almost immediately events taking place on the other side of the globe. According to commentator Alastair Cooke referring to the August 1991 coup in the former Soviet Union, [1]

They called the Russian Revolution of 1917 "the ten days that shook the world." This counter revolution took 60 hours to shake the world because the whole world saw every minute of it from the start. In other words it failed, I think, mainly because of something new that the eight plotters had not taken into account—satellite broadcasting. If that sounds glib or vague, let us go back to the beginning and recall, not so much how we felt as it went along, but why we people by the millions on five continents were able to make their feelings known. Monday morning, the pull of a knob, and we're on to the source which for the next three days will be the ever present source of the news for the great, the presidents, and the PMs, and the tyrants, and the not so great in 80-odd countries—Atlanta Georgia's 24 hour cable news service, CNN.

As communications satellites provide constant global communications, weather satellites, also in geosynchronous orbit, provide constant global monitoring of weather patterns. Weather satellites routinely provide advanced warning of impending hurricanes, typhoons, and other violent weather. By giving sufficient time to evacuate exposed regions, weather satellites have saved countless billions of dollars in property damage and many human lives. These satellites also enable a clearer understanding of the dynamics of the Earth's weather that is essential for day-to-day accurate and long-range forecasting. Such systems include the U.S. Goes, the European Meteosat and the former Soviet Meteor series spacecraft.

A number of satellites are able to peer at the land and sea in orbits nearer to Earth that pass over the north and south poles. The U.S. Landsat and the French Spot series spacecraft are able to observe areas through the use of multispectral imaging systems (Figure 1.2). These space-borne platforms are capable of distinguishing clearly between a variety of features including, for example, deforestation and the health of crops. They can also help locate mineral and oil deposits. Other spacecraft, such as the U.S. Tiros and NOAA satellites, are also equipped with instruments capable of studying the atmosphere in selective detail. A highly publicized result of this work was the discovery of an ozone deconcentration, or "hole," over the South Pole.

The military has given space systems top priority because of the vantage point of the "high ground" it provides. The Persian Gulf War in 1991 was a clear demonstration of the military effectiveness that space systems provide in being able to overwhelm an opponent and coordinate strategy. For example, the Defense Satellite Communications System (DSCS) was used to provide tactical communications services to the troops stationed in the Gulf, while the Navstar constellation of spacecraft permitted troops on the ground to locate their position with an error of just a few meters. The secret KH-11 satellites were able to provide high resolution imaging of the location of Iraqi positions and targets, and provided critical information on the progress of the war before the ground battle was joined.

Military space systems have also played a vital role in enabling the United States and former Soviet Union to agree to reduce their nuclear weapons. It is extremely unlikely that either side would have felt comfortable in reducing their stockpiles of weapons without an independent means to verify for themselves that the other was observing the agreements. Military spy satellites have been a vital element in assisting the disarmament process.

In the area of manned spaceflight, the former Soviet Union has demonstrated the ability to support cosmonauts in space for up to a year on the Mir Space Station without significant irreversible physiological effects and, as a result, has laid the groundwork for future long-term voyages to Mars. In the United States, the National Aeronautics and Space Administration (NASA) operates the first partially reusable space transportation system, the Space Shuttle, which has been able to recover stranded commercial communications satellites, such as the Westar and Palapa spacecraft (Figure 1.3); launched the largest optical astronomy telescope ever put in orbit, the Hubble Space Telescope; and placed in orbit a variety of other payloads including the European-built Spacelab manned laboratory for missions lasting about a week. The former Soviets have also developed their version of a Space Shuttle, Buran. Buran is launched on the side of the world's largest booster, Energia. Without Buran, certain configurations of Energia can place as much as 250 tonnes in low Earth orbit.

In the immediate future, the United States, Europe, Japan, and Canada are working to build the International Space Station, probably the largest and most expensive international technological and scientific program ever attempted. First element launch of the Space Station is planned for 1996, with permanent crewed occupation by the turn of the century. The United States plans to build the basic structure, Space Station Freedom, while Europe and Japan hope to

**Figure 1.2** Our fragile oasis in space (*NASA*).

contribute one pressurized module each, the Columbus Attached Module and the Japanese Experiment Module. Canada is developing the telerobotic Servicing Facility attached to the external structure of the station to support construction and maintenance activities. In addition, the European Space Agency (ESA) had hoped to provide a small man-tended space station, the Columbus Free-Flyer, intended to provide near weightlessness, or microgravity, conditions, as it will be free of disturbances from crew activities. The Columbus Free-Flyer was planned to be launched by the new European Ariane 5 booster, and supported once per year by the European manned spacecraft Hermes, also to be launched on top of Ariane 5.

In the area of remote sensing, the United States hopes by the turn of the century to launch various platforms capable of measuring the global environment with unprecedented precision and detail. This program is designated the Earth Observation System (EOS), the largest element of NASA's Mission to Planet Earth Program (MTPE). Like the Space Station program, MTPE is also an international effort and forms one part of the interagency Global Change Program. The European Space Agency is developing a large polar platform, ENVISAT-1 with instruments complementing those of the U.S. EOS platforms.

In the longer term, plans are being drawn for the establishment of permanently occupied bases on the Moon and

**Figure 1.3** Working in space (*NASA*).

crewed sojourns to Mars. U.S. President George Bush said on the twentieth anniversary of the first manned landing of Apollo 11 on the Moon, that the United States should go "back to the Moon, back to the future and, this time, back to stay," followed by "a journey into tomorrow, a journey to another planet, a manned mission to Mars." Later President Bush put a date of 2019 for the first U.S. footprint on the surface of Mars.

## *Success at Any Price?*

It would be difficult to deny that many of the spaceflight achievements over the last three decades have been anything less than staggering. Yet, against what standard can these achievements be measured? How do we determine success? For example, how much do we value pictures of Triton or evidence for the existence of black holes? Can we really quantify the value of the discovery of the ozone hole or the ability to warn coastal areas of impending hurricanes? What is the value of being able to verify adherence to arms treaties or monitoring enemy troop movements by reconnaissance satellites? What investment should be made in understanding human adaptation to weightless conditions or how protein crystals form in the absence of weight? Is endangering human life an acceptable risk to take when launching a space telescope or for the study of plant cell development? What is the value of having a human footprint on Mars, and ultimately perhaps, what value could be put on the discovery of extraterrestrial life?

The achievements of spaceflight have been extraordinary and, as many would argue, at only a small fraction of gross national products. But is it possible to rationally justify spending $500–800 million to launch a shuttle half a dozen times per year, $2 billion for a telescope that cannot be repaired when it needs to be, $9 billion over 12 years for a manned spacecraft that can fly only once or twice per year, $20 billion for a handful of Earth observation satellites, and $40 billion to construct a space station over 10 years?

Until now, these are costs that the major nations have seemingly been willing to pay. Has this investment been worth the price, and, importantly, is it *still* worth the price? Given a rapidly changing world swamped by increasing economic and social concerns, it is perhaps worthwhile to review these investments from the critical perspective of their value for money relative to alternatives, whether in space or on the ground.

In short, what price space?

## *Rationality Check*

In simplistic terms, when we go into space it should be because the benefits outweigh the costs. In this sense, space is just another domain for human activity, as are the land, sea, and air. Benefits are, of course, sometimes difficult to quantify in terms of dollars. Take unmanned exploration of the planets as a first example. Sending automatic probes to the planets is currently the least expensive way of exploring the solar system (Figure 1.4). The total cost of a typical planetary spacecraft is on the order of $1 billion spread out over 10 to 20 years—less than $100 million per year [2]. This is, by any standard, a significant amount of money just to obtain knowledge. Certainly, no one would expect to make any money commercially by dropping a probe into the atmosphere of Jupiter or sampling the surface of Venus, and there are many pressing concerns in society that could easily use an extra $100 million a year.

**Figure 1.4** Saturn from Voyager 1 (*NASA*).

Yet, exploration is, and always has been, a normal aspect of human behavior. It is an important facet that contributes to the completeness of human society. While, at one extreme, it might be difficult to justify spending many *billions* of dollars per year on an extensive solar system exploration program launching several probes every year, the other extreme of not spending *any* money would seem to be a poor reflection on humanity. Therefore, it is considered rational, and beneficial relative to the cost, to have some level of solar system exploration with robotic probes. Although developing technology is an important issue, for the most part solar system exploration is performed purely as science for the sake of expanding knowledge.

Inherently, there is no other choice but to launch spacecraft in order to perform in situ solar system exploration. This is not necessarily true for all space science activities. Consider the Hubble Space Telescope (Figure 1.5). Hubble was heralded as a telescope that would bring about a revolution in astronomy and our understanding of the universe. Orbiting above the blurring effects of the atmosphere, Hubble would be able to see clearer objects deeper in space than ever before possible. The $2 billion spent building Hubble over 13 years seems, by contrast, a small price to pay for an instrument that can see back in time almost all the way to the Big Bang 15 billion years ago.

When Hubble was conceived in the late 1970's, it was certainly a great idea and, at that time, the only means to overcome the blurring and absorptive effects of the atmosphere. Unfortunately, while it was a good idea then, technological advances in Earth-based astronomy have raised questions over the cost-effectiveness of Hubble today, or at least future optical space telescopes. This, it is important to note, has nothing to do with the defective mirror and other problems Hubble has encountered. Even if Hubble worked perfectly, the conclusions would be the same. Bruce Abell explains, "I recently saw somebody talk about the terrific astronomy you could do by having an optical array on the Moon—you could see back into forever—and God knows what the cost would be of actually putting a dozen large aperture optical telescopes on the surface of the Moon, and supporting them. Or you look at something like Hubble, even if the mirror were perfect, you're still talking about a $1.5 billion expense. Then you look at what is being done with the Keck observatory in Hawaii. As they're finishing that telescope, they're talking about building a second one, a twin nearby, so they can do optical interferometry and that will cost another $100 million. They may have invested all of $250 million in easily supportable ground facilities that will last 50 years and will enable them to probably do 80% of the kinds of things they can do with Hubble. That to me is a fairly clear tradeoff. There's no mystery about being able to do it from space. There's no special 'Columbus' kind of thing connected with it. It's strictly science and what you can discover."

These sentiments are reinforced by a report in *Aviation Week & Space Technology* which state, [3]

The ability to link large, ground-based optical telescopes to achieve optical interferometry is a technology that is also expected to enable some ground-based observatories to approach Hubble's spatial resolution capabilities. By linking the data of two or more observatories imaging the same object [simultaneously] the light gathering power of the combination is increased significantly. A second 10-meter (33-ft.) Keck telescope in Hawaii is being built to form such a pair. In Chile the Europeans are building four large interconnected observatories, providing the equivalent of about a 100

**Figure 1.5** Hubble deployment from *Discovery* in April 1990 (*NASA*).

meter (330-ft.) mirror compared with Hubble's much smaller 4-meter (13.2-ft) light-gathering capability, [sic—Hubble's mirror is actually only 2.4 meters in diameter]. With advanced SDI and other military technology enhancing developments by the astronomers themselves, telescopes like the new facilities in Chile are expected to make just as significant discoveries as Hubble, even with the spacecraft repaired, astronomers said.

There are some aspects of Hubble which cannot be performed on the ground, especially observations in the ultraviolet part of the spectrum, as this is radiation that is normally absorbed by the atmosphere. However, can these aspects alone really merit the enormous cost of continuing with Hubble? For example, the cost of the Hubble repair mission *not* including the launch costs is around $200–300 million. For that same amount of money, it would be possible to construct at least one or two additional 10-meter telescopes like the Keck and further improve the capabilities of ground-based observations—perhaps even well beyond those of Hubble. If the launch costs are included, a cost of at least $500 million, then the total amount of money that was to be spent to effect the repair is around $1 billion—enough to pay for a whole family of optical telescopes arranged in an interferometer with an equivalent aperture of hundreds of meters in diameter. Further advances in technology between launch and the repair mission—successfully achieved in December 1993—are likely to make the case for ground-based systems more compelling and much less expensive.

The *only* objective of Hubble is to perform astronomy. Even though Hubble is an outstanding technical achievement—and it has already provided many startling insights in astrophysics [4]—repeated launch delays appear to have allowed Earth-based technology to overtake it. Specifically, it was designed before optical interferometry and atmospheric compensator systems were practical. Viewed through purely rational eyes, today the high cost of Hubble—or more especially future large optical space telescopes—does not seem entirely justifiable as the objective is to perform astronomy. After all, it is the *function* of the telescope that matters, not the telescope itself, nor whether observations are done on the ground, in orbit, or on the Moon.

There are benefits to space-based optical astronomy, *but maybe not at any price*. By contrast, space telescopes like the Gamma Ray Observatory, the Infrared Astronomical Satellite, the Infrared Space Observatory, and the Advanced X-Ray Astrophysical Facilities are considered more rational because the function they perform is impossible from Earth as the atmosphere completely absorbs the frequency of the electromagnetic radiation they observe. Perhaps optical space-based astronomy can only be rational if the cost to build and launch space telescopes can be reduced to levels approaching equivalent ground-based arrays, *and* if they can be supported on a regular and timely basis. (This will be discussed in Chapter 18.)

## There Must Be Space Stations

If the space-faring nations ever intend to do more than launch one-shot automatic probes and satellites, there is a need for crewed space systems. As a consequence, it is rational and, indeed, mandatory to construct space stations. Deciding that a space station is needed is, of course, not the same thing as deciding what type of space station should be built and, more importantly, how much it is going to cost. Given the benefit of the doubt, the debate over Space Station Freedom is not really about whether America should have a space station. It is about whether the American taxpayer will really get an equivalent benefit commensurate with spending $30–50 billion over the next decade on Freedom, and a further $5 billion each year thereafter to operate it (Chapter 19). By the year 2010, Freedom could cost the U.S. taxpayer some $100 billion. Are the benefits worth this expense? Will Freedom give value for money?

In 1984 it was a different story. When President Reagan announced that the United States was going to build a space station, NASA immediately set to work to design its vision of what one should look like. After several iterations, the dual keel configuration was chosen in October 1985 [5]. This "all singing and dancing" facility was to simultaneously be a home for six to eight astronauts in four U.S.-supplied pressurized modules developing new space technologies and performing life science and microgravity research along with commercial development and materials processing work. Experiments were to be located on special attachment points all over the 90 meter by 40 meter rectangular truss structure, performing space science, astronomy, and Earth observations. The station design was also to incorporate various extra features—designated hooks and scars—to allow eventual expansion to a transportation and servicing node for crewed lunar and Martian space vehicles. The price: just $8 billion (1984) for the acquisition cost. This Space Station—it didn't receive the name *Freedom* until mid-1988—was clearly what most people would think a space station should be. This was to be the "space station of all space stations," and far superior to anything the Soviets had. Plus, first element launch was planned for 1992—the 500th anniversary of Columbus's discovery of the New World.

Ten years later, Freedom is still 4 or 5 years from first element launch. Budget realities and concerns about the Shuttle's availability have forced major whittling down exercises at intervals of roughly every 2 years. The Shuttle's unavailability has had particularly profound impacts. In 1984 it was still hoped that it would fly 24 flights per year. Today, the best that can be hoped for is just 8 flights per year

[6]. Freedom has grown up, but it is a pale reflection of what it was supposed to be (Figure 1.6). There is only room for 4–6 astronauts in two shortened U.S. modules. Major research activities will be limited to life science and materials studies. Gone are most of the external payload accommodations, along with the ability to expand the facility into a transportation node. This cut-down Freedom, sometimes referred to by U.S. congressional staffers as Space Station "Fred," is still claimed to be better than what has gone before. According to Senator Barbara Mikulski, [7]

> How does the space station compare in scientific capability? It is 20 times greater than the capability of a space shuttle with an extended duration orbiter. It is two times greater than Skylab ever was, and 54 percent greater than the Soviet Mir space station. That is pretty terrific.

There is no doubt that Freedom would be an impressive facility. However, quantifying Freedom's "greatness" through direct size comparisons can sometimes be misleading. A cost-effective analysis might be equally appropriate. Assuming Senator Mikulski's figures are accurate, greatness might be calculated per billion dollars spent. Freedom will cost about $40 billion to obtain the permanent manned capability, giving a "greatness index" of 2.5 (i.e., 100%/ $40 billion). Skylab cost about $7 billion, giving a greatness index of about 7 (i.e., 50%/$7 billion) The backup Mir, offered to the United States at a price of about $1 billion including launch, would give a greatness index of 65, as Mir is 65% of what station Freedom plans to be (i.e., 65%/$1 billion). In this simple analysis, Skylab was 3 times and Mir would be 25 times "greater" than Freedom *per dollar*. In other words, Skylab was and Mir would be more cost-effective than Freedom.

This very simple type of cost-effectiveness analysis cannot take account of the many semi-intangible justifications

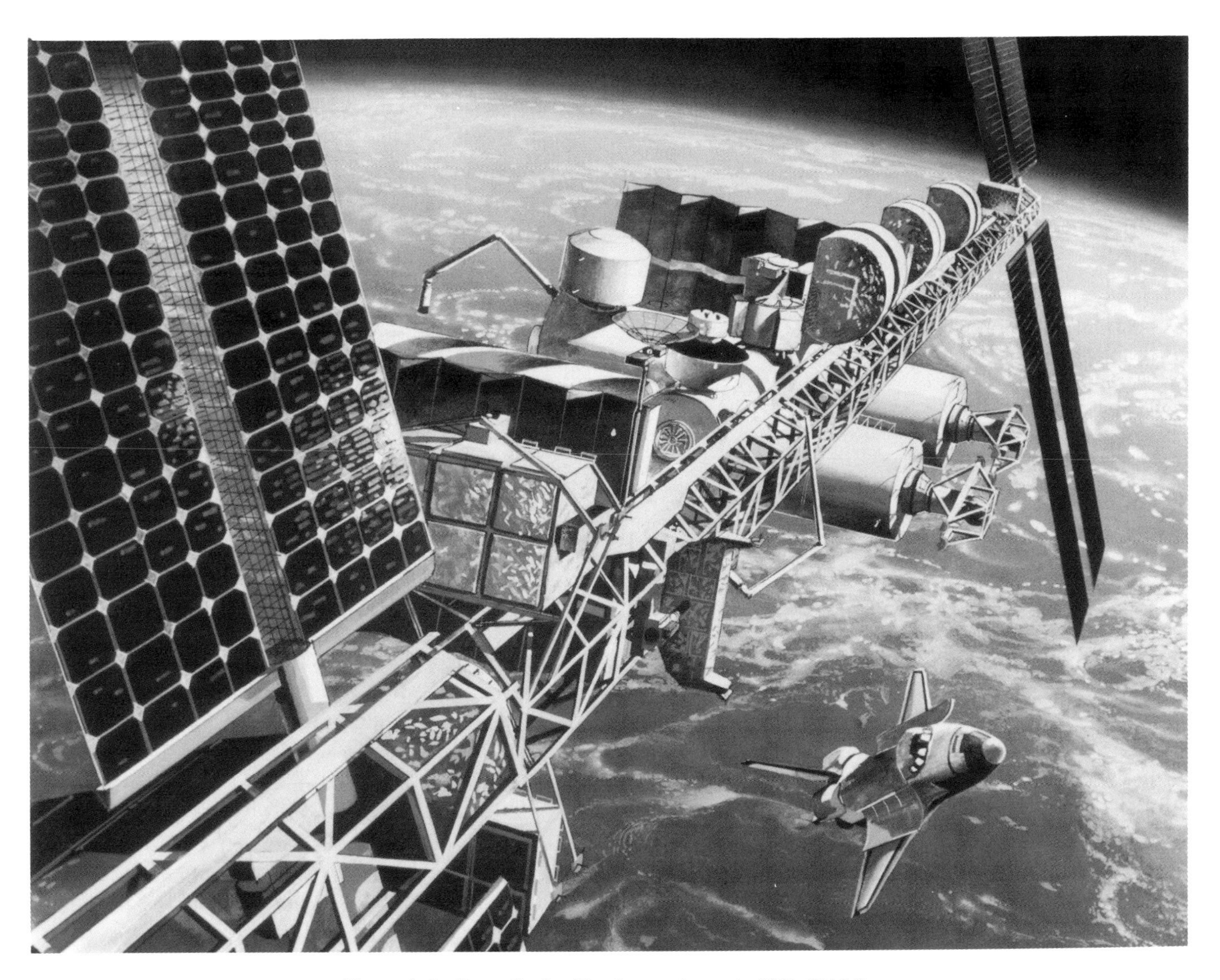

**Figure 1.6** Space Station Freedom as it was in 1991 (*NASA*).

for Space Station Freedom, including international leadership, technology development, jobs, and stimulating education. However, as Freedom is being built to support certain *functional* objectives, a cost-effectiveness technique is considered appropriate. As Freedom is presently defined, its primary function is to do two things: life science and materials research. These were explained by Admiral Richard Truly, during his term as administrator of NASA, [8]

> The space station will help provide answers to many critical problems that plague humans on Earth. For example, in studying the effects of microgravity on the human body and psyche, researchers will have unique opportunities to seek new insights into cardiovascular diseases. Freedom also will have laboratories for pharmaceutical research. The research aims to accelerate the discovery of new pharmaceuticals to treat such diseases as cancer and AIDS. Materials experiments on Freedom will attempt to produce stronger and lighter materials and new processes for recycling on Earth.

Many of these claims have, however, been strongly refuted on practical economic grounds. One perspective in the area of medical research alone is provided by John Pike, [9]

> Claims of the Station's contribution to medical research do not appear to be supported by those researchers and organizations working in these fields. Indeed, an examination of major documents outlining current approaches to these diseases fails to disclose a single reference to space-based research . . . The primary strategy document for the war on cancer is the *Cancer Control Objectives for the Nation: 1985–2000*. This 90-page report outlines the Year 2000 program to reduce American cancer mortality by 50% by the end of the century. The primary focus of this program is prevention through reduced smoking and improved diet, improved treatment through earlier detection, and more effective transfer of research results to clinical practise. There is no mention of space-based biomedical research in this document. Nor is there any mention of space-based research in the 337 pages of the National Cancer Institute's *1992 Budget Estimate*.
>
> References to space-based research are also conspicuous by their absence from major strategy documents on AIDS research. The 142-page publication of the Institute of Medicine, *The AIDS Research Program of the NIH* [National Institutes of Health] makes no mention of space-based research in its recommendations for research priorities. Similarly, recent statements by the National Osteoporosis Foundation make no mention of potential contributions of space-based research for treating this disease.
>
> The fact that Space Station supporters attach much greater priority to space-based biomedical research than those active in the field of treating diseases such as AIDS or osteoporosis would be of less concern were it not for the significant unmet funding needs for research on these diseases. The National Osteoporosis Foundation estimates that $100 million is needed for research on this disease in 1992, in contrast to the $33 million in the current budget request. And the National Organizations Responding to AIDS, which includes over 150 health, labor, religious, professional and advocacy groups, has called for $1 billion in additional funding for AIDS, over the $2 billion in the current budget request.

It might be argued that the reason why the various strategy reports listed above do not include references to space-based research is because it is such a new domain for this type of activity, and of course, Freedom is still a decade or so away. Further, in 1992 NASA and the National Institutes of Health are planning to sign an agreement on closer co-operation in preparation for Space Station Freedom [10]. Yet, the cost differences are so enormous that the benefits claimed for space-based research really must be very airtight indeed to justify a station which is likely to cost $100 billion by 2010. This is evidently far from the case, according to Senator Dale Bumpers, [11]

> Here is what science is: the United States put $1.5 million in NIH funding for research on Lyme's disease; it gets $5 million back [in reduced government costs for health care]. It gave NIH $11 million for breast cancer research and got $170 million per year back. It gave NIH $24 million to do research on kidney stones and gets $75 million a year back. For the space station, in contrast, the country will be starting on a $100 billion to $200 billion project that returns nothing.

Economically, it is difficult to make a justifiable rationale for Freedom in supporting the medical research. For example, if only one week per year on Freedom was devoted to osteoporosis research, the cost would be equivalent to about $95 million, or roughly triple the total annual U.S. national expenditure for this affliction. But economics is not the only problem. The other is time. Osteoporosis, cancers, cardiovascular conditions, and AIDS are problems that afflict people today, now. Space Station Freedom will not be available much before 2000. Diseases like AIDS already afflict millions of people all around the world and researchers cannot afford to wait another 10 years for the Space Station. If space-based research could really play as vital a role in AIDS research as some may think, then every effort should already be underway to do it now and not sometime in the next century.

## *How Important Is the Function?*

As the research functions of Space Station Freedom have changed so substantially, then it is perhaps reasonable to ask, does the current configuration best reflect these research objectives? Ignoring all the other semi-intangible rationale for Freedom for a moment, is there really the need to go to the vast expense and high risk of building a space station requiring some 25 consecutive Shuttle flights over 4 years just to obtain a permanent manned presence in space? Didn't the former Soviets achieve a permanent manned capability after just one launch? (see Chapter 2.)

The needs of life science and materials research strongly conflict. Materials processing work requires very low and sustained microgravity levels to grow delicate crystals, whereas life sciences generally need the presence of astronauts who have a tendency to bump off walls and jog on treadmills. Therefore, to best meet these objectives and to support cost-effective, near-term life science and materials work, it might be a good idea to split the station into two mini-stations, one dedicated to life sciences and the other dedicated to materials research. Such mini-stations would be much cheaper, demand less from the Shuttle, and could be available far sooner than Freedom.

Each station could be identical and about the size and mass of the Columbus Free-Flyer or the Mir Space Station core, thus enabling launch on one Shuttle mission. Two or three times per year, a Shuttle could visit the life sciences station to rotate crews and deliver large logistics modules that would double as extra work space. A Delta 2 or Atlas 2 booster could provide more routine logistics flow in between Shuttle visits. The un-crewed materials station would periodically return from a higher orbit and dock with the life science station in order for the crew to change out experiments, resupply the module with consumables, and perform repairs and maintenance. The life science station could be equipped with a multiple docking adapter, again like Mir, to allow the European and Japanese modules to be attached. The Canadian Servicing Facility would also still be necessary for station maintenance and for docking the Shuttle, as it is for Freedom.

An interim measure might be to launch just the life science station, and instead of launching a separate materials station, Shuttle Orbiters equipped with stretched Spacelab modules and modified to stay aloft for 2 or 3 months, could be used. Also, these long-duration Shuttle flights could plug into a powersat already in Earth orbit, thereby reducing the mass of cryogens that must otherwise be carried to power the Shuttle's fuel cells [12].

Using the Shuttle in this role is at least partially supported by J. R. Thompson, the former NASA deputy administrator, who has stated that, [13]

> We ought to exploit the use of the shuttle once we get it in orbit. It can stay up there a lot longer. Forget the space station; two, four months, we ought to push the capability with the shuttle [alone]. We do not fully utilize Spacelab, . . . for example. The space station can then take advantage of that. It would be a more natural evolution and allow you to accelerate the use of the station, so it answers the critics on using the station earlier for science.

The materials research on the Shuttle would still be disturbed from the presence of crews, although the fact that they are not conducting physical life science research would minimize these disturbances. This work would be left to the life sciences station. In addition, a large unmanned platform, such as the German Astro-SPAS, (see Chapters 17 & 20) could be deployed from the cargo bay, allowed to free-fly for a few days, and then be recovered by the Shuttle, loaded with different experiments, and redeployed. This platform would provide very low levels of microgravity for modest periods of time, as well as have the availability of a crew to perform any necessary repair and maintenance work. These types of materials research Shuttle missions could be flown once or twice per year, several years before the materials station could be ready for launch, allowing substantial experience to be accrued immediately.

Again, it is important to remember that the emphasis of this discussion is the scientific research *function* of space stations. Therefore, if the Shuttle can be readily modified to support materials research, then all that is needed is a very small space station for life sciences. Such a small station already exists: the backup Mir Space Station that has been offered for sale at a price of less than $1 billion, including launch and ground support operations [14]. Theoretically at least, it would be possible to modify the backup Mir with Western computers and other systems, and install special adapters to allow small U.S., European, or Japanese modules to be attached to the facility as required, as well as to allow the Shuttle to dock. While obviously technically challenging, it is not impossible and is significantly more practical and less expensive to use the backup Mir.

The dual Shuttle/Mir architecture would not match the single Freedom in technical sophistication and resources. Each element would probably only be able to provide one-quarter to one-third of average continuous power needs planned for Freedom, although the use of a relatively simple powersat would improve the situation significantly. As a result, neither would be able to support all the research activities of Freedom. However, they could certainly support most of them. More importantly, they could be available many years before Freedom, and perhaps an order of magnitude or more less expensive. Mir would not meet all the U.S. manned space station priorities, but it would certainly be a good interim facility, enabling the United States and other nations to begin research in long-term spaceflight almost immediately and at minimal investment. Such experience might be an essential stepping-stone before more lofty space station goals can be attempted, as will be discussed in Chapters 2 and 19.

Are the few research objectives that cannot be supported by the Shuttle/Mir combination really worth the extra expense and delay of building the $30–50 billion Freedom? Based on the earlier discussion about life sciences research, it would be difficult to answer this question positively.

## *Political Considerations*

Until recently, it has been politically inconceivable to use former Soviet technology in a U.S. space program. The former Soviet Union has been the "enemy." For the United States to buy C.I.S. technology could possibly be seen as an admission of failure on the part of the Freedom program, as it was originally sold on the basis of being superior to Soviet capability. However, buying the backup Mir could be turned into a positive foreign policy move, while simultaneously giving a rational function to the U.S. on-orbit life sciences and materials research program.

The cold war competition with the former Soviets for technological supremacy is clearly inappropriate in today's rapidly changing world. With the former Soviet Union now a Commonwealth of Independent States, the United States is actively seeking ways to improve relations with all of the Republics. Most importantly, the C.I.S. is in desperate need of assistance, and the United States is currently responding, along with the rest of the world, by working out a package of food and cash aid. What better deal than to give cash and/or goods (e.g., food) to the people of the C.I.S., while at the same time use their technology to meet a logical, functional need in the U.S. space program?

It would also avoid the irreversible loss of the enormous wealth of space technology the former Soviet Union has acquired over the last three decades, much of which is readily exploitable by Western space powers. It would be a tragedy if all that were lost, and especially if Western taxpayers eventually had to pay the enormous cost of developing such technologies indigenously (Chapter 9).

## Building Things Just for the Sake of It

The above rational solutions may seem simplistic in the face of the reality that the function of Freedom probably has little to do with its configuration, or indeed, its existence at all. More likely, it is being built because the "thing" itself is worth the expense. In this sense, Freedom has followed the same rationale used to justify the Shuttle, as explained by John Logsdon, "Why was the Shuttle built? It had very little to do with expected uses and a lot to do with maintaining an institution that was created during Apollo to build things. So, in the '69 to '71 period what was going on was a search for something to build that was interesting enough politically so the politicians would provide money to build it. A set of arguments was constructed to convince the politicians that this was worth building. The arguments turned out later to be fallacious, but they were effective in achieving the goal of getting enough money to have a big development project."

Building things just for their own sake is not necessarily wrong. Freedom is, after all, the largest international civilian program ever conducted. The problem is with the impacts on lesser, perhaps more rational, space programs. For example, Freedom and the Shuttle already (1992) consume somewhere between $7–8 billion of NASA's fiscal 1992 authorization, leaving $7–6 billion for everything else, including planetary exploration, space sciences, Earth observation, and technology research and development. Indeed, the saving of Freedom in fiscal 1992 has been to the detriment of many other smaller NASA programs. The Spacehab module received less than half the requested $58 million, or $20 million, the Commercial Experiment Transporter (COMET) capsule received $14 million of the $20 million requested, and the Comet Rendezvous and Asteroid Flyby and Cassini Saturn/Titan missions were reduced in funding from a request of $348 million to just $211 million. The Space Infrared Telescope did not receive any of the requested $15.9 million, as neither did the Orbital Solar Laboratory which requested $11 million or the Lifesat recovery capsule which requested $15 million [15]. By early January 1992, the budgets for Spacehab and COMET were upgraded to $39 million and $18 million, respectively, but only after intense lobbying activities [16].

It could be argued that big ticket items like shuttles and stations are needed to "boot-strap" funding for smaller, less expensive programs that wouldn't exist otherwise. This seems poor rationale. Programs should be justified on their own merits and benefits, and not just because it's "fair." Building major, hugely expensive space programs just for their own sake ensures that other programs will suffer.

## What About Europe?

Back in the early 1980's, France started to recognize that Europe "needed" a manned space launch capability in order to gain independence from the United States and the Soviet Union. In other words, they wanted autonomy. It embarked on the Hermes mini-space shuttle program that was to be launched on a future version of the Ariane series of launcher, Ariane 5 (Figure 1.7). At the beginning, the intent was to build an experimental vehicle with no operational functions, rather like the original version of the US X-20 Dyna Soar. (see Chapter 5.) This experimental vehicle would pave the way for later operational systems should Europe decide to pursue such a goal.

At the same time, Germany and Italy were jointly exploring options to use their Spacelab hardware and experience to build their own man-tended space station. This program is called Columbus. Because of the drive toward manned spaceflight autonomy—made clearer as a result of ESA's Spacelab experience—and the fact that Columbus needed a crew transportation system, an operational version of Hermes seemed an obvious possibility. This worked both ways because if Germany and Italy could provide political support for Hermes, France would be able to provide political support for Columbus—they all got what they wanted. By 1985 the symbiotic union was more or less complete. George van Reeth provides a perspective from ESA, "The operational [aspect of Hermes] wasn't meant from the be-

**Figure 1.7** Human spaceflight autonomy for Europe? (*ESA*).

ginning. It worked out that the French thought they could bring the Germans into the program this way because they could argue that the Germans wanted MTFF, [the Man-Tended Free-Flyer is the original name of the Columbus Free-Flyer] Hermes was essential. And it's true. Another characteristic of the optional program now is that it's much more interwoven than they were, say, eighteen years ago, when we started Ariane and Spacelab. You could have done Spacelab and never have even bothered about the rocket. The two were totally parallel. You could lose one of the programs, and the other one was not influenced. If now you lose Hermes, forget about MTFF, you really don't need it any more."

In 1985, this operationalized Hermes was configured with an unpressurized payload bay, complete with opening doors just like the Shuttle, and could carry a crew of four to six and 4.5 tonnes of payload into a 400 kilometer low inclination orbit, or a crew of two to four and 1.5 to 2.5 tonnes of payload into a 500–800 kilometer Sun synchronous polar orbit. The vehicle could launch and retrieve satellites, service man-tended platforms such as the original version of the Columbus co-orbiting space platform, and dock with Space Station Freedom. Hermes was to have an all-up launch mass of about 15 tonnes and could be launched by the original version of Ariane 5. The price: $2 billion in 1985 dollars, or $3.5 billion in 1991 dollars. First flight was planned for 1996 [17].

In 1987 Hermes became a full ESA program, but concerns about its costs and technical problems limited the funds for just the first 3 years, after which a review was performed. At the ESA ministerial level council meeting at Munich in November 1991, Hermes was not given the requested full approval, but it was put on hold for at least another year. The 1991 Hermes is, as with Space Station Freedom, a pale shadow of the original vision. After many design iterations, Hermes now has grown in both mass and cost and, consequently, in time. Hermes has a total launch mass of 21 tonnes, without a payload, will cost around $9 billion (1991), and won't attempt its first operational manned mission until around 2004. Further, Hermes has been split into two major parts, the reusable winged section and the expendable resource module.

While the mass, cost, and schedule increased, the function has been steadily eroded. With the current version of Ariane 5 undergoing final development, Hermes is so overweight that it can carry a maximum crew of just three. The unpressurized payload bay has been replaced with a small pressurized section, and the maximum payload lift capability is just 1 tonne, [18] although the design specification is 3 tonnes. The only mission of Hermes is to service the Columbus Free-Flyer, although it could probably be used to visit Freedom and Mir. In essence, Hermes still exists in name, but at twice the cost, Europe will get less than half the capability first promised.

In the midst of all this, Hermes has had profound impacts on both Ariane 5 and Columbus. The steady growth of Hermes from 15 to 21 tonnes has necessitated the stretching of Ariane 5, increasing the launcher's liftoff mass by 30%. This stretching may have seriously compromised the ability of Ariane 5 to be operated commercially. (The reasons for this are discussed in detail in Chapter 3.) The 15-tonne Ariane 5 was a commercially optimized configuration, but to suggest that the current bloated version is also commercially optimized might seem unrealistic, although this is yet to be proven. Ariane 5's main market is launching commercial communications satellites, but its performance into low Earth orbit was clearly optimized for Hermes. Interestingly, because first launch of Ariane 5 is planned for 1995, its design had to be frozen at the initial optimistic estimates of the Hermes mass. However, continued growth in the Hermes program now means that if Hermes is to meet its specified payload to orbit objective of 3 tonnes, Ariane 5 will have to be stretched again to lift a Hermes with an all-up mass of 24 tonnes. Indeed, at the council meeting in Munich, ESA requested around $700 million to upgrade the performance of the Vulcain cryogenic main-stage engine, but this proposal was also put on hold [19]. It is interesting to note, however, that the companies building the Ariane 5 main engine have started to spend their own money on developing the Vulcain Mark 2 [20].

The other major impact of Hermes was on the Columbus Free-Flyer program. The inability of the European nations to substantially increase their contributions to ESA made it necessary to stretch out the Hermes and Columbus programs by a few years. In this way the *yearly* program costs are less, but the *total* program costs are more. In the case of Hermes, this funding shortfall has meant that the number of orbiters has been reduced from two to just one. Europe hopes to build a second vehicle 2 or 3 years later, if approved. As a consequence, the goal of performing two servicing missions for the Free-Flyer per year has been reduced to just one, compromizing the utilization of Columbus.

It is also important to note that the Columbus Free-Flyer mission had a profound impact on Hermes. As explained earlier, Hermes was proposed initially as a small experimental vehicle. However, to win over Germany's financial support for the program, Germany insisted that the experimental Hermes be upgraded to an operational vehicle capable of supporting the Columbus Free-Flyer. This necessitated a larger crew size and a greater payload capacity, a rendezvous and docking capability, and the ability to stay in space for several weeks.

Hermes, in its present configuration, is an overweight, semi-reusable vehicle, that can carry just three astronauts. Initially at least, only one vehicle would be built, which would fly just once per year, carrying an internal pressurized payload of just 1 tonne [21]. It has stretched Ariane 5 to the point where it may be so large that Ariane 5 is commercially uneconomical to use, and has delayed and compromised access and use of the Columbus Free-Flyer. Hermes would take another decade to construct, at a cost to the European taxpayer of at least $9 billion up to the end of its first manned flight, and around $500 million [22] per year to operate thereafter. From a functional perspective, is the limited capability really worth the money? Obviously not. In November 1992, Hermes was drastically scaled back to a 3-year technology effort. In 1993 the decision was made to cancel Hermes completely.

## Europe and Manned Spaceflight Autonomy

As with the discussion on Space Station Freedom, Hermes cannot be rationalized based purely on its ability to provide a cost-effective function. Just as there are alternatives for Freedom, cheaper, quicker and less risky alternatives exist for Hermes and Europe. For example, Europe could opt to develop a manned semi-ballistic capsule similar to Apollo. Such a capsule could be launched on Ariane 5 and because it is much lighter than Hermes, it could also carry several tonnes of payload and a larger crew. Indeed, such a capsule could probably be launched on a suitably upgraded version of Ariane 4. A capsule would also be much safer than Hermes because an escape tower could be used to detach the entire capsule at any point during launch. By contrast, Hermes is limited to ejection seats which can only be used up to separation of the Ariane 5 solid rocket booster.

Further, this type of capsule would make an ideal crew escape vehicle for Freedom, as well as offer a backup personnel launch system in the event that the Shuttle is grounded again. Hermes is inadequate for both crew rotation and serving as an escape vehicle for a future European space station. Thus, a joint U.S./European program could be envisaged that would share costs and provide each with a much needed capability. Europe might still want to build the winged Hermes at a later date, but a capsule acts as an interim measure enabling Europe to gain necessary experience in manned spaceflight. Importantly, the objective of obtaining manned spaceflight autonomy would also be reached far sooner. In this light, if autonomy is so important as to justify the $9 billion Hermes, why should Europe wait until 2004 to obtain it? After all, for the next 12 years Europe will have to rely on the United States and C.I.S. for crewed access to space—and probably many years after that because of the operational limitations of Hermes.

Europe is not building a capsule. After all, why copy what the Americans and Russians did 20 years ago? It was argued that Hermes acts as a stimulus for high technology research and development, as, indeed, it already has. Other arguments are put forward that Hermes is a logical first step before a reusable launch system can be built. Both of these arguments, however, appear to be flawed. In the case of stimulating technology, the vast majority of the required advancements must be available long *before* the final design is frozen. Potentially, the program could stop at that point, without losing much more technology (Chapter 12). In the case of future launch systems, a better alternative would be to build the simplest possible unmanned sub-orbital demonstrator to provide all the necessary data. Such a vehicle could, of course, also act as the technology stimulator, if so desired.

Development costs are one thing, but operational costs are perhaps even more significant. This is particularly true for Europe which has not yet undertaken a large-scale operational space program. This is different from the United States which has done Apollo and supported the Shuttle for over a decade, and the Russians who have supported manned space stations since the 1970's. When the initial Ariane-proving flights were completed, the program was handed over to Arianespace and is now commercially self-supporting. The high cost of using Spacelab, coupled with limited launch opportunities has meant that Europe has only participated in a few Spacelab missions to date, although ESA is planning the Spacelab E-1 mission in 1995 as a precursor to Columbus. Hermes and Columbus are, by contrast, programs that cannot be started and stopped like Spacelab. Nor can they be handed over to the private sector like Ariane. They must be continuously supported on an annual basis for many years. As Roy Gibson observes, [23]

With Columbus and Hermes, however, we are dealing with major space infrastructure items which will not only determine the way Europe is to operate in space over at least the next two decades, but will also involve significant and unavoidable expenditures: annually not less than three times the amount ESA now [1989] spends on its scientific programme.

The operational costs are something of increasing concern to ESA, according to George van Reeth, " . . . there are ideas around in the Member States that once the programs [Hermes and Columbus] get operational in 2004 or so, we should pass it to some form of private initiative to run it. I don't believe in it, it doesn't stand to reason. I don't see it as a profitable organization. The problem for us is that operations are going to be expensive. You probably need up to a billion a year to cover the operations. If you do that, that's a substantial part of the budget, as the budget by that time is three and half, four billion, maximum. That's 35% for operations. That's better than NASA's doing, but, nevertheless . . . "

Combined, Hermes and Columbus will cost Europe some $15 billion between now and 2004, and then a further $1–1.5 billion per year for long as these programs are in existence. As Roy Gibson notes,

Are we [in Europe] really sure that the various elements of the ambitious Columbus programme and the value of the hastily conceived Hermes programme (with its attendant anti-commercial effects on the configuration of the Ariane 5 launcher) are really worth this rate of investment?

What price does Europe really put on manned spaceflight autonomy?

## The Institutionalization of Space

Freedom, Space Shuttle, Columbus, Hermes, and other programs all began as rational ideas. It is a logical next step for the U.S. space program to have a space station after the Shuttle, and it is logical that Europe should want an autonomous manned space launch capability. Yet, the logical origins for these programs have been turned on their heads to the point where the systems being developed are only partially capable of meeting the original objectives, and costs have soared beyond original estimates. Why is this consistently the case?

One reason can be put down to the institutional establishments formed in the wake of the space race, and the Apollo program in particular. The justifications for Apollo were primarily political: to beat the Soviets to the Moon and, thereby demonstrate technological superiority. As Lyndon B. Johnson noted in a memo to President Kennedy, [24]

It is man, not merely machines, in space that capture the imagination of the world . . . Dramatic achievements in space therefore symbolize the technological power and organizing capacity of a nation . . . The non-military, non-commercial, non-scientific, but "civilian" projects such as lunar and planetary exploration are, in this sense, part of the battle along the fluid front of the Cold War.

Throughout the 1960's, the United States and, as has become increasingly clear in recent years, the then U.S.S.R. were locked in an intense race to reach the Moon as quickly as possible. The political stakes were perceived as enormous. However, to achieve President Kennedy's objective "of landing a man on the Moon and bringing him safely back to Earth" also carried with it a massive investment in a government and industrial infrastructure costing in the region of $24.5 billion (1960's dollars) or $110 billion (1992).

America won the race to the Moon and in the process also gave birth to a new institution: NASA. At its height, nearly half a million people spread across the United States were employed on the Apollo program. However, after the race was won, the rationale and single-purpose motivation for Apollo vanished. Unfortunately, the United States, and presumably the U.S.S.R. as well, was left with an enormous institution that had no clear-cut purpose or, indeed, function. Although more than two-thirds of the people in industry who served the Apollo Program were unceremoniously discarded at its end, it became politically unacceptable to dismember NASA completely, such was the political clout of the institution that made Apollo a success in the first place.

At the same time, a growing attitude emerged that spaceflight was an activity of direct benefit to the "man in the street." The development of communications and weather satellites, as well as the perception that space programs were at the leading edge of technology, gave the feeling that a space program was a good thing to have, even if no clear rationale for the big ticket budget items existed.

The United States was left with a space agency with an extensive capability, but critically, only enough political support to ensure the continued existence of the institution, leaving little room to embark on Apollo-type programs. Maintaining an institutional structure like NASA, with all its field centers, is expensive. Therefore, this highly institutionalized means of doing business adds an enormous surcharge to any major project, such as building a shuttle or a space station. NASA and the U.S. administration do not lack direction, but know exactly what they want to do and where they want to go. Unfortunately, they just don't have the money to do it *and* keep all the institutions intact. In the opinion of John Logsdon, "Apollo created a particular kind of organization and a particular style of doing business that was wonderfully functional for Apollo but dysfunctional for anything else."

Europe's space institutions were not derived from an Apollo-style program, of course. Rather, they grew out of a political decision to gain autonomy to launch their own satellites and to develop space technology capabilities within Europe. The impetus for this can probably be pinpointed by two particular events. The first is the launching of the Franco-German Symphonie satellites on a U.S. Delta rocket. The second is the invitation to join the U.S. Space Shuttle program with Spacelab. With respect to the former, Joan Johnson-Freese writes, [25]

> Dependence on U.S. launch vehicles had also become a particularly acute issue after the first European (Franco-German) communications satellite, Symphonie, ran into problems with the U.S. government. As an operational communications satellite, Symphonie was viewed as competition to INTELSAT, so the United States mandated that Symphonie could be used for experimental purposes only . . . The French were particularly upset by what they perceived as the United States being able to in some ways dictate the direction of future European programs through launch restrictions. This increased their desire for autonomous European launch opportunities.

Another point to remember is that European space institutions are a fraction of the size of those in the United States. For example, the total ESA *plus* national spending on space (both civil and military) was slightly less than $5 billion in 1991, whereas NASA's budget was about $13.5 billion. In addition, the United States spends around $20 billion on military space applications per year [26]. In total, the United States spends about seven times as much on space as Europe.

## Programs With a Purpose

Apollo, Freedom, Shuttle, Columbus, Hermes, Buran, Energia, and others are not the only large-scale programs ever conducted. The Alaska oil pipeline, the Canadian gas line, the Trans-Siberian railway, the U.S. transcontinental railway, the North Sea oil platforms, the Panama Canal, the Suez Canal, and the Grand Coulee Dam are notable examples. All of these were enormous triumphs, and no less challenging than anything achieved in space. Moreover, they were of comparable cost to major space efforts.

However, in direct contrast to big ticket space programs, these terrestrial infrastructures were not built for their own sake, but to serve a functional purpose that met a real economic need. The U.S. transcontinental railroad was built because there was a clear need for affordable, reliable, and available transportation, making it possible to exploit the potential of the West. (see Chapter 16.) It was the establishment of such cost-effective infrastructures that enabled the industrial revolution and the continued growth of national economies. Few realized the potential impacts at the time, however, as Paul Czysz notes, "As late as 1917, the U.S. National Academy of Sciences said it was impractical to have a transcontinental highway because you would have to have a filling station every 100 miles. 'That's over a 1,000 filling stations, and we can't figure out having more than 1 or 2 filling stations per city.' They made these marvellous projections, and it is true, there is not a gas station every 100 miles on the transcontinental U.S. highway system, *there's one every five miles*."

Spaceflight today is the only field of human activity with programs that do not seem to need a clear-cut rationale to justify their existence. Some have likened the achievements of Apollo, and other programs, with being at the top of the technological pyramid. Although undeniably outstanding triumphs, these space programs were, and continue to be, about as practical from a functional point of view as were the Pyramids of ancient Egypt. If space is to become just another domain for human endeavors, then shouldn't major space programs be justified by the need to do them, as with any other activity on Earth?

*With all the potential that space holds, where is the equivalent of a railroad to space?*

## Summary—Not at Any Price

There have been many great triumphs in space. Yet, as long as programs like Freedom and Hermes are justified by intangible, uneconomic rationale, they will always be subjected to simultaneous reductions in capability, increases in total program costs, and stretching of their schedules. None of them have a single, absolute, functional requirement that must be met to justify their existence. Instead, the selling of major space initiatives is a balance between serving a limited range of technical needs, maintaining the existence of the institutions, developing technology, being an international partner, demonstrating leadership, and ensuring autonomy and independence. Although these can make good political justifications, they do nothing for space—except increase costs, constrain capability, lengthen schedules, slow technology advancement, hamper innovation, and prohibit growth-enabling commercial development. How we do things in space today seems to have little in common with how or why we do things on Earth.

Major manned space programs in particular are apparently conducted for their own sake, and not to incrementally build a cost-effective and functional space infrastructure. Further, big ticket programs tend to squeeze out other smaller, less expensive, but perhaps more purposeful efforts. The epitome of this is that the United States can afford to launch Shuttles at $500 million a shot, and spend $2 billion per year on a space station that will not be ready until the year 2000, but has had to "borrow" a relatively simple and inexpensive weather satellite from the Europeans [27]. This is ironic given that one of the principal justifications for space spending has been the great benefits to the public derived from weather satellites.

The truth of the matter is that launching people into space, building space stations, performing medical and materials research, constructing optical telescopes, monitoring the Earth's environment, establishing outposts on the Moon and Mars, and exploiting new commercial opportunities are basically sound ideas. If the cost and time necessary to conduct many of these activities could be brought more in line with how things are done on Earth, then perhaps they could be justified less on political intangibles and more on practical, cost-effective, and functional needs.

Suppose that . . .

*. . . launching people into space cost a few million dollars and could be performed many times per year, instead of several hundred million dollars a handful of times per year as today.*

*. . . a space station just as capable as Freedom could be built for a few billion dollars and take a few years, instead of many tens of billions of dollars and over a decade.*

*. . . the total cost of research to investigate a particular medical problem, such as AIDS or osteoporosis, in space was comparable to the cost of performing the same type of research on the ground.*

*. . . the cost to build a space telescope like Hubble was hundreds of millions of dollars and not billions of dollars, and it could be repaired and upgraded as and when required.*

*. . . a large number of modestly sized platforms to monitor the Earth cost hundreds of millions of dollars and a constellation could be deployed rapidly, instead of many tens of billions of dollars for a few satellites launched after many years of development.*

*. . . the cost and time needed to set up an outpost on the Moon was more in line with the cost and time needed to establish an outpost in the Antarctic or at the bottom of the sea, instead of hundreds of billions of dollars over several decades as proposed under current plans.*

*. . . the driving force behind manned spaceflight was to build a cost-effective, multielement, and frequently accessible space-based infrastructure to support a wide variety of activities, instead of the presently planned very expensive, unique, and infrequently accessible infrastructure which will support a limited range of activities.*

*. . . a wide variety of commercially self-supporting activities could be conducted beyond the relatively modest productless commercial communications satellite industry of today.*

*. . . access to space for crew and cargo was truly routine, demonstratively safe, and a factor of ten or hundred times less expensive than today.*

Would the rationale for these activities be more compelling with dramatic reductions in costs simultaneously coupled with a fundamental improvement in the way such activities are actually conducted?

## References


1. Transcript from Alastair Cooke's weekly "Letter From America" radio broadcast for the British Broadcasting Corporation on August 24, 1991.
2. NASA current fiscal 1992 budget for unmanned solar system exploration is about $500 million.
3. Covault, C., "Hubble Returns Good Data but Future Is Clouded," *Aviation Week & Space Technology*, October 28, 1991, p. 21. Courtesy *Aviation Week & Space Technology*. Copyright 1991 McGraw-Hill, Inc. All rights reserved. See also an article by Joel Kirkhart entitled "Adaptive Optics: Forget Hubble. Ground-based Telescopes Will Soon Open a Whole New Universe," *Ad Astra*, November 1991, p. 42. Kirkhart discusses the use of adaptive optics technology originally developed for SDI. Further, an article in the *International Herald Tribune*, August 8, 1991, explained that the U.S. Defense Department had recently declassified the adaptive optics technology, benefiting the astronomical community. The article states, "For infrared observations, . . . existing and soon-to-be-completed astronomical telescopes equipped with the system will far exceed the capabilities of the Hubble Space Telescope, even if that defective instrument is eventually fixed . . . "
4. See, for example, an article by James Asker entitled "Data From Hubble, GRO Likely to Shake Astronomical Theory," *Aviation Week & Space Technology*, January 20, 1992, p. 24.
5. "Space Station Redesigned for Larger Structural Area," *Aviation Week & Space Technology*, October 14, 1985, pp. 16–18.
6. "NASA Managers Will Review Work Needed to Ready Endeavour for Flight," *Aviation Week & Space Technology*, October 28, 1991, p. 26.
7. Commentary, Sen. Barbara Mikulski (D-Md.), *Space News*, "Scientists Need the Station: Project Improved by Clarifying Mission, Design," September 23–29, 1991, p. 28. With permission of *Space News*.
8. Commentary, Richard Truly, *Space News*, "The Challenge of a Dream Fulfilled," June 3–9, 1991, p. 15. With permission of *Space News*.
9. Pike, John, "Testimony Before the Committee on the Budget, Task Force on Defense, Foreign Policy and Space of the House of Representatives: Space Station Freedom," *Federation of American Scientists*, July 11, 1991.
10. Lawler, A., "U.S. Health Institute Steps Up Research With NASA," *Space News*, January 27–February 2, 1992, p. 17.
11. Commentary, Senator Dale Bumpers (D-Ark.), *Space News*, "Support Science That Will Pay: Station's Expected Results Do Not Match Costs," September 9–15, 1991, p. 20. With permission of *Space News*.
12. The NASA Marshall Space Flight Center proposed such a powersat called the 25 kW Power System in 1977. It would have been used for

Spacelab and other missions requiring high power levels over a long duration. Such a powersat could also form the resource module of a small permanently occupied space station.

13. Interview, J. R. Thompson, *Space News*, November 4–10, 1991, p. 23. With permission of *Space News*.
14. See, for example, Commentary, Steve Hoeser and Bill Haynes, *Space News*, "Buy a Station off the Rack: Why the West Should Consider Soviet Hardware," November 11–18, 1991, p. 16.
15. Lawler, A., "Congress Hits New Projects, Saves Station," *Space News*, September 30–October 6, 1991, p. 3.
16. Derived from a communication from Courtney Stadd formerly of the U.S. National Space Council.
17. There are many references to Hermes in the popular literature. One article of note was written by Jean-Claude Bouillot, head of the CNES orbital intervention division, entitled "Hermes Contract Decision," printed in *Satellite & Space Technology*, December 1985, p. 9.
18. The Hermes pressurized payload essentially consists of drawers that insert into the space station racks. Hermes does not have the volume capacity to carry a fully integrated rack.
19. de Selding, P., "ESA Readies 14-Year Space Plan for Ministers' Review," *Space News*, October 28–November 3, 1991, p. 5.
20. "Firms Gamble on Bigger Vulcain," *Space News*, January 13–26, 1992, p. 2.
21. de Selding, P., "Hermes Builders Claim Further Cost Cuts Not Possible," *Space News*, June 24–July 7, 1991, p. 3.
22. It was reported at the AIAA Third International Aerospace Plane Conference that 20% of the cost of each Hermes flight will be for the Ariane 5 (Paper AIAA-91-5004). Thus, if it is assumed that each Ariane 5 will cost around $100 million, this leads to a cost per flight of Hermes at about $500 million. This is equivalent to about $170,000 per kilogram of payload, assuming Hermes and Ariane 5 are eventually upgraded to carry 3 tonnes of payload. For comparison, the Shuttle rates at about $25,000 per kilogram.
23. Gibson, R., "Europe's Next Move," *Space Policy*, August 1989, p. 187. With permission of *Space Policy*.
24. Aldrin, B., and M. McConnell, *Men From Earth*, (Bantam Books, New York, July 1989), p. 67.
25. Johnson-Freese, J., *Changing Patterns of International Cooperation in Space*, (Orbit Book Company, Malabar, Florida, 1990), p. 25.
26. Budget numbers are detailed in the *European Space Directory 1991*, (Sevig Press, Paris, 1991).
27. See "GOES-Next Goes Next," *Space News*, September 16–22, 1991.

# Chapter 2

# The Achilles Heel of Space

## Analogy: A World Reliant on Expendable and Untestable Ships

Imagine a world in which trade between continents relies entirely on a few types of ocean-going vessels of various sizes. Suppose that because of technology limitations alone, a brand new ship must be manufactured for *every single voyage*, and the time required to actually construct the ship and verify its readiness is around 2 or 3 years. Further, because these ships are expendable, it is impossible to perform sea trials before setting sail, with paper calculations being one of a few ways to estimate the probability of each ship's success. Indications that each ship will perform as expected come from a small number of previous sailings of similar ships, some of which may have failed to make it to their final destination. Their success is also, in no small way, dependent on the ability of the shipbuilder to assemble each ship properly. Certainly, it is not possible to go for a test sail out of the harbor. Niggling technical problems with the ship's systems, and the need to reassure customers that it will work as advertised, cause departure delays of months or years. Once the voyage is underway, it is not possible to turn the ship around and return to port if a problem occurs and almost any single component failure results in total destruction of the ship and its cargo. Each ship must work the first time, *or not at all*. Finally, because these ships are expendable, the price of each voyage to the customers includes the total cost of manufacturing the ship (Figure 2.1).

In this somewhat ridiculous-sounding scenario, it is reasonable to assume that intercontinental trade would not exist, except for those missions where a small part of each expendable ship can return back to port with a payload of extremely valuable items, gold or diamonds perhaps. Even then, travel is unlikely because of the prohibitively high initial exploration costs needed to determine whether such treasures exist or not. If it is uncertain that sufficiently valuable materials actually can be found, then it is rather difficult to

*Would this ship have supported wide-scale industrialization ?*

**Principal Characteristics**

- Brand new ship for each voyage, (i.e. expendable)
- Cannot test sail the ship first, (i.e. not incrementally testable)
- Cannot abort the voyage if a problem arises, (i.e. ship drifts if the sails tear)
- Cannot save cargo if problem occurs, (i.e. cargo either safely delivered or lost)
- Cannot demonstrate reliability or safety, (i.e. judged by a success of other ships),
- Most models go in one direction only, (i.e. no return capability)
- Less than a dozen sailings per year ,
- Susceptible to delays of months or years,
- User pays the full cost of each ship,
- But, low cost per kilogram !!

**Figure 2.1** An imaginary ship with the same characteristics as today's space transportation systems.

justify such a high initial investment based on pure faith alone. Watertight guarantees of profitability are essential to leave port. Either that, or the only way to achieve a specific objective is to set out to sea with unambiguous justifications.

The makers of such ships insist that the cost of the ship could drop to more affordable levels, given the opportunity to build a great number on a production line. This may be true. However, the makers find it rather difficult to obtain the considerable up-front investment needed to build facilities for manufacturing ships in this way, simply because it is difficult to convince investors that the market is elastic enough to expand in response to lower transportation costs. It is argued that there isn't a need for a lot of ships because there just aren't that many treasures to find or there aren't enough missions that absolutely necessitate setting out to sea.

Even if reducing costs on a production line basis is possible, the users still have to struggle with a transportation system that is prone to random catastrophic failures. This means that the users find it necessary to purchase an absolute minimum number of ships and then try to achieve as much as possible on each voyage. The resulting complexity makes the cost of each customer's cargo very high, but nevertheless still significantly less expensive than building many cheaper payloads and purchasing a larger number of ships. In addition, the users have to restrict their activities to single, self-contained voyages which can be completed without the need of subsequent support missions. Apart from the high costs, another reason for this is because it is impossible to guarantee that each ship will arrive at its destination on time, if at all.

If only a handful of ships are purchased, and one sinks before reaching its destination, the financial loss, as a fraction of the user's total investment, is very large because of the combined cost of the ship and its cargo. Thus, unless the user's business is so certain, or the reason for having to set sail is so important that such failures can be tolerated, this lack of transportation robustness probably in itself spells death for many promising programs and ideas. Potentially, insurance coverage is obtainable, but the premiums are very high for an untried business critically dependent on a potentially unreliable ship. Most likely, because the risks are so high, such ventures are simply uninsurable.

After a catastrophic failure, all subsequent sailings of that particular make of ship are halted until the reasons for the incident are completely resolved. Unfortunately, the downtime of other ships can be very long, primarily because wreckage has to be recovered, if possible, and painstakingly pieced together. Further delays are then necessary so that any design changes can be made and the ships re-qualified. Few users want to risk their expensive assets soon after an accident, preferring to wait until the situation is properly resolved.

Although the above scenario may seem a little absurd, just such a world existed 3,500 years ago [1]. At that time, the Polynesians had perfected technology enabling them to build large double canoes capable of carrying sufficient supplies for voyages lasting several months and travelling hundreds of kilometers. They also developed relatively sophisticated techniques to navigate their canoes through the Pacific Basin. These canoes were probably just as expendable as the ships described in the scenario above, with the important exception that each could be tested first before the start of a voyage. Even though these canoes utilized the best available technology of the time, their performance was still insufficient to allow extensive commercial activities, other than trading small items or exchanging information between islands that were in relatively easy reach.

Today's global space industry is little different from the analogy of the expendable and untestable ships. Rockets such as the Space Shuttle, Ariane, and Proton are the modern equivalent of those Polynesian canoes. Currently, the only fully self-supporting commercial space industry is one not even involved in the exchange of materials or products—communications satellites. Instead, the industry revolves entirely around information exchanged through massless electromagnetic radiation, a situation analogous to the word-of-mouth exchange of information between Polynesian settlements. Similarly, the protein and other crystals grown in space today, although not directly commercially self-supporting, might be analogous to the exchange of taro, yams, bananas, dogs, chickens, and other small items between the different Polynesian groups. Other analogies between military reconnaissance satellites and spy canoes can probably also be made.

The Polynesians did the best they could with the technology available at the time, just as the space industry does today. However, for the same reason that canoes could not support the establishment of major seaports for trading large quantities of materials and products, neither are the users of today's launch vehicles able to bring about the widespread utilization of space.

## *Space Transportation and the User*

Access to space undoubtably poses difficult technical challenges, but space transportation is, after all, just transportation—a means to serve an end. What happens in space, and not how to get to space, is what matters. If the space-faring nations of the world are genuinely determined to have extensive space activities, then accessing space frequently, reliably, safely, and economically must be given the highest priority. According to former U.S. Vice President Dan Quayle, [2]

> A robust, reliable, available, and affordable launch capability is critical to success in space.

And the U.S. National Space Council's space transport strategy is, [3]

> To develop U.S. space launch capability—our transportation to and from space—as a national resource; the space transportation infrastructure will be to the 21st century what the great highway and dam projects were to the 20th. We will ensure that this infrastructure provides assured access to space, sufficient to achieve all U.S. space goals.

To meet this priority, the requirements potential users might desire from a space transportation system, regardless of the technical problems that must be overcome to build a vehicle capable of meeting these objectives, will include the following:

1. *Launch costs*: The lowest possible cost for use of the transportation system, either as a dedicated user or when shared with other users.
2. *Payload mass and volume*: Broad range of payload sizes, shapes, and masses that can be accommodated, with standard payload integration procedures for rapid installation of the payload into the launcher.
3. *Rendezvous and return*: The ability to rendezvous with spacecraft in orbit to support servicing activities, and to collect payloads for return to Earth.
4. *Launch availability*: Numerous launch opportunities available to the user every year.
5. *On-demand access*: Access to space as and when demanded by the user, even on very short notice.
6. *Launch reliability*: Very high probability that the payload will reach the intended destination.
7. *Abort capability*: Even higher probability that payloads can be safely recovered and returned to Earth or another stable location (e.g., a space station) in the event that a failure occurs in the transportation system that precludes successful completion of the mission.
8. *Alternative launchers*: In the event of the unavailability of the primary launcher (perhaps due to grounding after a loss), alternative means to access space always available immediately.
9. *Launch sites*: Options to choose from a number of convenient launch sites around the Earth.
10. *Orbit selection*: Wide variety of orbits, ranging from equatorial to sun-synchronous, accessible from a number of launch sites.

| | |
|---|---|
| Dedicated launch costs | : *$40 to $500 million.* |
| Payload mass into LEO | : *2 to 25 tonnes (OK!).* |
| Rendezvous and return | : *Shuttle and capsules (very rare).* |
| Cost to return payload | : *$25,000 to $250,000 per kilogram.* |
| Availability per vehicle | : *8 to 10 flights per year.* |
| On-demand access | : *Non-existent.* |
| Lead-time to launch | : *2 to 4 years.* |
| Typical launch delay | : *6 months to 2 years.* |
| Reliability | : *90 to 98%, not demonstratable.* |
| Abort capability | : *Non-existent expect Shuttle.* |
| Alternative launchers | : *Not easily interchangable, delays.* |
| Launch sites | : *Usually in remote locations.* |
| Orbit selection | : *Variety, but depends on launch site.* |
| Launch environment | : *Very rough, high vibrational loads.* |

**Figure 2.2** Current launchers do not meet the needs of the users.

11. *Launch environment*: Benign launch and return environment compatible with sensitive cargo.

The majority of these requirements are directly analogous to any other form of transportation. However, in contrast with today's cars, trains, boats, and planes, the capabilities provided by the current generation of launchers are, in almost every respect, diametrically opposed to what the user actually wants from a space transportation system. Typically, the capabilities of the world's current stable of launchers are as follows [4] (Figure 2.2),

1. *Launch costs*: Ranging from around $100 million for a dedicated Ariane 4 launch (approximately $20,000 per kilogram to geosynchronous transfer orbit or GTO) to about $500 million for a Space Shuttle launch (approximately $25,000 per kilogram to low Earth orbit or LEO). Prices for former Soviet and Chinese boosters are quoted as being around one-half those of Western vehicles. However, profound differences in labor costs and practices in these countries make direct comparisons impractical.

2. *Payload mass and volume*: Rockets such as Ariane 4 can carry payloads of between 1.9 and 4.3 tonnes to GTO, with a maximum size of 3.65 meters in diameter and 9 meters long. The Shuttle can carry payloads into LEO that are as heavy as 25 tonnes and 4.6 meters in diameter by 18 meters long. Energia has the potential to carry between 100 and 250 tonnes with a maximum diameter and length of about 5.5 meters and 35 meters, respectively.
3. *Rendezvous and return*: The former Soviets' Soyuz and Progress vehicles have demonstrated the ability to rendezvous with Salyut and Mir Space Stations for crew exchange and resupply. Only the Space Shuttle has demonstrated the ability to service unmanned spacecraft in orbit and return large payloads to the ground—a feat which has been achieved on two separate Shuttle missions up to 1992.
4. *Launch availability*: Launch opportunities are very limited and, typically, Ariane 4, Atlas, Delta, Titan, the Space Shuttle, and other vehicles are rarely launched more than 8–10 times per year. In the former U.S.S.R. launch rates for some types of vehicles have been two or three times higher, for reasons that are discussed later (Chapter 3), but large vehicles like the Proton are usually launched once per month.
5. *On-demand access*: In nearly all cases, planning for a launch may have to begin several years in advance due to limited availability. Further, launches are frequently subject to delays of weeks, months, or even years for a variety of reasons.
6. *Launch reliability*: Reliability is difficult to measure because the sample number of launches is so very small, and rockets tend to be upgraded and stretched at frequent intervals. Very approximately, the probability of a payload reaching orbit and not being catastrophically lost ranges anywhere from 90% to 99%.
7. *Abort capability*: All unmanned rockets have two possibilities following launch—either they reach orbit, or they catastrophically fail. There is no opportunity to return to the launch pad for another try at a later date. Only manned rockets such as the Space Shuttle and the Soyuz have some form of abort and crew escape capability, although only after loss of major elements of the rocket and a risky maneuver.
8. *Alternative launchers*: The capability exists to interchange modest payloads between vehicles such as the Ariane 4 and Atlas 2. However, alternative launch opportunities cannot be obtained immediately due to availability problems, and usually periods of a year or more must first elapse. Backup launch capabilities for manned missions no longer exist, nor are backup capabilities feasible for large payloads originally intended for the Shuttle or Titan, for example, without extensive redesign of the payload. The idea of assured access to space is clearly a misnomer.
9. *Launch sites*: There are about eight major launch sites throughout the world, and the majority, like French Guiana or Yichang, are in remote locations.
10. *Orbit selection*: A variety of orbit inclinations are available in most cases. Ariane can launch payloads into near equatorial orbits at 7° orbital inclinations and to polar orbits at 98.5°. The Shuttle can launch into orbits from about 28.5° to a little over 60°. However, the more flexibility required in orbit selection, the more remote the launch site tends to be.
11. *Launch environment*: The launch environment is comparatively hostile, and strong low-frequency acoustic and low-frequency vibrations are imposed on the payloads. Maximum static acceleration loads for Ariane are about 7 g's, and for Shuttle they are a relatively benign 3 g's. This hostile environment places extreme constraints on the design of payloads. If a major section of the payload were to become detached during launch, it would likely result in the catastrophic loss of the vehicle.

The current generation of launch systems cannot be described as user friendly in the same way as airplanes, railways, or ships. It just isn't possible to turn up at the launch site and buy a ticket for a ride into space the same day. Of course, this argument immediately raises some important questions. First, for the type of space operations undertaken today and those foreseeable in the future, is there really a need for a user-friendly space transportation system? After all, it might be argued, space activities are fundamentally so different from terrestrial activities that simplistic comparisons between airplanes and rockets just cannot be made. Second, if it turns out that there is a significant benefit from having user-friendly space transportation systems, then why don't they exist today? In this case, it might be argued that if space operations can benefit significantly from a more user-friendly launcher, then market forces alone should have brought about its existence already. Is the fact that this has not happened a sign that the benefits may not be as significant as has been suggested?

## Impacts of High Launch Costs on Space Operations

If a potential user—whether commercial or government—conceives of a business or project requiring the use of space, then the first level of concern is the cost to place a spacecraft into orbit. Launch costs range from $50 million for a modest spacecraft weighing 2 tonnes placed in geosynchronous transfer orbit, to as much as $500 million for a 13-tonne space telescope. Thus, the first impact of transportation on space operations is to eliminate potential users who cannot afford such an expense, whether because the commercial revenues are insufficient to cover these costs or the government budget is simply not large enough.

Cost is the overriding constraint for using any form of transportation, whether it is in the space launch business, the expendable ship analogy discussed earlier, or with conventional transportation systems. However, few forms of terrestrial transportation come anywhere near the cost of space transportation. To send 2 tonnes of hardware from Europe to the United States via cargo ship will cost on the order of $5,000, or about $2 per kilogram, while the cost to charter a complete 747 is about $250,000. This contrasts with using an Ariane 4 to launch a 2-tonne satellite, which might cost $40 million if the launcher is shared with a second payload, or about $20,000 per kilogram, while the cost of the dedicated Ariane 4 is about $100 million.

Arguments justifying high launch costs can be made. For example, it can be argued that launch costs are high, but this has to be looked at in context of the mission the launcher is performing. If a spacecraft lasts 10 years, then launch costs should be looked at in terms of how much they cost per year. Therefore, if it costs $100 million to launch a satellite that lasts 10 years, then this is equivalent to $10 million per year for transportation. The fundamental problem with this type of argument is, what is this relative to? For example, even though an airliner like a 747 may also cost in the region of $100 million to purchase, it can be used more than just once. Over a lifetime of 20 years, an individual 747 may make up to 10,000 flights, giving an amortized purchase cost per flight of $10,000. Similar comparisons can also be made for cars, ships, trains, and so on. All that these comparisons seem to do is reinforce how expensive space transportation actually is.

Another commonly expressed argument in defense of high launch costs is that it is expensive to get into space simply because the high technology involved, combined with the low frequency of launches, makes it expensive to do. Also, it is often expressed that to compare space activities with more common terrestrial doings is impossible, because of the entirely different environments experienced. However, if the technology problem can be overcome, then why shouldn't it be as easy and as inexpensive to do things in space as it is to do them on Earth?

Despite high launch costs, there will nevertheless remain some users who are in a position to purchase launch services. However, few of these users are likely to be able to purchase more than a handful of launches. The impact of this is that demand for launch services is minimized. The lower the demand, the smaller the launcher production run, and the less effect economies of scale can have in reducing launch costs. Space operations will remain at a minimum. Thus, launch costs not only remain high, but the market forces required to bring them down are very, very weak.

## *Limitations on In-Space Supportability*

Launch costs are a major inhibitor to space operations, but they are by no means the only ones. The next question the potential user must ask is, "What do I get for my money?" Some mission concepts under consideration may require repeated access to a spacecraft on orbit. For example, space stations like Mir or Freedom require a certain number of resupply missions each year to support the crew and operations. Other mission concepts may require a payload to be deployed in low Earth orbit for recovery at a later date, such as NASA's Long Duration Exposure Facility (LDEF) or ESA's European Recoverable Platform, Eureca. (see Chapter 17.) Unfortunately, the United States and the C.I.S. only have one means each to place people in orbit, the Space Shuttle and Soyuz respectively. In addition, only the United States has the means to rendezvous and recover large spacecraft from orbit, while the C.I.S. has the means to return smaller items only from the Mir space station in their expendable capsules. Their Buran orbiter is currently facing the possibility of cancellation and is unlikely to ever be used in such a role. Clearly, if the ability to perform such missions is limited –and so very expensive–then so is the opportunity to undertake them.

Western dependence on the expensive Space Shuttle has all but eliminated missions requiring in-space support (i.e., rendezvous, servicing, and return), even though this was one of its chief selling points. (see Chapter 17.) It is important to emphasize that the reason why so few support-type missions are undertaken is not because it is particularly difficult to rendezvous with a satellite. It is simply because of the unavailability of the Shuttle to perform this type of mission. By late 1993, the LDEF and Eureca spacecraft are the only spacecraft *designed* for launch and recovery that have actually been deployed on one Shuttle mission and then recovered by a later mission. For LDEF, the planned six months it was supposed to stay in space eventually became six years before finally being recovered. After several weeks of delays, LDEF was finally recovered, just 2 months before it would have burned up in the atmosphere [5] (Figure 2.3.).

The Shuttle has been used in an orbital support role to repair the ailing Solar Maximum Satellite (SM 41C, 1984), to recover the Palapa B2 and Westar 6 communications satellites (SM 51A, 1984), and, shortly thereafter, to jump start the stranded Leasat 4 spacecraft which failed to switch on following deployment on an earlier mission (SM 51I, 1985). In 1992, the new Shuttle *Endeavour* was used to attach a new rocket motor to the stranded Intelsat 6 satellite.

**Figure 2.3** Long Duration Exposure Facility (*NASA*).

This spectacular mission included the first three-person EVA. A few months later the dramatic Hubble repair mission was successfully completed. All of these were dramatic demonstrations of the potential of orbital support operations, although it might be argued that the actual cost of each rescue mission was more than the cost to build a brand new spacecraft and launch it on an expendable rocket. While this may be true from an isolated perspective, money was probably saved overall because the annual Shuttle operating budget is effectively independent of launch rates.

Several of these missions were performed prior to the *Challenger* accident. Today, stricter quality control procedures, coupled with the backlog of payloads, have made opportunities to perform these types of missions rather more difficult. It is surprising that even though the Hubble Space Telescope is a high priority and very valuable mission for the United States and Europe, it was not be possible to mount a repair mission until late 1993, 4 years after launch (Chapters 1 and 18). The repair of the Intelsat-6 communications satellite that failed to separate properly from its Titan 3 launcher in March 1990 was not done until mid-1992. Finally, the very same Solar aximum Mission spacecraft repaired by the Shuttle in 1984 was allowed to tumble out of orbit and burn up in the atmosphere in late 1989, simply because the Shuttle was unavailable.

These examples show that for as long as only one type of vehicle is capable of supporting spacecraft in orbit, it is extremely expensive to fly, opportunities to actually use the vehicle are scarce, missions have to be planned years in advance, the launch date is susceptible to significant delays of months or years, and only a few missions which require repeated access to systems already in space will be attempted. Missions requiring such support will be very limited while nations remain dependent on the current generation of launch systems.

## *Prospects for Supporting Space Stations in the Shuttle Era*

The above discussion raises serious issues for programs like Space Station Freedom and the proposed Columbus Free-Flyer. Even in Freedom's current scaled-back configuration—a design which was chosen to reduce concerns that the Station was already too dependent on the Shuttle—as many as 25 Shuttle missions will be needed, one after the other, over 4 years to assemble and support the Station [6]. After it is fully assembled, a further five or six Shuttle missions are required every single year for as long as the program lasts, just to support it in the planned permanently manned configuration. (see Figure 2.4.)

Since the United States has never assembled or supported a major structure in orbit over an extended period of time, the question has to be asked, "Is Freedom buildable?" Is it fair to ask such a question so late on in the program and after more than $8 billion has already been spent? What happens, though, if the Shuttle is grounded midway through Freedom's assembly sequence, or if one of the modules cannot be attached to the Station? (see Chapter 15.) According to David Webb, "I think the design of the Space Station has been a disaster from the start. It was a disaster in 1985 when I was on the Commission [National Commission on Space], but we weren't allowed to say so."

Examining the former U.S.S.R.'s experience in supporting space stations could provide an indication of the problems the United States and Europe may face. The Soviets have incrementally accumulated experience in supporting cosmonauts in space with the launching of eight practically identical space stations over the last two decades: Salyuts 1–7 and Mir. As these stations were built in a series, each one benefited from the former, allowing the experience obtained to be incorporated into subsequent stations through appropriate design upgrades. Thus, through actually building and supporting a string of relatively modest space stations, the former Soviets have been able to fully understand and appreciate the technical problems associated with building and supporting space stations. Although today's Mir is very similar in size to the Salyuts, it has benefited from new technologies, and its multiple docking adaptor is configured to allow expansion.

As the Soviet manned space station program advanced, the methods of supporting their stations also had to make a parallel advancement. Over the years, the Soviets incrementally upgraded their Soyuz manned capsule to take advantage of new technologies. Likewise, the methods to resupply and refuel the stations have evolved from the unmanned Progress vehicles first introduced in 1978 in order to allow crews to stay far longer on the stations, as well as to dispose of waste buildup. It is interesting to note that despite the small size

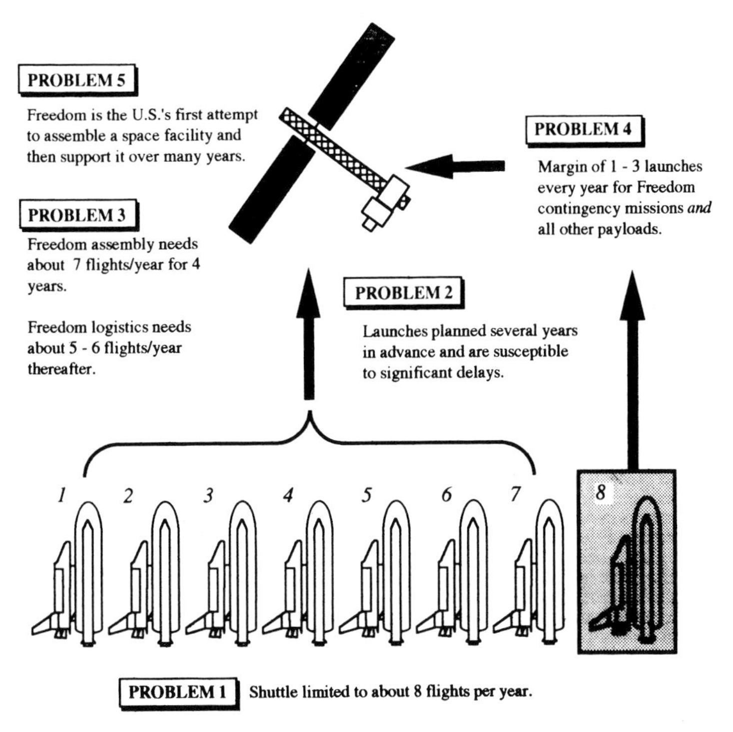


**Figure 2.4** Can Freedom be built with the Shuttle?

of Mir, supporting a permanent two-person crew requires about one Progress mission every month.

The C.I.S. space program has clearly benefited from the experience of operating eight individual stations. Nevertheless, 20 years after the launch of Salyut 1, the improvement in capability of the current Mir seems to be a surprisingly modest one over its predecessors. Importantly, despite considerable experience in docking Soyuz and the automatic Progress vehicles to space stations, when attempts were made to dock the first two expansion modules to Mir, serious problems occurred which could have led to the loss of either module. The first module was only finally secured to Mir after a space walk found that a piece of cloth had been mistakenly left in the Kvant docking ring. It was fortunate, or prudent, that there was a crew within the Mir core station to troubleshoot problems as they arose [7].

From a technical standpoint alone, the United States and Europe can clearly learn valuable lessons from the Soviet experience—*what it actually means to build and support the most basic of space stations.*

Just as space transportation follows the same basic rules as any other form of transportation, so do on-orbit operations such as the assembly of large space structures. When a bridge or a building is assembled, problems frequently arise requiring immediate action at the area of concern. Perhaps a particular section won't slot properly into place, or the plumbing system has been incorrectly installed. When such situations occur, access to resolve the problem is taken for granted. The same certainly cannot be said for space, especially as mankind's experience in building and supporting large space structures over an extended period of time is limited to three modules docked to Mir. By contrast, experience in constructing major buildings and bridges has been obtained over many thousands of years.

For all these reasons, the plan now is to build as much of Freedom as possible on the ground in order to minimize assembly activities in Earth orbit. This has led to the deletion of the large 5 meter by 5 meter square truss structure which was to be assembled strut by strut (Figure 2.5). It has been replaced by a smaller structure of six 4.5 meter diameter and 15 meter long sections which are pre-integrated on the ground [8]. Yet, even this does not mean that serious problems won't arise. The expansion modules docked to Mir were also fully integrated and tested on the ground before launch. Indeed, the two Kvant modules were fully autonomous spacecraft in their own right, containing all the necessary propulsion, guidance, and navigation systems needed to effect a rendezvous and docking with Mir.

Coupling ambitious space projects with an extremely expensive, unreliable, and occasionally available space transportation system has a profound impact on total program costs. In the case of Freedom, a major contributor to the cost of the program is the extra effort that must be put into the design of the hardware elements to ensure that each will function correctly when finally launched. NASA will not

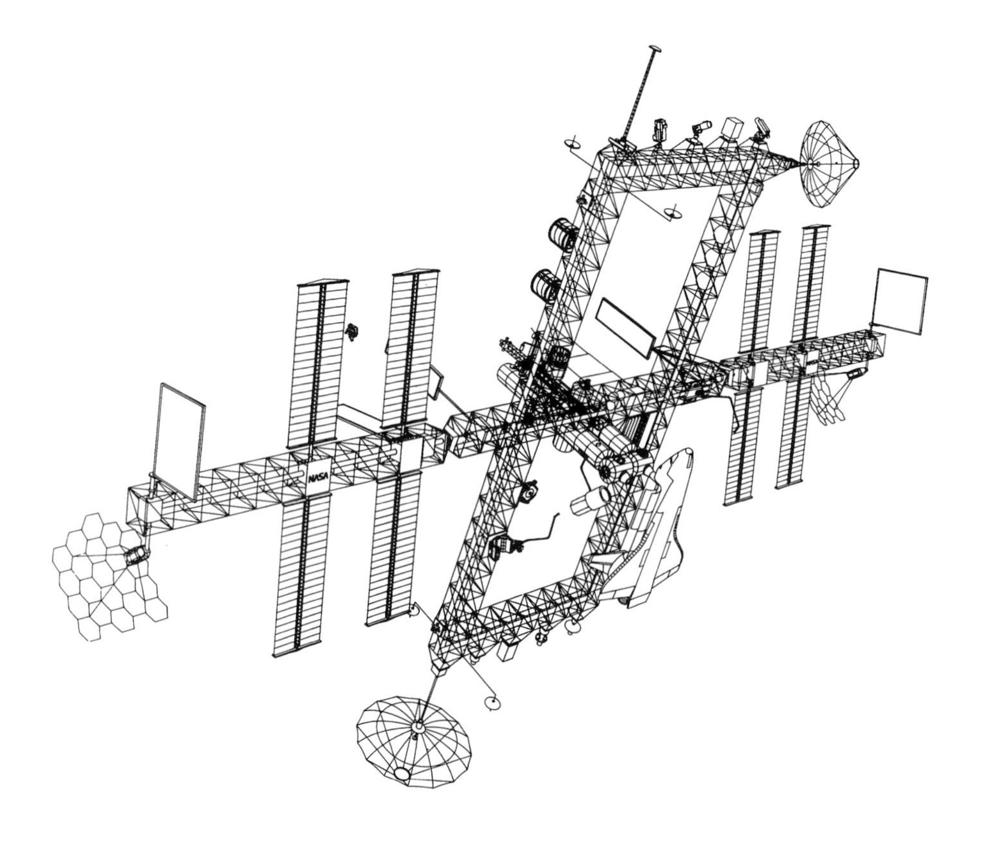


**Figure 2.5** Earlier version of Freedom with extensive astronaut-assembled truss structure (*NASA*).

launch a single space station element unless it has successfully passed an exhaustive series of ground tests. Unfortunately, these tests have to be limited to ground-based simulations. In spite of the clear advantages of testing complex systems—whether airplanes, cars, tanks, or boats—in the environment which they will eventually operate, Shuttle constraints preclude launching and testing of development models of major elements of Freedom. All major elements of Freedom must function properly *the first time they are launched*.

Terrestrial experience indicates that no matter how much money is spent, few major systems work perfectly the first time—nothing is ever 100% reliable. The same is true for space. Therefore, the inability to perform development testing of complex systems in space increases the probability that unforeseen problems will emerge once in orbit. (see Chapter 18.)

Because of the lead time required to prepare the Shuttle for launch, if a serious problem occurs, it will almost certainly be several months before a repair mission can be launched. The inevitable consequence would be the halting of Freedom's assembly until the problem is resolved. If problems like this occur every three or four flights, it might take somewhat longer than 4 years to build Freedom, if it can be built at all. According to William Haynes, [9]

> The history of space flight is replete with examples of systems which required significant initial "tuning," or even emergency repairs, before they were able to operate satisfactorily. Skylab almost went derelict before the first crew arrived because of the loss of one solar panel and some key thermal insulation. The first week of the crew's time was spent remedying the results of those discrepancies. It proved fortunate, indeed, that Skylab was ready for crew occupancy from day one.

Haynes also goes on to provide this warning for the periods between assembly missions when Freedom is unoccupied for 2 or 3 months at a time,

> SSF [Space Station Freedom] will undoubtably be the most complex space system ever orbited. To expect it to operate "hands off" from day one is naive, to say the least, and may be considered irresponsible. There are a number of malfunctions which can occur which would make the SSF unrecoverable if no one is aboard to intervene. They include runaway RCS [Reaction Control System] thrusters, failure of the thermal control system, seized control moment gyro, space debris collision, as examples. A system as complex as SSF will generate any number of other unforeseen possibilities. No one foresaw the Skylab near-disaster, and no one would have believed it if anyone had suggested it in advance.

**Figure 2.6** The now cancelled Columbus Free-Flyer (*Photo ESA*).

ESA does not have an independent means of returning anything from orbit. Therefore, the success of European space efforts relies entirely on high-fidelity simulations, exhaustive ground testing, thorough analysis, and basically, good engineering. The Columbus Free-Flyer is particularly vulnerable in this regard (Figure 2.6). None of the Free-Flyer's systems can be tested in space before it is launched on a single expendable Ariane 5 booster in 2003 to begin its 10-year mission. Thereafter, sole access to this man-tended facility will be through use of the Hermes spacecraft. Given that Hermes won't be available before 2004, can only fly once per year, and carries 1 tonne of pressurized payload, the Columbus Free-Flyer will have to function satisfactorily from the moment it is launched, and then survive for periods of up to 1 year untended. Options for recourse in the event of a Skylab-type incident just will not exist. Theoretically, the Shuttle could be used in the event of an emergency. More likely perhaps, the demands of Freedom, even in normal operations, will consume what limited number of Shuttle flights there are. If Freedom encounters the same type of problems outlined earlier, then the situation for the Columbus Free-Flyer would become even more questionable. In November 1992 the Columbus Free-Flyer was effectively cancelled, partly for budget reasons but mostly because of Hermes.

In all practicality, the Columbus Free-Flyer would not have been launched until engineers were confident that it would work properly and only if Hermes were available. Unfortunately, ensuring a high probability of success is an extremely expensive process, as Paul Czysz recalls, "When you worked for the 'Old Man,' by God, you knew who was responsible for success and failure. I can remember when we tested light bulbs ad infinitum, and we used to say 'Why the hell are we spending all this money testing light bulbs?' He said, 'Do you know what happens when you're up in space, and the light on the warning light that tells you that you've got a fuel problem, burns out?' He said, 'That's a 4 cent lamp, but it causes this entire project to fail—I'm not going to have a 4 cent lamp cause my project to fail.' "

No organization is in a position to take unnecessary risks when dealing with billion dollar investments. Of course, if Shuttles and Hermes could be safely launched every week, and at a fraction of the cost, it would be an entirely different story.

## Test and Test Again

There is no substitute for orbital testing when it comes to verifying systems, and the real benefits of space verification can only be realized through repetition. In particular, this means launching an element, testing it on orbit, returning to Earth for modifications, then relaunching it for further development tests, and so on, and so on, until the system characteristics are fully understood. Small components of Freedom, such as the heat pipe thermal radiator, flown on Shuttle Mission 28 (STS-29), are being tested in this manner [10]—but only as secondary payloads of some Shuttle missions.

An interesting example of the criticality of being able to test complex systems in the environment where they will eventually operate, is the Space Shuttle waste collection system, or toilet. Although several million dollars was spent over nearly a decade to develop this toilet, it partially or totally failed to operate on the vast majority of the early Shuttle missions. Clearly, getting a toilet to "flush" in zero gravity is somewhat more complex than it is in one g. Despite this, the Shuttle toilet was not and, moreover, could not be tested in a zero g environment—except for 20–25 seconds of microgravity on a KC-135 aircraft [11]—before it was placed on the Shuttle Orbiter and used operationally. The Shuttle toilet almost certainly would not have cost as much or taken as long to produce if it could have been developed in a zero gravity environment from the beginning, and it would have stood a better chance of working properly when installed for operational use. Today, the toilet performs nominally, but only because of what was learned from the many repeated failures that occurred during operational Shuttle missions.

The example of the Shuttle toilet is not unique. It highlights similar concerns in more complex systems such as Freedom's semi-closed loop environment control and life support system (ECLSS), which performs water reclamation and air recycling functions. The Shuttle toilet, which is only designed to be operable for the short Shuttle mission, is comparatively trivial in technical complexity compared to Freedom's ECLSS which must work for years, not just a few days. The ECLSS, even in its current scaled-back configuration, may be the single biggest inhibitor to the operational feasibility of Freedom. As with the Shuttle toilet, this is only because it cannot be developed in the environment in which it will be used. Ten years from now, assuming Freedom can be built, it may not be surprising to find that the majority of the crew's time is being spent on maintaining the life support system, at least during the first few years of operations. The two person crew of Mir is understood to spend about 80% of their time just nursing the station along [12], with the majority of time devoted to ECLSS maintenance.

Maintenance itself is vital to the proper operations and support of any complex system. Without it, most activities would quickly grind to a halt. The Golden Gate Bridge isn't constantly being repainted just to preserve its appearance, nor are a car's oil or tires changed periodically just because of the recommendations of the manufacturer. The situation is no different in space. However, as long as access to space is as expensive and restrictive as it is today, opportunities to perform essential maintenance will be limited. Instead, as much effort as possible will be put into ensuring that spacecraft can function for as long as possible before or even without maintenance being required. As has already been discussed, this extra effort costs dollars.

## One-Shot Missions

For all the reasons discussed above, it is not surprising that the only space station in orbit is the relatively simple Mir, and that Freedom is still another decade away and more expensive than first envisaged. It also comes as no surprise that, apart from Mir, LDEF, Eureca, and a handful of others, the vast majority of the world's space activities are little more than one-shot affairs. Not that these missions are particularly inferior because of it; quite the contrary in many cases. Planetary probes such as Voyager 1 and 2 could only have been one-shot missions. Nevertheless, it is interesting to note that after three and half decades in space, one-shot space missions are still so dominant. All of this is due to the transportation systems available—there has simply been no other choice.

The impacts current launch systems have on classical satellites are, in many ways, just as profound as on missions requiring repeated access and orbital support. In the West at least, communications satellites (comsats) (Figure 2.7), interplanetary spacecraft, weather and remote sensing satellites, astronomical platforms, and military systems, must:

- be designed so that they can carry enough propellant to last perhaps 10 years;
- have adequate redundant systems to ensure that when components fail the spacecraft can still function; and
- be built from sufficiently robust materials capable of withstanding many years of exposure to atomic oxygen (in low Earth orbit), cosmic radiation, and solar fluxes.

Building such systems into satellites incurs considerable expenditure, and, as with Freedom and the Columbus Free-Flyer, every effort is made to ensure they will work as planned. Nevertheless, problems and failures still occur. Hubble is one obvious example; another is the failure of the high gain antenna to deploy properly on the Galileo probe en route to Jupiter [13].

The drive to build a reliable and robust spacecraft clashes sharply with another requirement that hovers over all space programs: to make it as lightweight as possible. Gold-coated mylar blankets are used on some satellites simply because they provide the best thermal insulation properties for the lowest possible weight. Carbon fiber reinforced plastic materials are now being extensively used for major structural elements and antennas because they are lighter and stiffer than metals, even though they are more difficult to work with and are more expensive to buy. Why is so much effort devoted to reducing mass? The reasons are, once again, that the extremely high cost, limited availability, and low reliability of launch services forces users to make the most they can from a minimum number of launches.

Today, it is considered less expensive overall to cram as much capability as possible on a small number of satellites, thereby minimizing the required launches, than it is to build many heavy, less capable spacecraft which need more or larger launch vehicles. Today's satellites are invariably very highly integrated, as the benefits of modularity to the satellite are minimal compared with the extra launch costs incurred [14]. As a simplistic example, consider a typical 2-tonne comsat. It might be possible to reduce the comsat's cost by perhaps as much as a third through incorporating

**Figure 2.7** Current practice: Highly mass-optimized satellite designs exemplified by the Olympus satellite (*Photo ESA*).

modularity and heavy materials. (see Chapter 18.) This could translate into about $20 million off a spacecraft cost of around $60 million. However, this cost reduction might come at the expense of a mass increase of perhaps 100%. Such an increase would be large enough to require a dedicated launch on an Ariane 4, whereas for the more expensive, but lighter weight satellite, a shared launch would suffice. Therefore, as a dedicated Ariane 4 is about twice as expensive as a shared launch at about $45 million, the result is that the *total* cost to build and launch this "cheaper" comsat is higher by $25 million [15]. In addition, insurance premiums would increase by about $5 million. (This complex subject is discussed in greater detail in Chapter 17.)

In addition, because only a few launches can be bought, this obviously means only a small number of satellites are manufactured. GE Astro Space Company, British Aerospace (Space Systems) Ltd., Space Systems/Loral, Aerospatiale, and other companies typically manufacture about two or three commercial communications satellites every year, with series of identical satellites numbering seldom more than three or four. Consequently, even limited mass production techniques needed to bring spacecraft costs down cannot be employed because the investment costs of building the facilities are so high. The only exceptions to this are programs like Navstar (18 operational plus several spares) and the proposed Iridium (66 satellites plus spares) [16]. For comparison, a typical new car may cost on the order of several billion dollars to develop, or about the same as a large spacecraft. The difference is that many thousands of cars are manufactured, whereas only a handful of spacecraft are built. Critically also, if a car breaks down, it can be easily accessed and repaired, further reducing costs.

The net effect of the drive to build a few, highly capable, but lightweight spacecraft, is that such satellites are expensive. A typical comsat may cost between $50–100 million, while very sophisticated satellites like the Hubble Space Telescope and military reconnaissance satellites cost nearer to $1,000 million or more. In the case of military satellites, widespread use of space for military activity is effectively limited to the United States and the C.I.S. The primary reason is the extremely high cost of building and launching satellite hardware. The present U.S. military budget for space activities alone is about $20 billion per year, or about the same as the entire military budgets of Great Britain or France. Although France, Italy, and Spain are building the Helios reconnaissance satellite and have access to a modest military communications capability, such high costs are likely to restrict European and other nations from developing significantly higher levels of capability [17].

The impacts of mass optimizing spacecraft due to launcher restrictions are also profound for manned systems. The Columbus Free-Flyer originally began as a modular and fully serviceable spacecraft with the critical subsystems placed inside plug-in orbital replacement units (ORUs) similar to the ORU that was replaced on the Solar Maximum Satellite in 1984. In the early definition days, the Free-Flyer was originally intended to be launched and serviced by the Shuttle and to be periodically docked with Freedom. However, an ESA policy decision to launch the Free-Flyer on Ariane 5 eventually led to the deletion of most of the capability to service the external resource module. As the Shuttle could not be guaranteed to service the Free-Flyer, and Hermes could only carry a small internal payload, it obviously made sense to minimize the servicing capability and nonrecurring cost. In addition, cost escalations of the Columbus program and NASA opposition forced ESA to eliminate the ability of the Free-Flyer to rendezvous and dock with Freedom. Unfortunately, the net consequence of these decisions is that the Columbus Free-Flyer became expendable and it could not be expanded into even a basic space station without major redesign. After an expenditure of about $500 million the Columbus Free-Flyer was cancelled for all of these reasons.

## *Summary—Transportation Is the Key*

Space will be critical in helping understand how human industrial activities are impacting the stability of the global ecosystem. (see Chapter 18.) It will almost certainly allow new manufacturing capabilities not possible under conditions of weight. (see Chapter 20.) It holds unlimited potential for exploitation of energy and material resources. However, the extent to which these opportunities are realized depends fundamentally on the ability to frequently, reliably, safely, routinely, and inexpensively transport people and hardware to and from space. The current means of accessing space fall far short of ever meeting any one of these criteria, let alone all of them.

With today's launch systems only a few satellites will be built and, although most will be highly successful, this will only be due to the very substantial investments made in building a reliable spacecraft system. Such spacecraft will take years to construct and test and be subjected to frequent delays. Once in space, some may also suffer what would be considered on Earth as relatively minor failures. Because they cannot be repaired, such simple failures might lead to total losses of these very expensive systems. As a result, new commercial initiatives will be stifled, and government space programs will have no margins, no robustness, no stability, and no room for growth. When failure occurs, recovery will be ponderous.

In spite of all possible effort and generous funding, complex space platforms like Freedom may, at best, be very difficult or, at worst, impossible to build. Yet, all the Shuttle missions to assemble and support Freedom may only experience minor disruptions, and the Shuttle may meet the launch window for each element. However, based on past experience, is this a realistic expectation?

It is postulated that, should Freedom go ahead in even its present scaled-back configuration, problems may arise every few flights that will interrupt the flow of components to the Station, so much so that it may be impossible to complete Freedom in its intended form. The rationale for this is that as Freedom grows, maintenance and repair work will also grow, using all the available Shuttle launches, leaving none for further expansion. In addition, there is always the threat that another Shuttle loss, either during assembly or operations, will leave U.S., European, Japanese, Canadian and now C.I.S. hardware marooned in orbit.

Extrapolating into the future, proposed crew missions to establish a permanent base on the Moon or journey to the planet Mars may be even less practical, let alone affordable. Apollo successfully achieved its objectives because each mission was conducted using a self-contained, one-shot approach, and the program was very well-funded. A U.S. return to the Moon in the near future could probably only be achieved by eliminating those activities requiring assembly in orbit of components launched separately. Unfortunately, such Apollo-style missions are inevitably extremely expensive. (This is subject is more fully explored in Chapter 19.) In these times of budget deficits, is it really fair to expect governments to spend at least as much as was spent on Apollo for missions that are basically the same, except for longer stay times?

In the opinion of Bruce Abell, "I find it hard to get excited about [spending] multi-tens of billions of dollars [for] planetary exploration programs at this point, and I never thought I'd hear myself saying this. Mars isn't going anywhere, maybe we can figure out a way to get to it in good time at a cost that is more in line with what we feel like spending."

All of this is directly a consequence of the current generation of launch systems. *Today's Earth-to-orbit launch systems are the Achilles heel of space.*

## References and Footnotes

1. Comparing the spread of humans into space with the spread of the Polynesians is discussed in the excellent book *Interstellar Migration and the Human Experience*, Chapter 10, "Voyagers into Ocean Space," edited by Ben Finney and Eric Jones (University of California Press, Los Angeles, Calif., 1985).
2. "1990 Report to the President," National Space Council, January 4, 1991.
3. Ibid, p. 7.
4. Data for this information is available in a variety of sources. A particularly thorough data base is available in the *Interavia Space Directory 1991–92*, edited by Andrew Wilson, (Jane's Information Group, Coulsdon, Surrey, England, 1991).
5. Ibid, p. 90.
6. See, for example, "NASA Trims Costs, Complexity of Station," *Space News*, March 25–31, 1991, p. 21.
7. *Interavia Space Directory 1991–92*, p. 132.
8. "NASA Trims Costs, Complexity of Station," p. 21.
9. Memorandum from William Haynes entitled "Space Station Crisis" which was distributed to various members of the U.S. space community in August 1991. The purpose of this memo was "to cite a number of critical flaws in Space Station Freedom's Architecture and Operations."
10. The space station heat pipe radiator has since reflown and operated more successfully. However, this second test didn't occur until mid-1991, two years after the first experiment.
11. NASA Fact Sheet, "Waste Management in Space," Dec. 1979, p. 3.
12. From discussions with Keith Hindley.
13. At the time of writing, Galileo's main antenna is still stuck in a useless, semi-open configuration. About three of the antenna's ribs remain attached to the central mast. See "JPL Controllers Trying to Free Galileo Antenna," *Aviation Week & Space Technology*, January 20, 1992, p. 24.
14. In the case of large and complex satellites such as Hubble, the cost of incorporating modularity can be recovered through simplifying prelaunch checkout and repairs.
15. The calculation is as follows: $40 million (heavy comsat) + $90 million (dedicated Ariane) = $130 million, versus $60 million (lightweight comsat) + $45 million (shared Ariane) = $105 million. Hence, $130 million − $105 million = $25 million. Insurance @ 18% × $25 million = $4.5 million.
16. According to an article in *Space News*, "Iridium Launch Contract Delayed," (Jan. 27–Feb. 2, 1992) the Iridium satellite construction at Lockheed will be the "quickest in the space space business." The current plan is for parts to be delivered to the facility and integrated into a complete satellite which will be "rolled out the door at the end of seven days." This is very much the exception because of the large number (more than 100) of identical satellites.
17. Richelson, J., "The Future of Space Reconnaissance," *Scientific American*, January 1991, p. 18.

# Chapter 3

# Space or Bust?

## The Price of Failure

On February 22, 1990, the eight Viking engines of the largest version of the European Ariane 4 rocket ignited and began to lift the Japanese BS-X and Superbird-B communications satellites off the ELA 2 launch pad in French Guiana. Six seconds into the mission, and before Ariane had travelled higher than the launch tower, the thrust of one of four core engines suddenly dropped off and the Ariane rocket started to swerve off course with the exhaust plume of the engines singeing the top of the tower. The Ariane rocket was doomed; without nominal thrust from all four of the core stage engines and all four of the strap-on booster engines it would be impossible to reach orbit. Ariane continued to ascend for a few more seconds, and as it became clear to controllers that the seven remaining engines could not maintain the rocket's proper attitude, the launcher and its expensive payload were destroyed by the range safety officer. There was no alternative. Kilometers above the Guiana launch site, more than $250 million of rocket and satellite ruptured into thousands of pieces, falling back into the Atlantic Ocean below. The cause of the failure: a small cloth, measuring no more than 3 centimeters by 2 centimeters, was mistakenly left inside a water line, blocking a valve.

The chairman of the Ariane failure investigation panel pointed out in an interview in *Space News* that, [1]

> Human error can occur anywhere, even when people are careful . . . you find sometimes surgeons leave implements in their patients, which doesn't mean they are bad surgeons.

Although this is a true statement, leaving an implement inside patients seldom kills them. Likewise, when a foreign object—a grain of sand, for example—becomes lodged in the carburetor of an automobile, normally the car does not crash, nor is it thrown away and a brand new vehicle purchased to replace it. Yet, a "foreign-object contaminant," analogous to a grain of sand, was considered the culprit which caused one of the engines of a Centaur upper stage to prematurely shut down, leading to the destruction of the Atlas 1 rocket and Japanese BS-3H satellite in April 1991. Another $112 million of space hardware wound up in the Atlantic [2].

## Coping With the Unexpected: Demonstrating Failure Tolerance

Complex systems will inevitably be exposed to the potential for human error and a range of other events that cannot be foreseen. It is impossible to make something as sophisticated as an airplane or rocket that will not suffer failures at some stage in its use, especially when experience in operating such systems is still minimal. As this is always going to be the case, it seems fundamental to incorporate into sophisticated systems the ability to cope with unexpected failures. "There is no such thing as 100% reliability," says Paul Czysz, "but there is tolerance to failure. An airliner loses an engine on takeoff, he seldom crashes, but he comes around, lands, and then is given a substitute airplane and, an hour later, he is on his way again." If airliners weren't failure tolerant, there would be few if any today. Of course, failure tolerant also means human tolerant.

With expendable rockets such as Ariane (Figure 3.1), H1, Proton, Zenit, Long March, Delta, and Titan, building in an airplane-like level of failure tolerance is all but impossible because these vehicles are discarded after every single mission. As a result, the only way to understand the behavior of such launchers is through the limited telemetry radioed back, an analysis of the wreckage, or the simple knowledge that the mission was a success. In this sense, it is impossible to learn how to "fly a rocket." Much the same is true for the Shuttle, as Bob Parkinson notes, "The Shuttles that fly are not the same Shuttles that fly each time. It may have the same name on the outside, but the complicated bits on the inside are getting moved around all the time." The contrast between rockets and other forms of terrestrial transportation is striking.

The commercial success of new aircraft built by Boeing or Airbus is intimately tied to the ability to demonstrate that

**Figure 3.1** The first Ariane 4 launch (*Arianespace*).

the product is safe and reliable. This is achieved by meeting an intensive series of federally regulated airworthiness requirement. (see Chapter 14.) An airliner that suffers a major system failure every few flights is never going to be allowed to carry fare-paying passengers or cargo. As Maxwell Hunter notes, [3]

> *Continuous intact abort* is a *major feature* of the transport airplane business. In the transportation business, the ability to almost always *save* not only the crew and payload, but the *entire vehicle* in case of even very severe problems, is a central paradigm. This is accomplished during takeoff by the use of wings and long runways permitting the saving of the plane if it must abort prior to achieving flying speed. Once adequately airborne, the airplane can then maneuver to an emergency landing . . . Airplane flight test programs can proceed gradually to full performance, developing safety procedures as they go.

The same is true for automobiles. A new car undergoes many thousands of hours of road testing, under a full range of extreme conditions, to prove its reliability before it is put on the market. On paper, the actual design of the car could be perfect for all anyone knows. Advanced analysis techniques, the use of tested components, and the fact that many dozens of different cars have been built provides a high level of confidence. However, the only way to be certain that this is true is through actual demonstration. How many people are going to purchase a car which has only been road-tested a handful of times under relatively normal usage, and every time a road test was completed, the test car was discarded and a brand new car was used for the next test?

Demonstrating reliability is fundamental to the success of all terrestrial transportation systems. Achieving it obviously requires vehicles that are fully recoverable and fully reusable. These characteristics facilitate two critical capabilities: incremental testing and abortability. Incremental testing allows a test program to be conducted in a manner where a new vehicle is gradually exposed to more extreme operating conditions. Initially, a new aircraft will make its first flight on a relatively calm day. However, as experience is accumulated and confidence is gained in the handling the aircraft, it will gradually be exposed to more severe weather conditions. Abortability means that should a problem occur during a test program, the vehicle can transition back to less severe operating conditions and have a fighting chance of being safely recovered intact.

Further, demonstrating that the vehicle can routinely and safely abort is, of course, an extremely important aspect of an incremental test program. The test campaigns of all airliners require many tests where engines are *deliberately* shut down and then restarted while the aircraft is in flight. Other tests involve demonstrations of the ability of aircraft to safely lift off after an engine has been shut down late in the takeoff run. The flight test program of Concorde provides an interesting example. According to Calvert, [4] (Figure 3.2).

> As the [flight] tests progressed, confidence in the aircraft's ability to cope with the effects of failures at high Mach numbers grew. Most of the failures were deliberately induced, but in January 1971 an intake ramp became detached when the afterburner was shutdown at Mach 1.98. Pieces of metal were ingested by the engine, which continued to 'windmill,' although seriously damaged. The aircraft landed perfectly safely, using its remaining three engines, and the incident—unpleasant as it was—provided welcome evidence that the immensely strong wall between each engine and its neighbour would contain such damage, as it was designed to do.

When finally in service, the abortability aspect ensures the vehicle has a high probability of safely recovering itself, and the expensive cargo it may be carrying, should an unexpected problem arise.

An important question arises. If incremental testing and abortability are fundamental and mandatory capabilities of terrestrial transportation systems, doesn't it follow that they should also be fundamental to space transportation systems?

## *Expendable Launchers: Munition-Derived Vehicles*

The first launch vehicles did not originate as launch vehicles but were derived, essentially, from munitions. The development of the intercontinental ballistic missile (ICBM) made the launch vehicle a practical reality *almost by accident*. The energy needed to deliver warheads weighing many hundreds of kilograms over distances of several thousands of kilometers is only slightly less than the energy needed to put a satellite in orbit. Indeed, the first Soviet ICBMs provided more than enough energy to put a satellite in space if launched westwards with the Earth's rotation. In their simplest form, a launch vehicle is basically an ICBM with an additional small rocket motor, or upper stage, needed to provide the last little push to stop the satellite from falling back to Earth.

Prior to the development of the ICBM, nations had to rely on aircraft to deliver warheads. Because aircraft developed during and after the Second World War flew relatively slowly and were easily detectable on radar, they were viewed as very vulnerable to attack. The perceived success of the V-2 first hinted at the strategic military advantage of using expendable missiles that could launch from within national borders, travel great distances at great speeds, and deliver a warhead long before countermeasures could be launched. With the development of the nuclear bomb and the start of the cold war, ICBMs became a vital military capability.

However, an ICBM does not have the same mission as a space transportation system. Normally, when a military strike is ordered, the enemy's target is bombarded by a barrage of many missiles—not just one. The rationale is that even though some may miss their target, enough of the rest will be successful. This philosophy means that munitions systems don't need to have the capability to abort missions and return to base, nor do they even need to be incrementally testable. For widespread blanket attacks, it is also much cheaper to use many relatively simple missiles than to develop a similar number of more complex, fully recoverable vehicles able to provide the same rapid attack capabilities as ICBMs—assuming such reusable vehicles were technically feasible (see Chapter 5.) Therefore, to attack a target on the other side of the world, it is cheaper, easier, and more effective to use large expendable rockets. If some fail to reach the target, it doesn't matter because others eventually will. ICBM failures are tolerable.

With expendable launch vehicles, however, the payload being launched might be the only payload of its type. It may have taken many years and tens of millions of dollars—or even billions of dollars—to build, yet, it still sits on top of a launcher that is basically "just" a suitably enhanced version of an ICBM. It is not surprising, therefore, that launchers occasionally fail. Further, because ICBMs don't have need or reason for an abort capability, neither do launchers. However, given the high cost and uniqueness of spacecraft, there clearly would seem to be some merit to the notion that space programs would benefit by having an abort capability as a means to ensure against potential catastrophic failures. As Maxwell Hunter observes,

> If an expendable rocket has a 0.95 reliability, one will be lost every 20 flights. If a rocket of similar reliability has an intact abort capability, one in every 20 flights will have an abort if the probability of a successful abort is roughly the same as the basic flight reliability. In other words, a vehicle would be lost every 400 flights. Reliability will not likely be the same and arguments will abound

**Figure 3.2** The Anglo/French Concorde (*Air France*).

over the basic reliability differences of expendable and reusable vehicles, but the key point is very powerful: When a sure mission loss is turned into a likely successful mission abort, vastly improved operations must result.

Remarkably, the present selection of Western launchers are all based on technology that was state of the art more than 20, 30, and even 40 years ago, with most being direct descendants of the ICBM. The first stage of the Atlas 2 vehicle is, basically, a stretched version of the original Atlas ICBM, the first U.S. ICBM, that was deployed in 1959 and was also the launcher that placed the first U.S. astronauts in orbit in 1962 (Figure 3.3). The Delta 2 launcher uses technology first developed for the Thor ICBM, although the present configuration bears little resemblance to the original Thor because of a plethora of evolutionary upgrades which increased the payload.

Perhaps the first large space transportation system would have eventually been developed to be fully recoverable and reusable had there not been a need for a strategic munitions capability. Were space launchers developed before they should have been?

**Figure 3.3** Mercury 4 launch on a modified Atlas ICBM (*NASA*).

## Accessing Space

This now raises the issue of technology. Accessing space is a more technically challenging activity than all other forms of transportation. At the very minimum, to put a payload into orbit, the transportation system must be able to accelerate the payload to speeds in excess of 7.5 kilometers per second or 25 times the speed of sound and lift it to altitudes of at least 200 kilometers above the surface of the Earth. Physics and the Earth dictate that there is no other choice. This contrasts sharply with terrestrial transportation. For example, if someone really wanted to, a rowboat could be used to cross the Atlantic (as indeed has occurred) instead of a more convenient—and cheaper—alternative like a Boeing 747 or Concorde.

Finding an engineering solution for a vehicle that can overcome the relentless pull of gravity and push its way through the dense lower atmosphere, as well as meet the type of user needs discussed in Chapter 2, is not a straightforward matter. Technological limitations alone have, until recently at least, meant that the only engineering solution available for building a space transportation system was the use of one-shot ballistic missiles. As a direct consequence, the first expendable launch vehicles had to sacrifice design margins and redundancy just to be able to put a reasonable payload mass in orbit. The payload mass of the first Atlas launchers accounted for less than 2% of the total liftoff mass of 150 tonnes. As a consequence, the type of systems needed to recover the vehicle and perform incremental testing and safe aborts (e.g., wings, extra engines, a heatshield, and an undercarriage) could not be carried simply because the launcher would be too heavy to reach orbit. The Atlas booster is an appropriate example in this regard as practically the entire rocket achieves orbit, with only two of the three main engines discarded midway through launch.

The Space Shuttle was the first attempt to build a reusable launch system capable of reducing the cost to place spacecraft in orbit because the reusability aspect of the vehicle would allow it to be launched frequently; at least, this was the original objective. (see Chapter 11.) The Shuttle was also planned to be more reliable than expendable launchers as the Orbiter has four in-flight abort options: a return to launch site (RTLS) abort, a trans-Atlantic landing (TAL) abort, abort once around (AOA) the Earth, and abort to orbit (ATO). The only in-flight abort that has actually occurred was in 1985 when the center engine of *Challenger* shut down midway through ascent. The only ground option is called the redundant set launch sequencer (RSLS) abort, which can be ordered after the main engines are started, but before the solid rockets ignite. This has occurred on four occasions [6].

The Space Shuttle that is flown today, however, has as many as 2,000 single point failures—those parts of the vehicle that if they failed would lead to a catastrophic loss—the most significant of which are found in the propulsion system. For these reasons alone, it might be considered that the Shuttle is too risky to fly. However, in the opinion of

Bob Parkinson, "It probably isn't possible to get a stack of hardware that complicated that will not fail with a reliability of better than 99%. Airliners and military aircraft don't manage it, so why should I expect [vehicles like the Shuttle] to do it. That's not the point. The problem is that when an inertial navigation system goes down on a Tornado, they actually manage to land the thing again and take it out and replace it. When an instrument goes halfway across the Atlantic on a 747 they still manage to fly through it. The problem the *Challenger* ran into, and it would have run into had there been a *tenth* the amount of checking on it, was (1) it couldn't detect when it was in trouble, and (2) when it was in trouble, it had no way out anyway."

The inability to recover expendable launchers—and even the Shuttle—after major failures is fundamental to why current rockets are unreliable *and* expensive *and* unavailable. Incorporating backup or redundant systems in those parts of the launcher that are not particularly heavy, such as in electrical components, can improve reliability but only in a limited way. Building in higher levels of safety, redundancy, and using higher reliability systems for expendable launchers soon reaches a point of diminishing returns. This is because the benefit of that technology would quickly be outweighed by its cost, especially as it is thrown away after every single mission. An aircraft can afford to invest in expensive quadruplex-redundant avionics and hydraulic systems as this cost can be spread out over thousands of flights. Further, even though such systems are expensive, in the case of a passenger airliner, the cost of failure resulting from *not* having enough redundancy would be far higher. The same is true for partially and fully reusable launch systems. The economic cost of the *Challenger* accident was far greater than the cost of developing a propulsion system that could be continuously monitored and, if necessary, switched off at all points during ascent.

## The Cost of Testing

Two impacts of using a large, expendable rocket to put satellites and people in space is that their test programs are inevitably both expensive and time consuming, as a new vehicle has to be built for every single test. Rather than spending large amounts of money and many years on a test program, launch vehicle developers invest a fraction of that money and time into trying to engineer reliability through thorough design and analysis. In this way, the launcher can be pushed into service as soon as possible at moderate cost. Both Ariane and the Space Shuttle had only four test flights, although many of these missions actually carried important payloads. ESA's Marecs 1 maritime communications satellite, for example, was launched on the fourth and final "test" flight of Ariane [7]. The new heavy-lift Ariane 5 launcher will have just two test flights [8].

From the perspective of the launch vehicle developer, the cost of losing a launcher and payload will be considerably less than even a relatively short series of pure test flights. Performing several tens of test launches might seem an unnecessary luxury in view of the uncertain probabilistic nature of the failures and the several hundred million or even billion dollar price tags. Further, because the current market for launches is small, a thorough test program for a new launcher would either need governments to underwrite a massive test campaign or the addition of a large surcharge to the user's launch costs. In addition, because building and preparing a launcher for a test is a time-consuming process, the rate of test launches is very low—every few months for complex systems. For example, the second Shuttle test flight occurred 7 months after the first, [9] and the second flight of Ariane 1 occurred 5 months after the first [10]. By contrast the second test flight of Concorde occurred 6 days after the first [11]. So, even if a thorough test program were affordable, it could take many years to carry out.

An interesting discussion of the relationship between testing and cost is described by Bob Parkinson, "When I was at Wescott we had some two inch rockets which cost about 'ten bob each': qualification trials about 600. The rockets used on Blowpipes that go off next to your ear, they did 300 or 400 tests of those before they were qualified. The air-launched GW 'Skyflash,' now this motor costs a lot more than a two inch rocket but it's potentially just as damaging if it doesn't work properly: qualification tests 30. When they got the 18 inch diameter Raven, the booster for the Skylark sounding rocket, which are much more expensive still: qualification trials 10. The biggest solid rockets that you ever built for the Shuttle: qualification trials 3, and then they sat a man between two of them. Now, if I had said on Blowpipe, 'we fired three, let's take a gunner and put it on his shoulder for the next one,' they would have had hysterics no matter what sort of theoretical program I've done to prove that they inevitably would work. If I took an airliner and said we will do four flight tests and then we'll put 400 people on it and fly it across the Atlantic—I would be certifiable!"

Of the Space Shuttle's 2,000 tonne gross takeoff mass, only 25 tonnes of this is actually payload, although the crew could be considered payload. This is about four times larger than the largest aircraft that exists today, the Soviet Antonov AN-225. However, because the Shuttle is launched vertically, all the force it needs to get it airborne has to come from its three high-pressure main engines and the twin solid rocket boosters. This contrasts sharply with aircraft like the AN-225 where the engines are used only to provide sufficient forward velocity so that the airspeed over the wings can generate the required lift to enable takeoff. Indeed, the six D-18 engines of the AN-225 provide the equivalent of one-fifth of the total force needed to lift the aircraft off the ground [12].

The probability of the Space Shuttle—or indeed any launch vehicle—safely reaching orbit is intimately tied to the reliability of its engines. Logically, it might be expected that the more complex an engine is, and the more critical it is to the success of a mission, the more testing such engines should undergo. In practice this is not the case, as Bob Parkinson continues, "The reliability of an engine is fundamentally a question of how many times you start it. Unfortunately, the more expensive the tests become, the more we excuse ourselves from testing."

This discussion becomes clearer by comparing rocket engine testing with that of a typical jet engine for an airliner. According to Hickman and Adams, [13]

> Typically, a jet engine accumulates well over 7,500 test hours through over 11,500 endurance cycles before flight testing begins. This testing is conducted over a wider range of conditions than expected during operational use. Extensive data bases have been developed describing operating characteristics and failure modes associated with each engine component. Deficiencies that would result in an engine shut-down probability of $10^{-4}$ or a catastrophic failure probability of $10^{-8}$ are eliminated.
>
> Rocket engines on the other hand, are typically tested less than 6 hours before flight certification. Additionally, testing is conducted at near-nominal conditions to avoid risking the loss of the engine and test stand, since the costs of such a loss are considered unacceptable. The hazards associated with rocket engine testing create safety and environmental restrictions that drive up the costs of developing and operating rocket engine test facilities. This cost must be amortized over a relatively small quantity of engines, whereas by contrast, the certification cost of hydrocarbon-fueled jet engines can be spread over a relatively large number of production units.

Funding limitations seriously restricted the reusable Space Shuttle main engine (SSME) testing. Before the first mission, the 20 SSMEs were tested a total of about 1,000 times. This compares to the lower technology, lower performance, and expendable Saturn 5 F-1 main engine, where nearly 60 units were fired a total of about 1,500 times [14]. The Olympus 593 engine was ground-tested for more than 5,000 hours before the maiden flight of Concorde—and this was in addition to the many years of operational experience the same basic engine had had on the Vulcan bomber [15].

Even though the SSME was intended to be a reusable engine capable of lasting 55 start/stop cycles before a complete overhaul, it is actually partially overhauled and inspected after *every single mission* simply because insufficient development testing had been done to qualify life-critical components such as bearings and turbine blades (Figure 3.4). This situation is compounded by the fact that the SSME normally operates at 104% thrust, i.e., 4% higher than the original design specification for the engine when the program began in the early 1970's. By contrast, most aircraft engines cruise at a thrust level of about 25% of their maximum capacity, although the takeoff run usually pushes the throttle near to the maximum thrust levels. Automobile engines similarly run at about one-third their maximum revolutions per minute.

The only way to be confident that any engine will function as intended is by demonstrating it over and over again. This was clearly not the case with the SSME. Although considerable additional testing is now being performed [16] and new, high-wear SSME power heads are being developed for use in the late 1990s, this is only now being undertaken because it was avoided 15 year ago. Additionally, it is not preventing the Space Shuttle from flying. Today, the safety of the Shuttle is directly a consequence of the decision to limit funding for the SSME and SRB test programs. (See also the discussion in Chapter 11 on SSME testing.)

## *Impacts of Unreliability on Availability and Launch Costs*

In Chapter 2 it was shown that because launch costs are high, the user endeavors to utilize as few launches as possible. Taking maximum advantage of a minimum number of launches inevitably leads to payloads that are also expensive, and in the case of commercial payloads, it might also mean that insurance premiums will be high. As a consequence, the total investment in the launcher, payload and, possibly, insurance package will be significant—whether a communications satellite or a space station element is being launched. Clearly, users would like to be confident that their payloads stand the best possible chance of reaching orbit. Therefore, they will demand assurances from the launch service supplier that everything has been done to maximize the probability of success. Likewise, the launch supplier will be all too well aware that a string of failures could lead to the user looking elsewhere for launch services.

Providing the user with a high degree of confidence is no easy matter, especially if the launcher has a limited track record that is littered with past failures, as is the case with nearly all launch vehicles today [17]. Thus, launch service suppliers expend a great deal of effort throughout the production, integration, and checkout of every rocket. Specifically, during:

1. *Manufacture and production*: Strict levels of quality control and product assurance procedures have to be followed for all components. Every component has to be put through acceptance testing and inspection to assure quality of manufacture.
2. *Assembly and integration*: Care has to be taken to ensure the various stages of the rocket are bolted together properly and all electrical interfaces are correctly wired. Full end-to-end testing of the assembled launcher has to be performed to verify the quality of the stack.

**Figure 3.4** Re-installation of a Shuttle Orbiter's engines (*NASA*).

3. *Launcher checkout*: Care has to be taken to ensure that all the procedures to checkout the launcher are properly followed during the countdown to launch, and all measurements are within the predefined tolerances. Ordnance systems have to be carefully installed and tested to ensure that booster rockets and stages will separate properly.

All of this activity leaves open the possibility for human error. Rags can get left inside rocket engines, as in the Ariane 4 failure. It is possible to incorrectly wire pyrotechnic devices, as in the case of the Intelsat-6 comsat which failed to separate from its upper stage after launch into low Earth orbit by a commercial Titan in March 1990. Sometimes bad decisions are made, as in March 1987 when an Atlas/Centaur failed when it was launched into a thunderstorm and struck by lightning. The Space Shuttle *Challenger* was launched in weather that was outside the temperature limits of the rubber O-ring seals in the SRBs. In all these cases, if the procedures had been properly followed, the failures would not have occurred.

No matter how much care is taken, it is sometimes impossible to avoid human error. The crash of an Embraer EMB-120 regional aircraft in September 1991 is just one example. According to *Aviation Week & Space Technology*, [18]

> On-sight inspection of the horizontal stabilizer revealed that the deicer and leading-edge segment, about 9 ft in length, were missing, as were 43 attachment screws that secured the deicer and leading edge to the stabilizer . . . A maintenance crew assigned to replace the deicers was assisted by a maintenance inspector, who reportedly removed the screws. Workers on the next shift apparently were not informed of the unsecured leading edge, officials said.

These types of aircraft incidents are, of course, very, very rare compared with the tens of thousands of flights that occur every day—yet they still happen. However, few if any aircraft are partially disassembled or reassembled between every single operational flight. If major components of all aircraft were always being removed and reattached between flights, many more incidents like the example above would occur. This would clearly be unacceptable and would obviously necessitate stricter quality control procedures. However, stricter quality control requires more people and time between flights, resulting in higher per flight costs and reductions in flight rates. Few organizations would be able to afford the cost to use such aircraft. Therefore, few aircraft would be built—unless absolutely needed, such as for military requirements—and the businesses that depend on them today would simply not exist.

The above scenario is as absurd as the expendable and untestable sailing ship analogy of Chapter 2. Yet, it more or less reflects the situation of all launch vehicles today. It is important to note that technology limitations have, until recently, ensured launch vehicle designers had no other choice but to build expendable or partially reusable launch vehicles. And in this light, it is perhaps remarkable that launch reliabilities in the high 90s are in fact achievable—but only at the expense of availability and affordability. Nevertheless, the consequences of such vehicles *for the user* remain severe. As Maxwell Hunter notes with respect to the U.S. space program [19],

*Continually building ammunition will not create a spacefaring nation.*

Launch availability and costs are clearly a direct function of how much work must be performed to prepare the vehicle for its mission. One reason why the Space Shuttle will never fly much more than 8 flights per year is because of the time required to build and check out the external tanks, salvage the SRBs, and integrate the Shuttle stack. Another equally profound reason stems from the enormous cost of a catastrophic failure, as was the case in 1986. NASA cannot afford to take any unnecessary risks when it comes to flying the Shuttle. Unfortunately, the effort needed to minimize these risks means that the Shuttle is, and will remain, expensive to use. Achieving a higher flight rate, while preserving the same production, integration, and quality control procedures, would require an increase in facilities and manpower and, therefore, funds. In view of the high costs of the program and the current drive to reduce these costs [20], significantly increasing the funding is obviously not realistic.

*If the Shuttle could have been tested incrementally, if 40 test flights could have been performed in one year as opposed to one decade, and if the vehicle could safely abort at all times during flight, the United States would now have the shuttle it originally wanted,—a true space transportation system.*

This is also true for expendable launchers. Even though the Ariane series of launchers had by late 1993 performed over 60 missions, an anomaly in the third stage was discovered in May 1991 that kept the launcher on the ground until July of that year. Even though this anomaly has always been present and never before caused a problem [21], Arianespace decided to ground Ariane 4 for 2 months until a complete understanding of the problem was obtained. The reason for this conservative approach was partly because three previous Ariane losses had been due to third stage failures. However, it was also because of the high cost of the payload— Europe's first radar remote sensing spacecraft ERS-1, which had been in development since the early 1980's at a cost of about $800 million. Loss of this expensive and unique payload would obviously have had an enormous economic impact, and the time to build and launch a replacement would have been on the order of a few years. Europe wasn't about to take an unnecessary risk.

## The Soviet Experience: One to Copy?

The former Soviet Union (now the C.I.S.) has launched twice as many satellites on twice as many launch vehicles than the rest of the world combined. In 1981, for example, a record total of 125 launches were carried out, a rate of better than two per week. The former Soviets clearly have an incredible launch capability; they launch payloads more frequently, more quickly, more reliably, and at somewhat lower cost than anybody else [22]. Therefore, shouldn't the previous section's comments also apply to their space program?

The answers to these questions are relatively straightforward. The Soviet and C.I.S. space launch capability is a manifestation of the way the Soviet society functioned. If a project was deemed to be of significant national or strategic importance, then it was relentlessly and single-mindedly pursued. Space in the former U.S.S.R. was one pursuit which received the highest political priority because of the propagandistic value of launching satellites able to sail unchallenged over an enemy's territory. Such an act was also seen as a clear sign of some considerable technical prowess. As Khrushchev said, spaceflight was an inspiring example of achievements that only "socialist man" could attain. By 1970 the Soviets were already launching more than 80 vehicles a year, whereas the United States was launching around 30.

The Soviet launch capability was also a direct consequence of their ICBM program, much more so than it was in the United States. Soviet technology forced them to construct much *larger* ICBMs. In addition, and not insignificantly, to attack the United States, the Soviets had to launch westward *against* the Earth's rotation. Therefore, while higher U.S. technology allowed the United States to build small, high performance ICBMs to deliver lightweight warheads, the Soviets had no choice but to build large ICBMs to deliver their heavier warheads. Militarily, the smaller U.S. ICBMs were evidently more efficient than their Soviet counterparts. However, U.S. ICBMs could only place small payloads in space, whereas Soviet ICBMs had the ability from the beginning to launch heavy payloads in space. As Vasily Mishin explains, [23]

The dimensions and lay-out of rocket R7 were determined by the requirement set to a combat missile complex in which it was integrated, and namely: necessary maximum range and payload mass transported over this range . . . It became clear to the developers of this rocket at the initial stage of its conception implementation that it could launch an artificial earth satellite. Thus, simultaneously with the development of ICBM R7, its adjustment was carried out for launching a non-orientated space laboratory satellite being developed at the same time. But for a delay in delivering the equipment for a heavy satellite under development and because of widely advertising plans of U.S. connected with a launch of the artificial earth satellite by "Vanguard" program, S. P. Korolyov made a proposal to launch a tiny satellite [Sputnik] already in the course of flight tests of this rocket.

The R7 rocket in its ICBM form and with few modifications or additions was capable of launching a payload of up to 3 tonnes into low Earth orbit. The basic R7 with a small upper stage added was used to launch Yuri Gagarin and, remarkably, the same basic R7 is still used today to launch the Soyuz manned capsules and Progress space station resupply vehicles (Figure 3.5). Indeed, the vast majority of the launches today use vehicles that are direct descendants of the R7 ICBM. More than 30 SL-4s and SL-6s are launched from Plesetsk Cosmodrome every year [24].

**Figure 3.5** Soyuz launch in October 1991 using the same R7 ICBM first fired in May 1957 (*Peter de Selding/Space News*).

The U.S.S.R. ICBM program paid for the initial Soviet heavy lift capability. Specifically, the ICBM program required an enormous investment in infrastructure, much of which was directly applicable to the space program. This infrastructure also facilitated production line assembly and integration of the many R7s built for the initial ICBM program. As a result, the cost of each R7 space launcher was kept to a minimum. The R7 did not last long as an ICBM, however, but was quickly superseded by the more efficient SS-9 Scarp ICBM. Like the R7, the SS-9 eventually was modified into a launch vehicle designated the Tsyklon, and introduced in 1966 as the U.S.S.R.'s third launch vehicle type. The Soviets' second booster, the Proton, was not derived directly from an ICBM, but was partly a spinoff from their Moon program.

U.S. ICBMs were also manufactured on a similar production line basis. However, for the United States to launch heavy payloads, the ICBMs had to be significantly upgraded beyond the basic needs of the ICBM program. These upgrades came in the form of stretching the missiles as well as adding upper stages and strap-on boosters. Consequently, the initial U.S. heavy space launch capability came at a price higher than that of the Soviets.

As a result, the smaller U.S. ICBM-derived launchers were flying far less frequently than their Soviet counterparts. In addition, most of the money for U.S. space activities was being spent on Apollo, leaving more modest amounts for other payloads. Once Apollo was completed, the drastic reductions in funding meant that the United States could still only afford to build and launch a relatively small number of payloads and launchers. To follow the Soviets would have required either a new launcher or the continuation of the Saturn production line, together with an enormous investment in infrastructure to facilitate high launch rates. The Soviet space program, by comparison, did not have to make such a direct investment in infrastructure because much of it was already paid for by the military ICBM requirements. Budget austerity measures in the early 1970's meant the United States was not in any mood to keep Apollo alive, especially with Saturn 5 launches costing around $1 billion (1993) a shot.

In any case, the United States didn't feel a need for large quantities of boosters because its satellite technology was far superior, allowing the building of fewer, but longer-lasting and highly capable satellites. Thus, U.S. satellite manufacturers were forced to maximize the use of a small number of launches by making the payloads as capable as possible. As the demands on spacecraft performance increased, so did their costs. Therefore, more effort had to be put into producing and preparing each launcher to ensure the highest probability of the expensive payload reaching orbit. The costs of U.S. launchers and payloads began to spiral upwards, but this was offset by a declining requirement for launches.

Meanwhile, the larger and simpler Soviets rockets were continuing to pump a large number of heavy and relatively unsophisticated payloads into orbit. Even if they could, it was unnecessary for the Soviets to build highly sophisticated and highly capable spacecraft because they already had so much launch capability. They were able to build simpler, heavier, and cheaper payloads than their American counterparts and launch them on cheaper, production line-assembled boosters. For example, the vast majority of these spacecraft use pressurized containers to house most of the electronic equipment. This has a number of advantages, the most notable being that thermal control is easier as the gas circulating inside the pressure vessel can evenly distribute the heat. Such pressurized spacecraft are intrinsically cheaper and easier to manufacture than their Western counterparts. Their one obvious drawback is that they are much heavier. As an example, the current generation of C.I.S. Gorizont communications satellites has a mass of 2.5 tonnes and provides 8 transponder circuits. This compares to Intelsat 6 which also weighs 2.5 tonnes but has 38 transponder circuits. However, it really doesn't matter if C.I.S. satellites are not perfectly optimized for a launcher or if they don't carry as much capability as they potentially could, because they are not unduly restricted from accessing space (Figure 3.6).

This philosophy also means that the C.I.S. can manufacture parts of the satellites practically on a production line as many spacecraft share the same basic components. For example, the same basic type of capsule used by Yuri Gagarin is still used today for microgravity experimentation and Earth observation, this being the Resurs-F and Photon spacecraft. The Progress vehicle used to resupply Mir is basically a stripped-down version of the manned Soyuz spacecraft [25]. Although sharing the same components across vastly different spacecraft requirements leads to less than optimum spacecraft, it can also lead to much less expensive spacecraft.

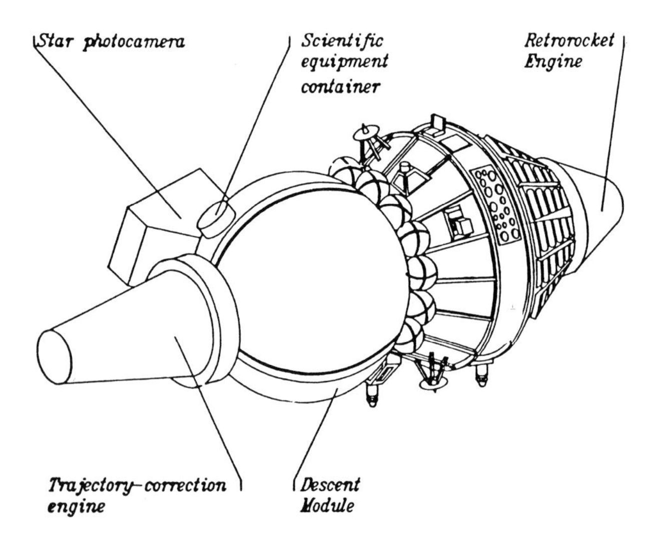


**Figure 3.6** The still operational Russian Photon spacecraft is derived directly from Yuri Gagarin's 1961 capsule (*Kayser-Threde*).

The difference in the C.I.S. and American programs is striking. C.I.S. boosters are cheaper and can be launched more quickly and frequently than their U.S. counterparts, but only because of the military investment in large ICBMs. Launchers are cheaper because so many ICBMs are built that production-line assembly techniques can be used. C.I.S. personnel can integrate a payload to a launcher in a matter of a few hours (compared with a few days or weeks for the United States or Europe) because they have done it so many times before. Likewise, countdowns also only last a few hours, and the amount of checking is minimized, as their experience in launching more than 100 boosters annually has taught them how to minimize these activities. According to one engineer, "They say 'we have payloads like U.S. TVs—we pick them up, plug them in, and if it doesn't work, we send it back. If it works, O.K., launch it.' "

From the very beginning, the United States has had to rely on a small number of higher performance and relatively expensive boosters to launch heavy payloads. This has automatically forced the improvement of satellite performance in order to get maximum benefit from a small number of expensive launches. Indeed, the drive to get the most out of a launch sometimes forces launch vehicles to be modified specifically to suit a particular payload. The German ROSAT high energy astronomy satellite launched in June 1990, for example, required the development of a specially stretched faring in order to be launched on a Delta rocket. All these activities push costs to the high level they are in the West today. Without a massive investment in infrastructure and a transition to building many heavier and cheaper satellites needing many more launchers than at present, the United States will not be able to significantly reduce launch costs or improve the reliability and availability of its expendable launchers. The same is, of course, true for Europe and Japan.

This situation was further compounded by the U.S. decision to phase out most of its expendable launchers in favor of total reliance on the Space Shuttle. Today, the expendable launcher industry in the United States has had new life breathed into it, due primarily to the military abandoning the Shuttle in favor of a mixed fleet of rockets. This has restored needed launch capability. However, it is still no match for the C.I.S. Then again, the United States doesn't really need to match the C.I.S. launch capability because U.S. satellites are far superior in performance.

Even though the C.I.S. has far greater flight experience than anybody else, it nevertheless still suffers failures, although how many is difficult to determine because of the reluctance to disclose past failures. In 1975, the Soviet Soyuz 18A spacecraft had to be detached from its malfunctioning third stage 9 minutes into the mission, and in 1983, the Soyuz T-10A capsule was the first to use the escape tower to separate it from its booster which caught fire 90 seconds before launch. The new Zenit booster—also the strap-on booster for the manned Energia/Buran combination—exploded 3 seconds after liftoff in October 1990, following an earlier failure in December 1985. The Proton launcher has an estimated success rate of about 86% (all launches) and 94% (last 50 launches), the Tsyklon is about 98% (all launches), and the family of R7-derived vehicles is about 98% (all launches) [26].

The Soviets have had their share of launch failures, but their experience has enabled them to bounce back rapidly. The Soyuz 18B was launched 6 weeks after the Soyuz 18A failure. The Soyuz T-10B took longer at 5 months to recover from the Soyuz T-10A failure, but only because the explosion knocked out two of the three launch pads. However, the first flight of the Zenit booster in August 1991 after the October 1990 failure also catastrophically failed. This was immediately followed by another failure—the third in a row—in early 1992. Thus, with 4 failures in 15 orbital launch attempts, the Zenit has a reliability of just 73% [27]. Even with all the Soviet experience with ballistic rockets, these failures demonstrate that the C.I.S. program, too, is far from perfect.

## U.S. Heavy-Lift Boosters: Business as Usual?

In 1987, the U.S. DoD initiated the Advanced Launch System (ALS) program following the results of its Space Transportation Architecture Study (STAS). Initially this was a program to devise an economic launch system capable of placing very large SDI payloads in space [28]. However, down-scaling of the SDI concept from the conjectured laser battle stations to arrays of tiny brilliant pebbles reduced the pace of the ALS program. This resulted in refocussing efforts, primarily toward engine development activities and specific launcher technology issues, now under the more broadly titled Advanced Launcher Development Program (ALDP) [29] (Figure 3.7).

The STAS, ALS, and ALDP programs have indicated that building a more cost-effective launch system might be achievable with new technology booster designs and, importantly, if the entire supporting launch processing infrastructure is completely overhauled. Critically, however, it must be appreciated that this cost-effectiveness is only from the perspective of the *cost per kilogram launched on a one way trip to orbit*. While reducing launch costs to under $1,000 per kilogram might be a realistic goal for an ALS-type expendable booster, it will only be achieved if a lot of mass can be launched all at one time. For example, suppose the dedicated cost of an ALS launch were as low as the cheapest medium-sized rocket today—the Delta 2 at about $40 million per flight. To achieve $1,000 per kilogram would require the ALS to launch some *40 tonnes* worth of spacecraft. In addition, achieving such low dedicated launch costs requires many missions per year over which the fixed operating costs can be amortized and take advantage of economies of scale. Since the vast majority of the payloads today weigh only a few tonnes, finding enough 40 tonne payloads to fill 10, 20, or 30 launches per year would seem unlikely at the present time.

Measuring a launcher's economic performance purely in terms of cost per kilogram is a poor standard. This is emphasized further by the experience of the C.I.S. Even with all the payloads it builds and launches each year, it hasn't yet found one payload to justify Energia's 100 tonne to orbit capability, except for Buran. This is probably because the cost to launch Energia is somewhat more than $40 million dollars. Alexander Dunaev said at the 41st IAF Congress in Dresden (October 1990) that the Energia/Buran launch cost was about "1 billion rubles" per mission, which, at the official exchange rate at the time, was about $2 billion. Further, according to Peter De Selding of *Space News*, the C.I.S. needs a "peak workforce of 10,000 people, including military personnel" to launch a single Energia/Buran. This is about the same number needed to launch the U.S. Space Shuttle.

Growing concern that the Space Shuttle will be unable to launch elements of Space Station Freedom, in particular, has given new life to the objective of building a new heavy-lift launch vehicle (HLLV). In the Augustine report released in December 1990, the advisory committee recommends, [30]

> Reducing our dependence on the Space Shuttle by phasing over to a new unmanned heavy-lift launch vehicle for all but missions requiring human presence.

Within 3 months was born the joint NASA-DOD National Launch System (NLS) program [31]. The NLS was envisaged as a modular launch system to launch payloads from 9 to 45 tonnes or more into low Earth orbit with various combinations of strap-on liquid and solid boosters. Thus, it was seen by some as a replacement for most existing launch systems. NLS was subsequently cancelled in October 1992, but "reborn" a few weeks later under a new name—Spacelifter.

The rationale behind the original NLS is interesting if looked at closely. While the ALS was intended to use advanced techniques, including automatic construction and checkout to reduce launch costs and increase launch rates,

**Figure 3.7** Advanced Launch System concepts.

the drive to get the NLS operational as soon as possible at the lowest *development* cost and risk led to a compromise solution using existing hardware. For example, the baseline NLS configuration would have used the Space Shuttle external tank and either simplified Space Shuttle main engines or a variant of the space transportation main engine (STME) being developed under the ALDP Program.

It is relatively easy to obtain an idea of the dedicated launch costs of an NLS-like rocket. The Space Transportation Propulsion Team—a consortium composed of Rocketdyne, Aerojet, and Pratt & Whitney—has calculated that by investing in advanced manufacturing techniques and building to cost rather than performance, the unit cost of the first 8 to 10 engines off the production line would be about \$8.5–\$9 million, and for the 500th engine the cost would be about \$4.5 million [32]. If an intermediate cost of about \$7 million is assumed, the total engine cost for a Shuttle-equivalent version of the NLS using four main engines would be about \$28 million. To this must be added the cost of the avionics, data handling, tanks, structure, shrouds, hydraulic systems, an upper stage and, perhaps, strap-on boosters. The cost of the Shuttle external tank, for example, is also around \$25–\$30 million. Once the cost of integration and launch processing has been added, the dedicated launch cost will be well in excess of \$100 million. This is significantly less than the Shuttle, but little different from other launch systems. It certainly cannot be described as "low cost." Just as importantly, it doesn't have the versatility of the Shuttle. It cannot launch people, travels in one direction, and cannot rendezvous or return payloads.

Further, the single engine configuration of NLS capable of lifting 10 tonnes into LEO was estimated to cost about

$40 million (1991) [33]. This "commercial NLS" was intended to replace vehicles like the Atlas 2 and Delta 2. However, to this $40 million must be added the cost of the upper stage required to place a comsat into geosynchronous transfer orbit, i.e., $10 million (solid) to $20–$30 million (cryogenic). Thus, the total dedicated cost of this vehicle would be $50–$70 million—little different from today.

For comparison, the European Ariane 5 is being built as a brand new launch system optimized to take advantage of new manufacturing, production, and processing systems—just like NLS and the current Spacelifter. However, even though it uses one main engine rated at a third the thrust of the NLS engine (i.e., 1,070 kN versus 3,000 kN), augmented by a relatively simple pair of solid rocket motors, the cost per flight target of Ariane 5 was originally about $100 million.

In June 1991, the Synthesis Group report was released, and its version of an NLS-like booster also used an external tank-type structure, but was powered by upgraded Saturn 5 F-1 engines [34]. However, regardless of which solution is finally adopted, it would appear that the drive to significantly reduce launch costs has begun to evaporate, as the Augustine report also indicates, [35]

> . . . a new unmanned launch vehicle itself can produce substantial savings . . . but not in the near term and in the longer term only if we change our processing philosophy and manpower . . . Future enhancements [to the proposed HLLV] would use elements derived from the Advanced Launch System technology program in progress.

If the United States is truly interested in reducing launch costs and easing access to space, then using existing facilities and components, and digging up former systems—all of which contribute to today's high launch costs—may not be the right way to go about it. Failing to invest in the near term along with compromising a configuration optimized for reducing costs is likely to strike back in the far term, just as it did with the Shuttle. (see Chapter 11.) Further, if a compromised NLS or Spacelifter is built, the cost to launch it might be so high that any thought of upgrading it with ALS-like systems might be pushed even farther into the future. As it is, the estimated development cost of the NLS or Spacelifter is anywhere from $10 billion to $12 billion. According to a prominent space industry official, "you can buy a lot of Titan 4s for $10 billion." [36]

Nothing is ever this simple. However, based on the arguments presented so far, it would appear that if an NLS-type program, or even the full ALS program, were started, then the end result would be a launcher that is fundamentally no different than today—it will be expensive to use, it will not fly very often, it will be incapable of returning payloads, and it will potentially be unreliable. Worst of all, perhaps, its size will not reflect the needs of the majority of the users who are just not interested in a launch capability much above 7 or 8 tonnes to low Earth orbit. There may well be a few payloads in the 20–100 tonne range. However, is it worth spending $10 billion over 10 years just to launch these few payloads? As Bennett and Salin note with respect to the commercial needs, [37]

> Only technologists operating without market feedback, such as NASA's upper management, Congressman operating in an isolated, politicized environment, and government contractors responding to the procurement process and with no capital of their own at risk could have believed that a few large vehicles such as the Saturn and the Shuttle were by themselves the most appropriate answers to the U.S.A.'s space transportation needs. Any commercial enterprise knows that picking too large and too complex a capacity for a new product leads to disastrous consequences . . . [Moreover] Giganticism crowds out alternative, more realistic, practical, inexpensive and rapidly attainable alternatives. Gigantic development efforts, although always justified by the great breakthroughs which they promise, instead tend to retard real progress by monopolizing financial and managerial resources which would have been more productively spent on a variety of smaller, less complex and much more rapid efforts . . . [For example,] the total development budget for Saturn and Shuttle was roughly $40 billion over the past 25 years [up to 1987]; whereas the total NASA and USAF R&D budget for improving moderate capacity launch vehicles such as Delta or Atlas probably amounted to no more than $2 billion over the same period.

Some form of heavy-lift capability may play an essential role in supporting more ambitious future space activities, such as for the bulk launch of propellants for crewed missions to the Moon and Mars. (see Chapter 19.) Yet, it does not seem realistic or rational to suggest that such launchers should become the *primary* launch system in a space transportation architecture in the immediate future, or ever. Perhaps, some of the $10 billion or so needed to build an NLS-like booster could be invested in upgrading existing launchers and ground processing facilities. Further, if the United States wanted a heavy-lift capability to assist in the bulk launch of Space Station Freedom elements, as noted in the Augustine report, a much less expensive and more readily available solution would be to build Shuttle-C (Figure 3.8). Although it could only be launched two or three times per year, this launch rate would be more than sufficient for the Freedom payloads in immediate need of such capability [38].

In essence, using heavy-lift rockets as the primary space access means will preserve a Shuttle-style status quo. As such, heavy-lift rockets will do nothing for the commercial community, and they will only launch very expensive payloads which the government has dictated they should launch. Heavy-lift rockets will do nothing to significantly reduce the

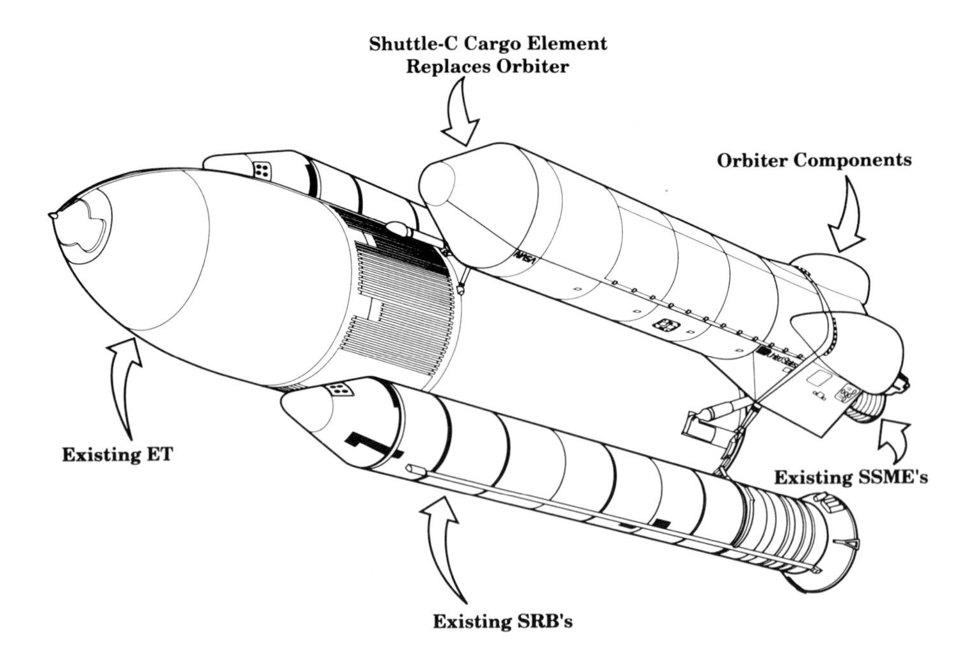


**Figure 3.8** Shuttle-C: A moderate cost option for the few heavy payloads (*NASA*).

cost of payloads, space missions, and space programs in general.

## *The European Ariane 5*

Europe is currently planning to launch its first heavy-lift launch vehicle, Ariane 5, in 1995 [39]. Ariane 5 is the latest in the European family of rockets, and apart from the name, bears little resemblance to its distinguished predecessors. For example, where Ariane 4 uses four low technology, hypergolic fuelled engines on the first stage, Ariane 5 uses a single, large, higher technology cryogenic engine. Ariane 5 was approved as a formal ESA program in 1987 with two distinct requirements. The first is to reduce the cost per kilogram to orbit by a factor of 2, while reducing the dedicated launch costs by 10% with respect to the current Ariane 4. Therefore, it will cost a little less to launch Ariane 5, and the cost per unit of mass will be about one half. The second requirement of Ariane 5 is that it must launch Hermes (Figure 3.9).

The Hermes requirement has had a considerable impact on Ariane 5. Originally (1985), Ariane 5 was to carry 15 tonnes into LEO and, with the assistance of a hypergolic stage, place up to three standard communications satellites in orbit simultaneously. However, the almost doubling of the estimated mass of Hermes from 15 tonnes to its current 23 tonnes (with 1 tonne of payload) has forced Ariane 5 to grow from a vehicle that was to have a takeoff mass of about 550 tonnes, [40] to one that now has grown to 716 tonnes, [41] with parallel performance growths in the cryogenic motor, propellant tank volume, and twin solid rocket boosters. In addition, Ariane 5 would have had to have been further upgraded early in the next century if Hermes had been built. (See Chapter 1.)

Under these conditions, is it realistic to expect that Ariane 5 can still be competitive with Ariane 4, and replace it by the year 2000 as currently planned? Consider the problems:

- The vehicle's growth leads to a growth in cost, as the cost for a like system is a function of its mass.
- Europe has no heritage or experience in building and operating large cryogenic motors or large segmented solid rocket boosters.
- Insurance for two large or three medium-class communications satellites on one unproven launcher will be difficult to obtain commercially.
- Commercial launches may eventually be interrupted periodically by crewed missions.
- There is only one launch pad.

Now that Hermes has been cancelled, Ariane 5 may meet its objectives. Otherwise, the complexity of placing people on "space or bust" boosters may jeopardize the chances of Ariane 5 every carrying forward the commercial work of its distinguished predecessors.

**Figure 3.9** Artist's impression of a Hermes/Ariane 5 launch (*ESA*).

## Summary—An Expendable Future?

Until recently, if an organization wanted to develop a space transportation capability, technology limited its choice to missile-like expendable rockets as the only way to access space. There has been no other option, and the existing launch systems have served their purpose as well as might be expected.

The types of failures that have occurred are not endemic to the launch industry but are a common, almost expected feature of any activity involving complex, sophisticated, and expensive machinery. But there are some crucial differences. Expendable launchers either achieve orbit, or they fail catastrophically—there is no way to turn around and come back to base. In addition, because current launchers are expendable or semi-expendable, testing a new launcher is an expensive and time-consuming activity that is usually minimized. Hence, high levels of user confidence cannot be established in the same way that people trust a Boeing 747. According to Paul Czysz, "What we've got now is a supertanker that goes out and a lifeboat that brings a couple of people back—are we going to make money on that? No! . . . Let's go with what technology we know, but let's not let the ship sink in the harbor if the sail tears." And Alan Bond also adds, "Any vehicle which you can't shake-down in service I think is doomed to be always very expensive."

The Ariane failure in 1990 discussed at the beginning of this chapter isn't the first time a vehicle has had a rag mistakenly left inside its plumbing. The following is a report printed in the July 15, 1991, issue of *Aviation Week & Space Technology*, [42]

> A piece of cheesecloth-like material inside a hydraulic line was responsible for a landing gear incident that cut short first flight of the USAF/Northrop B-2 No.3. The cloth evidently was inserted into the line to prevent hydraulic fluid from dripping while a repair was made prior to the aircraft's first flight on June 18 [1991]. Rejoining the high-pressure hydraulic fluid lines involved the use of cryogenic fluids and special processes, possibly necessitating a dry, clean surface at the joint. Quality control inspectors who checked the completed repair had no way of knowing the cloth was still inside the hydraulic line. In flight, the cloth apparently moved in the line and jammed a valve, preventing the gear from functioning properly during retraction/extension tests. Fuel was dumped and the aircraft landed safely on the dry lakebed at Edwards AFB, Calif. (AW&ST June 24, P. 17). Northrop managers are evaluating possible courses of action aimed at reemphasizing the need to eliminate foreign objects from production aircraft.

Constructing a vehicle which does not "sink in the harbor" and can be safely recovered after a failure; a vehicle which can be tested incrementally and repeatedly to demonstrate reliability; a vehicle which can bring back as much as it launches; a vehicle that can be launched rapidly and frequently; and a vehicle which simultaneously can drastically reduce launch costs; would obviously be desirable. This implies a fully reusable vehicle operated effectively just like an airliner is today.

If technology allows it, what seems to be needed is an *aero-space plane*.

## References

1. de Selding, P., "Ariane Accident Cause Found," *Space News*, April 16–22, 1990, p. 3. With permission of *Space News*.
2. Marcus, D., "Debris in Atlas I Centaur Stage Likely Cause of Launch Failure," *Space News*, July 8–14, 1991, p. 21.

3. Hunter, M., "The SSX: SpaceShip Experimental: Draft II," March 11, 1989, p. 2.
4. Calvert, B., *Flying Concorde*, (Airlife Publishing, Shrewsbury, England, 1989), p. 164.
5. Hunter, M., "The SSX-A True Spaceship," *The Journal of Practical Applications in Space*, High Frontier, Fall 1989, Vol. 1, No. 1, p. 50.
6. *National Space Transportation System: Overview*, NASA Kennedy Space Center, September 1988, p. 4.
7. See *Interavia Space Directory 1991–92*, (Jane's Information Group, Coulsdon, Surrey, 1991). Also, it is important to note that the second Ariane 1 test flight was a failure.
8. See "Target: 'Flight 501,' " *Aerospatiale*, April 1991, p. 24.
9. Wilson, A., *Space Shuttle Story*, (Crescent Books, Hamlyn Publishing Group Ltd., London, 1986), p. 63.
10. *Interavia Space Directory 1991–92*, p. 227.
11. See Calvert, p. 157–158.
12. Data obtained from a viewgraph presentation pack produced by British Aerospace (Space Systems) Ltd., and the Central Institute of Aero- & Hydro-dynamics (TsAGI) for the An-225/Interim-HOTOL presentation to ESA, Paris, June 21, 1991.
13. Hickman, R., and J. Adams, "Operational Design Factors for NASP Derived Vehicles," AIAA-91-5081, *AIAA Third International Aerospace Planes Conference*, Orlando, Florida, December 3–5, 1991, p. 7.
14. *Technology Influences on the Space Shuttle Development*, NASA Johnson Space Center, June 8, 1986, p. 5–11.
15. Calvert, p. 155.
16. According to a report in *Space News*, (November 25–December 1, 1991, p. 12) the total accumulated firing time of the SSME has exceeded 500,000 seconds or 140 hours.
17. It might be argued that "littered" is a little too strong a word to use. Five Ariane losses in 47 missions may not seem that much. However, it all depends on what it is relative to. For example, a low launcher failure rate of 1 in 40 (98%) missions would be equivalent to five crash landings at a major airport *every single day*. While such comparisons might seem unfair, five crashes per day per airport would definitely be described as a littered record.
18. "Investigators Believe Missing Screws On Stabilizer Caused EMB-120 Crash," *Aviation Week & Space Technology*, September 23, 1991, p. 65. Courtesy *Aviation Week & Space Technology*. Copyright 1991 McGraw-Hill, Inc. All rights reserved.
19. Hunter, M., "Tbe SSX: SpaceShip Experimental: Draft II," March 11, 1989, p. 19.
20. Isbell, D., "Budget May Lead to Mission Delays," *Space News*, November 11–17, 1991, p. 12.
21. According to a report in *Space News*, May 27–June 2, 1991, the launch delay was caused by a redesign to the third stage to correct a momentary drop of pressure in the hydrogen feed line just as the engine is ignited. Apparently, this problem was observed on several previous Ariane flights, but has been more pronounced since an Ariane launch in October 1990.
22. One space industry observer said, on returning from a trip to the then U.S.S.R., that the man-hours per Proton launch were greater than those for a equivalent Western rocket.
23. Mishin, V. B., "The USSR Launchers Programs," *Aéronautique & Astronautique*, Revue Bimestrielle, No. 150, May 1991, p. 40.
24. Covault, C., "Plesetsk Cosmodrome Gearing for New Heavy Booster Role," *Aviation Week & Space Technology*, September 16, 1991, p. 46.
25. *Interavia Space Directory 1991–92*, p. 341.
26. Ibid., p. 122.
27. Ibid., p. 345–347.
28. Branscome, D., and R. Harris, "Heavy-Lift Launch Vehicle Options for Future Space Exploration Initiatives," IAF-90-196, *41st Congress of the International Astronautical Federation*, Dresden, Germany, October 6–12, 1990, p. 4.
29. *Interavia Space Directory 1991–92*, p. 307.
30. "The Report on the Advisory Committee on the Future of the U.S. Space Program, Executive Summary," Washington, D.C., December 1990, p. 11.
31. Isbell, D., "New Booster to Draw on ALDP's Successes," *Space News*, February 11–17, p. 1991, p. 1.
32. "NLS Propulsion Team Shifts Engine Choice," *Space Exploration Technology*, January 17, 1992, p. 8.
33. Isbell, D., "Engine Redesign Aims to Boost NLS Advocacy," *Space News*, March 2–8, 1992, p. 3.
34. Synthesis Group, "Report on the Synthesis Group on America's Space Exploration Initiative," Chaired by Lt. Gen. Thomas Stafford, May 1991, p. 31.
35. "Advisory Committee" report, p. 8.
36. With Titan 4 launches costing around $150 million per flight, $10 billion would pay for 67 vehicles.
37. Bennett, J. & Salin, P., "The Private Solution to the Space Transportation Crisis," *Space Policy*, August 1987, p. 193. With permission of *Space Policy*.
38. According to *Interavia's Space Directory 1991–92*, (p. 318) the cost to develop the Shuttle-C would be around $1.5 billion over 4.5 years. Other estimates have put the cost higher at $3–4 billion. Launch costs would be about the same as the Shuttle Orbiter at about $400 million per flight. Thus, for a $10 billion investment—i.e., the same as NLS will cost to develop—the United States could develop Shuttle-C and fly it 15 to 20 times over a decade.
39. See, "Target: 'Flight 501,' " *Aerospatiale*, April 1991, p. 24.
40. See, an article in *Dornier Post*, published by DASA-Dornier in Friedrichshafen, February 1985, p. 37.
41. *Interavia Space Directory 1991–92*, p. 246.
42. Courtesy Aviation Week & Space Technology. Copyright 1991 McGraw-Hill, Inc. All rights reserved.

# SECTION 2

# AERO-SPACE PLANES IN PERSPECTIVE

*Nearly all great innovations seemingly must be preceded by prophets, men for the most part ignored or ridiculed, who have the vision to comprehend the implications inherent in a new discovery or a new innovation. Stating something novel, something that differs radically from what is currently believed, is dangerous business. It used to be common practice to feed such prophets a stout dose of hemlock brew, or simply burn them at the stake. In any period the role of prophet is no life of joy.*

—Stewart Holdbrook, *The Story of the American Railroads*
(Crown Publishers, NY, 1964)

*Civilized men simply took it for granted, for centuries, that if the wind blew against them they could do nothing but anchor, or "lie to" and wait for it to shift astern. Yet the mathematical and mechanical problem of sailing against the wind was simple, compared with that of building a stone cathedral roof more than a hundred feet in the air without steel girders . . . The men who created those miraculous buildings could have fashioned the clippers, if they had wished.*

—Alexander Laing, *Clipper Ships and Their Makers*
(Putman, NY, 1966).

*Vessels earn a return on their capital outlay only when plying between ports, and are a liability at other times. One of the advantages offered by a fully containerized transport system is the ability to load and unload cargo of various types and sizes encased in standard sized containers, thus avoiding the necessity of handling many small packages one at a time during loading and discharge. Vessels can therefore be turned around far more rapidly and savings will occur not only to the shipping company through less time being spent in port but also to all those parties involved in the handling and movement of the cargo between the dock and point of origin or consumption.*

—R. Whittaker, *Containerization*
(Hemisphere Publishing Corp., 1975)

*The novelty of long-distance passenger flight had ceased to be novel. There was now no other way to go. The miracle of flight, which had always depended on machines, had been triumphantly turned into a repetitive and unexciting routine. The adventure was over.*

—Carl Solberg, *Conquest of the Skies*
(Little, Brown & Company (Canada) Ltd., 1979)

# Chapter 4

# The Potential of Aero-Space Planes

*For decades before the advent of spaceflight, scientist and engineers pooh-poohed the sleek, needle-nosed, swept-finned spacecraft of science fiction and film, telling us that the idea of a streamlined, single-stage-to-orbit spaceship was nonsense . . . But these people were living in the Conestoga wagon era of spaceship design, and their assumption that what was state of the art then was as good as anything was going to get was fallacious at best and exceedingly narrow-minded at worst . . . something like assuming that all cars forever were going to look like Model Ts.*

—*The Dream Machines*, Ron Miller.

## A True Transportation System

Aero-space planes would enable a space transportation capability fundamentally different from anything that has gone before.

- Where current launchers cost several tens or hundreds of millions of dollars per flight, *aero-space planes might potentially be one or two orders of magnitude cheaper.*
- Where current launchers offer only one-way transportation, *aero-space planes can potentially bring back as much as they take up.*
- Where current launchers either reach orbit or catastrophically fail, *an aero-space plane should potentially be recoverable at all times during ascent, orbital operations, and descent.*
- Where current launchers are very expensive to test thoroughly, making reliability difficult to judge, *aero-space planes are potentially much cheaper to test, as a new vehicle does not have to be built for every single test, thereby allowing reliability to be demonstrated and experience to be accumulated with every flight.*
- Where current launchers can only be launched a few times per year from remote launch sites, *aero-space planes potentially can be launched many times per year from a variety of sites.*
- Where current launchers must be purchased many years in advance, *aero-space planes can potentially be launched on relatively short notice, within days or weeks.*

Aero-space planes have the potential to achieve all of these capabilities simultaneously (Figure 4.1). However, it is important to understand that launch cost reductions, reliability improvements, and availability enhancements are not achieved through having smarter engineers designing aero-space planes than those designing expendables. Nor is the number of personnel needed to turn the vehicle around decided upon by an arbitrary cost-saving measure. These basic aero-space plane capabilities are inherent characteristics of any true transportation system—specifically, vehicles that are fully reusable and incrementally testable.

Full reusability means that it is not necessary to build, stack, inspect, and check out a new vehicle before every single mission. By itself, full reusability does not necessarily lower costs; an aero-space plane that flies three or four times per year might be just as expensive and unreliable to use as an expendable rocket. An equal feature of aero-space planes is that they can be tested in a step-by-step approach. The first flight of an aero-space plane won't go supersonic. Indeed, it probably won't even pull up its undercarriage. Eventually, though, perhaps after many dozens of flights, an aero-space plane may attempt an orbital mission. In the intervening period, the incremental testing approach will allow operators to understand the individual characteristics and handling qualities of the vehicle. This is vital in developing the levels of confidence needed to reduce servicing activities in between missions and, as a result, increase launch rates, minimize delays, and reduce launch costs.

Together, full reusability and incremental testability have the potential to improve safety and reliability in a profound way. Essentially, aero-space planes should potentially have the ability to safely abort a mission at practically any point and return to Earth. Minor problems caused by rags left in

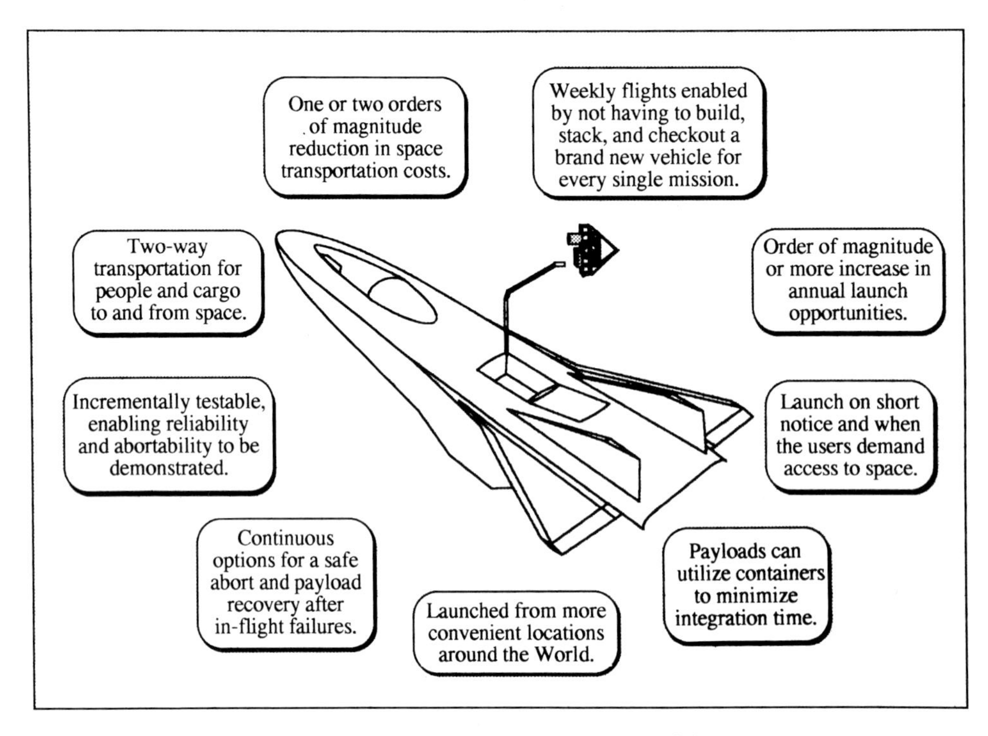


**Figure 4.1** The aero-space plane potential.

fuel lines or contaminates stuck in turbopumps might lead to an engine shutdown, but not the entire loss of the vehicle. Moreover, the inherent testability of aero-space planes should allow operators to deliberately cause failures in flight in order to learn how to safely recover these vehicles in various kinds of contingency situations. As a result, aero-space planes should be able to demonstrate that they can be safely recovered during the course of an incremental flight test program, and long before entering service, too. This is undoubtably the most important aspect of aero-space planes, just as it is with airliners. It is a capability with profound ramifications for all space activities in the future, because users will not necessarily lose their expensive payloads after an in-flight failure—as they frequently do today.

By the time aero-space planes enter operational service, such vehicles will have flown more flights in their test and qualification programs than most expendable launch vehicles will fly in one or two decades or, indeed, in their lifetimes. Aero-space planes should be able to demonstrate what they can do long before a payload is carried—a luxury that is impractical with expendable rockets.

Assuming all of the above is correct, the proposed aero-space planes have the potential to become a more affordable, responsive, and user-friendly means of travelling to and from space. If space becomes profoundly easier, safer, and cheaper to access, what impacts might this have on how space activities are conducted in the era of aero-space planes? (This is the subject of Section 4.)

While aero-space planes may indeed enable a profound improvement in launch capability compared with the current expendable rocket approach, it is nevertheless still only a potential. At the present time, no country or organization has made the political or economic commitment to actually build an aero-space plane capable of attaining orbit, although the DC-X suborbital experimental vehicle has been flight tested. (See Chapter 10.) Why should this be the case if the potential of aero-space planes is as great as supporters claim it to be?

Aero-space planes are not plying the 200 kilometer gap between Earth and space today for precisely the same reason the Polynesians did not voyage the Pacific in clipper ships or Christopher Columbus did not travel to the New World in a Boeing 707. Aero-space planes have not been built simply because they could not be built with the technologies available. Any major advance is always a hostage of technology. Indeed, it wasn't until the 1800's that the first clipper ship sailed out of the harbor and enabled widespread international colonization and trade. We had to wait until 1958 before the Boeing 707 brought the world even closer together by providing rapid, long-distance travel that, most critically of all, was affordable to many users, particularly the general public.

Yet, even though technology is the primary constraint, this does not necessarily mean that once the technology exists, aero-space planes will automatically be built. The actual benefits proffered by aero-space planes, in enabling more cost-effective space operations, must be weighed against the investment required. Also, what is the likelihood that high risks associated with this new technology really can be successfully overcome in a functional vehicle? More importantly, perhaps, their eventual construction may well hinge on whether we are *truly* interested in exploring and exploiting space more practically, routinely, and economically than we do today.

The potential of aero-space planes is enormous—but so may be the problems to overcome before they take flight.

## What Are Aero-Space Planes?

It is probably easier to explain what aero-space planes aren't, rather than what they are. Within the context of this book, aero-space planes are not high-speed passenger transportation systems that take people from one continent to another, crossing the fringes of space on the way; they are not a next generation Concorde. While this tends to be the popular view—and one that has been used for political reasons by most major aero-space plane programs to various extents—it is considered as a longer range objective that may or may not happen. So many alternatives exist for economically travelling around the world in a matter of hours that the benefits of shaving a few extra hours off the travel time might not be significant compared with the investment. Likewise, aero-space planes are not high speed reconnaissance aircraft like a Super SR-71. Although this certainly may become an important mission for aero-space planes, the almost constant availability of military reconnaissance satellites in space seems to make the need for reconnaissance aero-space planes less pressing. (See Chapter 18.)

In this book, an aero-space plane is simply a transportation system to low Earth orbit that is operated in a manner that mirrors airplanes. Specifically, it is a fully reusable vehicle that carries people or cargo to and from orbit. After returning from orbit, aero-space planes are serviced, loaded with another cargo, refuelled, and flown again—precisely in the same way a Boeing 747, Lockheed C-5A, Antonov An-225, and any other large aircraft operate. Even though aero-space planes bring up connotations of airplanes, this doesn't necessarily mean they have to look like airplanes. In this regard, probably no single word can describe these vehicles accurately. Names like "fully reusable space ships" or "non-expendable space ships" don't seem to evoke quite the right image. Other names such as "hypersonic vehicles" can, depending on the definition, exclude pure rocket-power vehicles or include expendable systems. Likewise, "single-stage-to-orbit" would exclude "two-stage-to-orbit" designs.

The term "aero-space plane" just about covers everything: "aero" because they are aerodynamic vehicles that use the atmosphere in some way during flight, "space" because they go into space, and "plane" because they are operated in a recurrent and economic manner like airplanes. The U.S. Shuttle, C.I.S. Buran, the now cancelled European Hermes, and the proposed Japanese Hope are not aero-space planes by this definition as their ground operations bear little resemblance to those of an aircraft. They could, however, assist in developing some technologies for aero-space planes.

Today, many different aero-space plane configurations are under study and technology development. Each reflects different approaches, varying requirements, and different levels of technical risk. Generally speaking, the more capable the vehicle is, the higher the technological requirements. The major programs are as follows:

- The U.S. National Aero-Space Plane (NASP) is currently planned as an experimental program, one which originated out of a desire to find a better way of accessing space (see Figure 6.5). Potentially, the experimental X-30 vehicles may pave the way for the later construction of airliner-sized, winged vehicles capable of using supersonic combustion ramjets—or scramjets—for air-breathing flight almost all the way up to orbital speed (Mach 25). A small rocket would then only be needed for the final push into space and to de-orbit the vehicle. (See Chapter 6.)
- The U.K. HOTOL, in its original form, is considerably less ambitious than NASP and aims to breath air only up to Mach 5, after which it would transition to a conventional rocket mode for the remainder of the ascent (see Figure 7.1). The joint U.K.-C.I.S. Interim-HOTOL is proposed to be launched off the back of the AN-225. It is powered only by rocket engines and, therefore, should be technically less complex and cheaper to develop than the air-breathing version. HOTOL was originated with the express goal of building an operational vehicle for no other reason than to significantly reduce the cost of accessing space. (See Chapter 7.)
- The German Hypersonic Technology Program has adopted the Sänger reference vehicle (see Figure 8.1). This two-stage-to-orbit launcher would be something of a mix between NASP and the Interim-HOTOL. It is proposed as having a manned first stage capable of flying hypersonically up to speeds of Mach 7 using a ramjet propulsion system. At such speeds, a manned or unmanned upper stage would separate and ascend to orbit through the use of conventional rockets. Sänger was initiated as an economic space transportation system and as an alternative to HOTOL. (See Chapter 8.)
- The U.S. Single-Stage-Rocket Techology program (SSRT) originally aimed at building a vehicle that can

travel to and from orbit using pure rockets in only a single stage. The McDonnell Douglas Delta Clipper concept lifts off and lands vertically using a ring of rocket engines around its base. In addition, and even though it does not have wings as such, the shape of the fuselage provides sufficient lift to allow the vehicle to fly when reentering at hypersonic speeds, thereby obtaining high cross-range (see Figure 10.1). Currently, the SSRT is the only program where a vehicle ihas been constructed—the experimental suborbital DC-X which flew in 1993. SSRT was initiated to reduce launch costs and provide an easier means of accessing space. (See Chapter 10.)

The programs of the United States, United Kingdom, and Germany are not the only aero-space plane activities. The former Soviet Union has a strong background in aero-space plane research and the C.I.S. is performing a wide variety of studies and major enabling technologies work on fully reusable launch system concepts. Its aim is to reduce the cost of accessing space. (See Chapter 9.)

The Japanese are performing advanced studies and technology work on various spaceplane proposals, including a horizontal takeoff vehicle that would use a liquified air cycle engine (LACE) and scramjets to help propel it into space. Their motivation for this research is part of a national strategy to reduce costs and find an easier way to access space.

France has begun a new national propulsion research program called PREPHA [1] which aims at investigating the feasibility of scramjet engines in lieu of a more distant goal to build a future international aero-space plane. (See Chapter 9.)

China, India, Italy, and Australia, and even countries like Sweden and Norway, have varying levels of interest in aero-space planes. These programs, like all the rest, are driven largely by the same goal—to find a cheaper and easier means of accessing space.

## *Back to Basics: Overcoming the Technical Challenges*

If the Earth were the size of Venus, then aero-space planes would not represent the challenge they do at the present time. Indeed, if Earth were the size of Mars, large expendable rockets would probably never have existed. As it is, the Earth makes life rather difficult for potential aero-space plane producers [2]. Understanding the problem is relatively straightforward; finding a workable solution, however, is an entirely different matter.

The following is a basic overview of the main technical challenges facing the development of aero-space planes. More detailed technical information is widely available in the literature.

### The Math

According to Newton's third law of motion, the force required to move an object is proportional to the mass of the object and the rate at which it is being accelerated. Therefore, to accelerate an object to a particular velocity, such as orbital speed, will take a certain amount of force maintained throughout the acceleration period. In the case of a launch system, this force comes from mass being expelled at very high speed through the engines of the vehicle—a consequence of Newton's first law of motion. The more efficient the engines are at expelling mass, the smaller the total mass or propellant that actually needs to be expelled. Hence, the desired final velocity of an object is proportional to the efficiency of the propulsion system, called the specific impulse or $I_{sp}$, and the total mass or propellant expelled by the engines. Explicitly, the relationship is expressed in the following form:

$$\Delta V = g.I_{sp} \log(M_i/M_f) \qquad (4.17)$$

where

$\Delta V$ = the increase or change in velocity of the vehicle

$M_i$ = the initial mass at launch, including payload & propellant

$M_f$ = the final mass when orbit is reached

$g$ = the constant of gravity that allows $I_{sp}$ to be expressed in units of seconds

This, in its simplest form, is the famous rocket equation. Regardless of the launcher configuration and its propulsion system, this equation always remains true. Its ramifications for placing objects in orbit, however, are rather profound. The $\Delta V$ is the total *equivalent* change in velocity that is required, not the final velocity on orbit. While the velocity of a spacecraft to remain in orbit is about 7.5 kilometers per second (km/s), in order to achieve this a launcher must also fight against the resistance of gravity and the drag of the atmosphere on the way to orbit. Consequently, the equivalent $\Delta V$ is larger than the ideal speed by a certain amount. For example, for a vertically launched rocket like the Delta Clipper, the $\Delta V$ is about 9.5 km/s, 2 km/s more than the required orbital speed. In this respect, for rocket-powered vehicles it pays to get out of the lower atmosphere as quickly as possible.

Achieving such high speed increments depends on the efficiency of the engine and how much mass is expelled, according to the rocket equation. Material thermal constraints and propellant mass properties limit the best rocket $I_{sp}$ to around 460–470 seconds (vacuum) using liquid hydrogen as the fuel and liquid oxygen as the oxidizer. Taking an $I_{sp}$ of 465 seconds and a $\Delta V$ of 9.5 km/s leads to a mass ratio ($M_i/M_f$) of around 8 for a vertical takeoff and landing, single-stage-to-orbit vehicle like the Delta Clipper. (see Chapter 10.) In other words, the takeoff mass of such a vehicle must be at least 8 times that of unfuelled mass. This means that 88% of the gross liftoff weight is propellant and the remaining 12% is structure, payload, subsystems, and so on. The problem can now clearly be seen. Until recent years, it has been extremely difficult to build a vehicle lightweight enough to achieve orbit, but also with the following characteristics:

- It has a structure and thermal protection system (heat shield) capable of withstanding the high acceleration, aerodynamic, and thermal loads during ascent and reentry.
- It can hold the large quantities of propellants at cryogenic temperatures.
- It is equipped with a relatively high performance, but lightweight, propulsion system.
- It holds additional extra propellant for a rocket-assisted vertical landing or has wings for horizontal landing.
- It has an undercarriage of sorts for a soft landing.
- It can hold all the avionics and other systems needed to control the vehicle.
- It can carry a reasonably heavy and voluminous payload to and from orbit.

If, for example, the total launch mass were 400 tonnes—a little more than a fully laden Boeing 747—this would lead to a dry mass including a 7 tonne payload on orbit of about 45 tonnes. The question then has to be asked, Can a vehicle that holds 355 tonnes of cryogenic propellants at launch actually be built for a mass of just 38 tonnes? Up until recently, a widely held view was that it could not. (Note: This calculation is for example purposes only and does not use actual Delta Clipper configuration data.)

Largely for this reason all launchers to date have been expendable or semi-expendable, and composed of more than one stage. If it is possible to throw away heavy parts of the rocket (e.g., tankage, rocket engines) after use, but well before orbit is reached, then less propellant is required compared with a single-stage vehicle simply because this expended mass does not have to be accelerated all the way up to orbital speed. Expendable, multistaged vehicles provide other advantages, too, such as allowing use of lower performance structures and propulsion systems that can be readily built with proven technology.

Until the last few years, building a fully reusable vehicle about the size and mass of a large airliner and capable of carrying a payload was considered difficult if not impossible to accomplish—as it still is in some quarters today. Actually, a one-shot *expendable* booster built with existing technology can get to orbit in a single-stage. In the mid-1960's, an Atlas booster achieved this feat, except it could not carry a payload other than a simple radio transmitter [3]. To place a payload in orbit, a much larger vehicle is needed. For example, the Saturn 5 S-II second stage alone could have placed a payload of about 5 tonnes into LEO, if it had ever been used for this task.

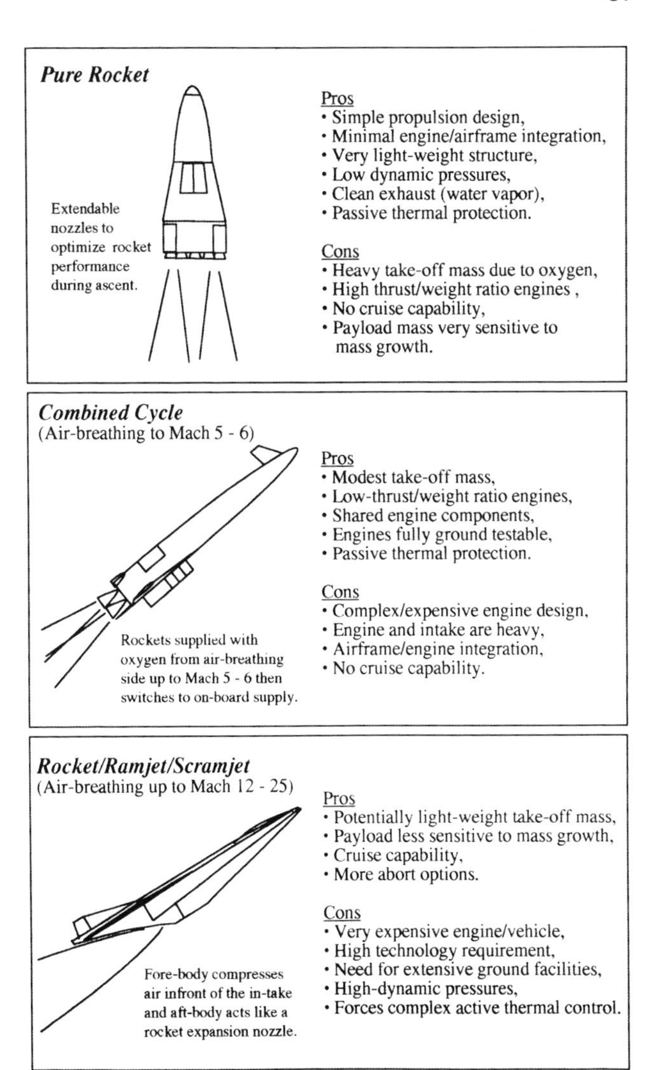


**Figure 4.2** Some propulsion system alternatives.

Clearly, for a reusable vehicle to ever be feasible, ways must be found to improve this situation significantly—in high performance propulsion systems, lightweight/high temperature/high-strength materials and structures, aerodynamics and flight control, and systems integration. These critical concerns are summarized in the following sections.

## Propulsion

Propulsion is key to the success of any transportation system, but it is particularly crucial to the future progress of aero-space planes (Figure 4.2). As mentioned earlier, the best specific impulse rockets can offer is limited to around 460–470 seconds [4]. However, if the engines of an aircraft had $I_{sp}$ equal to even the best rocket, then they would only be able to sustain flight for a matter of minutes, rather than hours. This is because a rocket-powered vehicle must carry the oxidizer on board, whereas an aircraft does not, as it is always surround by a plentiful supply. As a result, aircraft turbojet engines have efficiencies that are measured in *thousands* of seconds as opposed to *hundreds* of seconds for a rocket. For a cryogenic rocket engine, such as that used on the Space Shuttle Orbiter, the oxygen it carries in the external tank weighs about six times that of the hydrogen fuel. As a result, if it is possible to breath air for as long as possible on the way to orbit, then less oxygen has to be carried. Considered in this simple way, the value of air-breathing propulsion systems seems relatively obvious. However, when considered in the context of a system, complications emerge:

1. Even though oxygen mass is saved, new mass is added in the form of the air-breathing engine itself. Thus, rather like the Archimedes buoyancy principle, an air-breathing engine must be lighter than the oxygen it displaces.
2. The fastest air-breathing engine ever flown operationally on a large, reusable aircraft is the SR-71 which officially has achieved about Mach 3.3. Thus, a new unproven type of engine is needed that can breath air to even higher Mach numbers.
3. The higher the Mach number, the greater the thermal heating becomes due to atmospheric friction. Thus, advanced, high-temperature materials are required in the engine and for the vehicle itself, some of which may get so hot they need to be actively cooled, adding more weight to the vehicle.
4. Breathing air forces the vehicle to stay in the atmosphere longer. Thus, an even higher $\Delta V$ is needed to fight the drag compared with a ballistic rocket, meaning more fuel. The vehicle is also exposed to higher temperatures for longer, putting a further constraint on the materials used.

Despite these inherent disadvantages, air-breathing propulsion systems hold considerable potential, as the NASP, Sänger, HOTOL, and other programs are attempting to show. The French PREPHA program was started essentially to address this single issue. One of the main reasons why air-breathing systems are generally preferred over rocket-only solutions is because they make airplane-like horizontal takeoff vehicles more feasible. By contrast, a runway takeoff is relatively inefficient for pure-rocket powered vehicles, as will be discussed in Chapter 10. In addition, air-breathing propulsion can reduce the gross takeoff mass considerably compared to a ground-launched horizontal rocket, such as the RASV concept discussed in the next chapter. Chapters 6 through 10 discuss the various propulsion systems in the context of the major aero-space plane programs.

### Materials and Structures

Any vehicle flying to and from orbit will experience extremely high heating from atmospheric friction, far higher than aircraft like Concorde and the SR-71 which experience a maximum heating of about 100° C and 300° C, respectively [5]. An aero-space plane may be exposed to temperatures typically in the region of 700° C to as high as 1,800° C, depending on where it is measured on the vehicle and the trajectory taken.

At the beginning of the space age, the only way to protect against this frictional heat was to use an expendable thermal protection system (TPS) that chars or ablates to get rid of the heat. The Soyuz manned capsules still use this approach. Then, as materials technology improved, reusable TPS systems became feasible. The Space Shuttle is the first vehicle equipped with a reusable TPS: the thousands of silica tiles, blankets, and reinforced carbon-carbon panels that protect the aluminium skin of the Orbiter from high ascent and reentry temperatures. Although such an arrangement functions adequately for the Shuttle, it is structurally not very mass efficient for some aero-space plane concepts. The Shuttle's tiles take the heat loads, but they do not take structural loads, whereas the conventional aluminium skin takes all the structural load and no heat load. What is needed is a structural design that, first, is much lighter than that of the Shuttle Orbiter, perhaps by as much as 30 or 40%, and, second, shares as much of the structural and thermal loads as possible to ensure the lightest integrated structure and TPS (Figure 4.3).

These requirements are extremely severe and demand advanced materials that, until recently, existed only as small test specimens. The challenge has been to take these new materials and develop techniques that allow them to be worked into structural components as well as economically mass-produced in quantities sufficient to build aero-space planes. Other problems are explained by Terance Ronald in the case of NASP, [6]

> While progress has been rapid, the use of these new materials must be approached with caution. In addition to their high temperature and lightweight capabilities, they must have the reliability needed for use on piloted vehicles, and properties such as fatigue behavior, creep resistance, toughness, and ductility are especially important. Many of the newer materials will have some of these properties—such as ductility—that are at levels less than traditionally acceptable for long-lifetime conventional aircraft. In recognition of this fact, an important part of the NASP program is addressing behavior analysis, life prediction methods, and non-destructive evaluation techniques, in the belief that they will be key to the sucessful application of these newer materials on NASP.

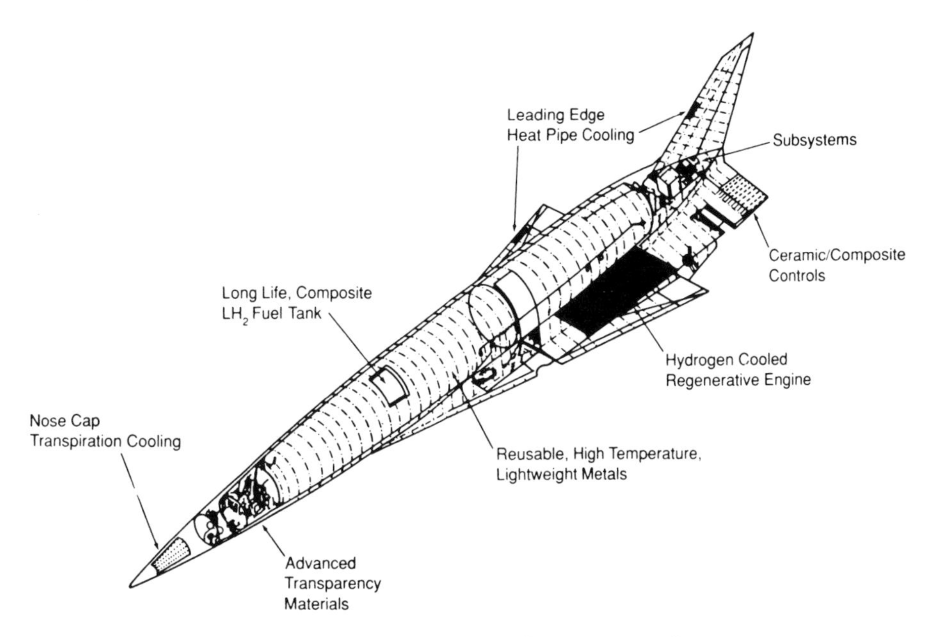


**Figure 4.3** Range of materials for aero-space planes.

Materials being pursued as candidates for current aero-space plane programs include titanium alloys, titanium-based metal-matrix composites, carbon-carbon composites, ceramic-matrix composites, and copper-matrix composites. Materials used on specific aero-space plane programs will be discussed in Chapters 6 through 10.

## Aerodynamics

Successfully designing a vehicle capable of flying through the atmosphere at very high speeds means that the aerodynamic performance must be extremely well defined. Even a small excess in drag, for example, might mean that aero-space planes will be unable get to orbit. For horizontal, air-breathing aero-space planes this is especially critical, as these vehicles stay in the atmosphere for longer periods than rocket-only vehicles.

Normally, wind tunnel testing using various combinations of continuous flow and instantaneous shock tunnels [7] are adequate for vehicles like HOTOL, Interim-HOTOL, or the Delta Clipper. For vehicles like the X-30, which hope to breathe air at very high speeds, other analysis techniques are needed. For example, wind tunnel facility limitations make it impossible to test fully functional air-breathing ramjets or scramjets much beyond Mach 8.

Consequently, aerodynamisists are forced to use computer modelling techniques to accurately analyze the airflow through the engines, including the actual combustion processes inside the engines and chemical species produced in the exhaust. This modelling, known as computational fluid dynamics (CFD), must be extremely precise as it is the only way of understanding what is likely to happen at these speeds. For example, at around Mach 20, the amount of energy that the propellant can add to the air is only about 5% larger than the relative energy of the air itself rushing over the vehicle. Therefore, when the air passes over the vehicle and through the engines, the maximum drag losses must be less than 5% of the energy of the air in order to maintain further acceleration.

Until the development of the super-computer, such as the Cray, solving the complex Navier-Stokes [8] linear equations of fluid dynamics flow over an aero-space plane was simply impossible (Figure 4.4). Aerodynamic issues specific to each of the major aero-space plane programs will be discussed in more detail in Chapters 6 through 10.

## Airframe/Engine Systems Integration

With normal subsonic aircraft like the 747, the shape of the wing and the fuselage are essentially independent of the engines, which are generally attached to the wings on an offset pylon. With many winged and air-breathing aero-space plane concepts such as Sänger and the X-30, the fuselage and wing are as much apart of the engine as the engine itself, and to maximize the performance it is vital that the aerodynamic configuration, structural design, propulsion system, and thermal management are all integrated together properly. Indeed, some people refer to the X-30 as a flying engine (Figure 4.5).

Deciding the best configuration for a vehicle like the X-30 is a complex problem. For example, the forebody is needed to compress the air before being ingested through the engine intakes, and the wider the forebody, the more air that can be fed into the engines. Increasing the fuselage size,

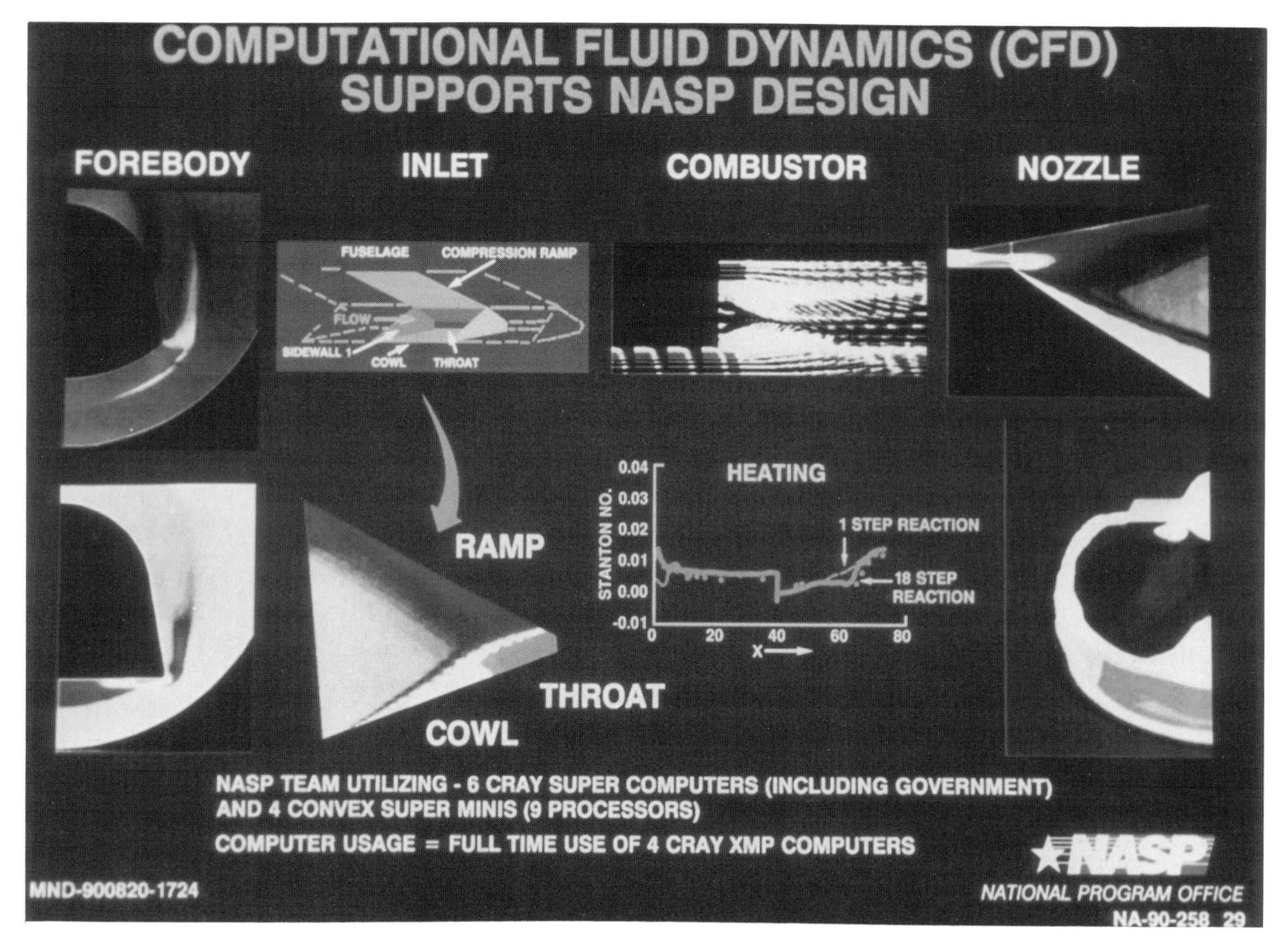


**Figure 4.4** Computational fluid dynamics for NASP (*NASP JPO*).

however, increases drag. In addition, the higher the level of forebody compression, the higher the lift that is also achieved. Therefore, a delicate tradeoff arises between the detrimental impacts of increased drag versus the positive advantages as a result of an increase in thrust and lift that is possible with a wider fuselage. A complete understanding of the entire system and how each area interacts with all the others is required to ensure that the vehicle has the best chance of working properly.

The X-30 is an extreme example because of its scramjet engines. For pure-rocket vehicles like the Delta Clipper or Interim-HOTOL, airframe/engine integration is obviously much simpler. Airframe/engine integration issues specific to each of the major aero-space plane programs will be discussed in more detail in Chapters 6 through 10.

## Can It Be Done?

As these discussions have highlighted, building aero-space planes is a difficult task requiring innovations and new developments in many critical areas, most notably propulsion and materials. The perception that aero-space planes face seemingly countless problems arises because nothing like this has ever been attempted before. It is not possible to say today with absolute certainty that aero-space planes can be built or how much they will cost. Until an organization attempts to put such a vehicle together, there will always be a question mark over their practical feasibility and economics. This raises a further question. If the feasibility of aero-space planes is so uncertain, is it realistic to speculate about their operational characteristics? After all, if current launch systems use relatively conventional tried and tested hardware and are as expensive and unreliable as they seem to be, why should an aero-space plane that uses much higher and riskier technology fare any better?

To understand why aero-space planes have the potential to fundamentally improve access to space, it is necessary to appreciate the differences between current launchers and aero-space planes that may make this possible. To under-

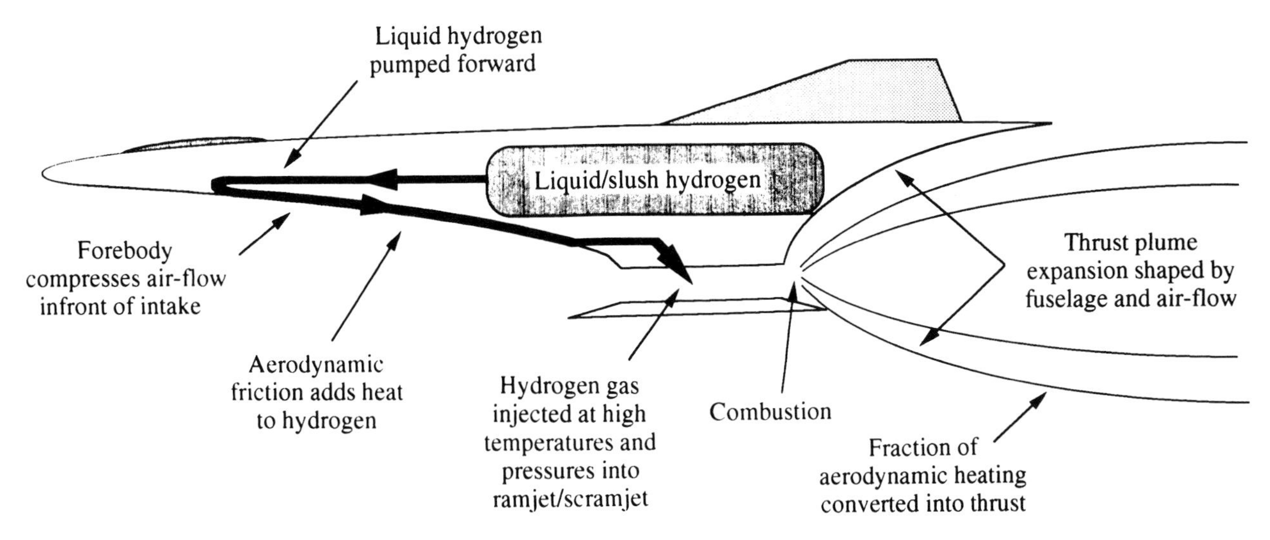


**Figure 4.5** The "Flying Engine."

stand the aero-space plane in operation is to understand the potential of aero-space planes.

## *The Aero-Space Plane in Operation*

Significantly reducing the cost of accessing space is the prime directive of operational aero-space planes. A reduction by a factor of about 10 is considered realistic by most aero-space plane programs today. Such reductions can be achieved if, first, the vehicle requires only a minimum amount of work to prepare it for the next mission and, second, the vehicle can be flown frequently relative to other launchers. The SSRT program believes that each Delta Clipper vehicle should be capable of a 7-day turnaround with 350 man-days of servicing work, compared to each Shuttle Orbiter's approximately 100-day turnaround schedule and about 1 million man-days of work. Turning around an aero-space plane in such a short period of time might seem a tall order. As Gordon Woodcock notes, [9]

> The cost targets described here are great challenges. As man is (geometrically) halfway between atoms and stars, these targets are halfway between shuttle and commercial jet-liners. Turnaround in a few days may seem incomprehensible in light of shuttle operations, but flying machines of comparable complexity are routinely turned around in an hour or less at commercial airports. Launch costs of a million dollars per flight do not defy laws of physics or even economics, but they place severe performance and life demands on launch vehicles and clearly can only be reached at launch rates far higher than published traffic forecasts.

Minimizing the activities needed to turn the vehicle around is a critical function of its configuration. In the case of large expendable rockets, the fact that a brand new vehicle must be manufactured for every mission is clearly a major contributor to high launch costs. In addition, each new launcher must receive the same standard of quality control during manufacture, and the various stages of the launcher must be properly integrated and verified. The high cost and uniqueness of the payloads sitting on top of the launcher make this necessary.

Why should this be different for an aero-space plane? Although the cost of manufacturing an aero-space plane might be significantly higher than the cost of an expendable rocket, a new vehicle does not have to be built for every mission. Ideally, the production costs of an aero-space plane would be amortized over several hundred flights. The other reasons are a little more subtle and are a function of the incremental testability aspect of aero-space planes discussed at the beginning of this chapter. Unlike current launch systems that must go to orbit on their very first flight, aero-space planes can be steadily pushed to greater speeds, temperatures, pressures, and altitudes until, perhaps after several dozen flights, orbit is achieved. A typical incremental test program might involve the following sequence of events:

1. Low and moderate speed taxis runs to verify brakes and ground handling
2. High speed ground runs up to and including nose wheel rotation to verify the speed required prior to takeoff (For vertical takeoff vehicles like the Delta Clipper, tethers might be used during initial hovering tests a few meters off the ground.)
3. First subsonic flight with undercarriage deployed to determine basic flying qualities and landing trim requirements
4. Longer subsonic flights including retraction of the

undercarriage, higher maneuverability demonstrations, trim verification, and crosswind landings
5. Demonstration of failure recovery including the shutdown and restart of an engine in flight
6. First supersonic flight to verify engine performance
7. Further supersonic characterization flights and demonstration of high speed abort capability
8. First use of rockets for air-breathing, combined cycle vehicles
9. First hypersonic flight for a short duration to verify engine performance and temperature predictions
10. Further supersonic and hypersonic flights including demonstration of higher levels of abort capability
11. Optimization trials of the ascent and reentry trajectory for orbital missions
12. First suborbital flights to verify vehicle characteristic in space
13. First orbital flight
14. Further orbital missions
15. Certification flight test program

This somewhat idealized sequence of events is not intended to be definitive, although it is interesting to note that the NASP program planned some 42 flights of the X-30s before attempting the first orbital flight. (See Chapter 6.) This list also helps to explain how the individual characteristics of the vehicle can be gradually assessed over a period of time. For example, if the instrumentation on the vehicle found that a particular area was being exposed to a higher level of heating than predicted, it would be possible to return the vehicle to the ground for analysis. If the problem was not fully understood, the vehicle could be reflown up to the point where the anomaly was first observed, and additional data could be collected with a more sophisticated instrumentation suite. The consequence of this activity might be a modification of the area of concern, perhaps with a higher temperature resistant material and changes to the vehicle trajectory.

With the Shuttle, the thermal protection covering the critical areas of the vehicle had to work perfectly on the first flight simply because the Shuttle had no choice but to go to orbit on its first flight. By the time the solid rocket boosters (SRBs) burned out, performing the difficult return to launch site (RTLS) abort—involving a 180° pitch over in the upper atmosphere and a rocket-assisted flight back to KSC before discarding the external tank (ET)—would have been almost as strenuous and demanding as actually going to orbit. In any case, the actual benefit of the data obtained from an RTLS-type abort might be small because the trajectory profile is not very representative of nominal operations. Even if useful, the many months of preparation time required between Shuttle missions mean that an incremental test program would have taken possibly as long as 10 years to achieve.

Other incremental testing options not involving an RTLS would have required landing the Orbiter in Europe, for example. This trans-Atlantic landing (TAL) option was also considered risky and expensive because of the lack of appropriate facilities available to support a landing at these sites. Likewise, the abort once around option, involving nearly one orbit of the Earth before a landing in California, would be little different from an actual orbital mission. The Shuttle was not incrementally tested because it could not be incrementally tested, with the exception of the five approach and landing tests performed by *Enterprise* dropped off the back of a modified Boeing 747 in 1977.

Clearly, if an aero-space plane can be flown in a manner that incrementally stretches the flight envelope of the vehicle over a large number of flights, each of which explores different aspects of the envelope, then this will provide confidence in how the vehicle actually behaves. This is, of course, how all aircraft, cars, boats, and trains are tested in order to demonstrate reliability before being allowed to perform an operational function.

The ability to perform incremental testing is a function of how frequently that testing can be performed. The fully reusable nature of aero-space planes will facilitate the rapid turnaround of the vehicle because of the minimum number of activities (compared with existing launchers) that need to be performed before each mission. In addition, the more frequently these tests can be performed, the quicker experience can be gained in learning how to optimize turnaround procedures. The Shuttle once again serves as an excellent example. One of the reasons why hydrogen leaks and hinge cracks still seem to plague the Shuttle is because 40 missions, of which only 4 were considered test flights, spread out over 10 years are simply insufficient to iron out these types of problems. If these 40 Shuttle flights could have been performed in the first 1 or 2 years, and each flight orientated purely toward achieving certain test objectives, the situation today might be different.

Comparing the flow of activities between the Space Shuttle and an aero-space plane is instructive in understanding why an increase in launch rates should be possible for the aero-space plane. Figure 4.6 takes an overview of the Space Shuttle processing cycle that occurs in each of the facilities at KSC, and indicates those aspects which are similar for a single-stage-to-orbit aero-space plane processing cycle. With an aero-space plane these activities are not necessary [10].

The stacking of the SRBs is a good example to highlight. After each booster segment has been stacked, gluded, and pinned together, a technician must be carefully lowered by

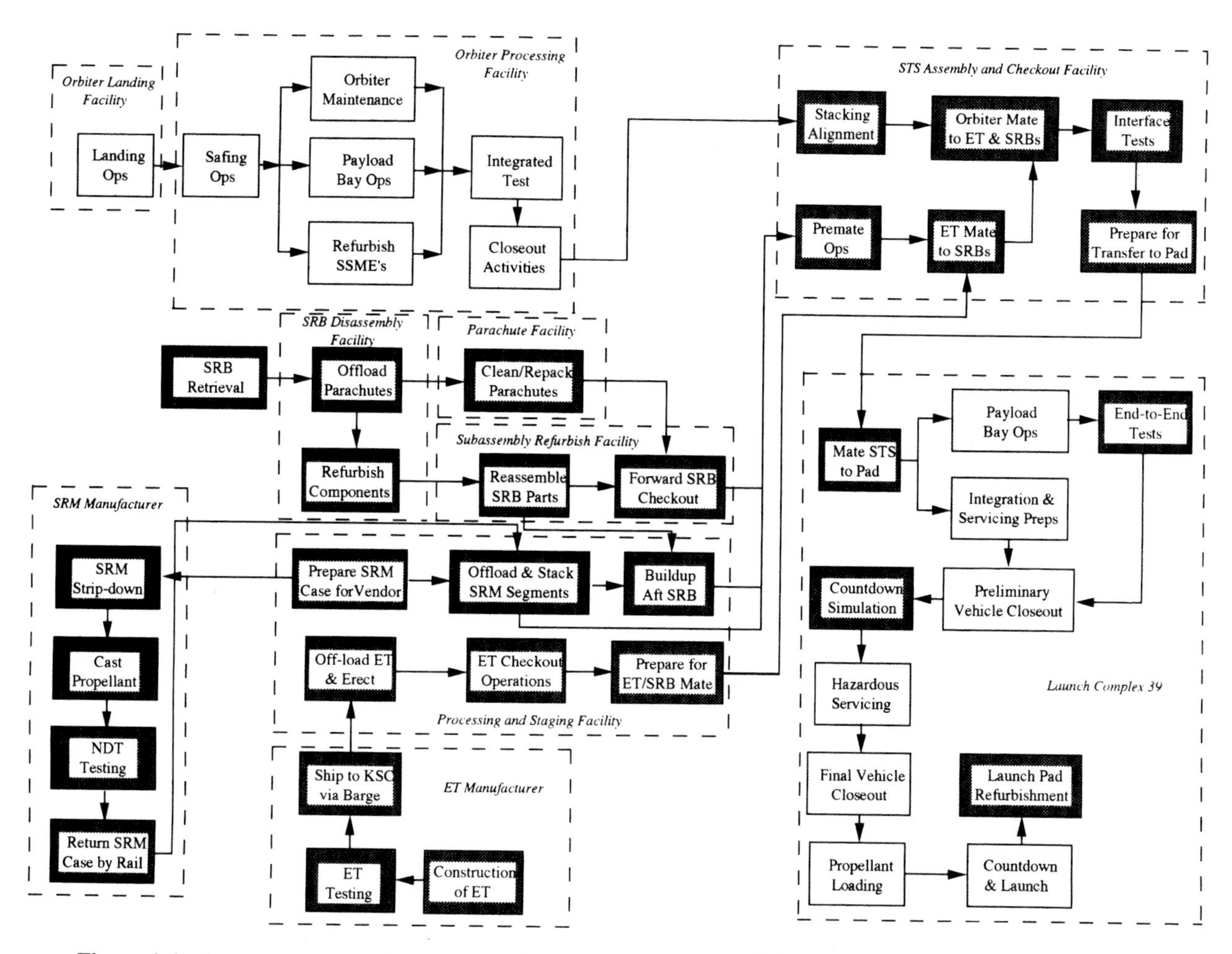


**Figure 4.6** Overview of Space Shuttle processing cycle showing areas (light grey) applicable to aero-space planes.

cable down into the SRB to inspect the quality of the joint (Figure 4.7). This is one of many time-consuming activities that must be performed before each Shuttle launch.

## Taking Advantage of Smart Technology

Aero-space planes are able to take advantage of modern methods of fault diagnostics, maintenance, and support—many of which weren't available when the Shuttle design was baselined in the early 1970's. HOTOL and NASP, for example, have proposed using smart skin technology, basically a lattice of fiber optic cables embedded into the skin of the vehicle and other major structural items such as the propellant tanks [11]. Smart skin technology allows nearly continuous measurement of the mechanical and thermal stresses within the fuselage. It also can show where the structure is experiencing greater than expected fatigue, as well as pinpoint the location of cracks as they occur.

On-board fault diagnostic equipment, including artificial intelligence systems, could constantly monitor the integrity of the structure, both on the ground and in flight, allowing prediction of when and where problems are likely to occur so that preventative maintenance work can be performed. Such systems would be of great benefit during the test program of an aero-space plane flying in an unchartered part of the envelope regime.

On-board fault diagnostics of the type described above could be applied to all systems on the aero-space plane in order to track the operations of the various subsystems and to indicate when unscheduled maintenance is necessary. One concept involves hooking the vehicle up, after landing, to a more powerful ground support computer that can interrogate the on-board computers in more detail. Such a system would also be able to compare the data with the performance of subsystems of previous missions in order to properly characterize their behavior. When problems are discovered at the appropriate scheduled servicing interval, units would be replaced with new systems.

Aero-space planes would be able to take advantage of line replaceable unit (LRU) type approaches more typical of the aeronautical industry than the space launch industry.

**Figure 4.7** A technician is lowered in the core of the Shuttle's solid rocket boosters (*NASA*).

LRUs contain elements of a particular subsystem and are built like modules that are easily plugged or unplugged from the vehicle, thus reducing the time required to replace a particular unit. LRUs also allow technology upgrades with minimum interference.

Current expendable launchers take little advantage of fault diagnostic systems. If a failure occurs during ascent, the only operation that can be performed is to switch over to a redundant system where available. The Shuttle was the first launcher to incorporate a few LRUs in certain easily accessible areas like the cockpit. In other areas they are little used, as Hickman and Adams explain, [12]

> It is important to note that the Shuttle's intricately integrated systems do not possess this [replaceable LRU] characteristic, and as a result, important Shuttle technology upgades are not economically achievable. Because the implementation of accessibility requirements can affect the vehicle's overall configuration, accessibility must be designed into the [X-30] research vehicle.

Limitations on Shuttle accessibility due to the vehicle's vertical, multielement configuration have caused problems in the past, necessitating rollback of the entire stack into the Vehicle Assembly Build for disassembly. In 1984, STS-9 carrying the Spacelab 1 payload had to be completely disassembled in order to replace the aft segments of each SRB. In 1990, both *Columbia* and *Atlantis* had to be destacked in order to access the Orbiter/ET hydrogen disconnect valves—a problem that grounded the fleet for 6 months [13]. With an aero-space plane, the impacts of such problems would be minimized because destacking-type exercises would be unnecessary, or minimal as in the case of two-stage vehicles.

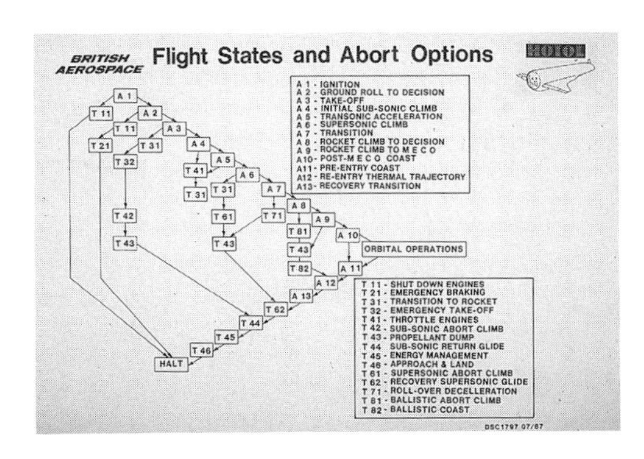


**Figure 4.8** HOTOL abort options (*BAe*).

## Failure Tolerance: A Second Chance

The ability to detect and diagnose problems in flight, coupled with the ability to abort at all points in a mission, is fundamental to building failure tolerance into a launch system—as it is with any form of transportation system. If an engine starts to overheat beyond what is expected in normal operations, then the engines can be throttled back, and an abort option can be selected based on what point the vehicle has reached in its ascent phase. If it happens while the vehicle is flying subsonically, then the vehicle turns around, dumps propellant if necessary, and lands. If the problem arises when the vehicle is more than halfway to orbital speed, it uses the remaining propulsion capability either to attain a lower than planned orbit, or if such a low orbit is useless to the mission objectives, the vehicle flies a suborbital trajectory around the Earth and returns to its original base. In all cases, the vehicle and the user's payload are safely recovered (Figure 4.8).

Assuming that aero-space planes can be built, an understanding of what the abort capability might actually mean can be obtained by comparing past launcher failures with equivalent aero-space plane (ASP) failures:

*Atlas 2: April 18, 1991*

Cause: Unknown contaminate in the engines, water or foreign object, resulted in failure of an engine to start (commercial comsat lost).

ASP: Unlikely that contaminates would get into the propulsion system if the vehicle had been flown many times before (experience). Even if they did, the remaining propulsion capability would allow an abort.

*Commercial Titan: March 14, 1990*

Cause: Incorrect wiring resulted in the satellite failing to separate from the launch adaptor (Intelsat 6 later reboosted by Shuttle).

ASP: If a payload could not be deployed, it could be returned to Earth and launched at a later date.

*Ariane 44L: February 22, 1990*

Cause: Small rag left in a water line blocking a valve, resulting in a premature engine shutdown (two commercial comsats lost).

ASP: Unlikely a rag would left in because a new propulsion system does not have to be built for every aero-space plane flight. Even if it did, the remaining propulsion capability would allow a return to launch site abort.

*Delta 178: May 3, 1986*

Cause: Hardware production deficiency resulting in premature signal given to the single main engine to shut down shortly after launch (GOES weather satellite lost).

ASP: Unlikely since any serious deficiencies would be detected during the test campaign and higher levels of redundancy (quadruplex). If it occurred, the multi-engine configuration would allow a safe abort.

*Titan 34D: April 18, 1986*

Cause: Debonding of the propellant from casing resulted in an explosion of one SRB seconds after launch (military satellite lost).

ASP: Propulsion system can always be monitored, and if outside acceptable limits, it can be shut down and the mission aborted and the vehicle returned to Earth.

*Space Shuttle: January 28, 1986*

Cause: As one factor in a complex situation, launch in weather that was unacceptably cold for the SRB O-rings to function properly (seven astronauts, TDRSS comsat, and SPARTAN lost).

ASP: Should be less constricted by weather restrictions and, in any case, propulsion system can always be monitored, and if outside acceptable limits it can be shut down, the mission aborted, and the vehicle returned to Earth (SRBs cannot be switched off).

Although this is not a rigorous analysis by any means, it can be seen in all these examples that an aero-space plane would either be unlikely to suffer these types of problems or would enable the payload to be safely recovered. Of course, an aero-space plane will have its own unique failure modes such as the burn-through of an intake. However, the ability to repeatedly test the same vehicle many times should contribute to the minimization of unexpected problems in the same way such problems are minimized on aircraft, automobiles, and most everything else.

## *Containerized Payloads*

So far in this chapter discussion has focussed entirely on the aero-space plane itself and the reasons why reduced cost launches, higher flight rates, and enhanced reliability should eventually be achievable. Aero-space planes might also fundamentally change how payloads are installed, or integrated, into the vehicle, further reducing launch cost and increasing space access. At first glance, the way an aero-space plane carries a payload would seemingly add little by way of improvement. To appreciate why the reverse is true, it is necessary to understand how payloads are currently integrated on today's rockets.

The process of integrating payloads to current launchers is complex, involving specially configuring the payload section of the launcher to accept the satellite. Ariane 4, for example, can launch single satellites, but it can also launch two payloads through use of a special adaptor (SPELDA) that come in three basic sizes. Ariane 5 will have an even greater range of payload adaptor combinations capable of supporting various types of satellites (Figure 4.9) [14]. The Shuttle also carries multiple payloads through trunion fittings placed at appropriate locations along the cargo bay. Almost every payload requires its own dedicated electrical power and telemetry/telecommand cable harness, new brackets to stop these cables flapping about during launch, new ducting pipes for thermal control (if needed), and payload-unique consoles and displays in the aft flight deck. Pyrotechnic devices and their initiators have to be carefully installed in order to separate the payload. After which, it is

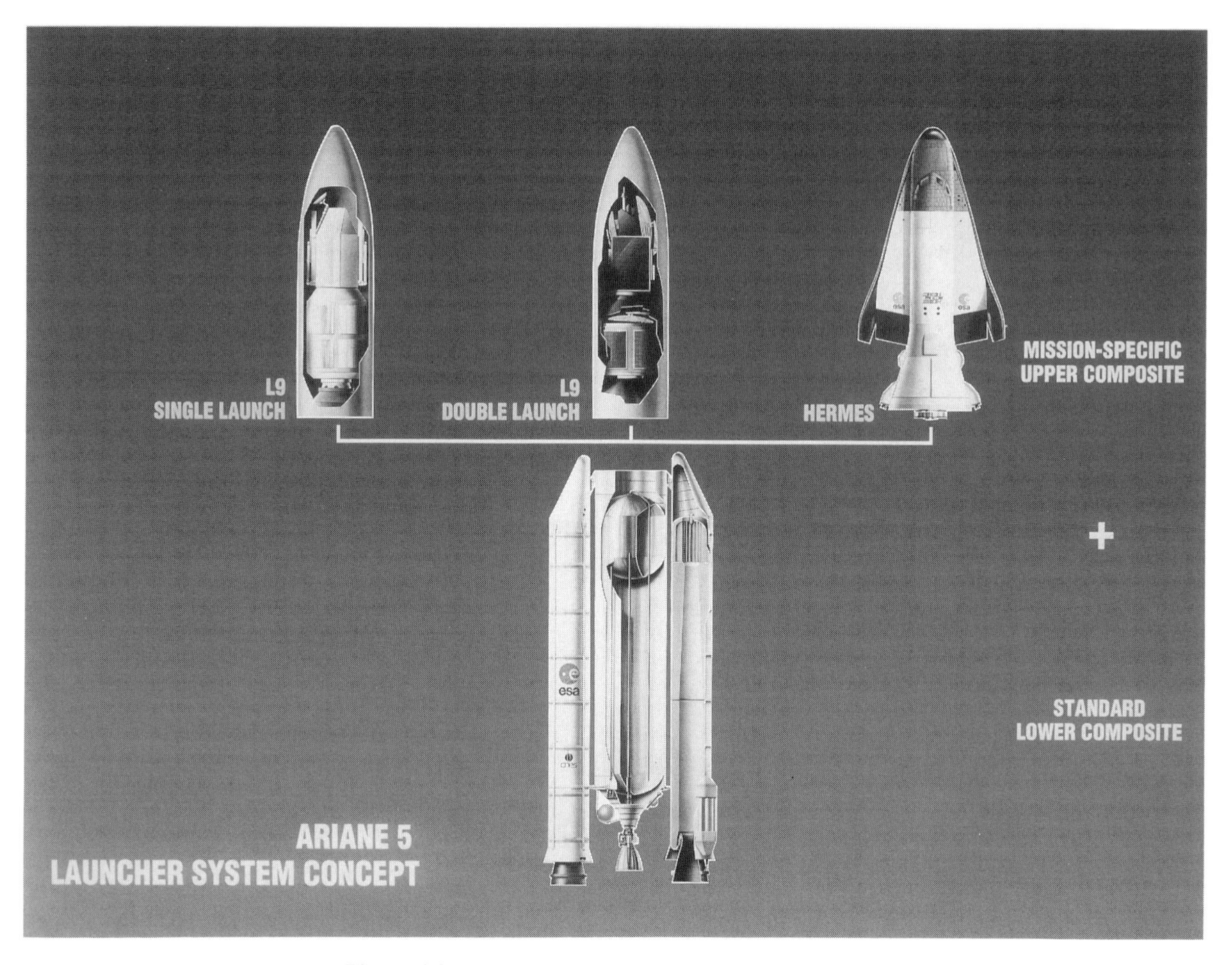


**Figure 4.9** Ariane 5 payload carrying options (*ESA*).

usually necessary to perform end-to-end testing to ensure that the payload has been properly secured to the launcher and is still functioning properly. Full countdown dress rehearsals for the deployment of Shuttle payloads are normally performed a week or more before launch.

Being complex and time consuming, current payload integration is, therefore, an expensive process. For Ariane, this process takes 1 or 2 weeks to complete, whereas on the Shuttle it can take more than a month. The uniqueness of today's payloads, coupled with the need to maximize the expensive launches, makes it actually cheaper overall to specially configure each launcher for each payload rather than standardizing everything. It is like having a tailor-made seat installed for every single airline passenger.

The difficulty is that if a payload suffers a problem, the launch is either delayed or the launcher must be reconfigured to accommodate the next payload in line. An Ariane 4 launcher was delayed several weeks in December 1991 because of faults with the Inmarsat 2 and Telecom 2 satellites. Several Shuttle launches have been also delayed, such as the last mission of 1991 due to the need to replace a faulty battery on the IUS upper stage.

Further, because the time required for payload integration is so lengthy, satellites must usually be installed many days or weeks before the major countdown activities begin. If a problem occurs with the launcher, the payload sometimes has to be removed to access the problem area. For example, during the final moments of the countdown to launch the Hubble Space Telescope, an auxiliary power unit (APU) on the Orbiter *Discovery* failed and the launch attempt was scrubbed. [15] However, to access the APU it was necessary to remove Hubble from the cargo bay—a delicate operation for this high-value payload. This activity contributed to a 2 week delay in Hubble's launch.

While launches of a particular vehicle occur every month or two, such payload problems can be tolerated, in the sense that the user has little other choice. Typically, nearly all satellite owners who book a launch 3 or 4 years in advance can usually expect a delay of several months to a year in the launch date due almost equally to the combined impacts of launcher problems and delays caused by other payloads. While this is tolerable for one-shot satellites, the ramifications for the construction and routine support of Space Station Freedom are clearly profound, as discussed in Chapter 2.

In the aero-space plane era this situation will have to change. If each aero-space plane were configured for each unique payload, then payload problems would mean that the fleet would be unable to fly as frequently as it could, and knock-on effects would delay subsequent payloads much as they do today. Likewise, if a particular aero-space plane suffered a problem forcing a launch delay, this would similarly delay the particular payload being carried. Overall, the cost per flight would be driven higher.

The answer is standardization of the interfaces between the payload and the launcher through the use of a payload container. A container acts as an adaptor between the payload and the launcher for all mechanical, electrical, and thermal interfaces. Specifically, although the internal aspects of the container can be tailored to suit the special needs of each payload, the critical feature is that *the launcher always sees exactly the same standard interface*. An aero-space plane wouldn't know what it is carrying. Whether a comsat or a space station logistics module, each payload looks and feels the same to the aero-space plane [16] (Figure 4.10).

The principal benefit of the container approach is that payload integration activities will not interfere with the aero-space plane servicing and preparation activities—and vise versa. The obvious disadvantage is the weight penalty of the container, with estimates varying at around 10% of the payload capability (about 750 kg). Such a penalty might be unacceptable today because of high launch costs, but in the aero-space plane era the benefits of containers would outweigh this penalty because launch costs would be so much less. It is also important to note that such containers don't necessarily need to be a complete enclosure, but could be a relatively lightweight frame-type structure or even a pallet.

Containers could be sent to the satellite manufacturers, where integration would be performed as leisurely as required without fear of compromising a launch schedule. Special kits could be installed inside the container with the payload to provide extra payload services as needed. Again, the aero-space plane would be utterly indifferent to the con-

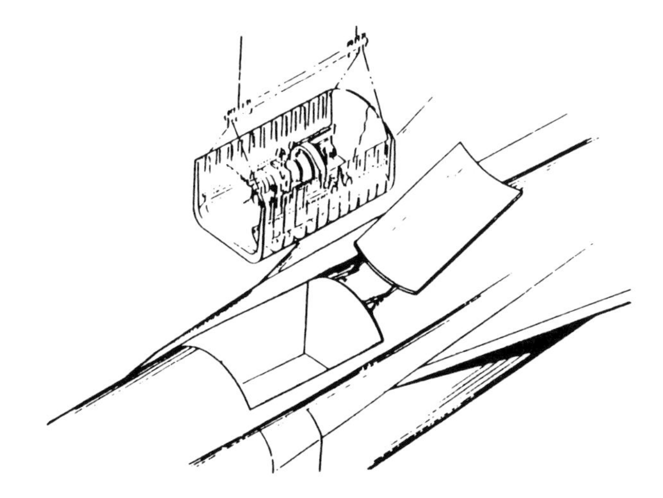

**Figure 4.10** Containerized payloads: A sign of a maturing space industry? (*Rockwell International*).

tents in its cargo bay. The container integration procedure at an aero-space plane launch site would be simple: once the launcher had been serviced, the container would be plugged into the bay, the door closed, and the vehicle rolled out, fuelled, and launched. If a problem then occurred with the aero-space plane, the container could be quickly removed and installed in the next vehicle available. Likewise, if the payload suffered a problem just before launch, the next payload in line could be launched. In both cases, interruption of the aero-space plane and payload activities could be minimized.

In addition, because the container's interfaces with the aero-space plane are always the same, operators could practice loading and unloading the container, and thereby learn how to install payloads rapidly, efficiently, and cost-effectively—something impossible today because of the uniqueness of each payload's interfaces.

McDonnell Douglas has baselined such a container for the Delta Clipper. Rockwell International—the builder of the Shuttle Orbiters—has done the same for their NASP-Derived Vehicle (NDV). Indeed, the same individuals at Rockwell who have worked on Shuttle payload integration since its inception have also defined the NDV container. According to Anita Gale, who states that Rockwell's NDV container work draws "on our unparalleled integration experience," [17]

> It is estimated that containerized installation of payloads will save weeks in the process of installing payloads in the vehicle. Assessment currently predicts that a containerized payload could be installed and checked out in an NDV in four hours—or less. Reduced vehicle turnaround times allow increased utilization of each vehicle, which leads to smaller fleet size and ground support requirements, ultimately reducing payload costs. Containerization will reduce payload integration costs as well . . . Standardization of payload accommodations for NDV also invites payload interface standardization between vehicle types . . . A sure sign of spacecraft industry maturity will be the development of standards for integrating payloads on diverse launch vehicles.

All other major forms of terrestrial transportation use standarized containers to economically and efficiently haul freight around the world. A short visit to an airport, harbor, or railway station is enough to confirm this. Pursuing space transportation concepts that allow the efficient use of containers, therefore, is a step in the right direction toward the goal of affordable, reliable, and routine access to space. Indeed, it seems totally obvious.

## Summary—Meeting the User's Needs

The above discussion illustrates some of the fundamental reasons why launch vehicles with the same basic characteristics of an aircraft would be a significant improvement on current launchers in terms of the cost of placing a payload in orbit, the payload launch schedule, and the probability that the payload will achieve its desired orbit.

Aero-space planes have the potential to reduce launch costs significantly below the best prices that current launchers can offer, first, because major elements of the launcher do not have to be built for every mission, and, second, because aero-space planes can be incrementally, repeatedly, and thoroughly tested. Such testing is critical to understanding the behavior of complex systems like aero-space planes and generating confidence in their performance. Clearly, the greater the confidence in the performance of the vehicle, the easier it will be to minimize turnaround activities important to the drive to reduce operational costs and to increase launch rates. In addition, the use of payload containers seems vital to achieving these objectives because of its ability to disentangle the payload integration from the launcher turnaround activities.

All of this is nevertheless theory. Until an organization attempts to build such a vehicle the potential of the aero-space plane will remain just that—a potential. The technical challenges are formidable, but only because nothing like an aero-space plane has ever been built before.

## References and Footnotes


1. PREPHA stands for *Programme de REcherche et de technologie sur les Propulseurs Hypersoniques Avancés*, roughly translated as: "Program of Research on Advanced Hypersonic Propulsion Technology."
2. According to Bob Parkinson, in his book *Citizens of the Sky* (2100 Ltd, England), "Unfortunately the Earth is about 10% too big to make single-stage-to-orbit easy. With the best rocket propulsion available, a vehicle has to use up about 87% of its mass as fuel to reach low Earth orbit."
3. See *Final Frontier*, "Mass Transit," March/April 1991, p. 14.
4. $I_{sp}$ is a function of the ambient pressure. The exhaust from a rocket nozzle can expand more evenly in a vacuum than if it must push through the air. While the Shuttle main engines have an $I_{sp}$ of about 455 seconds in space, at launch the ambient pressure reduces the performance to only 363 seconds.
5. Heppenheimer, T. A., *The National Aero-Space Plane*, (Pasha Market Intelligence, Arlington, Virgina, 1988.), p. 94A.
6. Ronald, T., "Structural Materials for NASP," AIAA-91-5101, *AIAA Third International Aerospace Planes Conference*, Orlando, Florida, December 3–5, 1991, p. 4.
7. An excellent compilation of the full range of wind tunnels (albeit non-U.S.) and their description and uses can be found in a U.S. General Accounting Office report entitled *Aerospace Technology: Technical Data and Information on Foreign Test Facilities*, (U.S. Government Printing Office, Washington, D.C., June 1990.)
8. There is no simple way of understanding the techniques of CFD. An overview of the CFD challenges for the NASP program can be found in McClinton, C., R. Bittner, and P. Kamath, "CFD Support of NASP Design," AIAA-90-5249, *AIAA Second International Aerospace Planes Conference*, Orlando, Florida, 29–31 October 1990. Also, Heppenheimer provides a more thorough treatment of the subject in

his book *Hypersonic Technologies & the National Aero-Space Plane*, (Pasha Market Intelligence, Arlington, Virgina, 1990.)

9. Woodcock, G. "Economics on the Space Frontier: Can We Afford It?" Space Studies Institute Update, Vol. XIII Issue 3, May/June 1987, p. 5.
10. This figure was based largely on data from a study entitled *Space Transportation Options & Launch Architecture Comparisons* performed by the NASP contractors for the NASP Joint Program Office, May 1990.
11. See Baldini, S., and W. Strange, "Development of Fiber Optic Sensors for Titanium Matrix Composites," AIAA-90-5237, *AIAA Second International Aerospace Planes Conference*, Orlando, Florida, October 29–31, 1991.
12. Hickman, R., and J. Adams, "Operational Design Factors for NASP Derived Vehicles," AIAA-91-5081, *AIAA Third International Aerospace Planes Conference*, Orlando, Florida, December 3–5, 1991, p. 7.
13. NASA Kennedy Space Center, "Information Summaries: Space Shuttle Mission Summary 1990," PMS 037, April 1991.
14. For a detailed description of the SPELDA and SPELTRA payload adaptors, see *Interavia Space Directory 1991–92*, (Janes Information Group, Coulsdon, Surrey, 1991), pp. 230, 247.
15. NASA Kennedy Space Center, "Information Summaries: Space Shuttle Mission Summary 1990," PMS 037, April 1991.
16. Each payload will not "feel" precisely same to the aero-space plane because of center of mass differences between payloads. However, provided each payload's center of mass is within a specified region, it will not matter to the aero-space plane.
17. Gale, A., R. Simberg, and R. Koeing, "Space Transportation and Payload Integration," AIAA-90-5272, *AIAA Second International Aerospace Planes Conference*, 29–31 October 1990, Orlando, Florida, p. 6. Copyright American Institute of Aeronautics and Astronautics © 1991. Used with permission.

# Chapter 5

# A Historical Perspective of Aero-Space Planes

## The Early Days

"Forever reaching beyond your grasp" epitomizes the history of the aero-space plane. Many of the first dreams of spaceflight, as far back as Konstantin Tsiolkovsky in 1903, centered around the development of spaceships that could carry people and cargo to and from space without discarding major sections on the way to orbit. Such concepts ranged from pure rocket configurations, rather like the Delta Clipper of today (Chapter 10), to others that involved the use of air-breathing, horizontal takeoff and landing vehicles, rather like original HOTOL (Chapter 7). Although many of the early designs could never have been built, each concept recognized the extent of the engineering challenges and attempted to solve them, at least partially, with the best available technology of the time. As a result, the pioneers of aero-space planes had to be content with building small —and relatively inexpensive—research vehicles capable of achieving only a fraction of the speed needed to orbit Earth. Even today, the technical practicality of building orbital aero-space planes has still to be demonstrated.

The evolution of the aero-space plane, as might be expected, goes hand in hand with the development of aeronautics and rocketry. Many of the earlier theorists believed that if it were possible to build aircraft capable of flying in the atmosphere, then it might also be possible to build aircraft-like vehicles that could fly into space. One of the earliest visions of aero-space planes was proposed by a Latvian in the 1880's. According to John Gutherie, [1]

In the 1880's, a Latvian theorist, influenced by both the development of rocketry in Russia and the work on kites and gliders then being conducted in Germany and Italy, proposed building an inter-atmospheric aero-spacecraft. This early Baltic "Spaceplane" concept would've used its wings to ascend through the Earth's atmosphere as it accelerated into interplanetary space, coasted to Mars (where it would fly in the Martian atmosphere) before returning to a gliding landing back on Earth.

This logical extrapolation of the aircraft into a spaceship was also recognized by Dr. Robert Goddard, the father of modern rocketry, and the first person to build and launch a liquid-fuelled rocket. Goddard's aero-space plane concept was published in *Popular Science* under the title "Robert Goddard's Turbine Rocket Ship to the Moon." His vehicle shown in Figure 5.1 [2], had a configuration that was, in many respects, similar to the DC-3, an aircraft that was still a number of years away. It incorporated elliptically shaped wings, as it was known that such a wing design is very efficient aerodynamically because it reduces losses resulting from air spilling over from the high-pressure undersurface to the lower pressure upper surface. For propulsion, his vehicle was powered by a combined air-breathing and rocket system. When out of the atmosphere, the vehicle used a liquid rocket motor, as would be expected. However, when in the atmosphere, two turbines were moved into the thrust stream of the rocket. The turbines drove two large propellers affixed to either wing via long prop shafts. Thus, instead of two entirely separate propulsion systems, Goddard attempted to integrate both into a single hybrid configuration to save mass. Although this design has obvious technical problems, it at least stayed within the reach of the best available technology, and some of the basic thinking behind the design of the vehicle is just as valid today as it was then.

The German Max Valier, like Goddard, was another of the many rocket enthusiasts dotted around the world interested

**Figure 5.1** Robert Goddard's aero-space plane.

in rocket-powered aircraft as a possible means to reach space [3]. He believed that aero-space planes would be developed from the hypothesized all-metal aircraft. Valier even suggested an evolutionary path for obtaining the necessary engineering experience. Initially, he proposed that low-thrust rockets be attached to a conventional G-23 Junkers airplane. Subsequently, the number of rockets would be increased and the wing span of the G-23 shortened to allow higher speeds. For flights into the upper atmosphere, Valier recognized that an entirely new vehicle design was needed, one equipped with three rockets under each wing and with a pressurized crew compartment in the fuselage to protect the crew from the low pressure at such altitudes. Between 1928 and 1929, Valier and his colleagues Fritz von Opel and Alexander Lippisch actually conducted small experiments with a rocket-powered glider. Interestingly, these experiments led Valier to discard winged vehicles in favor of wingless ballistic systems. The arguments used are as valid today as they were in Valier's time, as evidenced by the SSRT program. This also showed an awakening through actual experimentation of the technical problems inherent in the development of aero-space planes.

## *The Father of the Aero-Space Plane: Eugen Sänger*

If anyone can be considered the father of aero-space planes, it is Eugen Sänger (Figure 5.2). In 1928, Sänger enrolled in the Technical University of Vienna, Austria, to do his Ph.D. thesis. At first, Sänger wanted his doctoral research to be on a conceptual design for a stratospheric transportation aircraft, which he later dubbed the *Silbevogel* or Silver Bird [4] However, the university faculty felt that his proposed study was too esoteric, and he was advised to change the course of his research. He eventually received a doctorate for studying the structure of multispar wings. Nevertheless, the idea of the Silbevogel remained an important part of Sänger's later career.

**Figure 5.2** Eugen Sänger: The father of aero-space planes (*DASA*).

**Figure 5.3** Sänger and Bredt's Rocket Spaceplane (*DASA*).

By 1933, the design of the Silbevogel was published by Sänger in his book *Raketenflugtechnik*. The vehicle was powered by a rocket motor fuelled by liquid oxygen and kerosene and was intended to be capable of flying a suborbital trajectory, reaching an altitude of about 160 kilometers at a maximum speed of Mach 10, or about half the velocity needed to achieve orbit. Among its design features were wings with moderate sweepback, low aspect ratio, sharp leading edges, and a wedge airfoil section. In his book, Sänger considered the overall design to be relatively conventional, even though for the 1930's it was rather radical. Many of his design features were utilized in the 1950's and 60's. For example, the basic wing design described above is similar to that of the NASA/USAF X-15 vehicles. Sänger later revised his design to a vehicle that was capable of Mach 13, followed by supersonic flight at approximately Mach 3 at an altitude of 50 kilometers, allowing the vehicle to travel about 5,000 kilometers.

Sänger's work on his Silbevogel led him to collaborative research with mathematician Irene Bredt, whom he later married. This work led to the design of the Rocket Spaceplane, although his research assistants nicknamed it the "laundry iron" because of the flat wedge shape of the fuselage and its all-metal construction (Figure 5.3). Sänger and Bredt proposed that this vehicle should be launched from a sled travelling at Mach 1.5; its 100 tonne thrust rocket motor would the boost the craft into orbit. There, the Rocket Spaceplane would either perform missions with a 1 tonne payload lasting approximately two and a half orbits, a 4 tonne payload based on a single orbit, or an 8 tonne payload delivered to a point halfway around the world from the launch site. For the shortest mission, in particular, the vehicle would not travel in a conventional circular orbit, but would skip off the atmosphere the way a flat stone skips on water if thrown fast enough and at the right angle. The theory was that the vehicle could absorb the frictional heat generated as it briefly passed through the atmosphere, then the flat undersurface would allow it to skip off the denser lower atmosphere. It would then ascend back into space where the heat would be radiated before the vehicle once again plunged back into the atmosphere for subsequent skips. Although many technical problems would have precluded this approach for the Sänger-Bredt vehicle, many of today's aero-space plane concepts, including the Sänger upper stage (Chapter 8), have proposed the use of a skip-glide reentry trajectory.

At the outbreak of the Second World War, Sänger and Bredt were moved to the Rocket Flight Technique Center located at Trauen in North Germany, a facility built especially for Sänger by the Luftwaffe. Considerable competition existed between the German army and the air force, and it is believed that Sänger's research was an attempt by the Luftwaffe to have their own Wernher von Braun [5].

The couple's original work led to the *Raketenbomber* concept, or Rabo, which was intended to be a piloted global rocket bomber rather than a space transportation system. The Rabo is more widely known as the antipodal bomber, a name given to it by the German Ministry of Propaganda under Joseph Goebbels, because of its unique skip-glide trajectory that would allow it to deliver a bomb as far away as the antipode, a point halfway around the world from the launch site. The antipodal bomber was designed to deliver payloads as large as 10 tonnes to the U.S. East Coast and 5 tonnes to the West Coast, after which the skipping trajectory would settle down and the, vehicle would glide hypersonically back to the launch site.

A 1 tonne subscale version of the 100 tonne rocket motor required for the Rabo was actually tested in Trauen. Sänger was one of the first people to design and test the regeneratively cooled rocket engine. Because the majority of high-energy liquid rocket engines, such as those which use liquid hydrogen/liquid oxygen or kerosene/liquid oxygen, operate at temperatures beyond the limits of available materials, it is usually necessary to use one of the liquid propellants (e.g., liquid oxygen) as a coolant by circulating it through many tubes embedded in the exhaust nozzle and other parts of the engine before it is combusted. Sänger's 1 tonne engine was tested several times and it was so loud that the test engineers had to wear thorax microphones around their necks to communicate with each other during the dangerous test firings [6].

The *Luftfahrtfoschungsanstalt Hermann Göring*, the Hermann Göring Aviation Research Institute, not surprisingly dismissed a draft report on the antipodal bomber. It was seen as a project that was too distant, too expensive, and too risky; this was the last thing the Nazis needed as their war efforts headed for total failure. Sänger and Bredt transferred in 1944 to the *Deutsche Forschungsanstalt für Segelflug* (DFS), the German Institute for Soaring Flight, and spent the duration of the Second World War working on ramjets.

## The V-2 Rocket Team and the A-4b

In parallel with Sänger's work, Wernher von Braun and his team at Peenemünde were examining methods of increasing the range of the highly successful A-4 ballistic rocket, more commonly known as the V-2, by adding short wings to the basic vehicle [7]. Originally, the Peenemünde team planned to develop a much larger vehicle, the A-10/A-9, that could deliver a 1 tonne bomb more than 5000 kilometers. The first stage, the A-10, was a conventional rocket booster and the second stage, the A-9, was a winged vehicle that would glide supersonically after reentry before impacting the target. Like Sänger' s antipodal bomber, the A-10/A-9 was seen as too ambitious, and the concept never moved further than subscale models of the A-9, designated the A-7, that were dropped from the Heinkel bomber. All work on the A-10/A-9 was stopped in 1943 because of the burden of the A-4 program [8].

Work continued on a winged version of the A-4 that would allow its range to be increased, so that the launch site could be located farther away from allied striking range. The winged A-4, designated the A-4b, was to have a range of about 750 kilometers, three times that of the A-4 itself. Under the leadership of Ludwig Roth, two prototypes were actually constructed. The first test launch was performed on January 8, 1945, but failed moments after liftoff due to control systems problems. The second launch on January 24 was remarkably more successful. Following reentry, the A-4b transitioned to Mach 4 gliding flight, but one of the wings collapsed due to greater than expected dynamic pressures. In many respects, the A-4 and A-4b combined represented the true beginning of both ballistic rocket launch vehicles and aero-space planes. The A-4 was the first vehicle to fly into space, albeit on a suborbital trajectory, while the A-4b was the first winged vehicle to exceed the speed of sound. Indeed, the A-4b was the fastest winged vehicle until the start of the X-15 program two decades later [9].

It is also interesting to note that a small amount of work was conducted at Peenemünde, in secret from the Nazis, on a variety of space transportation systems. For example, there was a design for a manned derivative of the A-9 that was to be launched vertically and returned to Earth for a horizontal landing, just like the present day Space Shuttle. An even larger vehicle, the A-12, was conceived as a launch vehicle able to place as much as 30 tonnes into Earth orbit. This was intended to be a three-stage vehicle, with the top stage a reusable, winged reentry vehicle. All these conceptual and engineering accomplishments were seeds for the work that was to come.

The Second World War accelerated the development of aerospace and other high technology industries at a rate that could not have been foreseen. The momentum generated was not lost—if anything, it was accelerated as the result of the struggle for supremacy between the United States and the then U.S.S.R. This polarization of the world resulting from diametrically opposed ideological views carried with it the need to develop increasingly capable military systems. It led to the development of intercontinental ballistic missiles (ICBMs) to deliver the atomic bombs and also, of particular relevance to this discussion, the development of supersonic aircraft—the immediate forerunners of the aerospace plane programs of today.

## *The X-Vehicles*

Other than the partial success of the second A-4b flight, little was known about supersonic flight, especially for manned vehicles. Many questions remained. Doubts were expressed whether it would be possible for a pilot to safely steer an aircraft through the transonic flight regimes because of the compressibility effects of the shock waves on the aircraft's control surfaces. Nevertheless the United States initiated a number of bold programs to develop the technologies and systems needed to build supersonic and hypersonic vehicles. The most famous of these was the experimental X-series program, undertaken by the National Advisory Committee for Aeronautics (NACA) [10]. NACA was later to take on the role of supporting space research and development and was rechristened the National Aeronautics and Space Administration (NASA). The successes of the X-series are well known, including the first manned supersonic flight by Chuck Yeager who piloted the Bell X-1 rocket-powered aircraft to Mach 1.06 on October 14, 1947.

The X-15 is unquestionably the outstanding success of the X-series so far.The program was initiated in 1954, and the vehicles were built by North American, after a four-way competition with Bell, Republic, and Douglas Aircraft corporations. The X-15 was intended to be the first manned aircraft to fly at hypersonic speeds greater than Mach 5. Like its predecessors, the X-15 was dropped at high altitude from an aircraft (B-52) to provide initial assistance. After a few seconds the liquid oxygen/hydrogen peroxide rocket propulsion system was ignited for a burn time of a little more than a minute, after which the vehicle reentered the atmosphere and performed and unpowered landing at Edwards Air Force Base in California. The list of firsts is impressive. The aircraft was the first to fly hypersonically, achieving a maximum speed of Mach 6.7, demonstrating that stable flight at such speeds is possible. It was also the first manned aircraft to ascend above the boundary between the atmosphere and space, achieving a maximum altitude of 66.75 miles (107 km) [11].

The X-15 utilized the recently developed nickel alloy Inconel X to demonstrate the practicality of a hot external structure (i.e., no thermal protection system) to withstand the heat of the high speed ascents and reentries. The Space Shuttle uses a cold structure throughout as the Orbiter is constructed from aluminium protected by a high-temperature insulation system. NASP and Sänger are being configured to utilize hot structures where appropriate. The X-15 was also equipped with a hydrogen peroxide reaction control thruster system to ensure that the vehicle was properly orientated during reentry.

Near the end of the X-15 program, more advanced tests were planned that would take it to even higher speeds. Such tests were to utilize the recently developed supersonic combustion ramjet or scramjet air-breathing propulsion system [12]. The scramjet engine itself was the invention of Antonio Ferri (New York University) assisted by Frederick Billig and Gordon Dugger (both of The John Hopkins University) in the late 1950's [13]. Their work eventually led to the Hypersonic Research Engine (HRE) project which was managed by the NASA Langley Research Center, with Garrett Corporation building the flight hardware for the X-15 tests. Although a series of tests were planned that would culminate in a fully functional engine to raise the X-15's speed to about Mach 8, only two dummy HREs were actually flight tested. (The first flight is shown in Figure 5.4.) The second HRE flight occurred on the X-15 mission that achieved the highest speed of Mach 6.7. Unfortunately, shocks from the central spike of the HRE impinged upon one of the struts attaching the HRE to the X-15, causing localized and intense heating. The structural integrity of the attachments was so severely damaged that the HRE eventually fell off as the X-15 was nearing a landing [14]. Although the HRE flight experiments were a failure, valuable lessons had been learned about airframe/engine integration, now recognized as one of the most critical aspect of hypersonic air-breathing vehicles.

In all, the three X-15 vehicles flew 199 times between 1959 and 1968. As experience was gained in flying and operating the X-15s, including the recovery from catastrophic and other failures, many of the later flights were devoted to a wide variety of scientific research. Such research increasingly took over from engineering testing of the X-15 vehicle itself, and included characterization of the upper atmosphere, collection of micrometeroids, and microgravity experimentation for the 1 to 2 minutes the vehicle was in free fall. Indeed, to facilitate micrometeroid collection, a small door was located on the outside of the fuselage to be opened

**Figure 5.4** First flight of the hypersonic research engine on the X-15 (lower tail) (*NASA*).

when the X-15 was at the peak of its trajectory above most of the atmosphere. This type of research alone could not have justified the cost of the X-15 program—estimated to be a little more than $1 billion (1994)—although that was never the intention. It did demonstrate that useful science could be performed with aero-space planes that were able to fly on a regular basis. In the long term, this might be how the X-15 program is remembered [15].

## The X-20 Dyna-Soar

Prior to the advent of the current NASP X-30, the most ambitious of the X-series programs was the X-20 or Dyna-Soar. The goal of the X-20 was to build a winged and piloted vehicle to be launched on top of an upgraded version of the Titan ICBM, the Titan 3. The configuration is identical in many respects to Peenemünde's concept for the A-12 and, more up to date, to Europe's cancelled Hermes/Ariane 5 and Japan's proposed Hope/H-2 concepts. The X-20 was inspired by the Second World War work by Sänger and Bredt on the proposed antipodal bomber. The X-20 was apparently intended to demonstrate the concept of atmospheric skipping, followed by a long hypersonic glide at high altitudes. The "dynamic soaring" characteristics gave rise to the X-20's name—Dyna-Soar (Figure 5.5).

Although the X-20 never flew, its story is instrumental in the development of aero-space planes. The X-20 faced many problems from the start, not only technical but economic and political as well. First, its mission was never clearly understood. Initially, like all the previous X vehicles, it was intended as an *experimental* vehicle. However, as its cost rose, the X-20 had to evolve to fulfil a *functional* role, thereby justifying the enormous expenditure involved. For example, the X-20 was proposed as a vehicle that could perform reconnaissance missions, deliver a warhead, or take military crews to the proposed Manned Orbiting Laboratory (MOL), planned as the manned reconnaissance satellite, the KH-10. [16].

Unfortunately for the X-20, politics and technical constraints conspired to kill it. At the time of the X-20 program, NASA had just been established and was charged with the task of placing the first man in space. Needless to say, as the first manned space shot by the Soviets was imminent, the United States had to find the fastest way to catch up and win the race. The X-20 was not the best way to achieve this objective, and NASA instead adopted what was dubbed the "quick and dirty approach." This led to the development of the Mercury program: an expendable, ballistic capsule that could be launched on top of a much smaller and more readily available booster. The massive Titan 3 needed for the X-20 and MOL was still many years away.

Thus, the United States had two parallel manned programs, one civilian and the other military. This in itself probably was enough to kill the X-20, but technical realities put final nail in the coffin. The X-20 simply could not be built with the technology available in the early 1960's. Today, engineers quib that the X-20 required a new material they called "unobtainium." But even if the final version of the X-20 had been technically feasible, the utility of the system was questionable. For example, the crew size was reduced to just one and the payload capability was all but eliminated. The demise of the X-20 was complete: it could not be built, it had limited utility, and it was much more expensive than both the highly successful Mercury and Gemini programs combined. In December 1963 Secretary of Defense Robert McNamara cancelled the X-20, even though some $400 million (approximately $1.8 billion, 1992) had already been spent on the program [17]. Interestingly enough, the Hermes/Ariane 5 and Hope/H-2 programs appear to be repeating the experience of the X-20, as

**Figure 5.5** The X-20 Dyna-Soar (*NASA*).

both have nearly doubled in mass and cost while discarding utility. In addition, both have also been effectively cancelled. (See also Chapters 1 and 9.)

## The 1960's and 1970's

From the cancellation of the X-20 to the start of the Space Shuttle program in 1972, it would appear that relatively little work in aero-space planes was performed in the United States. This is, of course, not the case. Considerable work in the key areas of materials, structures, propulsion, and computation analysis techniques was undertaken, together with a range of systems studies. McDonnell Douglas was involved with Wright-Patterson AFB in the study of hypersonic gliders. They had a range of vehicles that could cruise hypersonically between Mach 8 and 12. One of these boost glide reentry vehicles (BGRVs) had a glide range of about 25,000 kilometers. Other BGRVs, including the ASSET and PRIME lifting-body reentry vehicles [18], were launched to further develop an understanding of hypersonic aerodynamic flight.

Aero-space plane work was not restricted to the United States either. The then Royal Aircraft Establishment (now Defense Research Agency) at Farnborough in England performed a number of classified studies of aero-space planes, and at one point the U.K. had an "operational requirement" for an orbital aero-space plane. One well-known concept was the British Aircraft Corporation's Mustard. This was a pure rocket, vertical takeoff reusable launcher involving the use of three large lifting body vehicles stacked together. During ascent, the two outboard vehicles of Mustard would pump propellant into the central orbital vehicle. Then, when these outboard vehicles had separated, the orbiter would continue the ascent starting with full tanks [19].

In 1962, Eugen Sänger reemerged to head up a space transportation systems project study conducted by a European consortium under the leadership of MBB/Junkers. Sänger's original vision of a sled-launched, single-stage-to-orbit concept had to give way to the constraints of technology, however. By the end of the study, his team concluded that a two-stage-to-orbit configuration held the greatest promise. This concept is today dubbed Sänger I, in tribute to its inventor (Figure 5.6) [20]. Later, in the late 1960's, Germany, also defined a reentry test program called ART that would have seen the launch of a reusable, boost glide reentry vehicle (BGRV) launched on top of a European rocket from Australia. This program never proceeded beyond wind tunnel testing and towing tests of a large model in Corsica in 1973. The program was terminated in 1974 [21].

In the early 1960's the Soviet Union had proposed the 50/50 concept for a small lifting body vehicle launched off the back of a hypersonic aircraft travelling at Mach 5.5–6. This vehicle was then to be boosted into space by a high-energy rocket stage rather like the French Star-H concept (Chapter 9). The Soviets actually built a full-scale version

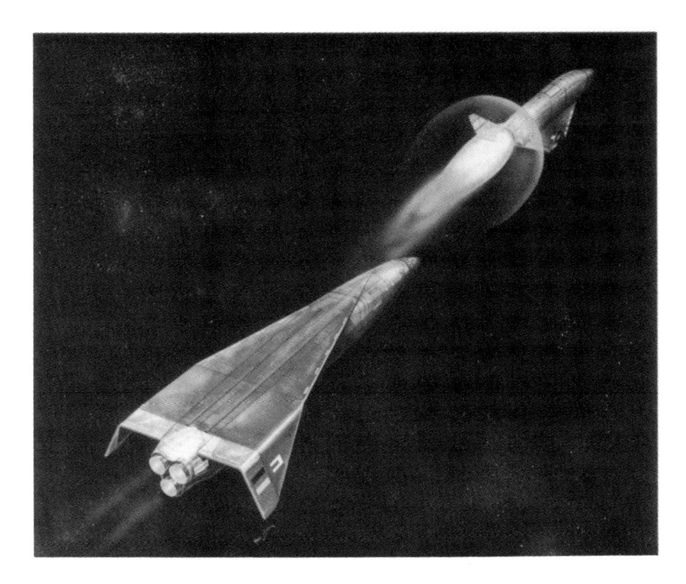

**Figure 5.6** Sänger I (*DASA*).

of the 50/50 orbiter, and it was initially dropped from the Tupelov TU-95 aircraft and later flew independently, propelled by a small turbojet engine. (More details on Soviet activities are presented in Chapter 9.)

## The X-24C, the Aerospaceplane Program, and RASV

Between 1966 and 1975, the USAF and NASA performed a wide range of flight tests of lifting bodies, which are wingless vehicles that derive lift from the shape of their fuselage. These included the M2-F2, HL-10, and the X-24A & B (Figure 5.7). Like the X-15, all were dropped from a B-52 and used rocket propulsion to raise their altitude and speed, followed by an unpowered approach and landing. The purpose of these tests was to study possible designs for orbital reentry vehicles. Originally, the X-24A & B program was to have continued through to the more advanced X-24C effort. The X-24C was intended to demonstrate many of the technical and operational characteristics of a manned, single-stage-to-orbit launch vehicle, although the vehicle itself was only to fly suborbital trajectories. This philosophy is also characteristic of the current Delta Clipper DC-X program (Chapter 10) and the Japanese Himes program (Chapter 9).

Importantly, the X-24C was to demonstrate scramjet propulsion systems to augment the built-in rocket system. Unlike the X-15/HRE, the X-24C was designed from the beginning to allow the proper integration of the scramjet with the vehicle, enabling air-breathing flights to above Mach 8. The X-24C was also to demonstrate aircraft-like supportability features, including the critical ability to turn around the vehicle relatively rapidly. Although the X-24C never passed beyond the feasibility design stage, the data collected during hypersonic wind tunnel testing has helped to verify the computational fluid dynamic codes for the NASP program [22].

The highly classified U.S. Aerospaceplane program started in 1965 and continued to 1969. A wide range of configurations were studied including combined air-breathing/rocket single-stage-to-orbit and two-stage-to-orbit vehicles, as well as rocket-only one and two stage vehicles. The technological agenda of the Aerospaceplane program did not, at

**Figure 5.7** The X-24B (*NASA*).

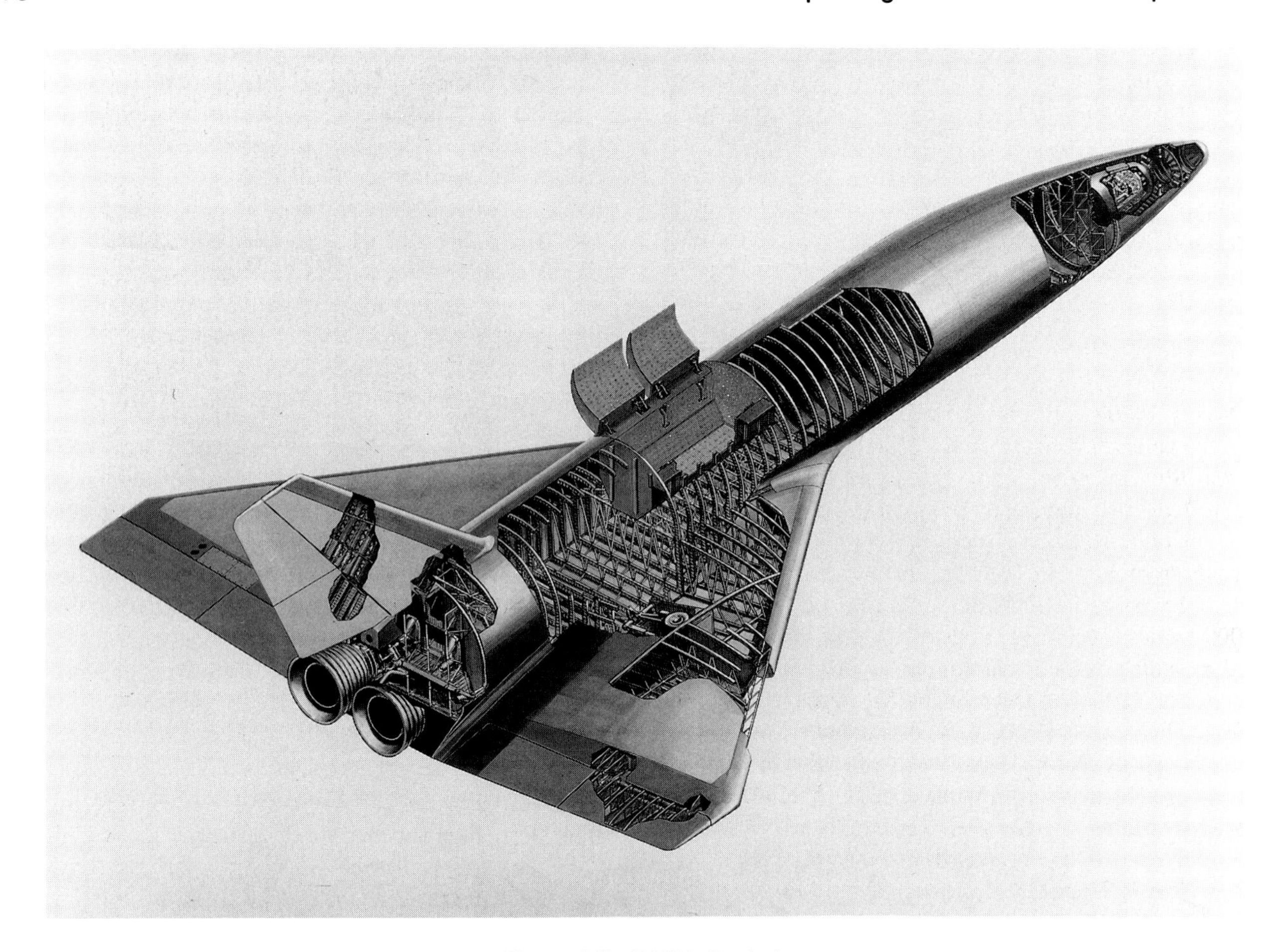

**Figure 5.8** RASV (*Boeing*).

first, call for the development of a full-scale orbital vehicle. Instead, technology efforts concentrated on the construction of a scramjet engine to be launched on top of a missile. Such an engine was under construction when the program was cancelled [23].

The Aerospaceplane program and similar studies led a number of organizations to conclude that developing air-breathing solutions for space launch vehicles would be difficult and impractical for some time. This led to a re-examination of pure rocket concepts. One which received considerable attention by the U.S. Department of Defense was the Boeing Aerospace Corporation's Reusable Aero Space Vehicle (RASV), with studies beginning in the late 1970's (Figure 5.8). The RASV was a piloted, fully reusable, rocket only, single-stage-to-orbit vehicle. The horizontal takeoff, though, was assisted by a sled system to save mass (as was Sänger's Siblevogel and the original version of HOTOL) and was conceived to place around 9–14 tonnes into Earth orbit. The RASV was the first aero-space plane design that took full advantage of advanced state-of-the-art materials and structural design technologies. The U.S. military invested modest money in the program for enabling technologies work, about $2–3 million by the early 1980's. These funds were used, in part, to construct a large section of the mid-fuselage to demonstrate that the proposed hot superalloy structure could be built and meet the specified performance requirements, as well as the thermal and mechanical stress of launch and reentry. Further, according to Gary Payton and Jess Sponable, [24]

> During the early 1980s the Air Force recognized the potential of using Space Shuttle Main Engines (SSMEs) with then-existing structural concepts and materials to achieve single-stage flight to low Earth orbit. In December 1982 Boeing Chairman T. A. Wilson committed his company to build a half scale single-stage-to-orbit (SSTO) rocket prototype at a price of $1.4 billion. But the Air Force elected not to award Boeing the contract.

Dana Andrews, the RASV program manager, provides a brief explanation as to why it wasn't pursued, [25]

> Boeing did indeed commit to build the single engine [half-scale] RASV for the USAF in 1982. The proposal was signed by the chairman of the board. It never happened because the B-2 Bomber got the money instead.

Although it is not certain whether all the technologies

were mature enough to support its development, the RASV work went a long way to provide confidence that at least the program was on the right track. If this contract had been awarded, perhaps vehicles with the Boeing label would be hauling cargo to and from space today. In 1991, RASV did have one last chance as Boeing's proposal for the SSRT program, but it was not selected. (The history of pure rocket, vertical takeoff SSTO concepts is presented separately in Chapter 10.)

The civilian community also became more interested in the operational potential of fully reusable aero-space planes. In the early 1970's, the United States became increasingly concerned about its reliance on the Middle East for oil and, after the OPEC oil embargo, the U.S. Department of Energy began to look at a number of energy alternatives. The development of large solar power satellites (SPSs) to beam electricity from the space to the ground was studied in some detail. The mass of the SPS was estimated to be a monstrous 100,000 tonnes [26]. Thus, to deploy such a high mass required a launch system that was economical to operate, could fly very often, and was able to lift large amounts per mission. This led to Rockwell International's Star Raker concept for a fully reusable, single-stage-to-orbit aero-space plane powered by scramjets and rockets. The similarity with the X-30 is striking, except when it comes to size. Whereas the takeoff mass of the proposed air-breathing NASP-Derived Vehicle is estimated at around 300–400 tonnes, the mass of Star Raker was estimated at about 2,000 tonnes so it could place a payload of about 100 tonnes into orbit [27].

## *The Trans-Atmospheric Vehicle Program (TAV)*

By the early 1980's it was becoming increasingly clear that the Space Shuttle would not fulfil the operational objectives planned for it. (See Chapter 11.) In 1982, the U.S. Air Force began to investigate possible successors to the Space Shuttle, initially under the command of Lt. Gen. Lawrence Skantze—a key player in the early success of NASP. This led to a major planning effort that culminated with the startup of the Trans-Atmospheric Vehicle Program, or TAV, which was later replaced by the NASP program. The latter completely absorbed the TAV team.

The requirements for TAV were little different from those for the Aerospaceplane program. The original studies were contracted to Batelle Columbus Laboratories and, in combination with major U.S. aerospace companies, produced 14 different TAV configurations. The best known of these was Boeing's Interim-TAV that utitilized a modified 747 as the platform for air-launching a small orbiter. Propellant for the orbiter was supplied by a discardable droptank and uprated RL-10 engines were proposed for propulsion. The 747 itself required a single Space Shuttle main engine (SSME) located in the tail in order to provide sufficient thrust necessary for a high angle of attack maneuver at the high altitudes that would be needed to efficiently separate the orbiter [28]. Interestingly, except for the SSME, the Interim-TAV proposal is almost identical to the Soviet's MAKS concept to be launched from the back of the Antonov AN-225 heavy-lift aircraft (Chapter 9). Rockwell also proposed an air-launched TAV concept using a special turbojet-powered launch platform (Figure 5.9).

**Figure 5.9** An air-launched TAV concept (*Rockwell*).

The Interim-TAV was eventually to lead to a fully reusable single-stage-to-orbit TAV configuration with supportability and maintainability features typical of military aircraft operations.One of particular interest was a concept developed by McDonnell Aircraft Corporation (McAir) that was similar to the RASV, except that it was to take off vertically. The McAir design included a modified SSME in the forward fuselage which would rotate the front end of the vehicle into a near-vertical position. At that point, the engines in the base of the vehicle would ignite and the vehicle would ascend to space. Like RASV, McAir also constructed a full-scale structural model of the part of the forward fuselage containing the SSME.

## *Summary—Technology Gains on the Dreams*

The dream of building a vehicle that can fly to and from space has been around for more than a century. However, the technology to fulfil these dreams has been slower to catch up. While there have been many ideas, the best that anyone has done so far is to build rocket-powered aircraft like the X-15 that could reach the fringes of space but were a long way from attaining orbital speed. Although the machines of Goddard, Valier, and Sänger could not have been built, these efforts all contributed to a better understanding of the problems and defined the direction of technological advancement. Today the technology for building aero-space planes is about to catch up with the dreams. The U.S. National Aero-Space Plane (NASP) and Delta Clipper, the U.K./C.I.S. AN-225/Interim-HOTOL, the German Sänger, and the programs and activities of other nations are the culmination of these past efforts. Each are discussed in the following chapters.

Will these vehicles be built? In the opinion of Paul Czysz, "I think in history there are windows of opportunity. In other words, the right people come together at the right time. That happened once before on hypersonic airplanes when, under Roy Marquardt, about 5 or 6 people, among a team of maybe 30 or 40, came together and said we can build an airplane that can go hypersonically, even to space. Even with the technology limitations they had, they could have come damn close to doing it—this was in 1956. By 1961 the whole team split up because they saw their dream crushed. The managers got into it, the bureaucrats, the politicians, especially the Air Force managers, and they created a massive program that just sank in the quicksand."

Perhaps the window of opportunity has cracked open once again for aero-space planes.

## References and Footnotes


1. Guthrie, J., " 'Spaceplanes' and the Rise of Ultra Tech," proceedings of the *First International Conference on Hypersonic Flight in the 21st Century*, University of North Dakota, September 20–23, 1988, p. 312.
2. This picture was kindly supplied by Paul Turner of Los Angeles.
3. Hallion, R. P., *From Max Vailer to Project Prime—The Hypersonic Revolution: Eight Case Studies in the History of Hypersonic Technology, Vol. 1 1924–1967, Special Staff Office, Aeronautical Systems Division, Wright-Patterson Air Force Base, 1987, pp. xi-xii.*
4. Ibid., pp. xiii-xvi.
5. " 'Spaceplanes' and the Rise of Ultra tech." p. 312.
6. Information on Sänger's rocket engine testing was derived from a video tape of a lecture by Robert Gutherie.
7. Ordway, F. I., and M. Sharpe, *The Rocket Team*, (Thomas Y. Crowell, New York), 1979.
8. From *Max Vailer to Project Prime—The Hypersonic Revolution: Eight Case Studies in the History of Hypersonic Technology, Vol. 1 1924–1967*, pp. xvi-xviii.
9. Ibid.
10. For a comprehensive review: of the X-series of research vehicles see Miller, J., *The X-Planes X-1 to X-29*, (Specialty Press, Marine on St. Croix, Minnesota).
11. Ibid.
12. For a comprehensive assessment of the early U.S. work in ramjets and scramjets see Gilreath, H., "The Beginning of Hypersonic Ramjet Research at APL," *John Hopkins APL Technical Digest, Volume 11, No. 3 & 4, 1990, pp. 319–335.*
13. Heppenheimer, T. A., *The National Aerospace Plane*, (Pasha Market Intelligence, Arlington, Virginia, 1987), p.10.
14. Ibid.
15. For a summary of the X-15 flights that achieved "spaceflight" (i.e., above 50 nautical miles) and descriptions of the science research performed, see Furniss, T., *Manned Spaceflight Log* (Janes Publishing Company, London, 1983).
16. Baker, D., *The Shape of Wars to Come*, (Hamlyn Paperbacks, Middlesex, 1982).
17. "Round Trip to Orbit: Human Spaceflight Alternatives—Special Report," U.S. Congress, Office of Technology Assessment, OTA-ISC-419, U.S. Government Printing Office, Washington, D.C., August 1989, p. 72.
18. *From Max Vailer to Project Prime—The Hypersonic Revolution: Eight Case Studies in the History of Hypersonic Technology, Vol. 1 1924–1967.*
19. Wilson, A., *Space Shuttle Story*, (Crescent Books, Hamlyn Publishing Group Ltd., London, 1986), p.14.
20. Högenauer, E., "Sänger: European Reusability," *Space Magazine, May-June 1988, p. 4.*
21. *Hypersonic Technology Programme*, Federal Ministry for Research and Technology, Bonn, Federal Republic of Germany, 1988, p.4.
22. Derived from discussion with U.S. aerospace officials.
23. Heppenheimer, T. A., *Hypersonic Technologies and the National Aerospace Plane, (Pasha Marker Intelligence, Arlington, Virginia, 1990).*
24. Payton, G., and J. Sponable, "Designing the SSTO Rocket," *Aerospace America*, April 1991, p. 40. (Copyright American Institute of Aeronautics and Astronautics © 1991. Used with permission.)
25. Letter from Dana Andrews of Boeing Defense & Space Group, January 14, 1992.
26. "Satellite Power System. Concept Development and Evaluation Program," NASA/DOE Report DOE/ER-0023, October 1987.
27. From a paper on Star Raker supplied by Rockwell International.
28. From discussion with U.S. aerospace representatives and the "Space Transportation Options & Launch Architecture Comparisons," presentation viewgraphs by the NASP Joint Program Office, May 1990.

# Chapter 6

# NASP: Pushing the Limits

## *What's in a Name?*

It is unusual for support from the highest political offices to be given to the development of an *experimental* aircraft. Even by experimental standards, the National Aero-Space Plane (NASP) is no ordinary aircraft, and it received no ordinary support from President Reagan. In his State of the Union address of February 4, 1986, he made the following announcement,

*We are going forward with research on a new Orient Express, that could, by the end of the next decade, take off from Dulles Airport and accelerate up to 25 times the speed of sound, attaining low-Earth orbit or flying to Tokyo within two hours.*

This announcement led to the formation of the NASP program in April 1986. NASP was in the late 1980's the largest of the aero-space plane programs around the world, and evolved from the secret Copper Canyon work conducted by the Defense Advanced Research Project Agency (DARPA) int the mid-1980's. NASP is generally regarded as one of the most technically challenging projects ever initiated, and the program is well known for its ground-breaking research in the areas of advanced materials, supersonic combustion ramjets, computational fluid dynamics (CFD), and systems integration aspects of hypersonic vehicle design. For example, during the late 1980's it was estimated that more than half the total U.S. supercomputer time (day and night) was devoted to NASP CFD analysis.

The NASP program was originally charged with the task of developing an experimental vehicle—not a prototype—able to test many of the key technologies critical to future operational vehicles, including hypersonic military reconnaissance aircraft and high-speed passenger transportation systems. Thus, a goal of the NASP program is to demonstrate the feasibility of sustained hypersonic cruise [1].

Although this objective is important, it is not the main reason for the existence of NASP. The primary objective and impetus of the NASP program is to build an experimental vehicle capable of flying directly to low Earth orbit, breathing air to as high a speed as possible before rocket motors must be used for the final burst to place the vehicle in orbit. Dr. Bob Williams, the originator of NASP, said that the "companies were all very eager [to work on NASP], and not just because of getting a contract, but because they were all pretty fed up with the Shuttle and the cost of launch systems. They knew there had to be a better way."

The combined requirements of single-stage-to-orbit through breathing air at the greatest possible speed is an immensely challenging goal. Indeed, it completely dominates the technical design of the vehicle, enveloping all other requirements including those relating to hypersonic cruise. Nothing like it has ever been attempted before. According to the former NASP Program Director Dr. Robert Barthelemy, [2]

The goal of the NASP program is to develop and demonstrate the feasibility of horizontal take-off and landing aircraft that utilize conventional airfields, accelerate to hypersonic speeds, achieve orbit in a single stage, deliver useful payloads to space, return to Earth with propulsive capability, and have the operability, flexibility, supportability, and economic potential of airplanes. In order to achieve this goal, technology must be developed and demonstrated which is clearly a quantum leap from the current approaches being utilized in today's aircraft and spacecraft.

These have been the NASP program requirements from the very beginning. Referring to a presentation the Copper Canyon team made to Dr. George Keyworth II, science advisor to President Reagan, Bruce Abell said, [3]

So imagine our shock, and delight, in 1984 to have someone walk in and describe a program that could offer a flexible, lower-cost alternative for space access. And this is something I must stress. It was the NASP's potential for *space access*, far and away above any other possible uses, that prompted our strong support for creating a highly visible national program . . . Virtually all our grand plans, as well as the smaller ones, are hostage to inadequate space access. Space access [today] is too expensive, it's too inflexible, and it's too unpredictable.

Throughout the short history of NASP, a remarkable amount of confusion has existed regarding the true purpose of the program. This is difficult to comprehend, although it is almost certainly a throwback to President Reagan's announcement of an Orient Express, added by his then speechwriter Dana Rohrabacher. The name was originally conceived by Gus Weiss of the National Security Council, A. Scott Crossfield, and Paul Czysz who said, "We thought up the name 'Orient Express' in the Aviation Club in Washington, D.C. one night. They wanted something catchy that talked about a transportation system that spanned continents." As Bob Williams adds, "The hypersonic airliner thing—[it] was basically Paul leading that. He was trying to get hypersonics going one way or another. When NASP came along, he just jumped on board and carried with him his vision of the hypersonic transporter, which I happen to think is great. There's a lot of truth and reality in that, if the country will ever move to it. But, that's kind of another tangent [from the space transportation requirement of NASP]."

The experimental vehicle that may eventually emerge from the NASP effort has been designated the X-30 although, as of late 1993, the prospects of the full-scale orbital vehicle being built in the forseeable future appear remote. Nevertheless, the X-30 concept maintains the distinguished tradition and heritage of many previous X-series programs, dating back to the X-1, the first aircraft to break the sound barrier in late 1947, and the highly successful X-15, the first manned aircraft to fly hypersonically and reach the fringes of space in the 1960's [4].

## Background: Copper Canyon

Surprisingly, the birth of the NASP program had relatively little to do with the Trans-Atmospheric Vehicle (TAV) program being performed at Wright-Patterson AFB. (See Chapter 5.) In fact, Dr. Bob Williams at DARPA spearheaded the idea of building an aircraft that could breath air up to nearly orbital velocity. His efforts led DARPA Director Robert Cooper to initiate in June 1983 the classified Copper Canyon program, with an investment of $6 million [5] (Figure 6.1).

By mid-1985, promising results from the Copper Canyon work and increasing enthusiasm by the contractors led Williams and Cooper to lobby for a national program to develop a vehicle that could reach orbit almost entirely on air-breathing propulsion. Meetings were held with a number of individuals including Edward Aldridge, undersecretary of the Air Force, James Beggs, NASA administrator, Lt. James Abrahamson, SDIO director, and the presidential science advisor, George Keyworth II, who advocated that trans-atmospheric research should be given a high national priority and was responsible for getting President Reagan interested in the program [6].

**Figure 6.1** 1985 government baseline configuration for the X-30 (*NASA*).

Of these meetings, perhaps the most critical was with Lt. Gen. Lawrence Skantze, the man who initiated the TAV program. Gen. Skantze was enthusiastic about the scramjet as a way to reduce the mass of rocket-only powered TAVs [7]. In addition, according to Bob Williams, "Skantze understood and he had that same vision. That's the reason I think Cooper introduced me to Skantze because he knew that Skantze understood. You kind of know in the first five minutes whether the guy understands the payoff or not. These guys implicitly understood the economic potential—you almost wonder if it is genetic or something. With other people, you try to show them fifty different ways, but they can't see over the next hill."

Bob Williams also tells about Gen. Skantze's first encounter with the scramjet, "We brought Skantze up to observe the engine tests, except we almost blew up the engine. What had happened, the technician was so excited to have this 'four-star' in there that he hadn't released the override on the igniters for the engine. There were two of them, and he hadn't released the second. So the chamber filled with hydrogen until it fed back far enough to get to the first igniter. It then went off with a kaboom! Well, anyhow, this fuel-rich plume went pouring out the back-end and so there was this building adjacent to the test facility with an asphalt roof. And the asphalt starts to catch fire. So we just finished the presentation to Skantze, and were going to walk out and change buildings, and there was this same technician standing outside spraying this hose on the roof, and we literally walked underneath the hose on the way out."

Skantze assisted Cooper and Williams in their efforts to build a national team. This secured support in the Pentagon and eventually led to DARPA being joined by the DOD Offices of the Air Force, SDIO, and Navy, and on the civilian side by NASA. Finally, by December 1985 Defense Secretary Caspar Weinberger gave authorization for the program to proceed, allowing the release of Requests for Proposals to industry [8].

At the same time, a suitable name had to be found. Prior to the formation of the multiagency team, Gen. Skantze had, in October 1985, launched the Advanced Aerospace Vehicle (AAV) program to evaluate the effectiveness of air-breathing propulsion for TAVs [9]. However, throughout the TAV and the previous Copper Canyon work, many of the participants referred to the vehicles they were studying as "aerospace planes," primarily because of the 1960's Air Force Aerospaceplane Program. In addition, because the estimated cost of their proposed program was already in the region of several billion dollars, it was recognized early on that any effort would have to be national in nature. By December 1, 1985 the National Aero-Space Plane or NASP program had been born, and 2 months later President Reagan made the first public announcement of its existence. By May 1986, the Reagan-appointed National Commission on Space made the following endorsement of the program, [10]

In view of the great promise of this [NASP] system we strongly recommend that: **The technology advances required for aerospace plane development and flight test receive the highest national priority.**

NASP got off to a flying start (Figure 6.2).

## *The Politics of NASP*

The NASP program is often compared with the spirit of the other experimental programs. One fundamental difference is apparent—the cost. The NASP program was going to be expensive with an estimated cost of $3.1 billion (1985) [11] through to completion of the flight test program. However, this estimated budget was based on an X-30 concept with an estimated gross takeoff mass of just 23 tonnes (50,000 pounds). Today, the current X-30 configuration has a takeoff mass in excess of 200 tonnes. Thus, the $3.1 billion is considered as a highly optimistic minimum. The true cost is more than $10-15 billion.

**Figure 6.2** NASP X-30 in Earth orbit (*NASA*).

The NASP program managers recognized that selling a multibillion dollar experimental program was not going to be easy. As a consequence, Bob Williams and the U.S. administration team pulled together an innovative arrangement between each of the five agencies in the program. According to Williams, "What we did was to tie the program up in a bureaucratic balance by constructing a Memorandum of Agreement between the four DOD agencies, and then between DOD and NASA. We had a situation there where nobody could back out. If one guy tried to reduce his funding, the others would all beat him up. If DOD tried to do that, NASA would beat them up. It worked marvelously; nobody was able to play games with the funding."

NASP's first budget request for Fiscal 1987 had the following breakdown: USAF $34 million, Navy $26 million, SDIO $30 million, DARPA $60 million, and NASA $62 million [12]. The appropriations committee did not agree with this arrangement and resolved to consolidate the budget. Bob Williams continues, "What happened was that staffers on the Hill said, 'This is too complicated for us to follow—we want you to consolidate everything.' So with a wave of a hand, they just set aside the administration's memorandums of agreement and everything was consolidated into basically a single DOD and NASA agreement. In retrospect, maybe we should have kept it in DARPA." The final budget for Fiscal 1987 consolidated DOD funds in DARPA ($100 million) and in SDIO ($10 million.) NASA received a total of $62 million, as requested [13].

For Fiscal 1988 and onward, all DOD NASP funds were put under control of the Air Force. DARPA funding was significantly reduced, although the plan at this time was for management control of the program to still remain with Bob Williams at DARPA. This situation was not particularly appealing to the Air Force. According to Heppenheimer, [14]

. . . Air Force officials were unhappy about putting in the lion's share of the funding—$245 million out of a total of $349 million in FY 1989 [proposed budget at that time]—and yet not having direct project authority. The Air Force traditionally runs what it funds . . .

Under the original Memorandum of Agreement, program funds were to remain broadly comparable. For example, the Air Force was to provide about 31% ($170 million) of the funds for Fiscal 1989, with the SDIO at 24%, the Navy at 19%, and NASA at 18%, with DARPA receiving about 8%

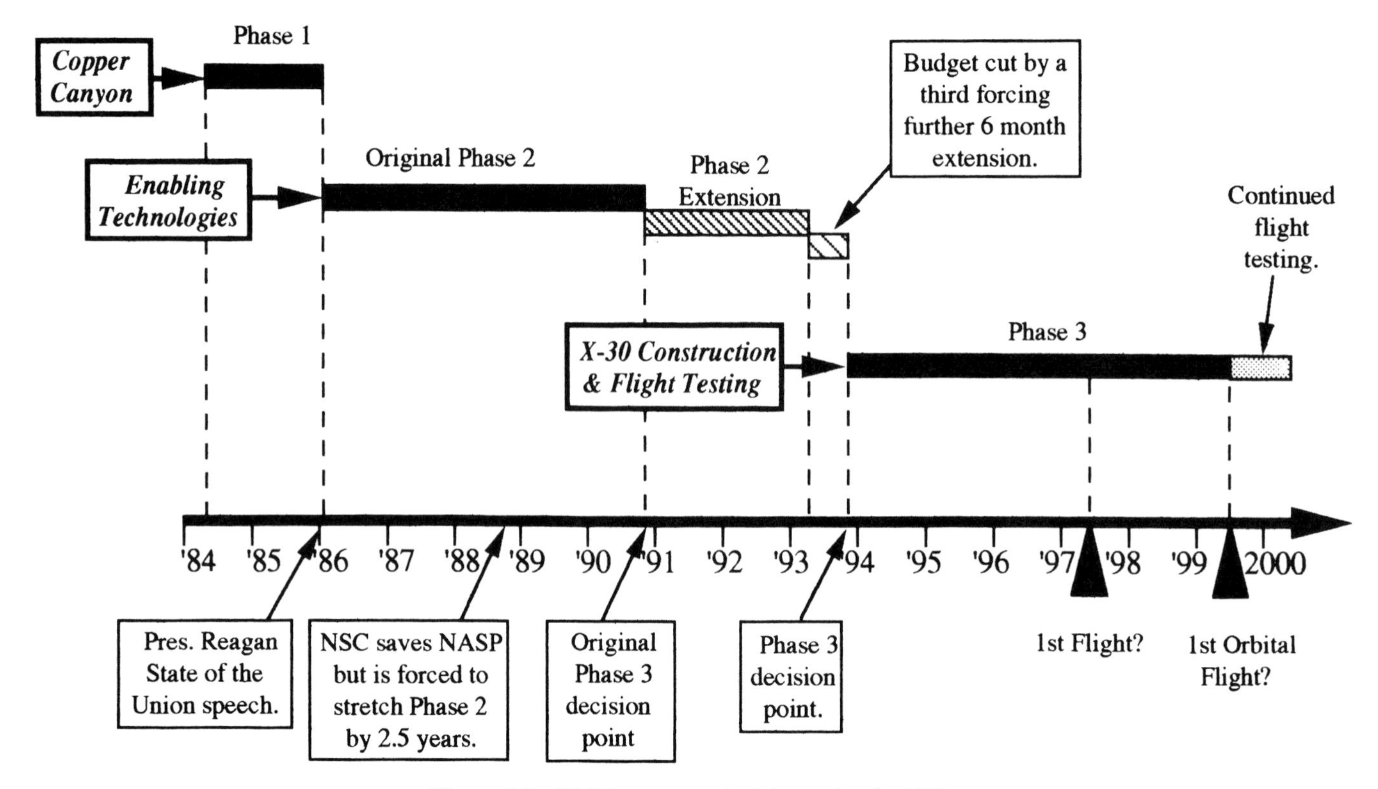


**Figure 6.3** NASP program schedule as of early 1993.

[15]. However, under the consolidation budget plan the Air Force found itself in the position of having to find around 70% of the funds in Fiscal 1988, 1989, and subsequent years—naturally it wanted to have management responsibility for spending its own money.

Management responsibility was intended to shift in September 1990 to the Joint NASA-DOD Program Office (JPO) at Wright-Patterson AFB for the start of the third phase of the program and construction of the experimental vehicles. With the change in the budget allocations, it was decided to bring this date forward. The JPO had been set up immediately after the formal start of the program in 1986 and was at this time responsible for executing the industrial contracts and supporting other technology research activities. After some reshuffling, Dr. Robert Barthelemy was appointed as the NASP program manager on September 25, 1987 [16]. This early move was allegedly instigated by DARPA Director Robert Duncan before he took up his new position in the Pentagon in early 1988 [17].

This is the current management structure. A NASP steering group was also assembled, made up of senior members from each of the government agencies involved with the program. The steering group is responsible for overseeing the programmatic and technical activities of NASP and reporting back to the members' respective agencies. The NASP program was able to establish the JPO very rapidly as it basically consisted of Gen. Skantze's TAV group in total. TAV became NASP with an air-breathing engine.

## *The Elastic Schedule*

The NASP program is split into three phases (Figure 6.3). Phase 1 was the initial Copper Canyon effort, aimed primarily at developing conceptual designs and identifying key enabling technology requirements in the critical areas of propulsion and airframe/engine integration. Phase 2 was the beginning of the NASP program, and the first seven contracts were awarded on April 7, 1986 [18]. For the X-30 airframe studies, one contract each was given to Boeing, General Dynamics, Lockheed, McDonnell Douglas, and Rockwell. For the propulsion system development, one contract each was given to General Electric and Pratt & Whitney. Rocketdyne later joined the NASP propulsion effort through developing and funding its own program. This bold move on the part of Rocketdyne was rewarded when on September 15, 1987, Rocketdyne and Pratt & Whitney were both selected for the continuation of Phase 2 of the program, while General Electric was eliminated [19]. Shortly afterward on October 20, Lockheed and Boeing were eliminated from the airframe work, leaving Rockwell, McDonnell Douglas, and General Dynamics [20]. These five are still the prime contractors and will remain so throughout the duration of the NASP program.

Originally, Phase 2 was to have been completed by September 1990. Assuming the requisite technologies were in place, a decision was to be made at this point on whether to proceed with Phase 3 of the program: full-scale development of the X-30s. However, the Phase 3 decision was initially delayed to April 1993, and later further delayed to September 1993. This delay stemmed from the decision in 1989 of Defense Secretary Dick Cheney to cancel the program in his first week in office. This was later reinforced by Secretary of the Air Force Donald Rice, who claimed that the Air Force didn't have a mission for the vehicle, despite a report published by the U.S. General Accounting Office the previous year that said there were significant military applications for future NASP-Derived Vehicles [21].

Certainly, few people in Congress, the administration, and the U.S. aerospace community really understood just why the United States should build NASP. This was—and still is—a problem that haunts the program.

The DOD's reluctance to support the NASP program came at a time when government spending restrictions were beginning to bite into a defense budget which had become used to large annual increases under President Reagan. Until this time, NASP had actually enjoyed significant budget increases with Fiscal 1988 being appropriated at $254 million and Fiscal 1989 at $316 million. Simultaneously, the pace of the program encouraged the private contractors to invest around $700 million of their own money into the effort. For Fiscal 1990, and the transition to the construction of the X-30s, NASP would have required a further $100 million increase in funding ($427 million), as indeed, it would have in Fiscal 1991 ($509 million) [22]. The need for sharp and continuous rises in funding began to draw unwanted attention to NASP.

It is not entirely clear whether Cheney was opposed to NASP. Cancelling the president's "favorite" programs was one way the DOD was able to show its displeasure at having to make large spending reductions. At the same time, the Senate backed the DOD and similarly zeroed its version of the NASP budget. In this case, the rationale for its action is remarkably simple; attempting to cancel NASP would provide it with leverage to overturn the House's decision to cancel the B-2. By this time, the House of Representatives had become the principle champions of NASP, and therefore it was a straight fight: the Senate wanted the B-2 so it used NASP as hostage, while the House wanted NASP so it used the B-2 as hostage [23].

It took the U.S. National Space Council, under the chairmanship of Vice President Dan Quayle, to step in and resolve the House-Senate conflict in 1989. Eventually, a 3-year budget agreement was struck: NASP would continue with Phase 2, but at a lower funding level of $254 million for Fiscal 1990 compared with the requested $427 million. For Fiscal 1991, NASP was to received $277 million, and for Fiscal 1992, $304 million. The compromise solution to extend the Phase 2 effort kept the program alive, but the decision to build the X-30s had to be put off until April 1993 at the earliest [24]. Although many of the people involved with NASP were unhappy about the schedule slip, most believe (in hindsight at least) that a 1-year extension of the Phase 2 research would have been necessary anyhow, because of the time needed to mature the critical technologies. Even though funding had been reduced, the stretching of Phase 2 actually increased the total amount of government money to be spent on the enabling technologies and provided more time for them to mature.

The 2 1/2 year delay had a considerable impact on the program's momentum. NASP was sold as a national program, and because it was also seen as a high-priority national effort, industries had to invest directly in the program with their own funds in order to be a part of it. Because only a relatively small number of major high technology military programs were likely to be initiated in the coming years, industries had no other choice than to invest in NASP to secure a foothold. If they had not, they would have been left out of new high technology systems and the potentially lucrative future development of operational hypersonic vehicles. It is a curious fact that up until Cheney's attempt to cancel NASP, the industrial contractors had actually invested more money in the program than the government. After the program was slipped and an agreement worked out which would keep the budget almost constant for the next 3 years, the industrial contractors not surprisingly decided to pull back on their investment. In the opinion of David Webb, [25]

> . . . three years leaves a degree of uncertainty in the program that is both patently unfair and perhaps damaging to the program as some of the best talent may leave it. Not to mention the added costs it will impose upon the American taxpayer should the program be finally completed.

## *The Roller-Coaster Budget Battle*

In the first 7 years of the NASP program, approximately $2.8 billion has been spent on the research phase, including the $700 million from the prime contractors during the first 3 years. As the program has progressed, the money has become increasingly difficult to win. The Fiscal 1991 budget is a good example of what a program like NASP must go through every year [26].

In August 1990, the NASA portion of the NASP budget was trimmed by the House Housing & Urban Development (HUD) Committee from a requested level of $119 million to $114 million. The Senate further reduced the NASA contribution to $75 million. There was also some concern that NASA would attempt to further reduce this amount by using some of the funds to bolster the Space Station Freedom program. In the middle of October, however, the NASA budget went to conference and the House and Senate authorization committees set the NASP budget at a compromise level of $95 million. This was a decrease of some $24 million over the administration request, but an increase over the NASA Fiscal 1990 budget of $35 million. By the last week of October, the NASP budget was approved by the Appropriations Committee.

While the NASA funding was relatively straightforward, predictably in early August 1990 the U.S. Senate Armed Services Committee deleted all funding for NASP from the DOD Fiscal 1991 budget request—a repeat of the 1990 budget situation. However, the U.S. House Armed Services Committee voted in favor of funding the DOD request to the full amount of $158 million. By early October, funds for the NASP program were in serious question in light of pressure on the government to reduce federal funding levels because of the recessionary climate, the savings and loan debacle, and the growing crises in the Gulf.

Concern was expressed that NASP would be cancelled simply because it was not an operational military priority. Congressman Dave McCurdy said, for example, "I always accused him [Mr. Chu, assistant to Cheney] of using the formula of dollars per pound of bomb dropped, and if you couldn't relate that to an aircraft then it wasn't worthy of support." McCurdy also explained the perception Congress had of the objectives of NASP, "Ask a member of Congress what Mach number is needed to achieve orbit and they can't tell you. 'What's Mach 8, so what, how does that relate to something, why is that tough, why do we have to spend for this, why are you building something that you'll never use and we're going to build two of these things and then scrap it? It's not going to be an airliner, it's not going to be a bomber, it's not a shuttle, then why are we putting money into it?' "

By mid-October, Defence Secretary Cheney reiterated that he did not believe the DOD had a need for the NASP research nor a mission for the follow-on vehicles. Cheney's statement prompted Congressman Tom Lewis to write directly to Vice President Dan Quayle and ask him whether his support for NASP had dissolved in light of the DOD's cold attitude to the program. Quayle had been a vocal supporter of NASP and was responsible for saving the program the previous year. He took the necessary measures within the U.S. administration that ultimately resulted in the Air Force assistant secretary alledgedly writing a letter to Senator Sam Nunn stating that the Air Force strongly supported the NASP program—apparently in direct conflict with Cheney's views. As a consequence, the House and Senate authorization committees went to conference in late October and jointly approved a compromised DOD NASP budget of $114 million, a reduction of $44 million from that requested.

But this was not the end of the story. Even though the total DOD budget for Fiscal 1991 was written into law at $289 billion, the NASP budget had not at that stage received its final appropriations. Normally, the appropriations committees approve funding at levels no higher than that approved by the authorization committee. According to Tim Kyger, staffer for Dana Rohrabacher, it was only theoretically possible for appropriations committees to increase funding over that which was approved by the authorization committees. However, a concerted lobbying campaign orchestrated under the guidance of former X-15 test pilot Scott Crossfield resulted in a last minute increase to the NASP budget. By October 29, the DOD NASP budget allocation was increased by an additional $49 million, bringing the total DOD NASP budget to $163 million. This extra money was provided to try and make up some of the shortfall in the NASA budget which, by this time, was already written into law. According to Dave McCurdy, "We were in the pattern of having to increase the budget over the past few years in order to keep one or both sides [DOD and NASA] alive [and] had to use unusual techniques to increasing the budget on authorization or increasing it on the defense component knowing that it's a joint program." By the end of October, after a barnstorming budget battle, NASP was funded to $258 million, a decrease of $19 million over the Bush administration request of $277 million, but a small increase of $4 million over the Fiscal 1990 allocation of $254 million.

That was Fiscal 1991, but Fiscal 1992 fared less well. By early 1991 the program seemed to have reached a reasonably secure funding level. In the opinion of Congressman Tom Lewis, "I think NASP will survive this year and next year as well, it's not going to be easy like any other program. We're going to have to do a lot of fighting to keep the funding available in order to do it . . . We have people who are more interested in killing other space programs, rather than NASP, and some of those programs many fall prey to the salvation of the space plane." According to Congressman Dave McCurdy, "I think we're at that stage now where we probably have a stable budget, not as high as we would like, but probably some stability for the future assuming the administration stays in."

It was not to be. By November 1991 (fiscal 1992), the NASP program was drastically cut back, primarily because of NASA. Up until this point NASA was apparently becoming a stronger supporter of NASP than at any time since 1986. However, Space Station Freedom took a higher priority in NASA's eyes than did NASP—as, indeed, it did over most everything else. (See Chapter 1.) When the appropriation committees met in conference, NASA's alledgedly tepid enthusiasm for NASP almost spelled the death of their involvement in the effort. It was only through a last minute, impassioned plea by Senator Jake Gam that the program wound up with any money at all—a mere $5 million [27]. According to David Webb, "It was only because it was ole' Jake Garn that Congress chipped in five million bucks—they would have eliminated NASA's involvement completely otherwise."

For a nation that sees itself as pushing back the frontiers of technology and intends to lead the way back to the Moon and beyond, this attitude toward NASP might seem terribly short-sighted. It is not that NASA doesn't care about NASP or doesn't recognize its potential applications. For example, according to a report from the NASA administrator's office in mid-1991, [28]

The program promises improvements in space-launch capability, with aircraft-like operations for future single-stage-to-orbit vehicles providing easier, more flexible and more efficient access to space . . . NDV operations can lead to cost reductions. Flexible, airplane-like operations, including in-space repair, allow for economies of payload launch and operations. The nature of ascent aborts should lower insurance and the cost of replacing payloads . . . Study results show that if payloads and manifests are not constrained to fit just Space Shuttle, then NDVs can capture about 70 percent of the payload types, over 80 percent of the launch events, and 30 percent of the mass.

NASA understands and supports NASP, but not at the expense of Freedom. In late 1991, NASA and congressional supporters of NASP made a last ditch effort to allow the space agency to reprogram $15 million from other agency efforts to NASP. These efforts were dashed in mid-January 1992 when the House and Senate budget appropriations subcommittee denied the request stating that, "the effort had merit but is not a priority in the tight budget climate" [29].

The defense component of the NASP budget fared better, receiving $200 million from a request of $230 million. This was despite concerns that the NASA budget would set a precedent for the DOD. The $200 million the DOD received, remarkably enough, was an increase of some $37 million over the Fiscal 1991 authorization [30].

The NASP budget for Fiscal 1993 faired worse still. NASA's contribution was eliminated, while the Air Force portion was reduced to $144 million. For fiscal 1994, DOD received $40 million and NASA $20 million. These lowering budgets have all but signalled the abandonment of plans to start construction of the orbital X-30 vehicles in the near future.

## *The National Program Office*

Traditionally, the United States has conducted its national and international business in a competitive manner. Although U.S. companies have been able to team with other U.S. or international companies in the pursuit of contracts, this teaming has usually been restricted to one or more companies providing subsystems and components to the prime contractor. Seldom have U.S. industries come together in the way Japanese and European companies have formed consortia working more as a single entity than the usual single company responsible for the management of the program. The European Airbus is a good and highly successful example. The NASP National Program Office (NPO) was the first major consortium established in the U.S. aerospace industry, and its management structure is being closely watched as a model for future U.S. consortia. NPO was originally located at the Rockwell facilities in Seal Beach, but in 1991 moved to Palmdale, California, in anticipation of the construction of the X-30 vehicles. Barry Waldman of Rocketdyne was put at the head of NPO [31].

The motivation to establish NPO resulted from two needs, technical and financial. As stated earlier, NASP is technically an immensely demanding program, and as Phase 2 progressed, JPO began to recognized that none of the five principal contractors had all the solutions. Indeed, because no one was certain NASP could even be built, it was impossible to know which solutions were best. This is different than with other industries, such as automobile and electronics, where the fundamental engineering principles are well established through experience accumulated over the years. According to Barry Waldman, [32]

For a new technology such as NASP, the bases for a downselect decision are not clear, and proceeding in a traditional manner requires a binary, yes/no, decision between technology packages that are unsuited for this sort of selection. An unfortunate result of competition is also the loss of the expertise of the unsuccessful competitors in the continuing work on the program.

The second problem, funding, was equally critical in shaping the decision to form NPO. By 1990, each of the three airframe contractors had one preferred design for the X-30 airframe design concept, while the two propulsion contractors each had one preferred engine design concept. According to Paul Czysz, " . . . we had six configurations and there wasn't enough money for all of them [to be matured equally] and so they tried to figure out a way to get consolidation of the effort." Bruce Abell also comments, "It was obviously done as a means of retaining the participants because they were going to fall off. The participants were no longer putting significant money into it because the program got stretched out. Given the fact that it started out as a means to retain the major players in this thing, it's probably working reasonably well in terms of actually developing a model of interest for a development program."

The formation of NPO was not the NASP program's first attempt at using the consortium model. The NASP Materials Consortium was the precursor to NPO and was, by all accounts, highly successful in its efforts to rapidly advance new materials. Originally, each contractor in the NASP program had responsibility for developing its own materials for its own design, which inevitably led to inefficiencies resulting from duplication. For example, at one point all contractors were developing the same type of carbon-carbon composites. According to Barry Waldman, [33]

Recognizing this, and realizing that materials selection should not be a competitive discriminator, the industry and government agreed to form a consortium to develop advanced materials and share the work done by the consortium . . . Each contractor was responsible for the leadership of the work done on one of the materials systems, but took part in specific tasks relative to the development of each of the others.

According to David Webb, "The National Research Council estimated that the development of suitable materials would take up to *fifteen years* and, as a result, were not supportive of initiating the NASP program. The Materials Consortium actually produced the required research breakthroughs and developed the needed materials in less than *three years*."

The National Contractor Team structure was first laid out in an Interim Teaming Agreement (ITA) that was originally signed by the contractors and submitted to the U.S. government for approval on January 22, 1990. After considerable wrangling, the Justice Department approved the ITA on May 25, 1990, allowing an immediate exchange of some technical information and the formal establishment of the NPO [34]. Even at this stage it was not certain that the Teaming Agreement would be formally legalized, and as a result, many of the team members withheld some of their most valuable research. Jeff Morrow of Rockwell International explains, "At the beginning, besides the 'not invented here syndrome' there was some people holding some things back because they weren't sure if the government would go for it. If all of a sudden the government turned around and said no teaming agreement and you've already released everything—well." And as Bruce Abell points out, "Ordinarily, when you get competitors involved in consortia, they hold back, they don't really share, and they don't trust each other. In this case it may be that because they're dealing with a fairly exotic technology, they have less reason to hold back. How many applications do you have right now for 4,000 degree materials, or Mach 12 engines?"

Finally, on January 30, 1991, the government formally approved the consortium and issued a single letter contract to begin the final part of the NASP Phase 2 effort (designated Phase 2D). The Teaming Agreement will remain in place for the duration of the NASP program. An interesting aspect of this agreement is in the way it allows the contractors to work with government laboratories without fear of giving other contractors an unfair competitive advantage.

A much broader motivation exists for the formation of NPO. According to Barry Waldman, [35]

From the contractors' viewpoint, an additional advantage of the team program is that each contractor will gain the expertise to be fully competitive in any follow-on hypersonic vehicle program.

And, in the words of John Welch, assistant secretary of the Air Force for acquisition, [36]

The five NASP contractors are sharing technology developments and each will be in a position to compete for future hypersonic systems . . . the NASP Program will help ensure that the U.S. continues as the world aerospace leader in the 21st century.

Another, subtler reason for forming NPO was the added base of lobbying power this would give to NASP. Having five major contractors with equal stakes in the program, backed up by more than a hundred subcontractors, all acting in unison, provides greater lobbying weight than if one prime contractor had been selected to carry the work forward.

It is important to appreciate that the NPO joint venture is organized entirely around the X-30. Any subsequent vehicles which may follow on from X-30 technology, and parallel developments not directly derived from the X-30, remain competitive areas. Effectively, as the U.S. law stands, it will be impossible for the NPO to build operational NASP-Derived Vehicles because this would mitigate participation from other contractors.

Finally, former Presidential Science Advisor Allan Bromley has voiced a note of caution by stating that, [37]

. . . this novel approach [to organizational management] is not without its critics. Some have been concerned that the teaming arrangement would work as a disincentive to creativity because competition is removed. That is something we will all want to watch and evaluate carefully. But if the NASP teaming arrangement proves the critics wrong, it may serve as a very fruitful model for future endeavors.

## *Contractor Management Philosophy*

The management approach adopted by NPO is indicative of the above intentions. Considered the most important aspect of this are the work-sharing agreements between the five major companies. This agreement differs radically from the current practice of cost-plus type contracts. Essentially, the contractor fee is divided equally among each of the consortium members and thus decouples the amount of money each company earns from the type of work being undertaken. Dividing the money equally eliminates the need for the contractors to flight over the work they want to do. The philosophy is, therefore, that no single company can be successful on its own. For a company to be successful, all the companies in the team must succeed. Likewise, the failure or lack of commitment of one company to the program will be detrimental to the rest.

The management structure (Figure 6.4) is such that each company has direct visibility of what the other company is doing, allowing monitoring of the progress of each. NPO, therefore, serves as the central focus of the NASP program for the five prime contractors, as well as being the single point of contact for the government with respect to contractual and technical matters. Barry Waldman stated that the three basic functions of NPO, in addition to its contractual responsibilities, are, [38]

(1) design of the X-30 airframe and engine, directed by a core technical team located at the NPO; (2) definition and administration of the Phase 2D program plan; and (3) development and documentation of the Phase 3 plan. The results of these activities are the basic products of NPO, while the supporting design and technology development are carried out at the CTM [Contractor Team Members] home sites.

Discussing how the contractor team has worked together to date, Barry Waldman said that the [39]

. . . smooth functioning of this National Contractor Team requires a considerable degree of trust among the team members, and this trust is rapidly being established.

This is probably the case, since the management philosophy requires all partners to succeed if the total program is to be a success. Given the intensity of the competition that preceded the formation of NPO, a certain amount of time has been needed for the team to become an efficiently operating organization. Although some people associated with the program have complained that the NPO simply added "another level of bureaucracy" and that few real decisions have been made, some members of the team believe the quality of the technical work has been enhanced.

## *X-30 Configuration: Design, Technology, and Testing*

Although as of late 1993 the NASP program has effectively shelved plans to build the full-scale orbital X-30s, it is nevertheless considered worthwhile to give an insight into the progress made in defining these vehicles.

One of the first tasks after the formation of the NPO was the selection of a single X-30 design concept based on the best ideas of the individual competitive designs. Although the NASP program is still a classified activity, some details of the X-30 composite configuration were first formally released at the AIAA *International Aero-Space Planes Con-*

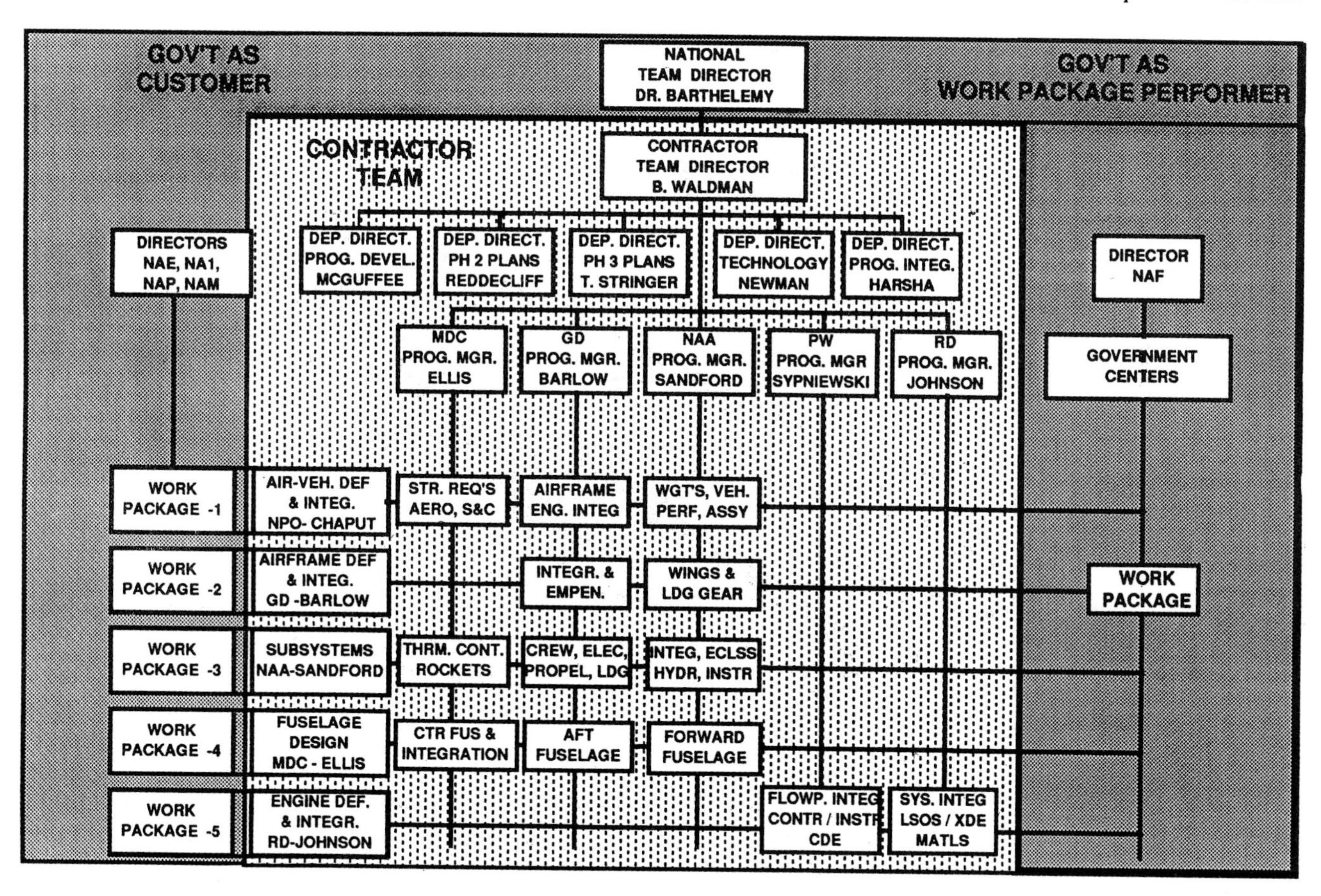


**Figure 6.4** NPO management structure (*NASP NPO*).

**Figure 6.5** The X-30 composite (*NASP JPO*).

*ference* in Orlando (October 29–31, 1990) [40] (Figure 6.5). The selection of a single configuration was considered critical to bolster the viability of the overall effort. It provided a strong focus, technically, psychologically, and politically, for the program. In addition, a single configuration allowed a rationalization of the activities and enhanced the chances of developing the needed technologies to support the NASP program goals. It eliminated any significant duplication of work done during the competitive phases of the NASP program.

The process of consolidating the work into one design is not straightforward, as described by Paul Czysz, "Then the problem came of how would we work on a single system. I think we lost 2 years, probably 3 because at the time the decision [to form NPO] was made we [McDonnell Douglas] basically had about 15 iterations, maybe 7 really good iterations on the complete resizing and sanity checks on the whole system, and now you had to go back to square one so the team could come up to the same level by a common configuration. It's not easy to take three separate hypersonic configurations that didn't look alike and two completely different engines and try to come up with one single engine and airframe. There are a thousand different configurations one can come up with, and they all had very little commonality."

Prior to the composite selection, JPO sifted through the data and made extensive evaluations of each design. As a result, JPO laid the groundwork for the selection of the composite design well in advance of the establishment of NPO. Once NPO was approved, JPO was in a position to brief each contractor on the strengths and weakness of its designs relative to the other contractors, and also to suggest some new concepts born out of JPO's evaluation activities. The composite vehicle is described as a directionally stable lifting body incorporating a twin stabilizer for hypersonic lateral stability, a very small wing platform, and a two person crew cabin. It is expected to be about 60 m long and have a gross takeoff mass of around 200 tonnes. For propulsion it has three engine modules located in a cluster for air-breathing flight. A small rocket motor with about 30 tonnes of thrust is incorporated in the X-30's tail to push the vehicle through the sound barrier, final injection into orbit, and the reentry burn. The rocket can also be a backup to the air-breathing engines. The X-30 is intended to be able to perform a powered landing if necessary [41].

The structural design and general types of materials for the X-30 have been selected. The airframe will use a hot structure for the most part, with a load-carrying skin made from Beta 21-S-based titanium-metal matrix composite panels. Three non-integral and conformal propellant tanks are held separately inside the fuselage. These tanks will not carry any significant structural loads and are constructed from graphite-epoxy. A test article of an earlier structural concept using a multibubble tank has already been constructed by McDonnell Douglas (Figure 6.6). This test article measures 2.2 meters high × 1.2 meters wide × 2.4 meters long and has been put through a series of flight-representative structural loading and thermal tests. Initial tests involved filling the tank with liquid nitrogen, exposing the external structure to a maximum temperature of 700° C, and subjecting the whole structure to the maximum bending loads experienced in flight, although not all these conditions simultaneously. In late 1991, this tank was filled with liquid hydrogen for the first time [42].

High heating areas (up to 1,700° C) including the nose, leading edges, and lower forward fuselage, which make up a quarter of the surface area of the X-30, will be covered by large advanced carbon-carbon panels that will generally take relatively limited structural loads. This material will also be used in certain strategic structural parts of the X-30. For example, a test article of the wing box was built by General Dynamics. This structure is constructed almost exclusively from advanced carbon-carbon composite materials (designated ACC-4) coated with silicon carbide and joined together with fasteners made from columbium and nickel alloys. One part of this wing box includes the largest piece of coated advanced carbon-carbon ever produced. This structure was successfully tested under represented flight conditions in 1992 [43]. For the most part, thermal control of these high heating areas will be passive, but for areas experiencing greater than 1,700° C such as the leading edges and intake ramp, an active cooling system is required [44].

While there is confidence that the materials and structural concepts for NASP could be ready in time for the Phase 3 decision, the primary concern remains with the propulsion system. Robert Barthelemy said that "If I had one wish to be granted, it would be to have some actual flight data of

**Figure 6.6** Integral propellant tank/fuselage test article (*NASP JPO*).

the scramjet performance at Mach 16. If I had that point, and knew that it was a good point, I'd be able to tell you exactly whether this vehicle was feasible and also what the weight was going to be. It is the performance of the propulsion system at high Mach numbers that's going to make or break this vehicle. I think it will be sufficiently good enough that we can have a single-stage-to-orbit vehicle, but the weight of the vehicle may be very, very large and not very interesting. If the performance is high, perhaps even higher than we're predicting, it's possible we could get Isp's of 700 or 800 seconds at that Mach number. And if that's the case, we'll never have to use rockets—we can go all the way to Mach 25 on scramjets alone, the vehicle could be very small and very, very interesting. You may still have a rocket on there for safety purposes, and to move around and get back."

The NASP propulsion system configuration is still highly classified, especially in the area of the low-speed power plant design. Once Mach 1 or 2 has been achieved, a ramjet propulsion system will be started and used to increase the X-30 to hypersonic speeds. At Mach 7 or 8, the airflow through the ramjet becomes supersonic, and the engine will convert to the so-called scramjet mode of operation for the remainder of the air-breathing part of the ascent. The NASP program is performing exhaustive engine testing in a range of continuous free-stream wind tunnels (up to Mach 8) and shock tunnels (up to Mach 25). This data helps to design the engine, as well as to validate the CFD codes needed to predict the engine performance above Mach 8. For this CFD work, the NPO has its own dedicated Cray YMP supercomputer in use virtually 24 hours a day, 7 days a week [45].

The structural design of the NASP propulsion system has also been selected, although there remains some uncertainty over its precise configuration. According to Barry Waldman et al., [46]

Engine structural design decisions have also been made, but in view of the very high thermal and pressure loads that the engines may see, the materials and cooling techniques chosen for the first prototype engines may change before or during flight testing of the X-30. This ability to modify the engine materials and, to some extent, the structure is enhanced by the fact that engines are designed as modules that incorporate all of the subsystems required for engine operation.

The first engine will be made from cobalt and copper alloys, and a molybdenum and rhenium compound known as MolyRhenium. However, there are concerns that the MolyRhenium material may not be able to withstand the rigors of the engine environment and especially the problems of atomic oxygen erosion. A carbon-carbon cowling will enclose the engine. The internal walls and the fuselage section in front of the intakes are to be cooled with liquid hydrogen.

After being used as a coolant, the temperature of this hydrogen is raised to very high levels. To take advantage of this, the super-heated hydrogen is injected into the engine, ensuring more efficient engine performance. As a result, the energy lost due to aerodynamic frictional heating can be partly recovered and used to provide higher engine thrust and performance. (See Chapter 4.) A full-scale engine model, using the actual materials and design of the flight propulsion system was constructed in 1993 [47].

One problem with all high-speed air-breathing propulsion systems is the almost immediate increase in drag that occurs if one or more engine fails. This is known as engine "unstarts" [48]. If this drag is high enough, it could destroy the vehicle almost instantaneously. Although this is considered a major problem, tests with representative multimodule engines performed during the competition phase of the program seem to indicate that the drag increase may not be as severe as first believed. Further tests of a modified version of the original Pratt & Whitney engine concept are being performed to validate techniques for minimizing the probability of unstarts and staving off their detrimental impacts [49]. If the problem of unstarts were to become more pronounced, then this might necessitate a complete revision of the current X-30 baseline design, and a return to a design more reminiscent of the original government baseline.

The X-30 will need a conventional rocket propulsion system, if only to circularize the orbit, manuever in space, and effect the reentry. The rocket system envisaged for the composite involves a single engine in the 30-tonne thrust class. Serious consideration has been given to using a linear motor or aerospike (Chapter 10) which can also act as a reaction control system by differential use of various sections of the aerospike. The rocket system also acts as a safety feature, as it will be able to take over from the air-breathing engines should they fail and provide a level of thrust needed for the vehicle to land safely. In addition, the X-30 is designed with an advanced heat exchanger capable of liquifying air and storing it on board. This may help to reduce the vehicle's takeoff weight because the oxygen needed for the final rocket ascent can be collected in flight.

The X-30 will need to use slush hydrogen for orbital flights, although liquid hydrogen can be used for the early low-speed flights. The NASP program has essentially pioneered large-scale slush hydrogen activities. Like the materials and engines, it is one of the key technologies that must be demonstrated. Slush hydrogen is essentially a mixture of liquid and frozen hydrogen, therefore providing a higher density compared with pure liquid hydrogen. At a 50:50 mixture ratio, it yields a 15% increase in density over liquid hydrogen and, therefore, a 15% decrease in tank volume and, it is claimed, as much as a 50% reduction in takeoff mass. Rockwell International and the NASA Lewis Research Center have performed much of this work and have been able to produce and transfer large quantities (about 2,000 liters) of slush hydrogen with a 66% solid content [50].

The weight of the X-30 is an area of some concern. Originally, the government baseline configuration of the X-30, developed by Tony DuPont for initial Copper Canyon activity, suggested an extremely small vehicle capable of single-stage-to-orbit with a total takeoff mass of just 23 tonnes (50,000 pounds). This particular design was heavily criticized as being impractical, although there are still strong pockets of support for the basic concepts embodied by the DuPont configuration [51]. Nevertheless, comparisons of the current X-30 with the original government baseline are often made because the program was sold on the basis of building a small and therefore relatively inexpensive vehicle. Today, the current X-30 design is nearly *ten times* heavier than the original baseline. Its cost has more than trebled from the original estimate of about $3.1 billion (1985), although a large contributor to this increase came from stretching out the program several years. This growth has caused concerns in political circles. For example, wording attached to the Fiscal 1991 budget requested the NASP program to reexamine the government baseline design in light of recent technical developments.

Robert Barthelemy said "one of things that really bothers me [is] the weight of the X-30 vehicle. [Weight] is a wonderful measure to see how you're doing, and we set goals and margins. But [I get the impression] every time I brief somebody [that] if we don't make a particular number, say 350,000 pounds [160 tonnes] for the X-30 it's like, 'forget it, there's no use doing it.' It's incredible to me to think that why wouldn't 500,000 pounds [227 tonnes] be a very exciting vehicle . . . as long as it fits into the category of an airplane . . . It doesn't seem to me that we should die just because we miss the [weight] by a 'couple' of pounds."

Prior to the selection of a single vehicle concept, the six competing designs were essentially combinations of the three airframe designs and two engine concepts previously selected in 1989. Each of these six competitive designs had different strengths and weaknesses arising from each contractor's approach to the X-30 requirements. According to Barry Waldman, [52]

> In fact, it was our growing recognition of these differences, which made it evident that no one contractor had all of the answers to the problems posed by the X-30 requirements, that was one of the reasons for forming the team.

The composite X-30 design concept, therefore, embodies many of the best features and none of the weaknesses of these previous competing concepts. Originally, Rockwell adopted a conventional wing and body design in order to minimize drag and, thereby, help facilitate air-breathing at extreme Mach numbers—this is known as the drag solution. The McDonnell Douglas design, concentrated on a configuration that would maximize the performance of the engine by adopting the much wider blended wing and body configuration—the propulsion solution. General Dynamics concentrated on a design that was influenced by the high thermal heating—the thermal solution (Figure 6.7). The end result of this activity, according to Barry Waldman is that, [53]

. . . the team X-30 is achieving in *combination* the values achieved *singly* by the best of the competitive designs. [Emphasis added.]

## Flying the X-30

The NASP program flight testing strategy is seen as very conservative in its approach, allowing the performance envelope to be expanded gradually while maximizing safety. The envisaged trajectories involve starting at Edwards AFB, accelerating up into the designated hypersonic testing corridors, and cruising at Mach 2–4 above normal air traffic. The vehicle would decelerate for a turn over the downrange base, such as the Kennedy Space Center, then re-accelerate to the required Mach number for data collection. After the appropriate data had been collected, the X-30 would perform a glide back to Edwards for an unpowered approach and landing. An unpowered recovery flight path is expected to place less stress on the vehicle, which could have received thermal damage during the data collection (acceleration) phase of the test flight. This trajectory keeps the X-30 over the United States, ensuring a reasonable chance for recovering the vehicle after an abort [54]. (See Chapter 16.)

The purpose of these tests will be to build up the speed of the X-30 gradually before attempting an orbital flight in order to fully understand the characteristics of the vehicle. This is in sharp contrast to all expendable launchers and the Shuttle which must fly directly to orbit on their first flight. An early X-30 test program concept would have attempted to reach single-stage-to-orbit after 2 years and about 43 flights. This is broken down as follows, [55]

| No. of Flights | Mach Range |
|---|---|
| 6 | 0–2 |
| 11 | 2–6 |
| 13 | 6–12 |
| 12 | 12–17 |
| 1 | 17–orbit |

In addition, the NASP program aimed to conduct about 70 missions over the subsequent 3 years. Overall, if this schedule were achieved, the X-30 would fly 113 test flights in 5 years. Even though not all will be orbital missions, this flight rate is considerably higher than the Shuttle's four test flights and 56 operational missions conducted to the end of 1993. The 113 X-30 flights would provide engineers with the opportunity to fully and incrementally learn how to fly true space transportation systems routinely to and from orbit.

The NASP program has also initiated environmental issues associated with flying the X-30. Assuring the public that the X-30 flights will have minimal impact on the environment is a mandatory requirement before the X-30 will be allowed to take off or, for that matter, be constructed, according to Reda, [56]

Before proceeding to Phase 3 of the NASP Program, the public, including all interested groups, and program decision makers must be informed of the potential environmental impacts of the proposed program . . . Safety and environmental concerns are being addressed upfront during the design phase of the program to have the maximum influence on the X-30 and the ground support system . . . The top three environmental topics are: 1) sonic booms, 2) stratospheric ozone depletion, and 3) hazardous materials.

The environmental impacts of the X-30 are being studied by the NASP environmental team using the standard US Air Force environmental impact and analysis process that all new military aircraft must adhere to [57].

## The NASP-Derived Vehicle: A Quantum Leap

The X-30 is an experimental vehicle which, in its last definition, cannot be easily retrofitted into an operational launch system, although it could probably carry small payloads located in the crew compartment, just as the X-15 did two decades ago. (See Chapter 5.) Incorporating a large-sized payload bay would necessitate a completely new vehicle. Air-breathing aero-space planes cannot be simply stretched like an airliner because of the high level of integration between the airframe and the engine. (See Chapter 4.)

Originally, the X-30 was to have had a small payload bay which could carry about 1 tonne of flight test instrumentation. Indeed, some members of the NASP program actually favored an X-30 with an even larger capability so that it could "easily" be converted into an operational vehicle. Robert Barthelemy was one of them, "I was one of the proponents of operationalizing the X-30, and thought that was a reasonable thing to do. I felt that you get to where you wanted to go quicker even though it might cost you a little more, but at least you'd be there quicker and that's what everybody wanted. I was pushing it really hard, but we got stopped because they thought it would add a lot of cost. The other reason I think we got stopped is that they really feel that the technology is so difficult to pull off, that they wanted this to be a pure R&D program. They're terribly afraid that we're going to get into something which we can't pull off from a technical standpoint and so the 'X' vehicle gives you

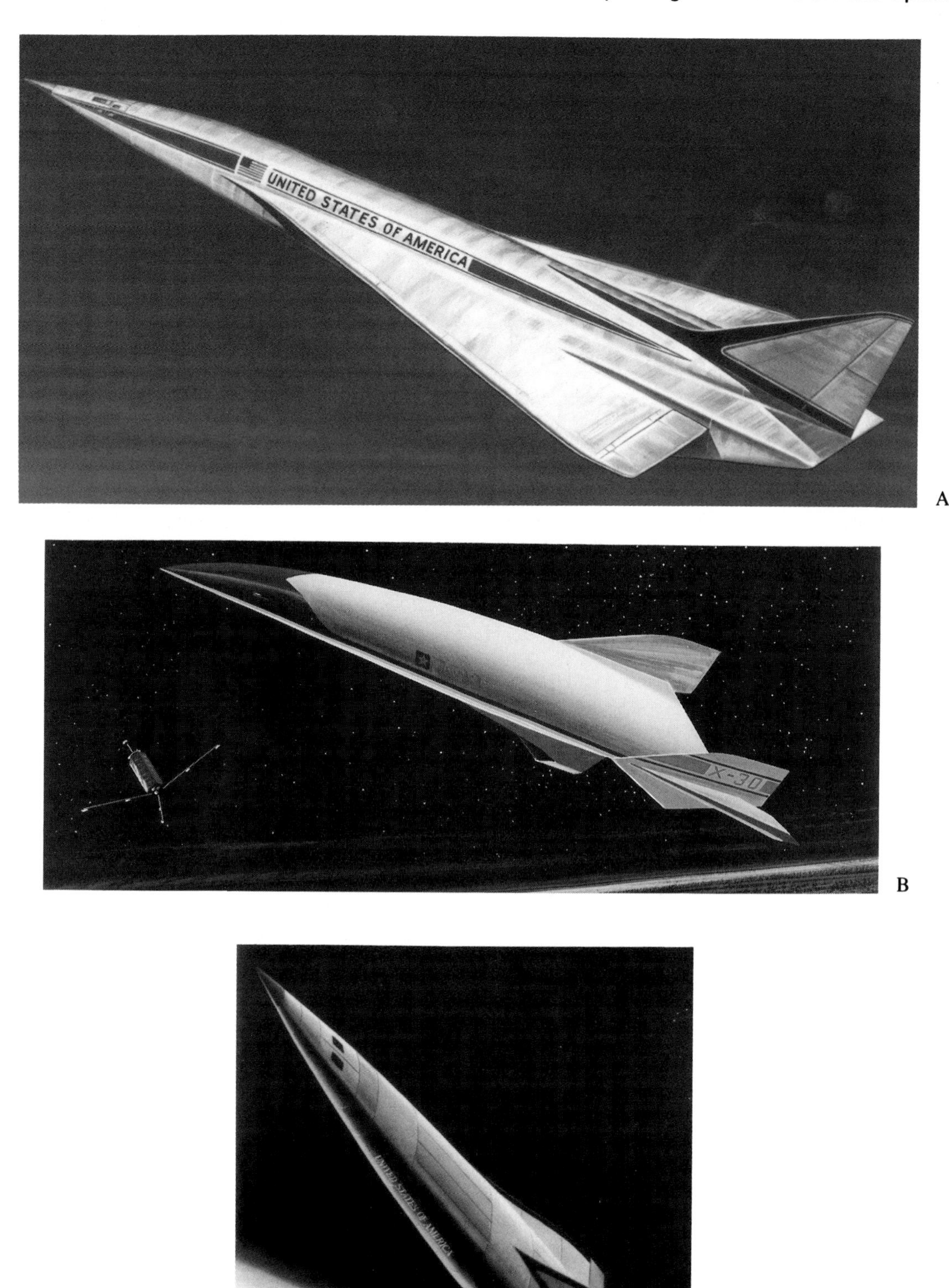


**Figure 6.7** Early X-30 concepts of (A) Rockwell, (B) McDonnell Douglas, and (C) General Dynamics.

another round of technology development which is quite separate from the prototype development."

The X-30 is intended to answer the fundamental questions regarding the practicality of air-breathing, single-stage ascent to orbit. After that, a practically brand new vehicle, usually called the NASP-Derived Vehicle (NDV) or S-30, would have to be built. This NDV could come in many shapes and sizes (Figure 6.8), including vertical takeoff and landing systems such as the proposed Delta Clipper (Chapter 10) and two-stage systems similar to Sänger. The ideal first generation NDV that the NASP program believes could emerge from a successful X-30 test program would have a gross takeoff mass of about 400,000 pounds (180 tonnes) and be capable of delivering a payload of 20,000 pounds (9 tonnes) to a Space Station Freedom orbit.

The choice of NDV after completion of the X-30 flights will be dependent on the success of the X-30, the technology maturity, and perhaps most importantly, the life cycle cost of the system. The NASP JPO and many contractor team members have already performed in-depth analyses primarily on the economics of an NDV fleet [58]. A major activity during 1989 and 1990 was the NASP study "Space Transportation Options & Launch Architecture Comparisons" [59]. This study concluded that the operational cost of an NDV fleet flying up to 100 flights per year would be in the region of $2–6 million (1986 dollars) per flight. If full life cycle cost recovery over 20 years were rolled into the equation, then this would lead to a cost per flight in the range of $10–20 million (1986 dollars). The study also concluded that while the NDV would be approximately two orders of magnitude (100 times) more expensive to use than a similarly sized aircraft, more importantly it would cost as much as two orders of magnitude *less* to use than any other similar capability launch system [60].

The Space Transportation Options & Launch Architecture Comparisons study attempted to calculate NDV costs directly, based on comparing the X-30 with other aircraft like the B-1B and C-5A. This approach has been criticized because the NDV is by no standard a traditional aircraft. Other studies have attempted to find a more conservative and therefore more immediately believable approach. One study

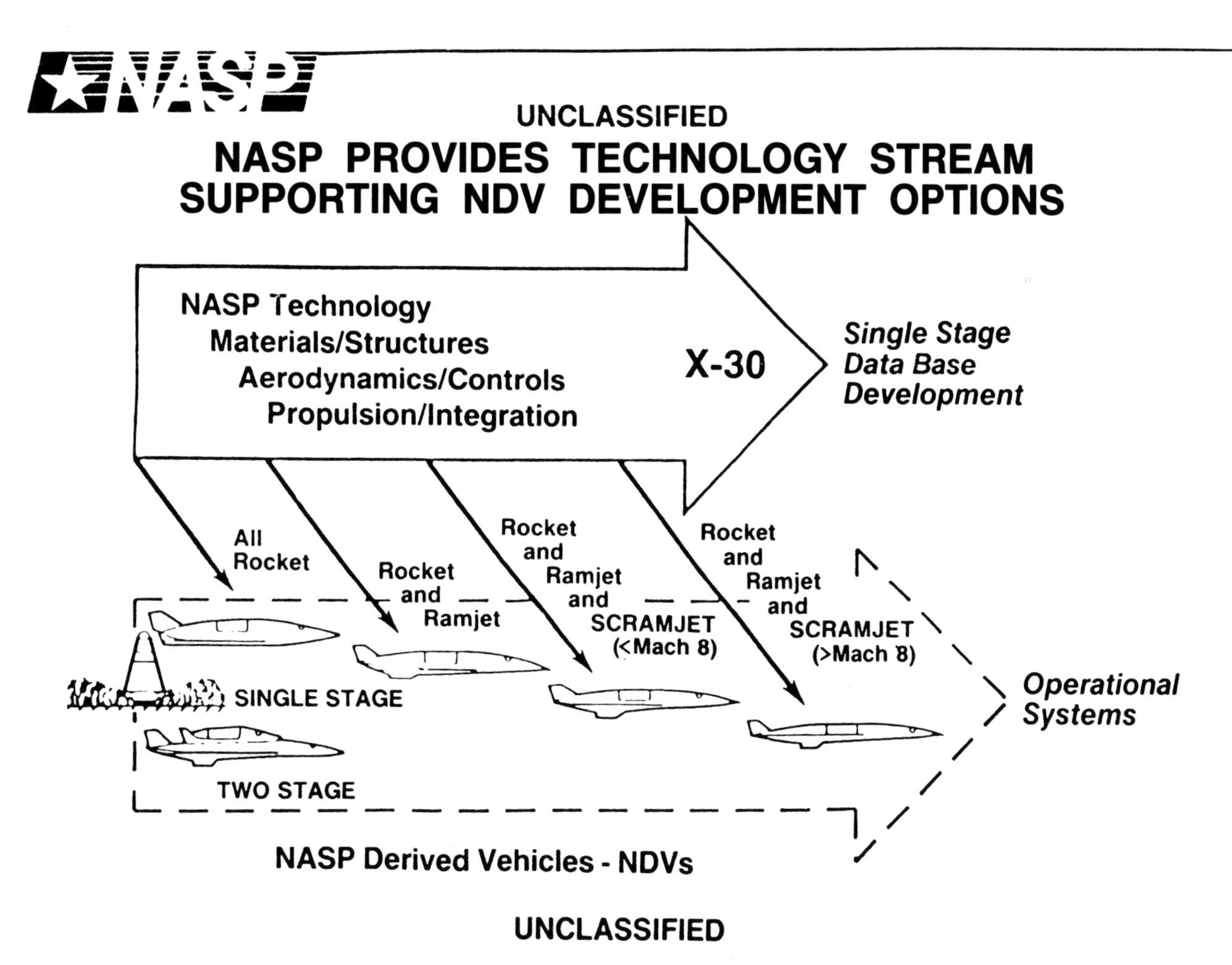


**Figure 6.8** Options for NASP-Derived Vehicles (*NASP JPO*).

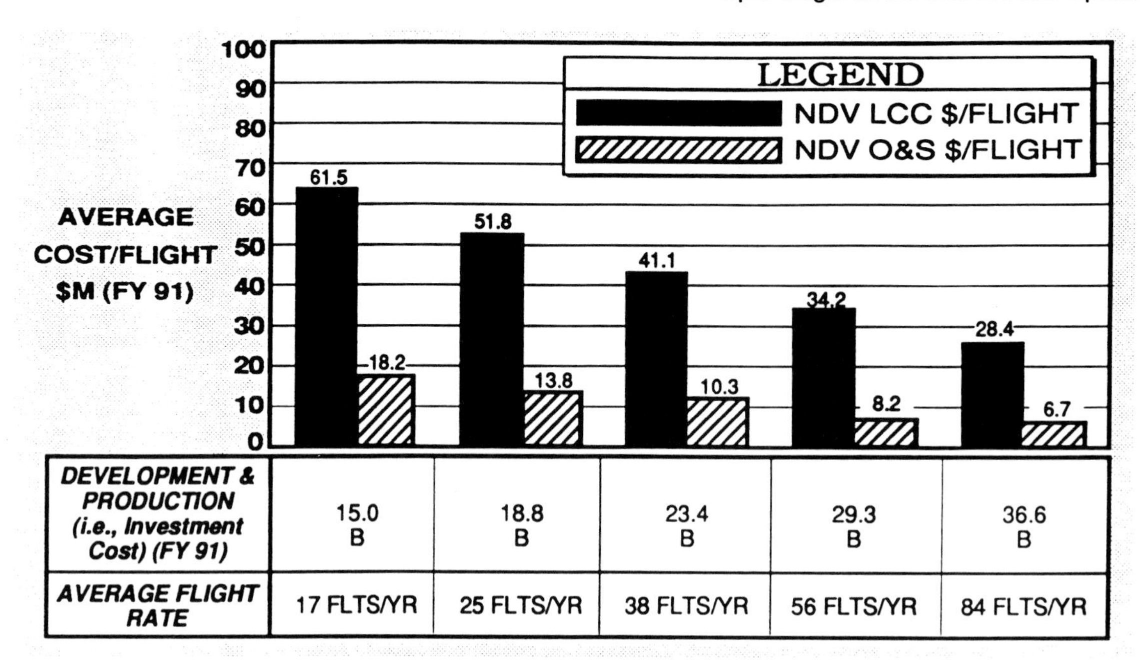


**Figure 6.9** A dramatic reduction in launch costs is the potential of future NDV fleets (*General Dynamics*).

uses what is termed the STS-Down cost analysis that analyzes NDV costs relative to the Space Shuttle operations costs. Toten et al. explain how this works, [61]

> The NDV is envisaged by the NASP contractors as an aircraft with designed-in aircraft-like characteristics which is also designed to routinely access low earth orbit. Nevertheless, there is a great deal of commonality between NDV goals and the original STS goals. There are also significant intrinsic differences between the NDV design, concept of operations, and technologies and those of the STS . . . NDV operating costs have been derived from STS costs by: (1) deleting those costs associated with the tasks or equipments that are clearly unique to STS design and/or operations, (2) adding in costs that are uniquely associated with the NDV design and/or operations, and (3) assessing the relative costs for common design features/operations. Adjustments for common features were made to account for new technologies and for new approaches stemming from application of STS historical lessons-learned.

For further checks, this study went to the length of having a working group, made up from representatives of the NASA Kennedy, Johnson, and Marshall field centers, periodically review the study progress. As an example of the conservative nature of the study, the refurbishment cost of the NDV engines was assigned the *same* cost as the Space Shuttle main engines simply because of the uncertainty over the servicing requirements of NDV engines.

The conclusions of this study are shown in Figure 6.9, with costs being in 1991 dollars. The results are comparable, if somewhat higher, to those of the Space Transportation Options & Launch Architecture Comparisons study, adding some credence to the accuracy of both techniques. For example, a flight rate of about 56 flights per year would lead to an operational launch cost of about $8 million per flight, or an operational cost of about $500 million per year. The launch cost jumps to $34 million per flight with a life cycle cost of about $29 billion amortized out over 20 years. Thus, the average total cost, with development and production costs rolled in, would be in the region of $2 billion per year at 56 flights per year.

The figures become more meaningful when compared to the Space Shuttle which, based on a maximum of 10 flights per year, would lead to an operational cost per flight of about $500 million, or $5,000 million per year. If the Shuttle development and production costs are included at about $35 billion [62], and it is assumed the Shuttle will average about 10 flights per year from 1991 to 2000, the average amortized life cycle cost *per flight* is in the region of $750 million. This gives the average total cost, with development and production costs rolled in, somewhere in the region of $6.75 billion annually.

Thus, in summary, the following conclusions can be drawn:

| | *Ops Cost/Flight* | *LCC/Flight* | *Ops Cost/Year* | *LCC/Year* |
|---|---|---|---|---|
| NDV | $8 million | $34 million | $500 million | $1,800 million |
| Shuttle | $500 million | $750 million | $5,000 million | $6,750 million |

(LCC = Life Cycle Costs)

Hence, over a 20 year period, an NDV could theoretically save an equivalent total cost of $80 billion compared with the Shuttle. More profoundly, where the Shuttle would fly only 140 missions during this period, the NDV fleet would be capable of 1,000 missions. This is the economic payoff potential that drives the NASP program today. As Robert Barthelemy notes, referring to the March 13, 1991 Joint Hearing, "For the first time I think, they [Congressional and Administration staff] began to appreciate the fact this was a revolutionary capability. If you look at space transportation in the future it could be radically different to what it has been in the past, even for military missions. Up until that moment it's been like 'well, yeah this is better than the Shuttle, maybe we can reduce launch costs or maybe we can fly fast.' It isn't that at all. It is a quantum leap, and they don't extrapolate from a capability that hasn't been available before, and what it might do. To me this thing could be like the discovery of the airplane, it opens up a different way of going to space."

## *Will the X-30s Be Built?*

In spite of this potential, the prospects of the United States moving forward into Phase 3 and the construction of the X-30 vehicles appear bleak. As mentioned earlier, it has been the strength of the House of Representatives, backed up by the National Space Council, that has repeatedly saved NASP. In particular, the man in Congress credited with this is Congressman Dave McCurdy, working in partnership with other congressmen including Tom Lewis, Dana Rohrabacher, and John Murtha. General Tom Stafford said, "I've been a big pusher of the NASP [and] Congressman McCurdy is in my home district. I advised him to send a letter to the vice president. McCurdy has personally carried the ball in the House of Representatives to save the NASP."

Congressman McCurdy has been able to keep NASP alive because he is a member of two critical House authorization committees, the first is the Subcommittee on Science, Space and Technology which oversees NASA's budget, and the second is the Arms Services Subcommittee which oversees the DOD research and development budget. Thus, as Congressman McCurdy explains, "for the past 5 or 6 years we've been having joint hearings with those two subcommittees, and they never before had been meeting together. The reason was because I was the one who sat on both and I chaired one of them. So, I was able to bring the two together and put some focus on NASP."

Congressman McCurdy's success is in no small part due to the efforts of A. Scott Crossfield, as McCurdy believes, "Scott Crossfield says every time I see him, he's not going to retire or die until he sees the NASP fly. He's probably the glue or the force behind this, if there's a single person . . . we ought to rename it [after him] at some point. Crossfield really has kept pushing for it." Combined, Crossfield and McCurdy have held NASP together as McCurdy continues, "We've been able to keep it at the level that a few number of people are controlling the outcome, good or bad. So far, we've got the right people helping and been somewhat successful. It gets harder and harder to keep increasing the budget [and] industry is going to have to show something for it pretty quickly to show they are making progress, and I think they are."

Keeping the program at a lower profile with only a few individuals has worked well for NASP while it received relatively modest funding during Phase 2. The problems in Phase 3 came because of the significant increase in funding that was required. The situation NASP faced for Phase 3 is summarized by Robert Barthelemy, "What happens is that we got these powerful groups with money that if they advocate something they're expected to pay for it. So, the DOD doesn't want to look too excited about the potential missions for NASP because they'd have to pay for the whole thing. NASA can't possibly say they need NASP, because they're going to have to pay for NASP. The Department of Commerce can't say that they need NASP because they've got no money at all. So, there's this terrible dilemma despite the fact that everybody really wants it. That's where the NSC [National Space Council] came in and they've said 'Look I understand why you guys are backing away but this nation needs this and you must give this much, and we're essentially taxing you to do that.' The big problem will come when the bill goes up higher for Phase 3. Then, it's no longer 'a hundred million from you and a hundred million from you'. It's maybe 'a billion from you and a billion from you' a year. Maybe that's what it's going to take to do the job. Now, if the country can really get that kind of overall vision, where it's good for the whole country, or the world for that matter . . . maybe that's what we ought to do; maybe that's what we're hoping will happen. But politics prevent them from saying 'I love this, I need it' because someone will say 'good—you pay for it.' No one wants to pay."

These sentiments are echoed by David Webb, "With the NASP decision coming in 1993 it's going to be very difficult if it's associated with NASA, and if it isn't, there's no other home for it. The problem that I see is that it's going to fall right between the cracks. The DOD certainly does not want a hypersonic vehicle at the present time because of the cutbacks in the DOD budget. It has enough problems just developing the Advanced Tactical Fighter. The Navy is desperate for a new airplane and there would be an uproar if the Pentagon decided that they would develop a hypersonic vehicle before they gave the Navy an new attack fighter. So DOD is not going to support hypersonics, and that means it would have to be NASA. But NASA has all of these other things on its plate and quite frankly, Congress is very doubtful indeed that NASA has the capability of doing what it has on its plate right now. I think that they would perhaps even shut down the hypersonic work as a result of that, or even some people will vote against it because they believe NASA can't do it. There are other people who will vote against it because they don't believe it should be done, and that it's taking bread from the mouths of babies, poverty and all of the other important issues NASP has to compete against."

It seemed for a time that NASA was gradually warming to the idea of NASP, where before it seemed somewhat indifferent. In the opinion of Robert Barthelemy, "They [NASA] would have zeroed the program a couple of years ago. If they had been given the program, I think they would have relegated it to a very small R&D program. They have made a complete turn around because they really began to understand what was going on. We probably weren't doing the right thing in not telling NASA what was going on. Once we got close to NASA, and they understood what was going on a little bit more, they came to support the program."

NASA support is likely to be crucial to the success of NASP, and the reason for NASA's emerging interest in NASP was a direct consequence of the involvement in the program of the former NASA deputy administrator, J. R. Thompson, who was also the deputy chairman of the NASP Steering Committee. Robert Barthelemy said that, "He is spending a lot of time with it; he believes it because he understands it. Probably of the sea of people in Washington D.C., he's the person who understands it the best, and takes a lot of time to pay attention to what's going on, and with that knowledge he backs the program. This is really encouraging. I think in the past NASP was sold mostly to people who didn't understand the technology aspects of it. They liked the idea and vision, but didn't really understand the technology and, when they found out about how risky the program was, they backed away from it. Here's a guy who understands the risk and is not backing away, and understands the implications. So we have a real friend in J. R. We tell him what's wrong, what's right, and how we're doing so as to maintain that trust he has in the program. Someday he may say he's looked this over and it's not technically feasible and say 'I want out.' That at least is the right reason to walk away from the program. To walk away because the politics change from year to year is the wrong reason."

Unfortunately, these comments were made prior to the resignation of J. R. Thompson in November 1991. NASP's strongest ally in NASA disappeared and, at the same time, the U.S. Congress slashed NASA's contributions to NASP from a requested $75 million to a mere $5 million. Whether this was directly a result of losing J. R. Thompson is unknown, but it almost certainly did little to help.

There is also the contentious issue of competition. NASP and NASP-Derived Vehicles are sometimes seen as something that will put other U.S. launch vehicle manufacturers and operators out of business. As Congressman Dana Rohrabacher explains, "You have got interest groups that are making profits in certain areas, and they are not experts on NASP. They will be left behind if NASP comes along. So their answer is 'NASP is undoable' or 'NASP is a pipe dream' or 'NASP is over-sold' or anything else that could deflect a real discussion which would reveal that these people are actually just trying to freeze the status quo so that they can maintain their status."

David Webb provides a sober view. "It's just the timing; I can't imagine the bad luck. If we had gone ahead with the hypersonic research engine in the '70s and it was 1983 when we were going to make this decision, I'd say history has finally coalesced. We had Reagan and 'go-goness' and everything else, and we would have gone for it. Now in 1993, I just don't know. It's one of those bad historical breaks."

Speaking in mid-1991 Bruce Abell said, "In spite of their relatively good progress in getting continuing support for the program, it's going to get a lot tougher when they go for big money after 1993. Again, my argument has been, you better make the case now for the kind of space access that NASP can give you, because if you wait until 1993 and say, 'OK, American public, we want $2 billion a year for an experimental airplane,' I think it's going to be a hard sell."

With all the technological progress the NASP program has made and, more importantly, having forged the way toward demonstrating that a true space transportation system might finally be within reach, it would seem a pity if the program ended prematurely. At the time of writing, the plan to build the orbital X-30 vehicles has been delayed indefinitely. Instead, the program is proposed to continue at a reduced funding level. The restricted NASP program is to include the flight testing of subscale scramjet engines and other critical components by boosting them to high speeds using Minuteman missiles (Hyflite). These tests will provide critical data on the performance of the scramjet and boundary layer transition, allowing future vehicles to be designed with greater confidence. By the late 1990's, the hope is that the United States will be in a better position to build the X-30, albeit a suborbital and unmanned configuration, prior to the construction of an operational vehicle.

How NASP has been sold has had a strong bearing on its progress. Perhaps it has put too much emphasis on the technology spin-offs and not enough on the applications. There is always a danger of overselling any new program, especially in light of the Shuttle experience. This is an understandable fear. Yet, with the original program priced at more than $10 billion—in addition to the $2.8 billion already spent—the people must have a crystal clear understanding of what they are being sold. To the average person, spending $10 billion plus for an experimental research vehicle might be considered an inappropriate use of taxpayer funds especially if its derivatives have only some vague applications in the next century. Apparently, it is this that contributed, more than any other, to the downfall of NASP.

However, investing $10 billion in a vehicle that is seen as leading directly to a usable space transportation system might seem an altogether more practical use of funds, if it could: save the taxpayer most of the money spent on the Shuttle and other rockets; save payload costs; eliminate Hubble-type fiascoes; enable cheaper and more accessible space stations; open up new commercial space business opportunities; and, as a byproduct, stimulate the nation's technology base.

*Access is the key to space. Realizing this is perhaps the key to building a national aero-space plane.*

## Summary—A Chance to Reach the Limits?

Whether or not the X-30 eventually flies, the NASP program has accelerated technology to the point where the feasibility of developing aero-space planes might be a realistic proposition. The technical advances have been ground-breaking and have laid the foundations on which future prototype or even operational vehicles could be built, although not necessarily scramjet-based configurations.

The NASP program has also been ground-breaking from the point of view of management and politics. The National Program Office represents the first time a major U.S. aerospace program has formed a consortium and, although it might not be ideal for experimental projects like NASP, it may well provide a model for other U.S. industries to conduct their future business. Politically, NASP has brought the prospect of an aero-space plane into the limelight for the first time. NASP has been able to test the political waters and help determine what actions are required before an aerospace plane can roll down a runway and fly into space.

Prior to the scaling back of NASP, one of the most remarkable aspects of the NASP program is the people who work on it. Despite the constant budget battles and the uncertainty over the fate of the X-30, many people have said that the excitement they have felt was at least as great as that which permeated the Apollo program a quarter of a century ago. When Apollo began, no one knew whether it would be successful, even with unlimited funding. This challenge made Apollo the success it was. In the same way, the challenge presented by NASP-type activities and the promise of routine access to space may well be the reason for its eventual success.

Bruce Abell provides a measured perspective that aptly sums up the present situation, "There are certainly people who are nervous about the enthusiasm, and they say we've seen it all before. In fact, we saw it all with the Shuttle; the same kind of prediction of technological success. But, in fact, to this point there just haven't been any setbacks. The setbacks have been political, they haven't been technological—the bloom is not off the rose for NASP."

## References and Footnotes

1. The definition of *hypersonic* is speeds of Mach 5 and above.
2. Barthelemy, R., "Tbe National Aero-Space Plane Program," AIAA-89-5001, *AIAA First National Aero-Space Plane Conference*, Dayton, Ohio, July 20–21, 1989.
3. Abell B., Speech given at *AIAA Second International Aero-Space Planes Conference*, Orlando, Florida, October 29–31, 1990.
4. Miller, J., *The X-Planes X-1 to X-29*, (Specialty Press, Marine on St. Croix, Minnesota).
5. Heppenheimer, T. A., *The National Aero-Space Plane*, (Pasha Market Intelligence, Arlington, Virginia, 1988.), p. 14.
6. Hallion, R.P, ed., *The Hypersonic Revolution—Eight Case Studies in the History of Hypersonic Technology: From Scramjet to the National Aero-Space Plane, Volume II, 1964–1986*, (Special Staff Office, Aeronautical Systems Division, Wright-Patterson Air Force Base, Ohio, 1987), p. 1362.
7. *The National Aero-Space Plane*, pp. 15–16.
8. Ibid., p. 17.
9. *The Hypersonic Revolution*, p. 1362.
10. *Pioneering the Space Frontier*, National Commission on Space, (Bantam Books, Inc., New York, May 1986), p. 115.
11. *The National Aero-Space Plane*, p. 27.
12. Ibid., p. 16.
13. "National Aero-Space Plane Program," Joint Hearing before the Subcommittee on Transportation, Aviation and Materials of the Committee on Science, Space, and Technology and the Subcommittee on Research and Development of the Committee on Armed Services, U.S. House of Representatives March 11, 1987, U.S. Government Printing Office, Washington, D.C., 1987, p. 23.
14. Heppenheimer, T. A., "How DARPA Lost the Aero-Space Plane," *Military Avionics*, June 10, 1988.
15. "Round Trip to Orbit: Human Spaceflight Alternatives—Special Report," U.S. Congress, Office of Technology Assessment, OTA-ISC-419, U.S. Government Printing Office, Washington, D.C., August 1989, p. 75.
16. "AFSC Announces Key Officer Changes," News release from the United States Air Force, Wright Patterson AFB, #87-18, October 25, 1987.
17. "How DARPA lost the Aero-Space Plane."
18. "National Aero-Space Plane Program Awards Contracts," News release, Office of Assistant Secretary of Defense, Washington, D.C., April 7, 1986.
19. "National Aero-Space Plane Propulsion Companies Selected for Phase II," News release from the United States Air Force, Wright Patterson AFB, #87-164., September 15, 1987.
20. "Three Airframe Companies Selected in X-30 Development," News release from the United States Air Force, Wright Patterson AFB, #87-194., October 20, 1987.
21. "National Aero-Space Plane: A Technology Development and Demonstration Program to Build the X-30," General Accounting Office, U.S. Government Printing Office, April 1988, p. 49.
22. "Round Trip to Orbit," OTA, p. 75.
23. Discussions with congressional staffers.
24. "Statement by the Press Secretary," The White House, Office of the Press Secretary, July 25, 1989. It is interesting to note that this statement explains that the NASP program was a "vital national effort" and that it also "provides the technological basis for greatly expanded access to space in the 21st century."
25. Webb, D. C., "Hypersonic Aerospace Planes: Launch Vehicles of the

Future?" *Third Pacific Basin International Symposium*, Los Angeles, California, November 6–8, 1989, p. 3.
26. Based on discussions with congressional staffers and congressman.
27. Lawler, A., "Congress Hits New Projects, Saves Station," *Space News*, September 30–October 6, 1991, p. 3.
28. "Civil Benefits of the National Aero-Space Plane (NASP) Program," NASA, *Report to the Committee on Science, Space and Technology, House of Representatives*, 1991. This report was written by the NASA Adminstrator's Office at the request of the House Committee on Science, Space and Technology.
29. "Washington Roundup: NASA Compromise," *Aviation Week & Space Technology*, January 20, 1992, p. 19.
30. Lawler, A., "Two NASA-Defense Projects Are Dealt Spare '92 Funds," *Space News*, November 25–December 1, 1991, p. 1.
31. See also a special report in *Aviation Week & Space Technology*, under the section entitled, "National Aero-Space Plane Team Selects Design," in which a full account of NPO is described, October 29, 1990, pp. 36–47.
32. Waldman, B., "When Competition Can't Deliver," *Space Technology International*, p. 83.
33. Ibid, p. 84.
34. "National Aerospace Plane (NASP) Fiscal Year 1992 RDT&E Budget Request," Joint Hearing before the Subcommittee on Technology and Competitiveness of the Committee on Science, Space, and Technology and the Subcommittee on Research and Development of the Committee on Armed Services U.S. House of Representatives, March 12, 1991, U.S. Government Printing Office, Washington D.C., 1991.
35. Ibid., Barry Waldman statement, p. 65.
36. Ibid., John Welch statement, p. 42.
37. Ibid., Alan Bromley statement, p. 15.
38. Waldman, B. J., and P. Hasha, "The First Year of Teaming: A Progress Report," AIAA-91-5008, *AIAA Third International Aerospace Planes Conference*, Orlando, Florida, December 3–5, 1991, p. 2. Copyright American Institute of Aeronautics and Astronautics © 1991. Used with permission.
39. "National Aerospace Plane (NASP) Fiscal Year 1992 RDT&E Budget Request," Waldman, B. J., p. 66.
40. The first pictures and descriptions of the Composite X-30 were printed in *Aviation Week & Space Technology*, October 29, 1990, pp. 36–47.
41. Kandebo, S., "Lifting Body Design is Key to Single-Stage-to-Orbit," *Aviation Week & Space Technology*, October 29, 1990, pp. 36–37.
42. "The First Year of Teaming," Waldman, p. 9. See also, Dornheim, M. A., "NASP Fuselage Model Undergoes Tests for Temperatures, Stresses of Flight," *Aviation Week & Space Technology*, February 3, 1992, p. 52.
43. Ibid., p. 12.
44. Ibid., p. 5.
45. Ibid.
46. Ibid.
47. Kandebo, S., "NASP Team Narrows Its Options As First Design Cycle Nears Completion," *Aviation Week & Space Technology*, April 1, 1992, p. 80. In addition, see also the GAO report, "National Aero-Space Plane: Restricting Future Research and Development Efforts," December 1992.
48. For a discussion of unstarts see Heppenheimer, T. A., *Hypersonic Technologies and the National Aero-Space Plane*, (Pasha Market Intelligence, Arlington, Virginia, 1990).
49. "The First Year of Teaming," Waldman, p. 7.
50. Ibid., pp. 12–13.
51. Based on discussions with NASP government and industry representatives.
52. "National Aerospace Plane (NASP) Fiscal Year 1992 RDT&E Budget Request," p. 61.
53. Ibid., p. 62.
54. "JPO Studying Flight Test Issues in Early Phase of X-30 Program," *Aviation Week & Space Technology*, October 29, 1990, pp. 46–47.
55. From a presentation delivered by H. Reda at the *AIAA Third International Aerospace Planes Conference*, Orlando, Florida, December 3–5, 1991.
56. Reda, H., "NASP and the Environment," AIAA-91-5051, *AIAA Third International Aerospace Planes Conference*, Orlando, Florida, December 3–5, 1991, pp. 2, 4.
57. "National Aerospace Plane (NASP) Fiscal Year 1992 RDT&E Budget Request," Welch, J., pp. 43–44.
58. See Appendix for list of NDV papers on the subject of operations and applications.
59. "Space Transportation Options & Launch Architecture Comparisons," Presentation viewgraphs by the NASP Joint Program Office, May 1990.
60. One of the problems with NASP is that it looks more like an airplane than a launcher. Therefore, when the economics of NDVs are discussed, comparisons with space launch economics are sometimes overlooked. For example, in *Space Watch* (July 1990) under the title "Multi-billion Dollar Price Tag Could Bury NASP," it was stated that the pure operational costs (ignoring fixed annual costs) would be between "$800,000" and "$4.4 million" per flight, or about $2.3–13.4 million per flight including fixed annual costs, assuming 50 flights per year for a fleet of five vehicles. It would have been enlightening to the reader if immediately after this, the article had made a comparison with the Shuttle's operational cost of between $400 million and $800 million per flight.
61. Toten, A., J. Fong, R. Murphy, and W. Powell, "*NASP* Derived Vehicle (NDV) Space Launch Operating Costs and Program Cost Recovery Options," AIAA-91-5080, *AIAA Third International Aerospace Planes Conference*, Orlando, Florida, December 3–5, 1991, p. 2.
62. According to an article by Gene Koprowski in *Washington Technology* (Sept. 26, 1991), a Congressional Research Service report concluded that the Shuttle's development and production costs were about $45 billion (1990). Therefore, over a nominal 20 year lifetime, this is equivalent to about $2.3 billion per year. (See also Chapter 11.) NASA maintains the actual figure is about $27 billion, or about $1.4 billion per year. Therefore, this gives an average of about $35 billion.

# Chapter 7

# HOTOL: The Pragmatic Approach

On August 23, 1984, the lead story of the Independent Television News took the United Kingdom by surprise, [1]

Secret plans have been drawn up for Britain to build a revolutionary space plane of its own-a sort of 'super Space Shuttle.' It would be the most advanced spacecraft ever constructed and the first in the world to take-off horizontally like a conventional aircraft. And it would be fitted with the very first rocket engine ever designed that can burn oxygen which it breathes from the air . . . Although it is still secret, ITN has learned exclusively from various sources some of the main features of the space plane—code-named HOTOL, for HOrizontal TakeOff and Landing . . .

Like NASP and Sänger, HOTOL has been plagued by misunderstandings of what it is and why it came into being. HOTOL is not a vehicle for flying passengers to Australia in 2 hours—although technically it probably could. It was not proposed as a program that would enable the U.K. to reenter and become more heavily involved in European space efforts—although it undoubtably would give the U.K. a much higher profile than in recent years. Finally, it is not a vehicle that was proposed with the primary objective of pushing back the frontiers of technology and engineering—although it most certainly would contribute to these fields in a profound way. HOTOL is none of these. The objective of HOTOL—the reason why it came into existence—was for one purpose only: *to significantly reduce the cost of placing objects in orbit [2].*

The HOTOL story is instructional in highlighting the interacting constraints of technology, economics, and politics encountered in the development of new programs like aero-space planes: technical constraints, because of the limitations imposed by technology; economic constraints, because of the drive to seek the most cost-effective solutions allowed by technology and the market; political constraints because of the way complex and original programs are appreciated by governments. Additionally, the HOTOL story also provides a lesson in how the British political system deals with high technology programs today. (Figure 7.1)

## History and Background

HOTOL was give a unique and, perhaps surprisingly for the British, unpoetic name. Indeed, some say the reason why this name stuck was because it was so awful that no one could forget it. Even though the amount of money spent on HOTOL was nowhere near as large as for NASP, the project's stature and its impact on thinking in the area of advanced space transportation systems were nevertheless significant. The HOTOL arguments have helped focus attention on the cost-effectiveness of European space programs, bringing attention to issues which some European nations have preferred to avoid. The emergence of HOTOL was in part responsible for the German Sänger program and the ESA Future European Space Transportation Investigation Program (FESTIP). (See Chapter 8)

HOTOL is the result of initially independent thinking by two individuals and long-time friends, who have been the force behind the project from its inception: Alan Bond and Dr. Robert Parkinson. Alan Bond, working on his own behalf recalls, "I'd basically been plugging away at my own thoughts on how to get economic spaceflight for a long time. In January 1982, there was a Space Transportation Systems symposium at the British Interplanetary Society. That was the one where CNES gave a presentation on ESA's next generation of launchers. I found it appalling that Europe was considering building twenty—year old technology and claiming they could make a commercial breakthrough with it. When [CNES] made this pronouncement, I decided to go away and see what could be done about it."

Alan Bond's first ideas were along a different line than a winged, air-breathing aero-space plane, and the evolution of his thoughts is interesting, as he explains, "Bob [Parkinson] decided independently of myself that what we wanted was a manoeuvrable aeroplane that could re-enter from orbit, be serviced on the ground and go back to orbit. I actually dismissed aircraft-type vehicles back in the 1970's—I didn't think they were practical. All my studies had been based on



CARL A. RUDISILL LIBRARY
LENOIR-RHYNE COLLEGE

A

B

**Figure 7.1** HOTOL of (A) 1985 and (B) 1989 (*BAe*).

rocket vehicles. In particular, I had become convinced by Phil Bono at McDonnell Douglas that the right way was to use a plug nozzle both for part of the ascent and as a heat shield for a ballistic re-entry. I spent the first half of '82 looking at higher energy propellants to see if there was some way out ot the trap I knew we were in—all the vehicles I looked at didn't quite make it [to orbit with a payload]. So on the 11th of June 1982, I decided that the only other way to improve the performance of the rocket engine was to use the atmosphere. I then started to look at how you can make use of the atmosphere and rocket propulsion in a single vehicle. In the literature, everyone in the past started off with an airbreathing engine and then found some way of making it into a poor-looking rocket engine for the latter part of its flight. When you look at the delta-Vs you've got to achieve [to attain orbit], that's clearly the wrong thing to do. Whatever you do in the airbreathing phase, when you go into rocket mode, the rocket part has got to be the best rocket you know how to make. That was the starting point in my own studies, to have a look at how you make a rocket engine that is at least as good, if not better than, current closed-cycle rocket engines, and then to find a way to graft the airbreathing engine onto the rocket engine using as much of the same machinery as possible. In other words, not to increase the mass more than necessary through the airbreathing phase. By September of 1982 what I finished up with was a Phil Bono-type vehicle [see Chapter 10], but with an airbreathing enhanced capability in order to do the mission a lot more effectively—such configurations may still be a possibility. I was having some problems in finding a satisfactory trajectory for it because the lack of lift of the vehicle means you've got to follow a rather unoptimum trajectory that throws away part of the advantage of the airbreathing."

At this stage, Alan Bond contacted Bob Parkinson, who had recently moved to British Aerospace (BAe) to study, among other things, future launch systems because of Europe's increasing interest in building a new launcher to follow Ariane 4. Alan Bond recalls, "Needing some ideas on vehicles, I got in touch with him, and needing an expert on pumps and turbo machinery [for this combined-cycle engine concept] I got in touch with John Scott-Scott at Rolls Royce. That led to a series of meetings which, in March 1983, caused me to stick with Bob's [ideas of using winged vehicles]." Until this point Alan Bond had dismissed winged vehicles as being impractical, "I did a lot of studies when I was with Rolls Royce back in the early 1970s on winged vehicles, and I never finished up with a positive payload. As soon as you started to put wings on a rocket and change its shape to make it more aeroplane-like, it just didn't have a payload. Although all the calculations were based on quite ancient technology, what was also going on in the 1980's, which I hadn't realised until we started work on HOTOL, was the materials revolution. This led to the second part of the practicality problem: one part is the propulsion, and materials contributes the second part—that's why space planes are actually possible now, not through any one single breakthrough but progress on all fronts."

By mid-1983, British Aerospace, supported by a small group of individuals working for Rolls Royce, had started some initial study work of a fully reusable launch system based on propulsion system concepts proposed by Alan Bond and a winged vehicle configuration by Bob Parkinson. The winged concept that was to become the shape of HOTOL did not happen by accident, but was the result of extensive study of future launcher configurations. As Bob Parkinson explains [3],

> Choosing a launch vehicle design is no easy matter. In a recent study at British Aerospace we identified 35 different types of launch vehicles without coming to the end of them. Just which you choose depends on two factors, economics and politics. The problem is that launch vehicles—any launch vehicles—are expensive to develop.

Alan Bond examined a number of different engine cycles, and by December 1983 he was confident that one particular concept was sufficiently promising, therefore he applied for a patent [4]. Almost immediately, the Ministry of Defence (MOD) slapped a prohibition order on the concept. Shortly afterward Alan Bond was informed that his engine, and all the work that had gone into it, was classified secret, primarily to stop it from falling into the hands of the Soviets. Although this was to become one of a number of reasons for the program's limited international success, the classification was initially welcomed for two reasons. The first was that it brought a certain amount of credibility to the concept. The second, and rather more important reason in the opinion of Alan Bond, was that it prevented the Germans or French from "running with it, which they could have done quite easily."

The basic feasibility studies conducted within BAe (Space Systems) were largely a result of the interest of Peter Conchie, the company's business development director, who also brought in some of the engineers originally associated with Concorde at BAe in Bristol. Integrated configuration studies were pursued on a commercially confidential level at this point, and as Alan Bond says, "Even using conventional materials, there had been sufficient progress made that it did show a credible real payload. There were some factors that had been dropped, and a genuine $\pi$ had been left out of one calculation as it turned out! But nonetheless it was sufficiently encouraging and it looked like a feasible vehicle. This was the first time I started to think about vehicles which had wings and took off horizontally . . . By mid 1984 to early 1985, the transition was complete, and I could see all the advantages, although my own calculations suggested that the payload was still marginal."

While the HOTOL feasibility studies were accelerating during the latter half of 1984, BAe also began work on an unmanned space platform proposal that was to be the U.K.'s contribution to the International Space Station, and an element of the proposed ESA Columbus program—the forerunner of what was to evolve into the ENVISAT-1 platform for Earth observation. As Bob Parkinson recalls, "When we went to talk to [Sir Geoffrey] Pattie, we actually went to talk to him about the Polar Platform, but he also wanted to know about HOTOL because he was on his way to [the ESA Ministerial meeting in] Rome. We told him we were concerned about the way the launch vehicle program was going in Europe and out of that meeting came the famous moment when the Admiral [Raymond Lygo, former Chairman of BAe] popped up and said 'this is worth a million pounds of my money, how much is it worth of yours?' And Pattie said 'well I think we ought to have a 50:50 split, it would be worthwhile just to find out what's going on.' and we proceeded on that basis."

From this, and subsequent meetings with Rolls Royce, who were brought on board to support the critical engine technology development work, the U.K. government decided to fund a 2 year Proof of Concept study for £1.5 million ($2.6 million) and British Aerospace and Rolls Royce contributed £1 million ($1.7 million) and £0.5 million ($0.9 million), respectively. The work began in mid-1986 and, at its peak, there were about 100 people working on the project within the Space Systems Division in Stevenage, the Civil Aircraft Division in Bristol, and the Military Aircraft Division in Warton of BAe, and the Ansty Division of Rolls Royce. In addition, a number of other companies (e.g., ICI) and universities (e.g., Bristol) were performing a wide range of enabling technology work. In a very short period of time, HOTOL had spread throughout various sectors of the U.K. industrial and academic centers. HOTOL also had a high profile with the general public, who seemed to adopt it as the flagship of Britain's high technology industry, even if they weren't quite sure exactly what it was; HOTOL was often confused with "Hotel." Everyone had heard of HOTOL, and even Prince Charles was known to be a fan of the project [5].

## Politics: The Wrong Place at the Wrong Time

During the 2 year period of the HOTOL Proof of Concept studies, a number of events occurred which could have worked to the benefit of HOTOL. In the end, though, circumstances worked against it.

Both 1985 and 1987 were critical years for the European space industry. In January 1985 the first ESA Ministerial Council meeting for a decade was held in Rome. This meeting was intended to help define ESA's long-range plan and policies for its next 10 years in space. Because the programs seeking approval were more ambitious than ESA had attempted before, requiring a doubling or even tripling in the budget, a second meeting had to be held 2 years later. The meeting at The Hague in 1987 was to approve the plan adopted in 1985, and commit the agency to carrying out certain programs.

Prior to the first council meeting, the U.K. Department of Trade and Industry made some briefings to ESA and indicated that it planned to propose HOTOL for "Europeanization," but only in the form of a technology program. The proposal was made, and HOTOL was included in the Rome communique, which stated [6],

> [ESA] takes note of the studies underway in the United Kingdom of the future generation HOTOL project and, following Annex IV of the Convention invites the United Kingdom to keep the Agency informed.

The U.K. now had 2 years to establish HOTOL as a European program to stand a chance of it being considered for approval at The Hague in 1987. This was a daunting prospect in light of France's already considerable financial and political commitment to Hermes. According to Sir Geoffrey Pattie, then Junior Minister for Trade and Industry, and a strong supporter of HOTOL and the U.K. space objectives, "In 1986, I went around most of the significant ESA partners to brief them on HOTOL—but obviously not the French—to build up a sort of a coalition which would be useful on our side of the table. Everybody liked [this] because they were fed up with being 'babooned' around by the French. Even so, having said that, there wouldn't be any European space program without the French, without any doubt at all. So one has to salute them even if one might be jealous, which is probably what it is . . . We never reckoned [HOTOL] was going to be a British program. We thought that it might have 30% UK funding in it. It was always going to be competitive, in *funding* terms and not *programmatic* terms, with the French Hermes, which is a payload and not a launch vehicle."

Not since the U.K.'s involvement in the Europa launcher program, the predecessor of Ariane, did it have such a high profile within the European space community, and much to the surprise of Europe, as Sir Geoffrey Pattie continues, "it was all of a sudden like Britain reappeared, and we came [to the Rome meeting] and said we will do the Polar Platform and we're going to do this and that." The U.K. gave the impression that it was serious about getting back into the space business. Despite the good words, it was not to be.

Britain's apparent renewal of interest in space, after considerable lobbying on the part of some British space companies, led to the establishment of the British National Space Centre (BNSC), the administrative body for coordinating the U.K.'s space efforts. Roy Gibson, former director general of ESA, was made director general of the agency, and his first task was to define a space policy for the U.K. Up to this point, Britain's participation in ESA programs had been on an ad hoc basis. Bob Parkinson explains, "Roy put together an 'all singing, all dancing' kind of campaign of which HOTOL was an aspect, although it wasn't the most important aspect. What we were looking for was a technology program and not an infrastructure program. [Unfortunately, from the Government's perspective] I believe Roy Gibson's plan was probably too enthusiastic and too much of a steep increase to be reasonable." At that time, the U.K. was spending about £135 million ($230 million) per year on space, and the BNSC plan proposed a range of options that could have doubled or tripled the budget.

The U.K. was actively trying to promote HOTOL and was pursuing a new commitment in space. Unfortunately, the situation then started to deteriorate rapidly, even though the HOTOL studies were proceeding satisfactorily. As Alan Bond recalls, "By the time the project stopped, we really had a superb atmosphere on how to resolve problems. We all used to take our stances and were confident everybody else was wrong, although deep-down we knew that wasn't true. It was sort of like a car tyre with a bulge. You'd shift the bulge to one place and it comes out somewhere else, but gradually all the bulges were getting smaller. I've got to say it was the most terrific working atmosphere I have ever worked in my whole career—those couple of years on HOTOL when we really got it moving."

One of the principal problems seemed to be the classified nature of the HOTOL engine, designated the RB-545 by Rolls Royce (Figure 7.2). As Sir Geoffrey Pattie explains, "The MOD classified the engine technology for some mysterious reason and declined to declassify it." Many nations and organizations were curious about HOTOL, perhaps for no other reason than the elusive secret status of the RB-545. Officials close to the program believe that HOTOL's classified engine status was used as an excuse to dismiss the concept out of hand as there were many ways around the problem. Further, although of less significance, the ESA convention's stipulation that the agency should promote activities "for exclusively peaceful purposes" conflicted with what was seen as a military interest in HOTOL. However, Ariane was developed by ESA and has launched military satellites.

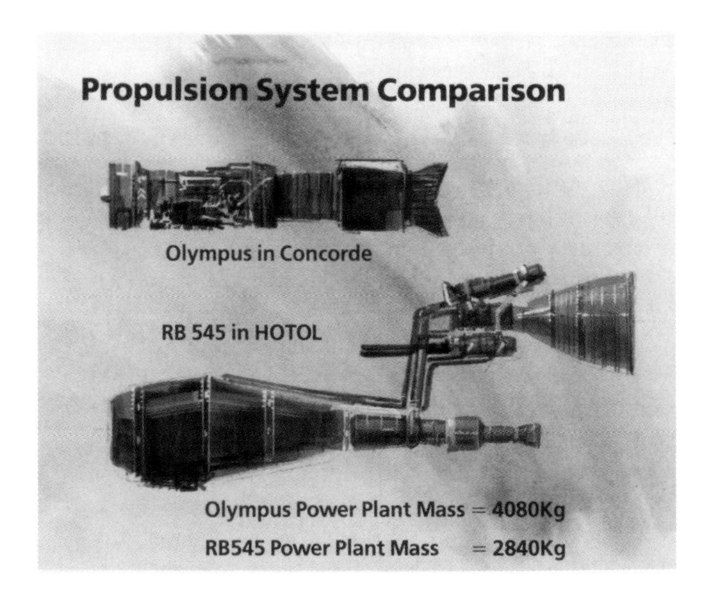


**Figure 7.2** RB-545 comparison with Concorde's Olympus engine (*BAe*).

The stigma of the engine classification threatened the program's future. Yet, as Sir Geoffrey Pattie explains further, "I think we could probably have mounted a major campaign to get [the engine] declassified without too much difficulty. These things are not completely immovable. You've got to be able to go to the PM and explain the situation. The problem was that Rolls Royce were not particularly interested in actually doing it. They didn't want to get involved, so who would be campaigning for it? If a government department comes along and says 'we would like to have that engine declassified and, mark you, the engine makers don't want to have anything to do with it' they would say 'go away, stop wasting my time.' So it wasn't worth the effort, and we didn't seriously try. Rolls Royce had their own reasons for not wanting to do it."

The rationale for Rolls Royce's lack of interest was never made entirely clear, although the company was being privatized at the time. Efforts to make the company as profitable as possible alledgedly forced a reduction in the funds available for long-range R&D projects. Rolls Royce was concentrating on those R&D programs with a better chance of making money in the near-term [7]. If the U.K. government had funded further work on HOTOL, Rolls Royce would probably have been asked to reciprocate with its own funds, personnel, and facilities which could otherwise have been devoted to nearer-term projects.

The engine status and Roll Royce's lack of interest were the first problems, but not the last. In June 1987, Margaret Thatcher performed her periodic reshuffling of the government and, unfortunately for HOTOL and the U.K. space program, Sir Geoffrey Pattie was removed from his post. The new Minister for Trade and Industry Kenneth Clarke quickly proved his distaste for the European space programs. The day before the 1987 International Astronautical Federation (IAF) meeting in Brighton, England, Clarke was interviewed on national TV and, demonstrating his masterful lack of subtlety, said he thought ESA was a "hugely expensive club." His appointment was a hammer blow to the U.K. space efforts. Indeed, his statements before the IAF led Alan Bond to say, in a widely publicized interview on British TV, that he might be prepared to leave the U.K. and risk prison by telling other nations about the engine just to keep HOTOL alive.

Kenneth Clarke also put an end to Roy Gibson's space plan. Bob Parkinson takes up the story, "I think Clarke came in [to his Department of Trade and Industry position] feeling somewhat insecure and recognising that to make his political position more secure, he was going to have to be seen as a somewhat harder man than he had been previously. So he said no to space and discovered that gave him a political reputation. So he continued saying no to space, not that he was against space, not because it did his PR image any good, but because it did his political image good because he was seen by Maggie [Thatcher] as someone who stood up to people and said no. It didn't matter what it was about, it just happened it was space that got it. Essentially, that left us rudderless."

The U.K. government gave the impression that it wanted to put Britain back in the forefront of European space. However, it rapidly became clear that the U.K. government could not find the funding and, in any case, had no such desire. When The Hague meeting finally came around in November 1987, Clarke ensured that the U.K. would not contribute any funds to Ariane 5, Hermes, and the manned elements of Columbus. A last minute effort saved the Polar Platform and Britain's leadership of that program, but only because it would now be dedicated to Earth observation. As for HOTOL, if the U.K. government had little enthusiasm for pursuing it, neither did anyone else in Europe. HOTOL's demise was complete—in its original form, at least.

Sir Geoffrey Pattie provides another perspective on the situation from the viewpoint of Margaret Thatcher, "She wasn't interested in HOTOL. She didn't care about HOTOL, quite frankly. She just wasn't into it at all. The fundamental problem in her eyes, and it was always a tricky area of ambivalence here, was that she had a scientific background where she was quite fascinated by science, including remote sensing, for example. And then there was a big rock sticking up out of the water and that was called Concorde. She would say 'it's not going to be like Concorde is it?' Because Concorde represented to her all that was wasteful, irrelevant, evil about technology for the sake of technology, scientists doing their own thing, engineers having fun at the public's expense, etc. This [HOTOL] looked to her like a kind of passenger-less Concorde, taking billions of dollars for a limited purpose."

Part of the reason for her seeming lack of enthusiasm and understanding of HOTOL was the departure of Robin Nicholson, the Chief Scientist to the Prime Minister and a supporter of HOTOL. As Sir Geoffrey Pattie explains, "Actually, I think one of the key things was that the previous Chief Scientist [Nicholson] was very much the apple of the PM's eye. The Chief Scientist and I [went to the ESA Rome meeting] on the *express* instructions and support of the PM. By the time a year had gone, he had gone, and the new guy had come, and he didn't have the same stature—what happens is that people take a while to play themselves in. If the Chief Scientist had actually said 'PM, I think this [HOTOL] is something we ought to do' she may have said 'OK.' If she had [then] asked for £5 million ($8.5 million) from everyone [Departmental Heads] around the table she would have got it."

The demise of the original HOTOL, therefore, happened in stages. First, the continued classification of the HOTOL engine made it impossible for Europe to believe HOTOL was practical. Without pan-European technical confirmation of the feasibility of HOTOL, it was impossible to provide positive inputs to the policy level. Second, an alledgedly small interest on the part of Rolls Royce ensured there was no way the engine could be declassified. Third, the Minister responsible for pushing HOTOL was replaced, and the new DTI Minister killed HOTOL purely as a tool to enhance his political stature in the eyes of Thatcher. Fourth, the loss of the Chief Scientist cut off any positive inputs to the PM. As Pattie said, "Things need a champion at the very highest level." By the end of 1987, any champions of HOTOL had disappeared. On July 25, 1988, Kenneth Clarke made the following announcement, [8]

> The scale of funding required to develop Hotol or any similar concept to eventual production would be far too great for the UK to contemplate on a national basis, particularly since only a relatively small number of launch vehicles are likely to be required. I have therefore concluded that further development must take place on the basis of international cooperation. The Government will support efforts by UK companies to find suitable collaborators but will not be providing any further financial support for the foreseeable future.

The only vacuum HOTOL ever found itself in was political.

## *Evolution of the HOTOL Design Philosophy*

Throughout the progress of the HOTOL studies, maximum emphasis was placed on optimizing the life cycle costs of every single aspect of the vehicle and its operations, i.e., the total costs associated with development, production, and operations over the vehicle's expected lifetime, including a contribution resulting from the cost impacts of failures. Technical solutions were never incorporated purely for the sake of technology—this would have defeated the object of the exercise. Every design decision ultimately was based on life cycle cost tradeoffs, and the evolution of HOTOL was carefully costed all the way through the design process [9].

For a vehicle like HOTOL to be economically attractive, the life cycle cost must be substantially better than existing launchers over a reasonable timeframe—20 years, for example. The sooner a vehicle is able to pay for itself, the better its chances of being funded in the first place. The philosophy can be most simply explained by Figure 7.3. Because an aero-space plane is technically more difficult to construct, it will cost more to develop than an expendable launcher, typically by a factor of about two. However, if the

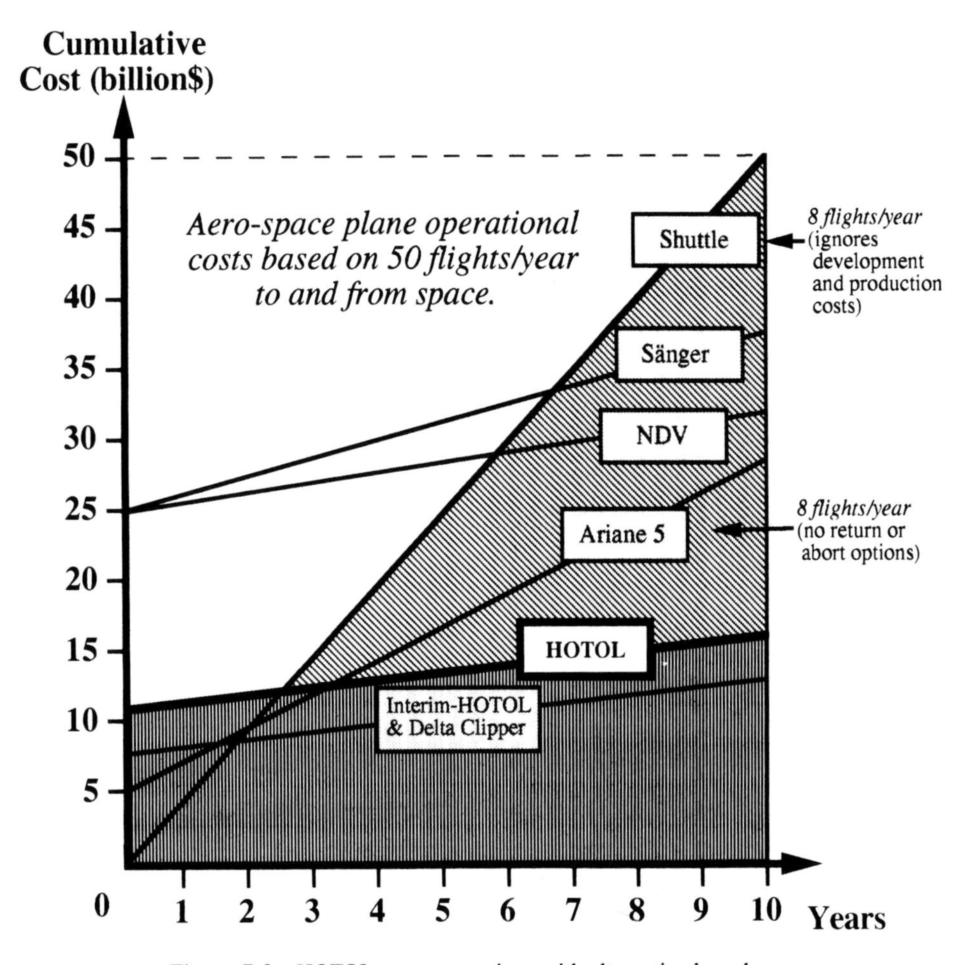


**Figure 7.3** HOTOL cost comparison with alternative launchers.

proper investment is made in critical support infrastructure needs (e.g., spare parts, logistics, facilities) at the beginning, the airplane-like characteristics of the aero-space plane should enable lower costs per flight compared with an expendable rocket, as discussed in Chapter 4. Eventually, after a certain number of payloads have been launched, the total cumulative costs of the expendable rocket will exceed those of an equivalent aero-space plane. Of course, this basic explanation doesn't even take into account the positive impacts on payloads of having a more affordable, available, and reliable means of accessing space. This conservative philosophy entirely underpins the HOTOL concept.

The market size that a vehicle like HOTOL is expected to serve is of critical importance because of its relationship to technology. An aero-space plane that must fly 400 times per year will place tighter constraints on technology and hence cost more to develop and procure than one only required to fly 40 times per year. For HOTOL, to maintain as much realism as possible, a market of around 12 to 25 payloads per year was all that was anticipated to be economically cost-effective [10]. This is about the same as the current Ariane 4 annual market. Over a 20 year lifetime, a fleet of vehicles would fly a total of about 250 to 500 missions. This means that investing in higher technology solutions that could achieve higher flight rates might be of little benefit because the market would not justify it, and the payback time would be longer. As Bob Parkinson explains, "There is a tradeoff between development and recurring costs, which is that you can reduce development if you do not use the vehicle vastly. The real question with NASP is that the development cost of the NDV, including NASP itself, maybe very large. So, the cross-over point compared with a simpler vehicle [like HOTOL] will be a long way downstream. In comparison with conventional launch systems, the break-even point [for HOTOL] is fairly close to the origin."

## Pleasing Some of the People All of the Time

"When we set out with HOTOL," according to Alan Bond, "the object was to take the expendable launcher and replace it with a recoverable vehicle that did everything the ELV did, but not a lot more because there isn't the payload traffic to justify a lot more at the present time. We were aware that there are lots of things you would like to do if launch costs were cheap enough. However, it would really be quite nonsensical to expect operations in space to increase by an quantum leap, because things just don't happen that quickly. So HOTOL doesn't have a pilot, it's developed using missile-type technology, it doesn't have masses of on-orbit manoeuvring delta-V, and it provides minimal services to the payload. HOTOL does everything an ELV does with the one exception that you've saved some money in the process, because at the end of the day the entire vehicle comes back for refurbishment." (Figure 7.4).

Alan Bond's description of HOTOL shows clearly that it was not designed with the intention of being all things to all people. One of the problems the Space Shuttle encountered was that it had to launch all sizes of payloads from small communications satellites to large space station structures. The Shuttle also had to put people in space and keep them alive for missions lasting several days. This requirement helped to make the Shuttle as expensive and unavailable as it is today. HOTOL, however, was optimized to place payloads in low Earth orbit that were no heavier than 7 to 10 tonnes (Figure 7.5). This mass range was chosen because the *existing* market indicated that around three-quarters of all payloads weigh less than 10 tonnes. (Chapter 16.) For the remaining quarter, vehicles like Ariane 5 could be used. Keeping Ariane 5 in service would also ensure continuous access to space in the event of a HOTOL standdown. (Chapter 15.)

The aim of HOTOL was to design a vehicle as small as reasonable while still achieving the defined 7–10 tonne payload performance. As a consequence, HOTOL is unmanned because the mass of a crew compartment large enough to house four to six people is close to that of the payload itself, about 5 tonnes [11]. An approximate doubling of the effective payload mass also leads to a doubling of the takeoff mass (i.e., 275 tonnes up to about 500 tonnes) and potentially a similar increase in the development costs. With a crew compartment heavily integrated into the vehicle design, servicing activities become more complex to ensure that the crew compartment and the vehicle are safe for the next mission. Such factors push up operational costs and reduce flight rates. With HOTOL, the philosophy from the beginning has been to carry crew or cargo as required, not both. For crewed missions, a capsule is to be serviced offline from the HOTOL launchers to avoid launch processing interruptions, and then inserted into the next vehicle just like cargo. The capsule itself is equipped with an escape system for contingency situations [12].

Because HOTOL is an unmanned launcher designed to stay in orbit for about 2 days, and not a mini-space station like the Shuttle, its electrical power needs are substantially less. HOTOL was configured to provide about 2 kilowatts [13] of continuous power for payload use (more than enough for a typical comsat deployment), compared with the Shuttle's 12 kilowatts of continuous power capability over a week in orbit. This low power requirement has profound impacts, and one of the most interesting is that HOTOL does not need a complicated deployable thermal radiator like the Shuttle has running the length of each cargo bay door. Instead, relatively simple fixed radiators are built into the skin of the fuselage on either side of the payload bay. Heat can be adequately ejected provided one of the radiators is not pointing at the Sun or the Earth.

As another impact of the lower power needs, HOTOL uses lithium-based high energy batteries designed to be replaced after every flight. The alternative was fuel cells which generate power by the chemical reaction between liquid hydrogen and oxygen, as on the Space Shuttle. Although fuel cells are extremely efficient and reusable, the decision not to use them was taken for a number of reasons. First, the cost to develop fuel cells is greater than batteries. Second, each fuel cell requires at least one pair of liquid hydrogen and oxygen tanks, and additional costs are incurred due to the need to carefully integrate them into the vehicle system. Third, fuel cells usually need maintenance between missions, and their installation makes them difficult to access. The cost and time to service them were expected to be at least as great as the purchase of a new battery, which has less critical installation requirements, only needing to be plugged/unplugged from the power supply system. Finally, fuel cells have to be loaded with cryogenic liquids before launch and the temperature and pressure of these reactants must then be carefully monitored, complicating the launch preparation activities compared with the solid state batteries. Overall, while fuel cells could have been used, for the objectives of the HOTOL mission they were considered more trouble than they were worth. For those few missions requiring more power and thermal control than the standard HOTOL is able to provide, the payload would have to make up the difference.

## *The RB-545: A Secret No More*

The power supply system is just one demonstration of the HOTOL design philosophy, but the propulsion system is a better one. Whereas NASP aims to develop a vehicle capable of breathing air to as a high a Mach number as possible before switching to rocket propulsion, with HOTOL, air-breathing only occurs up to about Mach 5.5 before the combined-cycle engine transitions to the pure rocket mode using the onboard liquid oxygen propellant. The principal reason for Mach 5.5 is because at or below this speed, incoming air can be shocked down to subsonic speeds prior

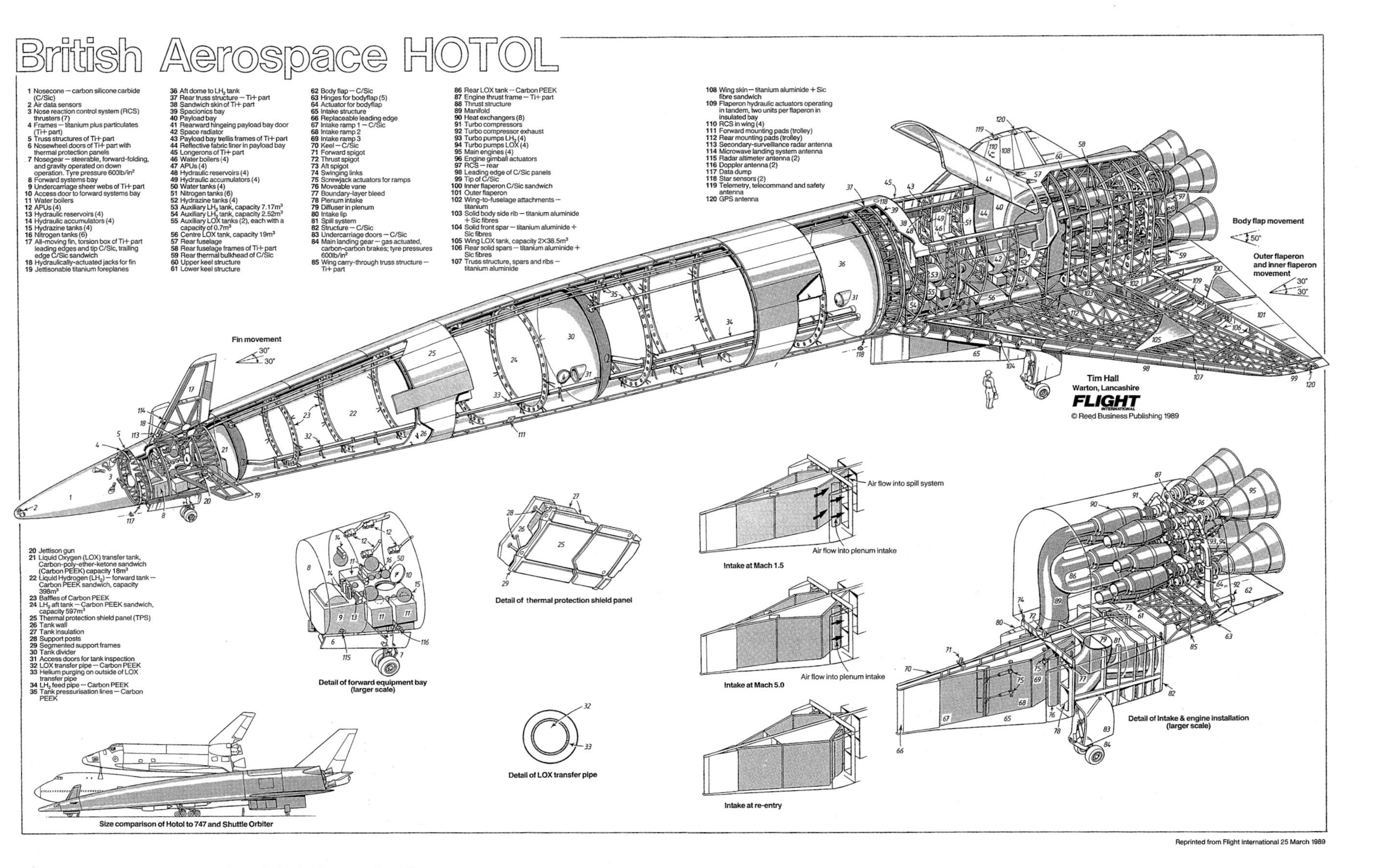


**Figure 7.4** HOTOL cutaway (*Flight International*).

**Figure 7.5** Deployment of a communications satellite (*BAe*).

to being combusted. If much higher speeds were chosen, HOTOL's engines would need to handle and control supersonic airspeeds. Developing such engines presents a much higher level of technical risk (See Chapter 6). More importantly, breathing air significantly below Mach 8 ensures that the engine can be fully tested on the ground and in existing facilities without the intermediary step of building an experimental vehicle.

As Alan Bond explains, "since the 1960's I have been very well aware of scramjets and ramjets—these were the love of my life in those days. One of the great things that happened in my life was meeting Val Cleaver [former manager of Roll Royce's rocket propulsion work in Ansty], because he had an enormous background, and one of the dangers is that unless you have contact with people who are amongst the pioneers you don't understand how things today are the way they are. Val Cleaver was one of the pioneers, and I still bear the scars where he kicked me all over the office one day when I was doing a rather clean-shaven, young whizz-kid act with him about why we weren't going scramjet. And he said 'While you draw the ramjet and scramjet on the piece of paper and that looks extremely simple, on the bit of paper next to it you got to draw all the facilities that are required to develop that system. It may be very intellectually satisfying to have a nice clean-cut aerodynamic vehicle that can do everything, the fact is at the end of the day somebody is going to have to pay to develop it.' When I started on the HOTOL engine, I deliberately—and I have said this many, many times—kicked all that stuff to one side for the simple reason that to do a development program on a ramjet or scramjet for this sort of application, you either have got to build a test facility or you've got to build test aeroplanes [like NASP]. Now, back in those days we didn't have CFD models, but it has not changed the picture all that much. Those [test facility] programs, even before you get to the thing you really want to build, are enormous programs in their own right. It would be great if you could just go and build the wind tunnel and have it work, but the first thing you've got is a development program on building the wind tunnel. You can then put engine modules in and get them to run. Roll Royce have a high altitude wind tunnel test facility at Derby but cannot afford to run it [more than] about twice a year, and maybe about six times during the course of an engine development program, because the sheer cost of running it is astronomical."

Criticism, most notably from the United Stated, has sometimes been levelled at the HOTOL engine approach. Alan Bond explains further, "What [some people] fail to recognize—and it's not because we're all brain dead and don't know about all these other engines [e.g., scramjets]—is that the first requirement for developing an air-breathing engine [is that it] must be capable of being taken outside, sat on an open-air test stand like a jet or rocket engine, so that you can run it initially in straightforward atmosphere, and then maybe start to plug on some rig facilities that simulate the high altitude operation. The HOTOL engine is based entirely on those principals, that's where it came from and those were the starting guidelines."

The divergence between the HOTOL and NASP programs is striking. HOTOL uses a far more conservative approach in the design of the airframe, propulsion system, and subsystems to minimize life cycle costs over modest mission models. This approach was adopted because HOTOL was not an experimental program like NASP, but was intended to lead to an operational launch system by way of normal prototype testing, just as with any other aircraft development program. Thus, by restricting the design to more available technologies and, perhaps more critically, to systems that can be fully ground tested in existing facilities, it is possible to minimize development risk uncertainties and, most importantly, cost. NASP pushes technology to the limit because that is one of the main reasons for its existence, but for HOTOL to be economically successful, such luxuries were considered intolerable. According to Bob Parkinson, [14]

The key technology issue appears to be with regard to testing. An engine which can be wholly demonstrated within existing ground test facilities has significant development advantages over an engine which requires flight testing onboard a flight test vehicle in novel flight regimes, or extensive new ground facilities. Subsonic combustion cycles therefore offer cost advantages over supersonic combustion systems. The design of the HOTOL engine allows the engine to be developed on ground test facilities without the need for a flight test demonstration vehicle before the airframe prototype is produced The prototype can then be progressively flight tested and "tuned" for best performance in the manner of a high-performance aircraft.

Another important reason for choosing to breathe air up to Mach 5.5 is because it enables the use of an entirely passive thermal protection system. Much beyond Mach 5.5, aerodynamic heating increases significantly to the point where overcoming it requires either more advanced materials or the incorporation of an active thermal control system into critical parts of the vehicle, such as in the intake, nose, and leading edges. Considering that no one has yet built a fully reusable launch system, the choice of keeping HOTOL's complexity to a minimum appears to be a prudent one.

The feasibility of the original HOTOL launcher hinged on the use of the new RB-545 combined-cycle propulsion system. Although the RB-545 uses an original cycle, its successful development was always also going to be a function of good engineering. Alan Bond explains, "When you did all the calculations, the bit that you finally finished up with as being a real problem was the pre-cooler, and I devoted 3 months to simply looking at how to get a light-weight pre-cooler. The thickness of the raw material in the heat exchangers [used to cool the incoming air] is extremely thin—on the order of 100 microns or less. You don't have a lot of room to maneouvre with materials of that sort of thickness. When it was first recognized what size of tubing the pre-coolers were going to have to rely on, I realised that we were going to have a lot of problems selling people we could actually make tubes that size. So one of my first tasks was to actually manufacture the tube of the final dimension. I actually manufactured a tube in the end by taking a tube of the appropriate bore and chemically milling the material away. I remember well in a meeting at BAe, considerable concern was expressed when we first talked about the heat exchanger tube size. I was then able to do my 'showman' stunt by saying 'I just happen to have one I made earlier,' which [unfortunately] Don Stott promptly broke. It took a long time to get to grips with all the factors involved with that pre-cooler."

The demonstrated ability to manufacture these tubes was a critical factor in the credibility of HOTOL. Another was the problem of frost control. Controlling the buildup of frost had always been argued as a major stumbling block to engines that use pre-coolers [15]. This is because moisture in the incoming air would freeze the moment it came in contact with the heat exchanger tubes through which super cold liquid hydrogen was flowing. With the RB-545 it was particularly serious because the pre-coolers cool the air to a temperature just above the liquid state. In addition, the system had to cope with the hot and humid climate of the Kourou launch site in French Guiana.

A central feature of Alan Bond's engine design was its ability to cope with the buildup of frost on the heat exchanger. Due to the short study period this problem was solved in a relatively "brutal fashion," as Alan Bond notes [16]. Essentially, the incoming air is sprayed with liquid oxygen. This has the effect of freezing the moisture in the air before it reaches the heat exchanger. In this way the small frost crystals pass through the pre-cooler without clogging it up. Since such a technique was unproven, one of the first activities of the HOTOL Proof-of-Concept work was to perform rig-testing of a scale version of the pre-cooler. These tests, coupled with the manufacture of the pre-cooler tubes, provided confidence that the RB-545 design was sound (Figure 7.6).

Although this frost control solution proved feasible, it consumed a total mass of about 3 tonnes. Since the HOTOL Proof-of-Concept study was completed, more elegant frost control techniques have been proposed that would eliminate most of the 3 tonne mass penalty and, thereby, significantly add to the payload performance.

The heat exchanger tubes were one of only a handful of aspects of the engine that HOTOL team members were allowed to discuss because of its classified status. However, this all changed when in April 1991 the RB-545 was suddenly and unexpectedly declassified, although the actual details of the engine remain proprietary with Rolls Royce. The rationale for this is not clearly understood. Alan Bond and others believe that part of the reason was that the then Soviets had published a paper describing an engine cycle similar to the HOTOL approach. It would certainly be ironic if the Soviets eliminated the classified engine problem, as the RB-545 was classified to stop it from falling into their hands in the first place.

If this had happened in 1987 or 1988, the story of HOTOL might be somewhat different. By the late 1980's, however, HOTOL had evolved into a new form, taking a direction that made the need for an air-breathing engine practically irrelevant.

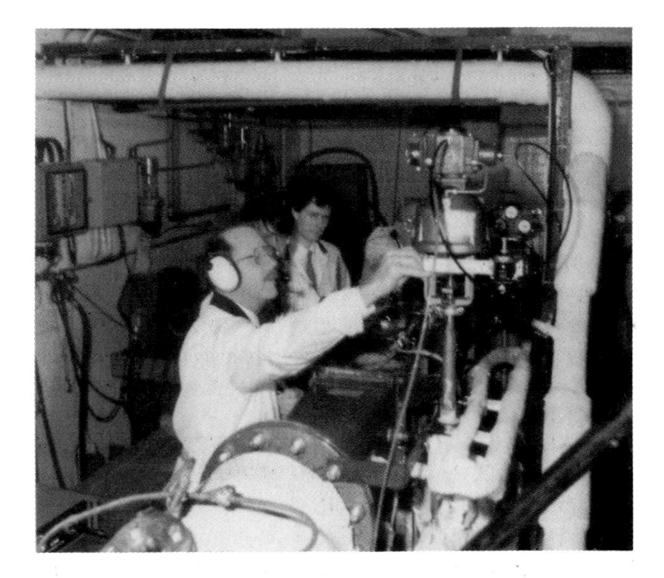

**Figure 7.6** Rig testing of RB-545 components (*BAe*).

## *On the Back of a Big Brother: Interim-HOTOL*

The original version of HOTOL used a combined-cycle airbreathing/rocket engine because it was considered the most practical way of minimizing life cycle costs and because it was believed to enable single-stage-to-orbit. This has led to a common misconception, as Bob Parkinson explains, "When HOTOL came up we said it was possible because it uses airbreathing propulsion, and because of this, we can bring the costs down. So everybody [implied this as meaning] airbreathing makes it cheaper. This is not actually true. If we built an airbreathing vehicle as difficult to operate as the Shuttle, it would actually be more expensive."

By the late 1980's it was clear that HOTOL was going nowhere quickly, even though the technical studies were proceeding well and were maintained into 1989. The U.K. government suggested it might be willing to fund additional HOTOL enabling technology work, but only if BAe could persuade other countries to join in the project first. Unfortunately, this was a Catch-22 situation because few organizations had any interest in joining a program that the U.K. government was unwilling to fund initially. Further, during this time the other European nations seemed decidedly less enthusiastic about the need or feasibility of fully reusable aero-space planes. Germany insisted that a hypersonically separating, two-stage-to-orbit vehicle was the only viable approach. (See Chapter 8.) France was pushing Hermes and expressed little interest in looking at something it hadn't invented. Not only did these nations believe HOTOL's engine had little credibility, they also believed HOTOL-type single-stage-to-orbit vehicles weren't credible—period. If the goal of making space easier and cheaper to access was to stay alive, British Aerospace needed an alternative that eliminated most of these problems. Almost by accident, a solution appeared.

Bob Parkinson takes up the story, "We always had a fight with Ivan Yates [Deputy Chief Executive (Engineering) BAe PLC] who didn't like the trolley and always said HOTOL should be launched from an aircraft. Then, when the AN-225 was unveiled, BAe Warton were asked to do some sums around it. But another thing happened. In April 1989, I was telephoned and told that the Russians were bringing the AN-225 to Paris [Le Bourget] and wanted to join up with people with major projects. So, I went to Paris and looked around the AN-225, and the Russians knew they may have a tiddler on the line. The following spring, Ivan went for a tour of Soviet aircraft factories on the basis that they might want to collaborate and discovered that HOTOL was one of the items on their agenda. He came back and said 'you better go and discuss about doing this.' We discussed whether it was reasonable to do and how we might do it. Eventually in July [1990] we went out there to discuss

**Figure 7.7** AN-225/Interim-HOTOL (*TsAGI/BAe*).

whether we could have such a cooperative program. When we got there they rolled out the red carpet for us."

Just over 1 year later at the Farnborough Air Show in September 1990, BAe announced an agreement with the Soviet Ministry of Aviation Industry for an initial 6 month joint study of a pure rocket version of HOTOL launched off the back of the Soviet AN-225 aircraft [17]. This was the first major aero-space plane international cooperative agreement to be signed. Interim-HOTOL is shown in Figure 7.7.

## The Trolley Problem

The trolley, or launch assist system, Bob Parkinson refers to was always an interesting problem for HOTOL. Because HOTOL weighs above five times as much on takeoff as on landing, a fully integral airliner-like undercarriage would need to be five times stronger for takeoff than landing. As a result, it would be heavy. Therefore, it was argued that a simple and lightweight undercarriage—or even a combination of a nose wheel and skids like the X-15—should be sized for the landing mass only, and the takeoff run should use a heavy piece of machinery that remains on the ground. In this way HOTOL would only need a light undercarriage and, as a result, it would be able to carry additional payload.

There are many advantages and disadvantages to using a launch assist system. If, for example, a takeoff run had to be aborted before vehicle rotation, the launch assist system would be able to arrest the motion in a short distance using brakes and reverse thrust. A vehicle with an integral undercarriage, by contrast, would need to be equipped with brakes, parachutes, or thrust reversers in the engine which, assuming they were practical, would be heavy and add considerable complexity. The disadvantages are that a launch assist system has to be developed and paid for, and it fixes HOTOL to particular launch sites. The latter disadvantage isn't particularly critical, since the first objective of HOTOL is to reduce the cost of space access. Therefore, even though a multilaunch site capability is desirable, it is not mandatory to achieving reduced—cost access to space. Further, a top level requirement forcing a multisite launch capability might preclude construction of a more economic launcher.

Another problem never completely resolved was how to deliver or return the vehicle to the test or launch site. Potentially, HOTOL could have flown itself to its own launch site. However, because of support infrastructure requirements (Chapter 16), such a flyback capability option may have been difficult to achieve. Thus, airlifting HOTOL, like the Shuttle orbiter flown on the back of a 747, might have been the best solution. However, if an aircraft is needed to airlift HOTOL, then could it also be used for launch and to provide a better payload-to-orbit performance? An aircraft able to lift a fully fuelled HOTOL just did not exist—until the AN-225 was unveiled.

The Interim-HOTOL exists only because the AN-225 exists. Building such a large aircraft from scratch was considered more expensive than developing a launch assist system and an air-breathing engine for a ground-launched HOTOL. Although, as Bob Parkinson notes, "The AN-225 is not actually available, because it's not a certified aircraft, and the certification program would have to be paid for. If you don't deal with the AN-225 properly, then some of the abort options won't be available, and other operational problems are going to be there. So you're going to have to spend money on the AN-225 to get the aircraft to the same standard you want the orbiter to be in anyway. From that point of view, while it is a good vehicle, there is still money to be spent on the AN-225, that's clear."

## The Benefits of Air-Launching

The benefits of having the AN-225 are considerable, according to Bob Parkinson, "There's not as much difference between them [HOTOL versus the Interim-HOTOL] as I had first thought there was. Part of the reason [is because] with the Interim-HOTOL I'm adding a new vehicle and integration problems, and that puts up costs. But, at the same time I have a simpler set of engines and a smaller airframe which means there are fewer riggers, effectively. Under those circumstances, I actually reduce the amount of work I need on the vehicle itself. There's a degree of balancing out, and I've removed the trolley problem. So the availability of the AN-225 does a lot of good for me, but again it's not a simple equation."

The first and most important benefit of air-launching for HOTOL is that the air-breathing engine is not required. Unlike the original HOTOL, the Interim-HOTOL does not need to expend propellant to lift itself off the runway and climb through the dense lower atmosphere. If launched at an altitude of 9 kilometers and a speed of Mach 0.8, the ΔV required to attain low Earth orbit is about 13% less than if launched from the ground. Specifically, the gains in ΔV compared to a vehicle like the Delta Clipper are shown in Figure 7.8 [18].

| | | |
|---|---|---|
| FOR VERTICALLY LAUNCHED SSTO (e.g., Delta Clipper) | | |
| Total ΔV required for LEO | 9,361 | meters/sec |
| FOR AIR-LAUNCHED VEHICLE | | |
| • Speed supplied by the AN-225 (Mach 0.8, 9 km) | −235 | meters/sec |
| • Increase drag loss | +67 | |
| • Difference in gravity losses | −670 | |
| • $I_{sp}$ underexpansion losses at low altitude | −180 | |
| • Thrust vectoring demands increase | +10 | |
| • Improved engine $I_{sp}$ due to altitude start | −214 | |
| Total ΔV change | −1,222 | meters/sec |
| Therefore, ΔV for Interim-HOTOL | 8,139 | meters/sec |
| **Total ΔV advantage for air-launched vehicle** (1,222/9,361) | **13%** | |

**Figure 7.8** Comparison of ΔV between a vertically launched SSTO and an air-launched vehicle.

Even though the ΔV requirement is reduced by 13%, this actually translates into a 24% decrease in the energy requirements and, therefore, a similar decrease in propellant mass. This is because energy (i.e., propellant mass) is proportional to velocity squared. As a result, the reduction in energy is sufficiently large enough to allow a relatively small vehicle (250 tonnes) that can rely entirely on conventional and essentially existing rocket engines.

Deleting the RB-545 and cooperating with the C.I.S. has also removed a principal stumbling block to European cooperation—credibility. As Bob Parkinson notes, "It's a bit like being the smallest boy in the playground, but you have this great big brother! It's no longer 'Bob Parkinson' talking about it, it's *the Russians* talking about it. This is no longer a vehicle that is designed by someone who has only designed things using pieces of paper. It's actually been approved by the people who designed Buran."

The only significant technical problem standing in the way of the Interim-HOTOL is materials. According to German Zagainov, "We have the possibility to accelerate this project, and the first flight may be organized by the end of this century. But, before this time we must organize the production of some materials, but not materials for the next century. Great Britain and the Soviet Union now have some new materials, but just samples. We must develop these materials further in construction and organize the production of this material—this is very important. We need some time for exploration, for preparation of our technology, of our production of this material. It needs time, it's not so simple. However, it seems to our specialists that Interim-HOTOL is realistic, with technologies available by the end of the century, but not current technology."

By current technologies, German Zagainov means steels, and the same type of aluminium and titanium alloys used in current airliners and military aircraft. The C.I.S. has an extensive advanced materials capability, but does not have suitable mass production techniques (Chapter 9). Zagainov continues, "If we have money, there is no problem to organize the next generation of new materials. It seems to me that Europe now has no real problem for developing aerospace [plane] systems."

## The Thrust Behind Interim-HOTOL

Materials and advances in structural concepts are critical to the development of all fully reusable aero-space planes because of the need to reduce the structural weight fraction. Propulsion is also an area where advanced development is usually required, especially with air-breathing vehicles. As a result, the combination of developing new materials and structures in parallel with a new engine pushes costs steadily higher. This high cost of developing aero-space planes is their fundamental impediment.

Using the AN-225 has provided British Aerospace with the opportunity to split the parallel development of materials and propulsion and still obtain a useful operational capability. Materials advancements are mandatory, but it would appear that an engine derived from the engines used on the Energia could be adequate. According to Bob Parkinson, [19]

With a launch mass of 250 tonnes, and an initial acceleration of 1.3 g, a four engined vehicle requires an engine thrust at ignition of about 800kN. [The figure] shows an engine concept, based on the Soviet RD0120 hydrogen-oxygen engine, designed for the *Interim Hotol* Orbiter. The engine is designed for reuse 20–25 times, and would incorporate a two-position nozzle to accommodate the large area ratio without flow separation at ignition. Performance gains in this engine design have been gained from the large area ratio nozzle, not by increasing the assumed combustion efficiency over the RD0120 measured result.

Using an engine derived from the existing RD0120—whose performance is known and has been tested for over 165,000 seconds—is clearly cheaper and less risky than developing a high speed air-breathing engine. Using rocket engines also reduces the amount of integration between the engine and airframe (See Chapter 4). With most air-breathing aero-space planes, it is necessary to perform careful airframe/engine integration to ensure that the fuselage forebody is correctly sculpted so as to properly compress the incoming air before it is ingested by the engine intake. Likewise, the aft fuselage is similarly sculpted to assist in expanding the engine thrust. The rocket-powered Interim-HOTOL has none of these problems.

Figure 7.9 compares the original HOTOL with its air-launched counterpart. Whereas HOTOL requires a slippery aerodynamic shape (fineness ratio), the Interim-HOTOL can get away with being much stubbier, fatter, and more structurally efficient. This is because of the differences in the ascent trajectory. The original HOTOL follows a flat trajectory that keeps it in the atmosphere longer than a pure rocket vehicle so as to optimize the use of the RB-545. Consequently, it requires a high aerodynamic fineness ratio in order to minimize drag during the air-breathing phase. For the Interim-HOTOL, the fineness ratio is less critical because as soon as the orbiter separates from the AN-225, it follows a trajectory that takes it out of the atmosphere as quickly as possible—just like a rocket. Indeed, the wing is used purely to assist in the pullup maneuver after separation and to facilitate a horizontal landing. With the original HOTOL, the wing had to be large enough and strong enough to lift the entire vehicle off the runway.

The only really significant concern is the safety issue associated with separating the two vehicles in flight. Studies have indicated that it may be possible to idle the upper two engines of the Interim-HOTOL immediately before separation (Figure 7.10). According to Bob Parkinson, [20]

The separation manoeuvre currently preferred . . . involves the AN-225 first entering a shallow dive about 9.4 km in altitude, increasing incidence briefly at about T-15 sec to release the Orbiter, before continuing the dive to avoid the engine exhaust plume. The four engines of the *Interim Hotol* vehicle would be started in pairs immediately before and after release, with the engine sequence being completed 4–6 seconds later.

Separation testing of a scale model of the current Interim-HOTOL 0068 configuration from the AN-225 was successfully performed in 1992 to verify the predictions. These tests were in addition to the static wind tunnel testing conducted

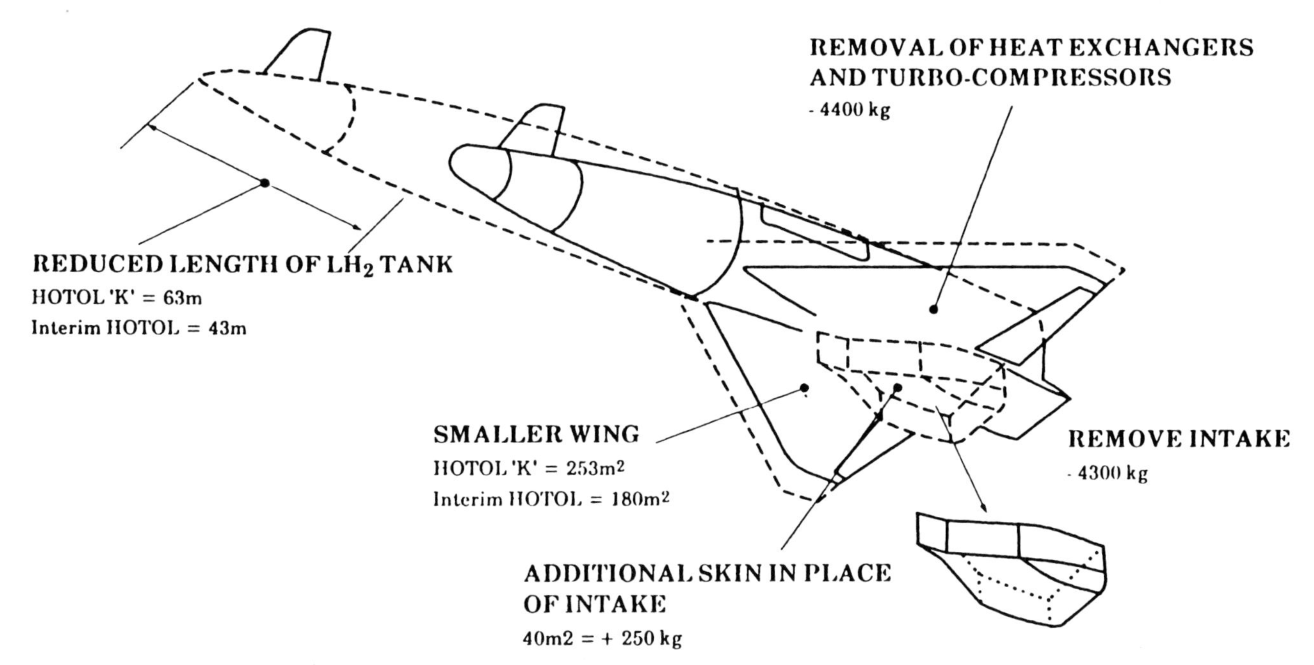


**Figure 7.9** Interim-HOTOL compared with its air-breathing predecessor (*BAe*).

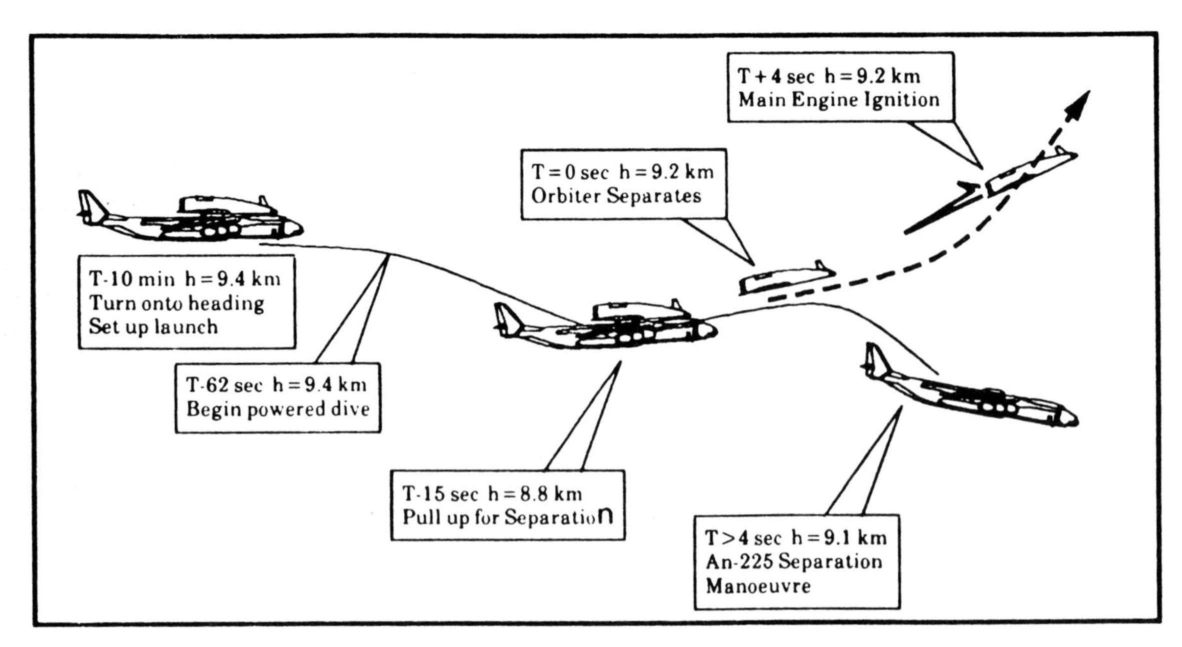


**Figure 7.10** Interim-HOTOL separation strategy (*BAe*).

in the Central Aero-Hydrodynamics Institute (TsAGI) facilities throughout 1990 and 1991 (Figure 7.11). Additional, wind tunnel testing at speeds of Mach 10.4 and Mach 14 was also undertaken in 1992 to determine the aerodynamic and heating interactions with the vertical stabilizer on the nose of Interim-HOTOL [21].

Tests in the C.I.S. have determined that the AN-225 will need an additional two Lotarev D-18 engines, one more on each in-board pylon. Surprisingly perhaps, the reason for these extra engines is not because more thrust is needed to attain the required Mach 0.8 speed and 9 kilometer altitude. They are needed to safely takeoff from a 3,000 meter runway under the worst-case hot and humid conditions in Kourou. Much like the original RB-545, if Interim-HOTOL could takeoff under European conditions, the existing six D-18 engines would be adequate (Figure 7.12).

The D-18 is the current baseline, although the use of the higher thrust and more fuel efficient Rolls Royce Trent engine has been studied. According to German Zagainov, "Six Trent engines are enough, and it seems to us that there is no problem to interface these engines to the Antonov."

## What Will It Cost?

The total acquisition cost of the Interim-HOTOL fleet and AN-225 modifications is about two-thirds that of the original HOTOL concept, around $7.5 billion (1990), primarily because it is a simpler and less risky concept [22]. A third of this cost would be a Western equivalent of a C.I.S. contribution. This also has the advantages of speeding up the initial operating date which further reduces costs. Flight testing could begin within 7 years of the completion of an enabling technologies program. This is about 3 to 5 years quicker than for HOTOL.

One advantage of a lower development cost is that it reduces the number of flights required for full cost recovery. From the operational cost analysis British Aerospace has been performing in conjunction with the C.I.S., it believes that full cost recovery would be feasible at about 15 flights per year—about the same number of payloads Ariane 4 carries annually—over a 20 year timeframe, giving a launch cost of about $40 million. For 24 missions per year, launch costs could be significantly less, at about $28 million. If life cycle cost recovery is not mandated, as is the case with all previous large launch systems, then the dedicated operational costs are estimated to be $12 million (1990 dollars) per flight, including fixed annual costs, spares and inventory, management overhead, servicing, and insurance [23].

## Politics: The Right Place at the Right Time?

Where HOTOL was stymied by its classified engine and a lack of political and international interest, the Interim-HOTOL teams believe their project could be in exactly the right place at the right time. First, there are no defense-related issues to worry about, providing the opportunity for other nations to scrutinize the technical feasibility closely. Second, the U.K. government has stated that it might be interested in participating in the new ESA FESTIP program once it is finally approved. FESTIP is intended to study a range of reusable launcher concepts including Interim-

**Figure 7.11** Wind tunnel testing at TsAGI (*BAe/TsAGI*).

HOTOL type vehicles. Third, the extensive experience of the C.I.S. in propulsion systems, materials technology, and launch processing is considerably in advance of what Europe could offer alone—this experience certainly carries with it considerable technical pedigree. Fourth, a cooperative program with the C.I.S. may well be one of a few possible ways to save its space program from total collapse. Potentially, Interim-HOTOL could save scarce financial resources that otherwise would be needed for expendable launchers, as V. T. Ivanov, then deputy minister of the Soviet Ministry of Aviation Industry, said at Farnborough in 1990, "None of us are indifferent to the cost of space activities" [24].

It would seem that the former Soviets are genuinely interested in a cooperative program like the Interim-HOTOL by the way they devoted large resources to the study work. Initially, they were to focus primarily on Interim-HOTOL/AN-225 subsonic wind tunnel testing, separation analysis, modifications to the AN-225, and cryogenic propulsion system development. According to Bob Parkinson, "The Soviets actually gave [the Interim-HOTOL design] to two independent teams. The Soviets said 'Yes, we've had this weighed by two independent groups and we've got two answers' and they were both positive. I think the Buran team gave the lowest [payload mass-to-orbit] estimate, and the fact that they were still positive makes me feel very confident because the Buran team built the last vehicle and, I reckon, they are damn well not going to underestimate anything this time." According to German Zagainov, "We used our specialists from TsAGI, and we had some specialists from Gleb Lozino-Lozinsky of the Buran team and some specialists from Central Institute of Aviation, and the Antonov Design Bureau. But the main experts were from TsAGI. It's our duty to estimate new projects in the Soviet Union. It's our duty to make special conclusions about any project in the Soviet Union: aeronautics, aerospace, military, helicopters, missiles, etc."

An example of the former Soviets' enthusiasm for the project was apparent during one meeting when the BAe team had a handful of people on its side of the table, while on the Soviet side there were 40. "It's embarrassing how many people they put on the other side of the table!" says Bob Parkinson.

## Summary—What Next for HOTOL?

The HOTOL program began for the single purpose of developing a launch system capable of significantly reducing the cost of accessing space. All technical decisions on the program were made with the express desire to build a cost-effective launch system. HOTOL was never intended as a technology or experimental program for its own sake, but was to lead directly to an operational launch system. While the technical arguments for the original air-breathing HOTOL concept were founded on solid rationale, the politics could not keep pace. The combined effects of the classified status of HOTOL's engine, the U.K. government's reluctance to increase space spending, and perhaps most importantly of all, the lack of a champion at the highest political level, led ultimately to its downfall.

The vision of a HOTOL-like capability remained. And, almost by accident, a second chance emerged courtesy of the U.S.S.R.—the country originally responsible for the secret classification of the HOTOL powerplant. This led to the Interim-HOTOL program, made possible purely due to the existence of the AN-225, the largest aircraft in the world. The principal benefit of air-launching was that the Interim-HOTOL would not require the development of an advanced and unproven air-breathing engine. Instead, it could rely on pure rocket engines derived from the existing Energia RD0120 engines. Thus, unlike most other aerospace plane programs that must simultaneously solve propulsion and materials problems, Interim-HOTOL would, in theory at least, only need to concentrate on materials. In the future, it may then be possible to develop an air-breathing engine and return to the original version of HOTOL.

The HOTOL program has revealed many of the problems that must be overcome in the development of a true space transportation system. Commenting on the original HOTOL,

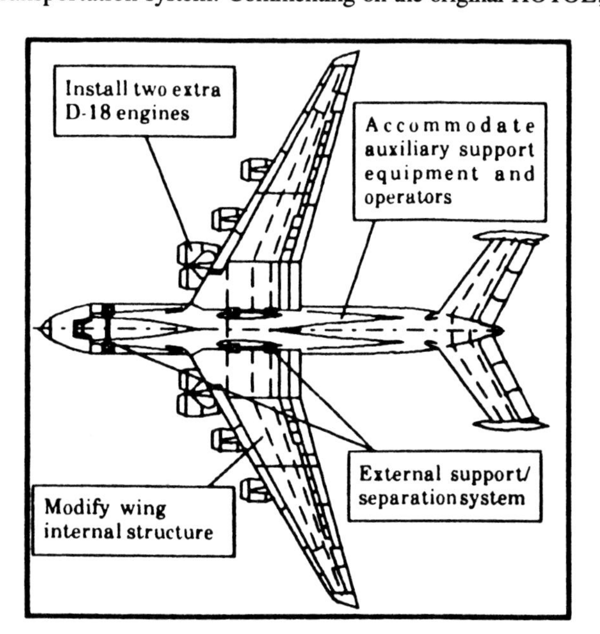


**Figure 7.12** Modifications to the AN-225 (*BAe*).

Alan Bond believes, "HOTOL to my mind is still the most convincing SSTO space plane in the world which stands up to engineering scrutiny. You can see the layout, you can see figures on temperatures, pressures, flows, and you cannot find that for any other space plane program in the world, other than possibly Sänger which isn't a comparable program. I think in those six years from 1982 to 1988 for about £10 million ($15 million) we achieved quite a lot."

HOTOL has undergone a traumatic evolution since its birth in 1983. However, despite the setbacks, the proponents have maintained their commitment to the basic rationale for the project—significantly reduced-cost access to space. As Bob Parkinson explains, "The most important thing is to get something flying along these lines—not to get the best one."

## References and Footnotes

1. This story was compiled by Frank Miles, a self-proclaimed supporter of British space efforts. Alan Bond and others have noted that HOTOL would probably not have progressed as quickly as it did had it not been for Frank Miles's efforts.
2. Parkinson, R. C., "A Total System Approach Towards The Design of Future Cost-Effective Launch Systems," IAA *Symposium on Space Systems Cost Methodologies and Applications*, San Diego, California, 10-11 May 1990, p. 1.
3. Parkinson, R. C., *Citizens of the Sky*, (2100 Ltd, Stotfold, England, 1987), p. 31.
4. Alan Bond's original engine design, designated "Swallow" in honor of Sir Frank Whittle, the inventor of the jet engine, was given the British patent number 8334308.
5. According to a front page story in the London *Evening Standard*, December 17, 1987, under the title "Spacecraft Plea by Charles," Prince Charles reportedly said, "We are in danger of losing another typical British and brilliantly simple aerospace concept to foreign interests unless we are extremely careful . . . Why is it that others see the commercial value in our ideas, but so often we do not?"
6. "Resolution on the Long-Term European Space Plan," ESA *Bulletin*, No. 41, Feb. 1985, p. 9.
7. "What Future for HOTOL?," *Space Markets*, Autumn 1988, p. 128.
8. "Hotol Funding Shock," *Spaceflight*, Vol. 30, September 1988, p. 345. Courtesy British Interplanetary Society.
9. "A Total System Approach," Parkinson, pp. 1–9.
10. These were the figures quoted in a BAe/Rolls Royce brochure originally perpared for Prince Charles. In addition, BAe had undertaken a range of other market studies. One of the earliest was requested by the BNSC and jointly conducted with General Technology Systems in June 1986.
11. Parkinson, R., and R. Longstaff, "The United Kingdom Perspective on the Applications of Aerospace Planes in the 21st Century," AIAA-91–5084, AIAA *Third International Aerospace Planes Conference*, Orlando, Florida, December 3–5, 1991 p.1.
12. Parkinson, R. C., and C. M. Hempsell, "Cost Impacts of Supporting a Future European Space Station,"IAF-90-602, 41st IAF *Congress*, Dresden, Germany, October 6–12, 1990.
13. Parkinson, R. C., "HOTOL System Requirements," British Aerospace (Space Systems) Ltd., March 1987, p.3.
14. Parkinson, R. C. and P. Conchie, "HOTOL," AIAA-90–5201 AIAA *Third International Aerospace Planes Conference*, Orlando, Florida,

October 29–31, 1990, pp. 6–7. Copyright American Institute of Aeronautics and Astronautics © 1990. Used with permission.
15. See, for example, Heppenheimer, T. A., *Hypersonic Technologies and the National Aero-Space Plane*, (Pasha Market Intelligence, Arlington, Virginia, 1990).
16. Alan Bond's new engine design, called SABRE, uses a more subtle and mass efficient means of controlling the frost problem.
17. Shifrin, C., "British Aerospace, Soviets to Study Launching Hotol From An-225," *Aviation Week & Space Technology*, September 10, 1990, p. 23.
18. From viewgraphs produced by British Aerospace (Space Systems) Ltd., and the Central Institute of Aero- & Hydrodynamics (TsAGI) for the An-225/Interim-HOTOL presentation to ESA, Paris, June 21, 1991.
19. Parkinson, R., "The An-225/Interim Hotol Launch Vehicle," AIAA-91-5006, *AIAA Third International Aerospace Planes Conference*, Orlando, Florida, December 3–5, 1991, p. 2. Copyright American Institute of Aeronautics and Astronautics © 1991. Used with permission.
20. Ibid., p. 4.
21. Kandebo, S., "Soviets to Conduct Wind Tunnel Separation Tests on Hotol Launch System," *Aviation Week & Space Technology*, January 6, 1992, p. 58
22. Data obtained from the Interim-HOTOL presentation to ESA, June 21, 1991.
23. Ibid.
24. "British Aerospace, Soviets to Study Launching Hotol From An-225," Shifrin, p. 23.

# Chapter 8

# Sänger: A Lead Concept

## *Two Steps to Space*

The aero-space plane world is benefiting from another major program—one that approaches the problem of access to space from a fundamentally different direction than either NASP or HOTOL. Germany does not believe the technology for single-stage-to-orbit launchers, or even for vehicles that are air-launched at subsonic speeds, will be mature for at least another decade or two. Therefore, the Sänger concept was conceived as a two-stage-to-orbit launch system. The first stage uses relatively conventional turbo-ramjet technology to accelerate to a little under Mach 7. Then a small manned or unmanned second stage separates and is propelled to orbit using similarly conventional rocket technology. (This second stage was formerly called Horus.) No new technology breakthroughs are required, unlike NASP, although very significant advances are still necessary (Figure 8.1).

Although Sänger has become a well-publicized concept throughout the aerospace industry, it is important to emphasize that it is, at present, only a lead concept in the German *Förderkonzept Hyperschalltechnologie* or Hypersonic Technology Program. In this respect, it is difficult to compare it directly with NASP, Interim-HOTOL, or the Delta Clipper. Whereas NASP aims purely at developing an experimental vehicle that can breathe air to very high speeds before attaining orbit, and Interim-HOTOL and the Delta Clipper are particular concepts for an operational launch system optimized to reduce launch costs, Sänger appears to fall somewhere in the middle. The Hypersonic Technology Program aims to develop the enabling technologies for future aero-space planes, but uses the Sänger concept to precisely focus their activities. In this way it is possible to ensure those technologies developed can be integrated into a working and, in the opinion of the Germans, realistic system. Politically, it also helps to have an identifiable focus for the effort.

Dr. Dietrich E. Koelle, of DASA in Munich, is credited with originating the current Sänger concept, although the idea of a horizontal takeoff, two-stage-to-orbit aero-space plane is not new. The name was chosen in recognition of the pioneering work of Prof. Eugen Sänger during the 1940s. (See Chapter 5.) Prof. Sänger was actually pursuing single-stage-to-orbit concepts, and only reluctantly agreed to the two-stage alternative because of technology limitations. In this sense, the Sänger of today is not entirely a resurrection of Prof. Sänger's dreams.

**Figure 8.1** Sänger 1991 (*DASA*).

## *History: Challenge and Response*

Germany, like many other spacefaring nations, has been acutely aware of the high cost of placing spacecraft in orbit and of the limitations these high costs place on the utilization of space. In the opinion of Sänger Program Manager Dr. Heribert Kuczera, "We need a next generation space transportation system. We need a system that will lead to significantly reduced transportation costs, as well as is inherently safe, eliminates orbital debris, and is environmentally benign. I think that is clear, otherwise we won't be able to do anything [in space]."

Never has this been more clearly demonstrated than by the Spacelab program. Germany was the principal sponsor of ESA's contribution to the Space Shuttle program and, at its inception in 1973, it had been hoped that Spacelab modules would fly as many as five or six times per year [1]. Because of the limited Shuttle flight rate and high launch costs, the pressurized modules have been used only six times by early 1992, only one flight being sponsored by Germany. The second German Spacelab D-2 mission occurred in 1993, although the launch was originally planned for 1988. Because of the high cost of this mission and the infrequent launch opportunities, Spacelab D-2 is likely to be the last German-sponsored flight of Spacelab.

Reducing launch costs was not the only motivation for the existence of Sänger and the Hypersonic Technology Program. HOTOL and, to a lesser extent, U.S. efforts provided the initial impetus, although something like Sänger would probably have emerged regardless. Heribert Kuczera takes up the story, "The British came up with HOTOL and submitted this proposal to ESA [in late 1984]. We knew a little about this concept—maybe not everything—but at this time our Ministry said 'what is your reply, what is your suggestion?' We said we would think about it. So we [Dr. D. E. Koelle and I] did some work over a weekend and on [the following] Monday we sat together, discussed our ideas, and then we made a few pictures and we gave it to our Ministry and they said, 'O.K., we have your ideas—fine.' We did not make a proposal, but just a brochure with a few pictures and just a few comments on it. We were saying [basically] it shouldn't be a single stage, it should be a two-stage-to-orbit. We were surprised that they took this brochure, a few pages, and submitted it to ESA and said 'that's the German proposal.' We didn't know this. I have only heard that Dr. J-B Mennicken [Director of Department 5 of the BMFT] was happy to have a very beautiful color brochure and I think it was, maybe, his first visit to ESA HQ [when] he turned up and said 'here is our proposal.' " The original version of Sänger used for the front cover of this color brochure is shown in Figure 8.2.

**Figure 8.2** The original Sänger configuration (*DASA*).

DASA was, apparently, delighted that the Federal Ministry for Research and Technology (BMFT) could take such a positive action without further consultation and analysis. What was its reason for making such a rapid decision to back Sänger? Prior to the ESA Ministerial Council meeting in Rome (1985), France had been the dominant member of ESA programs and, in particular, space transportation systems. "When an Ariane successfully launches a satellite, the newspapers in France declare it a great *French* triumph, but when an Ariane dumps a couple of satellites in the Atlantic, the same newspapers declare it a *European* disaster," or so the infamous joke goes. The French have dominated space transportation, and Germany has played an important support role in the Ariane consortium. However, this is not a situation Germany would like to see continue with future launch systems. Dr. Horst Hertrich explains the BMFT's perspective, "Within the Ministry, the main rationale for the Hypersonic Technology Program was to give Germany an opportunity to play a major role in future space transport systems, and not leave this area to the French. The European Ariane launchers have been dominated by France, and, apparently, they also want to continue with Hermes. However, the strategy of future space transportation systems is important to Germany, and that is also where the most challenging technology is." The strategic importance of space transportation to Germany is explained further in the official booklet of the Hypersonic Technology Program, [2]

> The Challenger disaster at the beginning of 1986 and the resultant delays affecting many space programmes clearly indicated that space transport systems are being assigned a key strategic role. Furthermore, the prolonged development periods of space transport systems such as SPACE SHUTTLE, ARIANE 5 and HERMES have proved that such systems are technologically very demanding and complex.
>
> As defined, the SÄNGER concept developed in the Federal Republic of Germany is an especially challenging task from a technological viewpoint. This offers the Federal Republic of Germany the opportunity of making an outstanding contribution to the development of future, less expensive and more reliable transport

systems, as well as assuming a responsibility on a par with the country's general economic and technological capabilities.

When the U.K. proposed HOTOL effectively as an alternative to Hermes (only in terms of funding, not function), this seemed to take Germany by surprise. From its perspective, if anybody were to challenge the French dominance, by all rights it should have been the Germans first and not the British, who up to this point had a mere 3% stake in the Ariane 1–4 series, compared with Germany's 18.65% and France's 56.6% [3]. The Germans had been complacent, and were forced to respond with haste.

It is generally recognized in Germany that mistakes in space policy have been made in the past, as Heribert Kuczera notes, "We now see and we now understand what we did wrong in the past, so there is a good chance to change it in a better way. What we obviously did wrong was we contributed to the ESA space program in a way that we only gave them our money, and they made the decisions, and that was wrong."

Originally, Germany wanted to provide about the same funds to Hermes as France intended to provide and also wanted to integrate one of the two Hermes vehicles in its own national facilities. Essentially, Germany wanted to benefit from Hermes technology. France, however, refused to allow the participation in the technology aspects of specific interest to Germany, particularly reentry systems. This provided further incentive for Germany to pursue alternative launch systems, leading to the Sänger decision [4].

In the end, the only probable reason why Germany agreed to fund Hermes was because it was the only way to get France to support Columbus, even though France's involvement in Columbus was relatively small compared with Germany's contribution to Hermes. (See Chapter 1.) Specifically, for Hermes Germany planned to contribute 27% (about $2.4 billion) with France providing 43.5%. For Columbus, Germany will provide 38% and France 13% ($646 million). Effectively, the price to the Germans of winning French support for Columbus was four times the cost to the French of winning German support for Hermes [5]. However, by late 1992 Hermes was effectively cancelled as was Columbus Free-Flyer.

Speaking in 1991, Heribert Kuczera clarifies, "I think the problems we have are related to Hermes. Hermes is mainly a political program, and the French will do something like Hermes, and they will do it successfully, and it's not so much a question of whether Hermes will really become an operational system or not. It is a demonstration of technology, even if there is only one astronaut and even if there is no payload. Our government representatives didn't like the way the Hermes program developed and were unable to get the responsibilities they requested."

The challenge of the British and the way Germany accepted a weakened role in Hermes effectively without condition in order to keep its Columbus aspirations alive, were evidently important factors in Germany's decision to begin Sänger studies. Another of the rationales for funding Sänger comes from Germany's limited experience in developing supersonic aircraft, as noted in the Hypersonic Technology Program booklet, [6]

. . . the situation in the Federal Republic of Germany can be described as follows: to date, the involvement of the German aircraft industry in the development of supersonic aircraft has been limited to the TORNADO [an Anglo/Italian/German] programme and a few guided missile projects.

Politically, Sänger is also seen as something that would help Germany establish its position as a competent partner in the important domain of high-speed flight. From this perspective, Sänger was viewed as a sound investment for technology reasons alone, bringing German industry and academia capabilities more in line with the United States, France, and the U.K. Importantly, the rewards of the investment would not depend on the eventual development of Sänger many years downstream. Sänger is only one contribution to this technology effort; Germany also participates in the U.S. X-31 program to test highly maneuverable aircraft.

In 1986 and 1987 a study funded by the BMFT entitled "Determining Key Technologies as Starting Points for German Industry in the Development of Future Hypersonic Transport Aircraft with a View to Possible Hypersonic Projects" was conducted jointly with industry and the German Aerospace Research Establishment (DLR) [7]. According to Horst Hertrich, "We had many long discussions with our experts before we agreed to Sänger as the lead concept for the Hypersonic Technology Program. I still think it is a realistic and worthwhile concept."

## The Hypersonic Technology Program

With this background—reducing cost to orbit, the HOTOL challenge, responding to French dominance and the need to push aerospace technology—the Phase 1 Hypersonic Technology Program was approved by Parliament on June 30, 1987. Originally intended to last from 1988 to 1992, funding was set at about DM 360 million ($210 million), of which DM 220 million ($130 million) comes from the BMFT, DM 30–40 million ($18–22 million) from private industry, DM 80–90 million ($47–53 million) from the DLR general budget, and DM 20–30 million ($12–18 million) from university establishments [8]. Approximately 200 people work on Sänger. Political support for the program is strong and bipartisan, says Horst Hertrich, "Sänger is equally supported by all political parties in Germany."

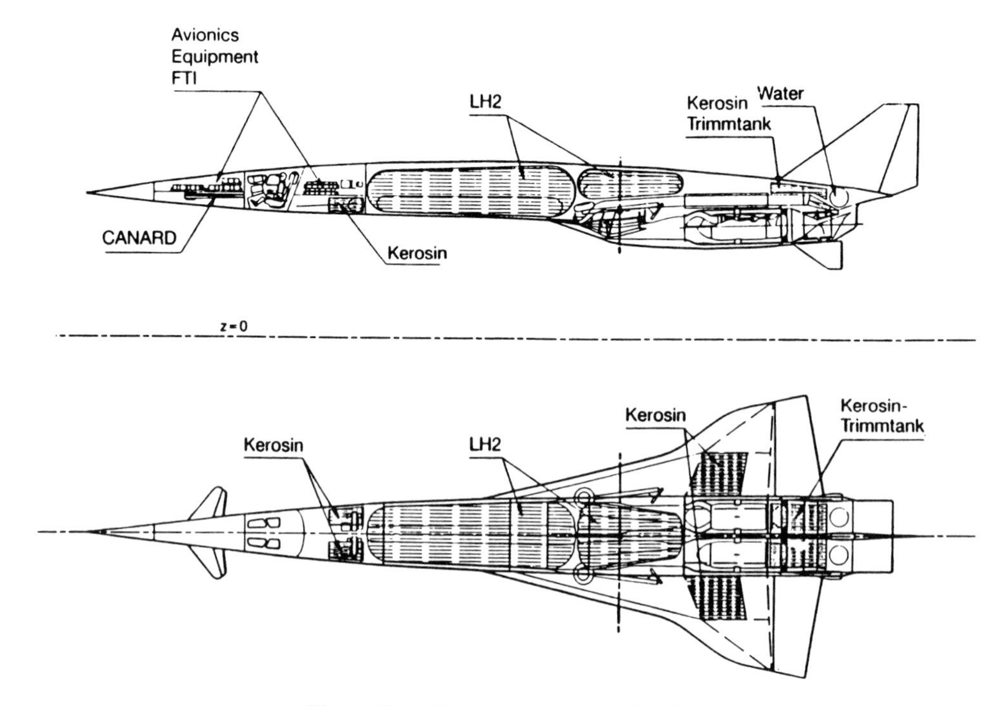


**Figure 8.3** Hytex demonstrator (*DASA*).

Like the NASP program, the Hypersonic Technology Program is split into three distinct phases, [9]

- *Phase 1* aims at performing extensive system studies of the Sänger concept, with emphasis on the air-breathing first stage. Basic enabling technology development is to be performed in areas of materials, structures, and aerodynamics. Test facilities are to be constructed.
- *Phase 2* aims at continuing and maturing the technologies developed in Phase 1, but under an international framework. In addition, an experimental demonstrator vehicle, designated Hytex, is planned for construction and test flight.
- *Phase 3* will depend on the success of Phase 2, but ideally it aims to construct a prototype version of the Sänger first stage and possibly second stage.

In 1992 a decision was pending on whether to extend the Phase 1 effort 2 or 3 more years rather like the NASP program. Horst Hertrich explains, "Funding is available for the rest of this year [1992]. The decision on whether to extend the program has not been made yet. I am trying to obtain funding for the extension phase. If I don't succeed, the program will stop at the end of 1992." The reasons for extending the program are considered to be due to the high cost of German reunification and the sluggish development of interest in hypersonic launch systems in Europe. However, this latter reason appears to be changing, as Horst Hertrich notes, "There is a good chance we will receive funding up to 1995–96 because of the [startup of the] French PREPHA program and the ESA [FESTIP] program." By late 1992 it was finally agreed to extend Phase 1, but at a slightly reduced budget.

Both the French PREPHA and the ESA FESTIP programs include exploring the possibility of building small hypersonic experimental vehicles as an intermediate step before construction of an orbital system. These prospects seem in line with Germany's Hytex (Figure 8.3). In January 1990, the BMFT initiated Hytex concept studies aimed at building a manned or unmanned hypersonic technology test-bed able to fly for brief periods up to a speed of about Mach 5.6 at an altitude of 28 kilometers, verifying the applicability and performance of the critical propulsion and materials [10]. Hytex is an activity within the overall program effort and an important first step in the eventual development of a Sänger-like vehicle. The original Hytex proposal is in many respects equivalent to a "poor man's NASP program," although at an estimated cost of $3 billion it is certainly not cheap. Germany is currently reviewing concepts for a smaller and less expensive, unmanned version of Hytex. According to an article in *Aviation Week* [11], the Sänger team is holding preliminary discussions with NASA about using the SR-71 to launch a small, unmanned drone that could achieve many of the basic objectives planned for the original Hytex concept.

The Hypersonic Technology Program strategy is outlined by Heribert Kuczera, "We are working on a technology program with a lead concept, so it is not a project or anything like this . . . We have a Sänger concept study, and we have a concept study of a hypersonic demonstration vehicle, and the results of these two studies always will be the reasons for modification to our technology programs, so if we change something in the concept we need to look at what we should change in the technology program." In essence, while the technology program aims at developing technical capability in a variety of areas, this work is restricted by the needs of the Sänger and Hytex concepts, ensuring that all technologies can function alongside other technologies within a single vehicle concept.

As an example of this process, in early 1991 it was thought that the Hytex propulsion system needed to be fuelled by a mixture of liquid hydrogen and kerosene. However, studies have revealed that while kerosene is needed for the turbojet, pure liquid hydrogen can be used for the ramjet phase at higher speeds. Thus, according to Heribert Kuczera, "So this part of the technology program was just deleted [and] we think that we are able to run during the ramjet mode with only liquid hydrogen [but we] are still considering the basic concept of using kerosene for the turbojet engine. This is our strategy, to have a technology program with a reference concept and so we will always modify our technology program and activities as soon as we learn something out of the concept studies."

## The Managment of Sänger

The Hypersonic Technology Program organization is shown in Figure 8.4. The BMFT in Bonn is the responsible government ministry for the program. The Steering Committee reports to them and is responsible for coordination and control of the Hypersonic Technology Program. The Steering Committee is composed of representatives from BMFT, the aerospace industry, and DLR. Its role is to assess proposed technological activities, establish study priorities, and define the funding limitations. Reporting to the Steering Committee is the Technical Program Committee (TPA) which is responsible for coordinating and monitoring all study and technology activities. The TPA is made up from industrial program management, representatives of all lead companies, and a coordinator from the DLR. The BMFT, Steering Committee, and TPA are assisted by a project monitoring organization under the auspices of the Industrieanlagen-Betriebsgesellschaft (IABG). The IABG acts essentially as an external monitor to the BMFT [12].

The BMFT is currently responsible for handling the Sänger program for the duration of Phase 1, after which responsibility will probably switch to the German Space Agency, DARA. Heribert Kuczera explains, "DARA has a lot of problems in their own organization. They have a lot of problems with the European space programs [Hermes and Columbus], and so they should not add additional tasks. Right now it is really OK that the BMFT is doing this, and

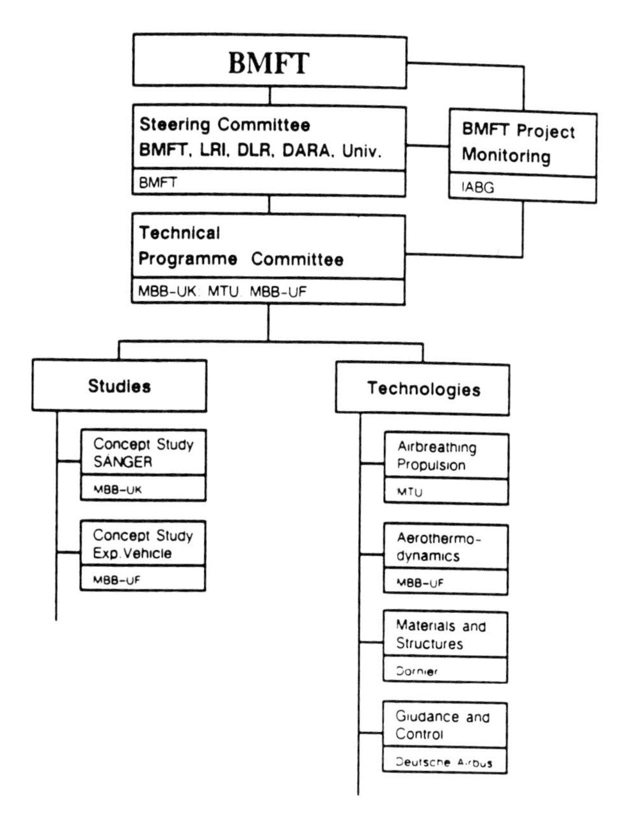


**Figure 8.4** Management of Sänger (*DASA*).

I think there is good reason to then switch over from BMFT to DARA because in Phase 2 it will be a European program and, therefore, a task for the German Space Agency."

At the industrial level, originally the Sänger work was being conducted by the Integrated Project Coordination group comprised of project directors from the leading companies involved in the program and the DLR, and organized by the former MBB-UK in Ottobrunn. With the formation of Deutsche Aerospace (DASA), a conglomerate of the major German industries in 1990, the situation has simplified. Heribert Kuczera became the Hypersonic Technology Program director, supported by two deputies, one from the DASA Military Aircraft division and the other from MTU. Even though the program functions as a single entity under the control of DASA Space Communications and Propulsion Systems Division, the individual lead company responsibilities are:

- DASA, Military Aircraft Division in Ottobrunn: Aerothermodynamics, propulsion integration, and hypersonic experimental vehicle studies
- Deutsche Airbus, in Hamburg: Guidance, navigation, and subsystems
- Dornier GmbH and Dornier Luftfahrt GmbH in Friedrichshafen: Materials and structures
- Motoren- und Turbinen-Union München GmbH (MTU) in Munich: Air-breathing propulsion

DLR is participating in the Sänger program through its own independent contribution funded by the BMFT, and it also receives some relatively small contracts. Its primary role is in performing hypersonic wind tunnel testing including separation tests of the second stage from the first stage, a critical area of concern, as will be discussed later. A number of technical universities and institutes are participating in the research effort, including those in Munich, Stuttgart, Aachen, Braunschweig, and Berlin. The Max-Planck Society and the Fraunhofer Society are also involved [13].

## The Sänger Concept

The Sänger system is a concept for a space transportation system—not a high-speed passenger airliner. As Horst Hertrich explains, "The [Hypersonic Technology] Program has always been the development of a technology base for future space transportation systems and not hypersonic transports. This has always been the position of the Ministry [BMFT] from the beginning, and it has never been any different." The size and performance of the first stage has led some to suggest that it could form the basis of a hypersonic passenger airliner in the distant future. Indeed, the first stage was sometimes referred to as the European Hypersonic Transport Vehicle (EHTV). In an MBB paper presented at the First International Conference on Hypersonic Flight in September 1988, it was stated that [14],

> The German Sänger Concept of an advanced space transportation system comprises the synergetic idea of designing the first stage vehicle with maximum commonality to a later hypersonic passenger plane.

The earlier EHTV concept is a long-range possibility driven by the economics of supersonic and hypersonic flight, as noted in the Hypersonic Technology Program book, [15]

> Future air transport will see an increasing demand for transport capacity at high flight speeds (above Mach 2). However, as flying becomes faster and the productivity of individual aircraft grows, demand in terms of numbers of aircraft will fall. Civil passenger transport at speeds of Mach 4 or 5, although conceivable from both technical and operational viewpoints, would probably result in such a limited demand that no economically viable operation is possible due to the high development and production costs. Current estimates therefore do not anticipate civil transport aircraft flying at speeds well in excess of Mach 3 in the foreseeable future.

The similarity between the first stage and a conventional airliner has led some people to think that airliners derived from Sänger, or indeed Sänger itself, could be flown from conventional airports, a point which Heribert Kuczera dismisses, "some important people, I do not know whether they really knew at this time what they did, tried to convince the public that [Sänger] could take off from every airport." Like NASP and HOTOL, Sänger has fallen into the trap of suggesting the use of such vehicles for airliner operations in order to heighten public interest and attract political attention, only to later deny it was ever a serious intention.

The Sänger reference configuration is first and foremost a space transportation system. To best understand the concept, it is worth outlining a typical mission scenario, as shown in Figure 8.5 [16]. Sänger takes off on its own integral undercarriage from a runway located in a remote region of central Europe and accelerates using its turbojet propulsion system. At an altitude of about 10 kilometers, the afterburners are ignited to push the vehicle through the sound barrier. The vehicle continues to accelerate and at about Mach 3.5 the ramjet propulsion system takes over from the turbojets to allow further acceleration. The current baseline is a turbo-ramjet system using a conventional two-dimensional nozzle, although studies of a plug nozzle are being pursued [17]. The propulsion system of Sänger is highly integrated with the vehicle's airframe, as with the X-30, and tests have already been conducted of a 1-tonne thrust hydrogen ramjet up to a simulated speed of Mach 7 (Figure 8.6).

Sänger then cruises at an altitude of 26 kilometers at an average speed of about Mach 4.4. The distance travelled during the cruise phase will depend on the desired orbit parameters. For example, for a space station located in a 28.5° orbit, Sänger would have to cruise more than 2,000 kilometers down to a latitude of 28.5°, while the launch of a communications satellite may necessitate a cruise all the way down to a 15° latitude or lower.

On approach to the desired latitude, Sänger turns due east, and the ramjets are throttled up to accelerate the vehicle to Mach 6.8 at an altitude of 31 kilometers. Maximum heating of the first stage is experienced, with temperatures as high as 2,000° C on the exhaust expansion ramp aft of the engine and other parts of the engine, up to 1,300° C on the nose, leading edges and intake ramp, and between 500–900° C everywhere else. Thermal protection in the highest temperature regime would be provided by carbon-carbon or other composites such as carbon silicon-carbide. For temperatures up to 1,300° C, Sänger would use fiber-reinforced ceramics covering the titanium intermetallic skin. For the wings and twin stabilizers, where the maximum surface temperature ranges from 700–900° C, a thermal protection system may not be needed. In such a case, the wings and stabilizers would be hot structures made from titanium alloys. Over the main fuselage body, even though the temperature never reaches much more 500° C, a cold structure design is necessary in order to protect the liquid hydrogen tank, propellant, and vehicle equipments. This cold structure is made from conventional titanium alloys and covered with a separate metallic thermal protection system [18].

At this point, a pullup maneuver of 6° is executed, slowing the vehicle to Mach 6.6 but raising the altitude to 35

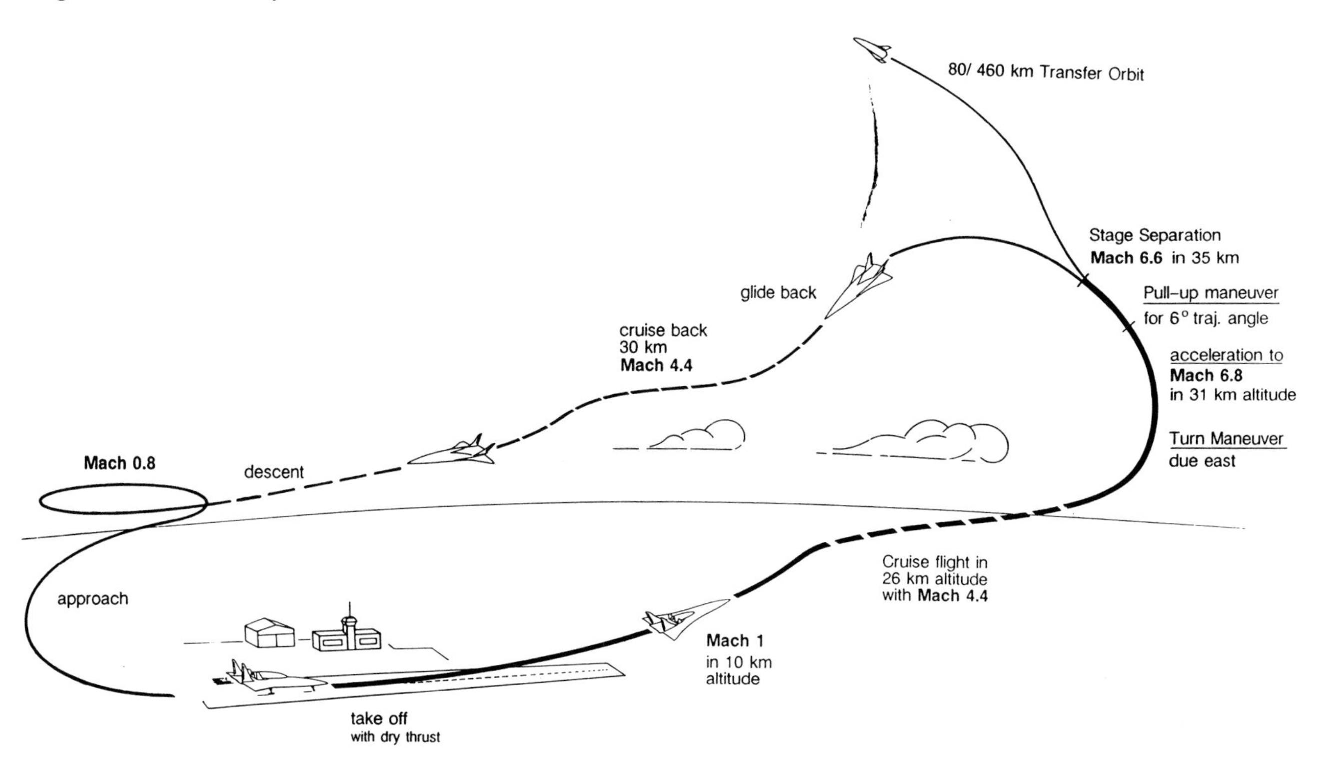


**Figure 8.5** Sänger mission profile (*DASA*).

km. The most critical maneuver of every mission is performed at this point with the separation of the second stage. The second stage is cranked up on its three point mounting to an angle of about 2° relative to the first stage, at which point the two small outboard orbital maneuvering engines (OMEs) are ignited—the thrust of which is just enough to counteract the increased drag of the elevated second stage. The single main engine is also ignited but initially kept at an idle setting, somewhat like the current Interim-HOTOL. Once the second stage is in position, it is immediately released, and the positive pressure between the two vehicles pushes them apart. All of this occurs in just a few seconds (Figure 8.7). Moments later, the second stage main rocket engine is throttled up and it ascends directly into an initial transfer orbit, with the OMEs used to later circularize the orbit. Meanwhile, the manned first stage descends down to its original cruising altitude and flies back to the launch site for servicing.

The second stage actually comes in two similar configurations, one for unmanned cargo missions and the other for manned missions (Figure 8.8). The basic difference between these vehicles is that the manned second stage has a crew compartment in place of a cargo bay. These vehicles can fulfil a range of missions, including the automatic deployment and retrieval of free-flying spacecraft like Eureca or Astro-SPAS, or the deployment of communications satellites. The unmanned second stage can also be used to supply 7 tonnes of logistics to manned space stations, while the manned second stage is specifically configured to deliver three person crews to a space station, as well as supply 3 tonnes of pressurized (internal) logistics payload.

On completion of the mission, which lasts no more than 50 hours, the second stage fires its two OME engines to change orbit for reentry. To return the second stage back to a central European launch site, the vehicle reenters in a

**Figure 8.6** Testing of a subscale hydrogen-fuelled ramjet engine (*DASA*).

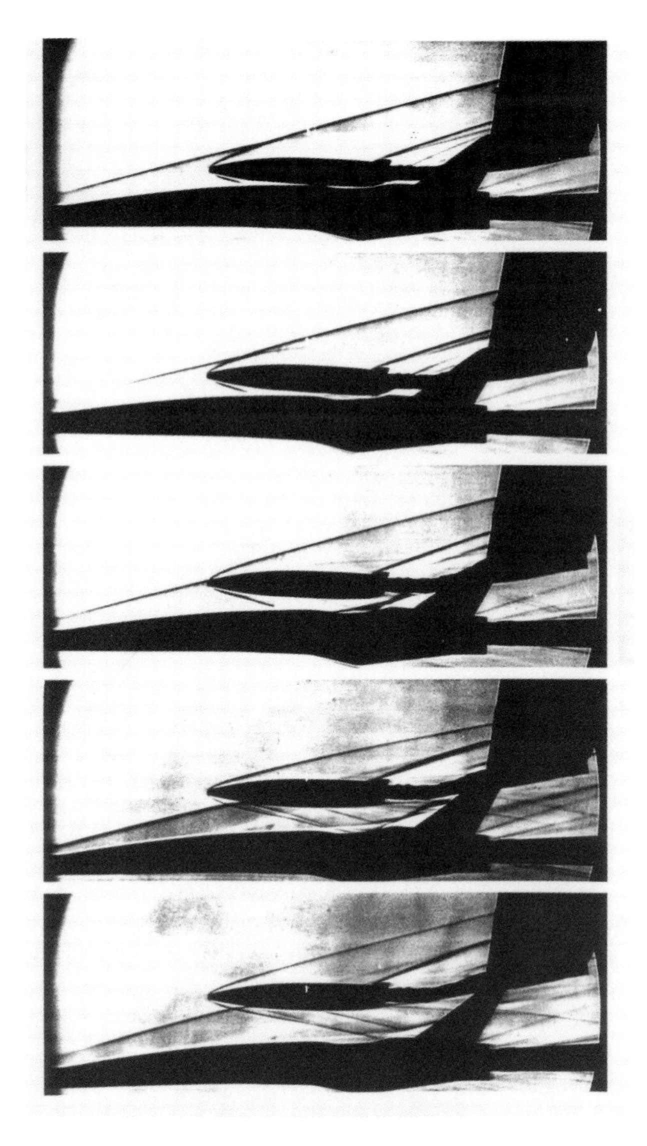

**Figure 8.7** Wind tunnel separation testing (*DASA*).

manner allowing it to fly 2,700 km north of its orbital ground track. To achieve this high cross-range, the Sänger team is considering a skip-glide trajectory strategy similar to that proposed by Eugen Sänger for his Silbevogel in the early 1940's. (See Chapter 5.) If this reentry approach is used, the second stage will repeatedly skip off the atmosphere in order to gradually slow the vehicle down before a final descent through the atmosphere. This skip-glide trajectory also helps limit the maximum heating experienced by the second stage, compared with the more usual direct reentry strategy, because during each re-ascent into space, the vehicle can radiate heat absorbed during the previous skip. The second stage then performs an unpowered landing to complete the mission.

## Is Cruise Capability Really Necessary?

Sänger is the German proposal to reduce the cost and increase the ease of placing payloads in Earth orbit, and it aims to achieve this from a launch site in Europe. The Sänger team believes this latter capability could be important because it will reduce dependence on other countries for a launch site. As discussed in Chapter 7, the ability to launch from Europe is a desirable option, but it is not considered fundamental to space access, especially if incorporating such a capability significantly impacts the vehicle configuration and development costs. It might be argued further that the objective should be to reduce launch costs first, and if there are any other benefits like cruise capability then all the better. The original HOTOL concept adopted this philosophy.

In the case of Sänger, the Germans do not believe that building a single-stage-to-orbit vehicle is feasible, as Heribert Kuczera explains, "I am convinced that this technology is not available and will not be available during the next 10 or 25 years. I am also convinced that an SSTO vehicle for a mission just accelerating, no cruise capability, is still not feasible. A pure rocket system is feasible, there is no question about this, but it's more a question of the takeoff weight, and of the level of technology achievements." At the same time, he also recognizes that these conclusions were, originally at least, the result of "first assessments, [and] we do a certain level [of work but] maybe not in very much detail." Currently, the Sänger team is performing more detailed assessments of other reusable launcher concepts.

The Germans are committed to the Sänger concept. Even though this two-stage concept appeared to be thrust upon them by the BMFT which was looking for an idea sufficiently different from HOTOL, there is reason to believe they genuinely thought that the Sänger approach was the best route to take. During the definition studies of 1986 and 1987, a number of other concepts were investigated, including a vehicle almost identical to HOTOL, called *Luftatmender Raumtransporter* or LART (Figure 8.9), proposed by the DASA-ERNO Orbital Systems and Launcher Division in Bremen. Heribert Kuczera says, "We had a working group led by the BMFT discussing potential concepts, [including] Sänger and LART, and the final result was that Sänger was selected as the reference concept. It was a vote of many experts that the Sänger concept was the most convincing concept. There were some activities [on LART] for a short period of time, and then everything was dead, because the weight was given to Sänger."

The selection of a two-stage-to-orbit concept thus enabled consideration of a launcher that could both reduce cost to orbit and achieve considerable cruise capability. This is

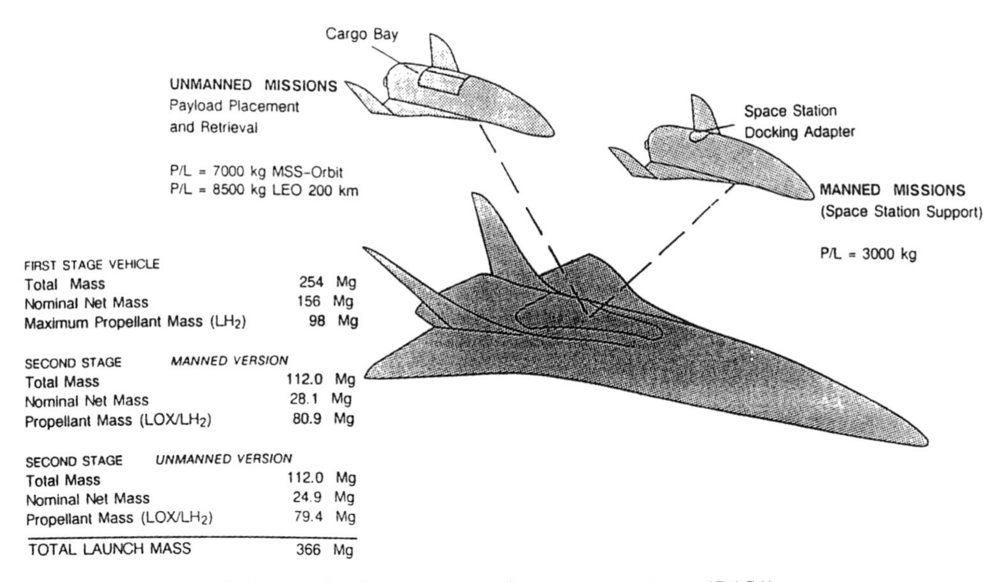


**Figure 8.8** Sänger crew and cargo upper stages (*DASA*).

practical because the cruise part of the mission can rely on air-breathing propulsion systems contained in a vehicle that does *not* go to orbit. Thus, it is merely a matter of ensuring that the fuel (hydrogen) tanks in the Sänger first stage are large enough (by about 12%) to contain this extra fuel needed for the cruise phase. This is not a serious problem for two stage vehicles as the lower stage has to be large enough anyway to physically support the orbital stage. With a single stage vehicle, the extra structure needed to hold the cruising fuel has to be taken to orbit. It is because the mass of this structure is heavy that single stage vehicles like the original HOTOL have limited cruise ability, unless they are very large, typically 500–600 tonnes. If this is indeed the case, as the Sänger team believes, then their reasons for opting for the two stage approach become clear: successfully building a single stage vehicle that weighs 600 tonnes (larger than any other aircraft), of which only 7 tonnes is actual payload, might be extremely difficult, given the uncertainty surrounding the practicality of building aero-space planes. A 1% margin for error is not much to play with, given the risks. The two stage approach makes Sänger feasible and, almost coincidentally, it enables a cruise capability. In addition, it allows the Sänger first stage to fly itself to the Kourou launch site.

Of course, this is not the end of the story. The NASP X-30, for example, could achieve a considerable cruise range, but only if it can breathe air efficiently at extremely high speeds—perhaps greater than Mach 18 or 19. This is because the mass of oxygen that needs to be carried is much less than HOTOL or Sänger. As discussed in Chapter 6, though, this demands a far higher level of technology and involves greater risk. Given that Germany wants to be involved in the development of an operational space transportation system, coupled with its limited experience in high-speed flight, a NASP-type approach is clearly too risky.

## Goodbye Cargus

The original manned second stage, Horus, was configured purely as a crewed vehicle optimized to take two pilots, three or four passengers, and 3 tonnes of pressurized logistics to a space station. (Less seriously, it was proposed that this Horus could be modified to carry as many as 36 people for space tourism type missions [19].) For missions involving the deployment of satellites, an expendable second stage vehicle designated Cargus was to be used [20] (Figure 8.10). The rationale for Cargus was to take the existing (at that time) Ariane 5 core stage and launch it off the back of the Sänger first stage. Effectively, Sänger's first stage would replace the *function* of Ariane 5's solid rocket boosters. Thus, Cargus would have been able to achieve approximately the same payload capability of Ariane 5—about 14 tonnes to LEO. Sänger's cruise capability would also have allowed payloads to be launched from Europe. Thus, Cargus would have achieved the same objectives as Ariane 5, at lower launch costs and from a rather more convenient launch site. Further, Cargus would have maximized the use of existing systems and kept the Ariane 5 production line open—an extremely important rationale, given the nature of European space politics.

Although this was the theory, technical realities, coupled with a better appreciation of payload requirements, resulted in the demise of Cargus. Analysis showed that Cargus would have required an active delta wing at the back of the vehicle to ensure aerodynamic stability during the critical separation maneuver at Mach 6.6. In addition, performance short-

**Figure 8.9** LART (*DASA-ERNO*).

falls, and the increased mass due to the wing, showed that Cargus had to be stretched and more propellant added to maintain the required 14 tonnes payload capability.

As the design evolved, Cargus began to look less and less like the Ariane 5 core stage. According to Kuczera et al., [21]

Originally it was planned to use the core stage of Ariane V as basis for CARGUS. However, the required modifications in particular for structure and propulsion turned out to be very extensive. They would have amounted to almost a new development, of course with corresponding cost.

None of this is particularly surprising in view of the problems being encountered in designing an optimum aerodynamic shape for Horus. The optimization of Horus had necessitated a flatter and more tapered nose shape to minimize drag. This resulted in a significant shrinkage of the Horus pressurized section and some loss of future capability, as noted in an earlier paper presented when Cargus was still an option, [22]

The more compact design [of Horus] became feasible by reduction of the pressurized volume, giving up some future growth potential. The more restrictive design . . . is limited to accommodation of a 2 + 3 crew and the payload of 3100 kg . . . The airlock has been eliminated since no EVA maneuvers are expected to be required from Horus.

If Horus was rapidly becoming flatter and more tapered, so must Cargus. As a result, by early 1990, Germany had a Sänger design that basically consisted of three entirely different and unique vehicles, and this meant three independent development programs would have to be paid for at an estimated cost in excess of $20 billion.

This three-vehicle Sänger became unacceptable because the goal of reducing launch costs was being compromised by the cost of making Sänger technically feasible. Cargus had become a very complex winged vehicle and, because it was expendable, it also became more expensive than first thought. Indeed, as the cost of each Cargus mission approached that of Ariane 5, the question was asked, Is it worth spending up to $20 billion on Sänger to develop a launch system that is basically little different from the existing Ariane 5? Likewise, as Horus continued to shrink, its capability was rapidly approaching that of Hermes, although it would still be at least an order of magnitude cheaper to use, fly more often, and be significantly safer. In view of what this configuration seemed to deliver for the money needed to develop it, the case for Cargus—and Sänger as a whole—did not look convincing.

At this point, it is important to bring into discussion the payload market needs. In the three-vehicle Sänger configuration there remained a very large, and very obvious, hole in the payload capability of Sänger. Horus was designed to launch and return a small crew and small payload, while Cargus was designed to launch very heavy payloads on a one-way journey into orbit. The capability to launch modestly sized payloads and, importantly, transport them back to Earth did not exist. After all, isn't the objective of aero-space plane launcher to provide a transportation service for taking the *majority* of payloads *to and from orbit*? If Europe wanted true autonomy from the United States, and wanted to do more in space than it does today, then the Sänger concept had to change. For example, Europe has already spent about $350 million developing the Eureca recoverable platform which was intended to be launched by the Shuttle in 1987 and recovered 6–9 months later. Because of Shuttle delays, this launch didn't occur until July 1992. (See Chapter 17.) Spacelab-type missions are another example. If the original Sänger concept was kept, Europe would remain dependent on the United States for such missions.

Heribert Kuczera describes the rationale for the switch away from Cargus, "Under the assumption that we will have beyond 2010–15, some kind of space station, we will need a vehicle like Sänger just for transportation of people and small payloads to the space station. It is not intended to be a flying laboratory like the Shuttle or Hermes, but is just a transportation system just to go to the station, to do something there very immediately, in a short time, to exchange payloads, and crews, and then return to Earth. That's one major task, the second task is to deploy payloads, maybe satellites, platforms, on unmanned missions which would use an unmanned upper stage. It would deploy GEO satellites, and platforms like Eureca or Astro-SPAS. The third task is a very important one, to retrieve these platforms—and I think that is new and it is a good reason for changing our strategy of switching over from the Cargus to the unmanned second stage, now a reusable system."

By mid-1991, Germany abandoned Cargus and Horus and replaced them with a second stage vehicle that comes in a manned and unmanned configuration, as discussed earlier. This decision reduces the Sänger development costs through the deletion of Cargus, while maintaining the distinction between manned and unmanned missions. In addition, the size of the crew compartment of the manned second stage can now be somewhat larger and more flexible than that planned for the original Horus because it is in the same location as the cargo bay. It has not yet been decided whether the manned second stage is to be a dedicated vehicle or if it will simply be the unmanned second stage with a crew module in the payload bay, reminiscent of HOTOL and the Delta Clipper. Before, the crew compartment was positioned in the nose which, because of the aerodynamic tapering, was volumetrically very inefficient and made growth difficult, as noted earlier. This relocation of the crew compartment allows the potential for larger crews to be carried.

Even though the unmanned second stage would be able to launch only half the amount envisaged for Cargus, this was not considered particularly important for three reasons. First, the demand for heavy-lift boosters is relatively limited because few heavy payloads are built. Second, the second stage has abort capability, allowing it and its payload to be recovered, unlike Cargus which would either achieve orbit or fail (i.e., no different from expendable rockets). Finally, it would not be acceptable to have Sänger as the sole means of accessing space. This third point is particularly important, as demonstrated by the U.S. Shuttle-only debacle. (See Chapters 11 and 15.) Europe would need a backup to Sänger to assure access to space, and as Ariane 5 would be available then, it would seem to be logical to keep it in service, albeit at a reduced launch rate. Thus, Sänger would take care of the bulk (80%) of the crewed and cargo missions, while Ariane 5 could take care of the occasional oversized payloads, as well as be on standby to support a possible grounding of the Sänger fleet—an event that might be likely in the first few years in service.

The recognition that Ariane 5 must remain in service to complement or back up Sänger is also somewhat inconsistent with the rationale for a European launch site. If Kourou is required for Ariane 5 anyway, then couldn't this same launch site be used for an aero-space plane like Sänger?

Overall, the transition from the Cargus and Horus concept to the twin second stage configuration has been welcomed because it is now a true, fully reusable space transportation system that allows both manned and unmanned missions to take equal advantage of the lower launch costs, higher reliability, and higher availability that are inherent capabilities of such a recoverable configuration. All of this also fits into the politics of European space. Whereas before Sänger might have been viewed by the French as a threat to Ariane

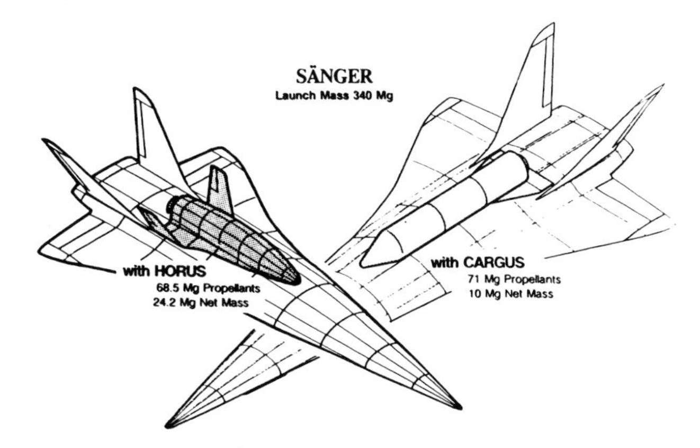


**Figure 8.10** Original Cargus and Horus upper stage concepts (*DASA*).

5, such a dual launch architecture could allow Europe to keep the Ariane 5 production line open.

In the opinion of Heribert Kuczera, "We think that we can already recognize today a trend which is the following: for heavy cargo we will need a transport system which is based on a vertical takeoff rocket. This may be expendable, partially reusable, or fully reusable—we don't know. So we will have Ariane 5, and maybe Ariane 5 will be modified or there will be an Ariane 6 or whatever. We have told these arguments to our French colleagues and now there [appears] to be a situation [where] they are trying to do something to accept it by saying 'we are now number 1 in Europe on the Ariane line, so if we can continue this line, or Ariane 5 and beyond, then everything is OK. Now the Germans are coming up with the idea of getting the leadership of a new system, which is this small taxi for the daily business that does not yet exist.' I would say they are trying to accept this step-by-step approach [because they can keep] the Ariane line open and we can do something else. It is at least an attempt."

## How Much Will It Cost?

The drive to minimize life cycle cost is as important to the Sänger program as it was to the HOTOL program. Unfortunately, because Sänger requires two totally different vehicles, the development, production, and operational costs might be somewhat higher than for a single-stage-to-orbit vehicle. For example, in 1987 an economic assessment of the LART concept showed that its development costs were nearly half that of Sänger: $7–8 billion versus $14.5 billion [23]. Presently, however, Heribert Kuczera does not believe it is clear that two "conventional" vehicles are necessarily more expensive than one "complicated" single-stage vehicle. Of course, and as discussed earlier, Germany does not consider a single-stage approach to be feasible within the next one or two decades, making this argument irrelevant.

Like HOTOL, the basis of Germany's work is to optimize its vehicle for a conservative future market, and if the market turns out to be larger, then Sänger will be able to meet the growing demand. Heribert Kuczera explains, "I think it is also clear that the development costs for a fully reusable system will be very high—higher than for current systems. We say that a system like Sänger will cost, in terms of development costs, maybe about twice as much as Ariane 5 and Hermes [combined]. That's an assessment that has been based on many calculations. On the other hand, we think with a realistic mission scenario, something like 10–20 launches per year, you can show that the development costs can be amortized during 12 to 15 years. Normally people do not talk about development costs because they think that governments will pay for this."

Based on a fleet of two first stage vehicles, each flying a total of 500 missions over their lifetimes, and each manned and unmanned second stage flying 120 missions each, DASA estimates the launch costs as follows, [24]

- Manned $25 million @ 12 launches per year
- Unmanned $23 million @ 12 launches per year

Of these costs, about one third ($7.6 million) is the cost associated with the amortized life cycle cost of the individual vehicles themselves. *Importantly, these figures also indicate that the cost difference between launching people and cargo might be marginal in the aero-space plane era.*

It is always difficult to estimate the costs of a new, risky technology like aero-space planes, and the Sänger team has put considerable emphasis on establishing a realistic basis for making such estimates. Perhaps the greatest uncertainty is with the operational costs for refurbishing the vehicles after every mission. A directly applicable data base does not exist. Instead, comparisons with the Space Shuttle, Concorde, and the X-15 have been made. According to Dietrich Koelle, [25]

> The vehicle refurbishment effort is one of the major uncertainties in the launch cost estimate since little reference data exists. Nevertheless, an estimate has been made as a compromise between some X-15 and Shuttle Orbiter data, and aircraft values, based on cost per flight (Manpower and spares) in % of the new vehicle recurring costs . . . For HORUS, one third of the initial Shuttle Orbiter cost has been assumed due the experience gained, improved technology and reduced launch and re-entry loads for Horus. For the first stage—closer related to high-speed aircraft—3.5 times the refurbishment and maintenance cost of CONCORDE have been assumed although extensive use of expert systems and a continuous diagnostic data system is planned.

Certainly, the fact that prototype versions of the Sänger first stage can be test flown many times before a second stage is put on top of it will go a long way to ironing out problems associated with refurbishment and spares requirements—again, just like an aircraft. Once the first stage is certified—a necessary and likely very expensive activity for this hypersonic aircraft (see related discussion in Chapter 14)—an unmanned prototype second stage can been flown attached to it several times at increasing speeds before separation tests are undertaken. Potentially, such separation tests can be performed at lower than optimal speeds, and the second stage can glide back to Earth (Europe) after a short engine burn. Gradually, the second stage envelope can be expanded until orbital flight is achieved. All the while, experience in operating and turning around (including mating) the vehicles can be steadily gained, and the individual characteristics of the system identified. Eventually, many flights and perhaps a year or so into the test program, a manned flight of a second stage would be attempted.

Understanding how a fully reusable vehicle like Sänger behaves is vital to reducing costs associated with refurbishment, maintenance, and spares. It is also just as critical to reducing the ground crew, as Heribert Kuczera says, "It is a lot of optimism when we say a fully reusable system like Sänger will lead to a significant decrease in transportation costs. But, I think there are a few good obvious reasons why there is a trend towards this direction. The first one is it's a fully reusable system, you have it and you just have to fly it. The second one is, it's an aircraft-like system, so the whole handling and ground operations are more comparable to aircraft handling [than that of a expendable rocket] . . . We think, as a result of comparisons with Concorde and other aircraft, we should be able to reduce the ground crew from the 14,000 people working in three shifts for the Shuttle to let's say 1,000. Then I think it is very easy to convince everybody that this must lead to a significant cost reduction."

## *Internationalizing Sänger*

Like all major European aerospace projects, the success of Sänger will depend critically on international cooperation. The Sänger team recognizes this and has stated that the Phase 2 effort cannot be initiated without significant foreign participation. According to Heribert Kuczera, "From our point of view, it is not completely clear how we could proceed in Phase 2 . . . Can ESA handle the development of a Hytex-like vehicle? Maybe they cannot. Could it be handled like the Tornado program? I think that this might be a good solution. This is a very delicate question, but we could do Phase 2 without ESA. We could do it on a bilateral basis with some others, maybe we are more efficient doing it this way like *Panavia*. For example, you could do the work maybe with Italy or the U.K. or with France—three partners who will do the development and flight testing of the Hytex. But the basic question is, if we do this in Phase 2 what are the impacts on Phase 3? Then will come the next step, and I think this is very clear, the implementation of Sänger can only be done on an ESA basis."

One of the concerns expressed by the Sänger team is that if Phase 2 is performed on a trilateral basis, for example, how would the teaming arrangement change when transitioning to ESA? In the opinion of Heribert Kuczera, "I think we have to prepare for cooperation for Sänger and so it would be really worthwhile to consider establishing a team for Phase 2 which might be fortunately the same for Phase 3. [Perhaps] ESA does not yet have the experience to lead the development of a hypersonic demonstrator. So they probably have to do something [similar] to Hermes." Potentially, a Euro-Sänger consortium would have to be established first somewhat like the Euro-Hermespace team put together in November 1990 by the principal Hermes contractors at the request of ESA.

Germany has been preparing for European cooperation practically since the inception of the Sänger activities, and serious efforts to attract European and other partners began mid-1989. The BMFT has made about a dozen official presentations to the U.K., Spain, Norway, Italy, the Netherlands, Sweden, Austria, Belgium, and Switzerland [26]. According to program officials, a number of requests were sent to France to discuss possible cooperative efforts. Since Germany never received a reply, no official briefings in France have yet occurred. Australia was the first country to participate in Sänger, as it was with HOTOL. British Aerospace (Australia) has been performing work under funding from the Australian space office on high-speed air data sensors. Presently, no formal government-to-government agreement exists, nor is Australia being paid by the BMFT or DASA. Cooperation is limited to information exchange [27].

The first formal cooperative agreements for the Phase 1 effort were approved in late 1990. In September, the Swedish National Space Board signed a Memorandum of Understanding (MOU) with the BMFT, and will commit around $5.5 million to the effort [28]. According to Per Nobinder, [29]

> Sweden is not big enough to carry out studies and work of this kind on its own, so we have to find friends willing to cooperate and, preferably, also take the lead . . . I am convinced that, if it is decided to start such a [full-scale aero-space plane] programme, Sweden will embark. The Swedish competence and capacity in the aerospace field is high, and work in a space plane development programme could certainly be performed at a much higher financial level than today.

In December, the *Norsk Romsenter* or Norwegian Space Center signed the second MOU and committed a similar amount as Sweden [30]. Although these contributions are small compared with Germany's, in the opinion of Heribert Kuczera, "From our point of view it is more important to have a lot of countries and companies all over Europe."

How many countries participate in and support Sänger is, presently, far more important than how much money is actually being spent. Further MOUs are expected to be signed with Austria, Belgium, and Spain, and potentially with the Government of Quebec. Apparently, the U.K. has also expressed some interest in cooperating on the Sänger program, despite continued reluctance to fund its own HOTOL. According to Heribert Kuczera, "we have had discussions with the BNSC who were very positive. What they said is they would be very much interested in participating in the technology program. We really want to get the British involved; we see that there is at least a chance, and we will do everything to support them and keep them informed." At the moment, cooperation with the U.K. is limited to information exchange.

These first contributions are relatively small, although politically significant. Germany hopes to sign an MOU with Italy soon. Potentially, Italy could contribute a more sizable amount to the Sänger Phase 1 activities than either Sweden or Norway. (In early 1991, the original target contribution was about 20%.) Currently, the agreement has been stalled, apparently because of the Italian industry's unwillingness to contribute its own funds to the effort. The concern is that this will set a precedent for other programs [31]. Should this conflict be resolved in the near future, this large contribution is just what will be needed for the Sänger bandwagon to roll into Phase 2.

Cooperation within Europe alone might be insufficient to raise the about $20 billion price tag for the Sänger Phase 3 effort, assuming the Sänger solution is adopted as a program. As Horst Hertrich notes, "We don't think that Sänger could be built in Western Europe alone. Its high development cost and limited market will probably require that it is done as a 'global project' around 2010." With this is mind, Germany has sent official delegations to the United States and Japan, resulting in limited agreements to exchange information.

More substantive talks were held with the then Soviet Union. After a major delegation of German industrial and government representatives visited Moscow in May 1991, Germany and the U.S.S.R. began planning for cooperative activities that would have led to wind tunnel testing and other component level testing [32]. These activities would have been on the same scale as the Soviet Interim-HOTOL work. (See Chapter 7.) Just as the agreements were being finalized, the August coup in Moscow quickly brought these efforts to a halt. As Horst Hertrich explains, "We were on the very edge of a cooperative agreement. Now we must start all over again."

## Summary—Only A Reference Concept

Germany has responded strongly to the challenges of Britain's HOTOL, and France's continued dominance in European launchers. The Hypersonic Technology Program now appears to be the world's second largest aero-space plane effort after NASP. Germany chose the two-stage-to-orbit approach with separation of the second stage at hypersonic speeds because it is, and remains, firmly convinced that single-stage-to-orbit vehicles are not yet practical and will not be for 10 to 20 years. This selection also enables a cruise capability to be incorporated into the Sänger vehicle, allowing launches into low inclination orbits, starting from a European base. Although all the work is scoped around the Sänger two stage configuration, Sänger itself is just a lead concept acting as a focus for the German Hypersonic Technology Program. In this respect, it is important to appreciate that the work presently being undertaken is purely to develop the critical technologies for hypersonic space transportation vehicles like Sänger.

The Sänger configuration, as presently defined, requires the development of two entirely different types of vehicle, and while Sänger may be technologically much more conservative than a single-stage vehicle like the X-30, perhaps more on a par with Interim-HOTOL or the Delta Clipper, it is likely to be more expensive than all of these. To pursue the Sänger route, developers must be absolutely certain that Sänger is the only practical approach because of both the massive investment required and the threat of another organization developing a more cost-effective operational system.

Germany clearly has little desire to allow the French to dictate the direction of any future European effort to develop a next generation launcher. Whether it will be a Sänger-like vehicle, or something more affordable, remains to be seen. Technology may be advancing more rapidly than expected at the inception of the Sänger program in 1987, making less expensive aero-space plane options feasible. If this is not the case now, it almost certainly will be by the end of the 1990's. Further, if Germany's goal is to make space easier and less expensive to access, then it will choose a technical solution that will best meet that end—one stage, two stage, horizontal, or vertical. For the present, it believes the Sänger route is the best. Still, as Heribert Kuczera reminds the space community, "We say [that Sänger] is only a reference concept and this can be changed. Why not? We are completely open and flexible enough to change it. This is really a very important point."

## References and Footnotes

1. According to *Jane's Spaceflight Directory*, R. Turnhill, Editor, (Jane's Publishing Co. Ltd., 1984, p. 232), Spacelab was planned for 50 mission in 10 years. It is not clear whether these missions also include single Spacelab pallets, such as when they were flown for the first time on the second and third Shuttle missions.
2. *Hypersonic Technology Programme*, Federal Ministry for Research and Technology, Bonn, Federal Republic of Germany, 1988, p. 7.
3. *European Space Directory 1991*, (Sevig Press, Paris, 1991), p. 193.
4. Based on discussions with European space officials.
5. Wilson, A., ed., *Interavia Space Directory 1991–92*, (Jane's Information Group, Coulsdon, Surrey, England, 1991), p. 153.
6. *Hypersonic Technology Programme*, p. 6.
7. Ibid., p. 15.
8. Kuczera, H., P. Krammer, and P. Sacher, "SÄNGER and the German Hypersonics Technology Programme: Status Report 1991," IAF-91-198, *42nd Congress of the International Astronautical Federation*, Montreal, Canada, October 1991, p. 2.
9. *Hypersonic Technology Programme*, pp. 16–18.
10. "SÄNGER News," DASA-MBB, München, No. 1, March 1990, p. 2.

11. "New Breaks," *Aviation Week & Space Technology*, December 9, 1991, p. 15.
12. "SÄNGER," IAF 1991, Kuczera et al., p. 2.
13. Ibid.
14. Koelle, D. E., "On the Optimum Cruise Speed of a Hypersonic Aircraft," *Proceedings of the First International Conference on Hypersonic Flight in the 21st Century*, University of North Dakota, September 20–23, 1988, p. 279.
15. *Hypersonic Technology Programme*, p. 2.
16. Koelle, D. E., and H. Kuczera, "SÄNGER Space Transportation System, Progress Report 1989," IAF-89-217, *40th Congress of the International Astronautical Federation*, Torremolinos, Spain, October 1989, p. 5.
17. "SÄNGER," IAF 1991, Kuczera et al., p. 5.
18. Ibid., pp. 6 & 8.
19. Koelle, D. E., and W. Kleinau, "Man-into-Orbit Transportation Cost—History and Outlook," IAA-89-695, *40th Congress of the International Astronautical Federation*, Torremolinos, Spain, October 1989, p. 4.
20. *Hypersonic Technology Programme*, pp. 7–8.
21. "SÄNGER," IAF 1991, Kuczera et al., p. 4.
22. "SÄNGER," IAF 1989, Koelle et al., p. 7.
23. "Aerospace Plane Technology: Research and Development Efforts in Europe," GAO/NSIAD-91-194, U.S. *General Accounting Office*, July 1991, p. 124.
24. Koelle, D. E., "Sänger Advanced Space Transportation System—Progress Report 1990," AIAA-90-520, *AIAA Second International Aerospace Planes Conference*, Orlando, Florida, 29–31 October 1990, pp. 6–7. Copyright American Institute of Aeronautics and Astronautics © 1991. Used with permission.
25. Ibid., p. 7.
26. "SÄNGER News," DASA-MBB, München, No. 2, July 1990, p. 1.
27. "SÄNGER News," DASA-MBB, München, No. 3, April 1991, p. 2.
28. Ibid., p. 1.
29. Nobinder, P., "Sweden and the Space Plane—A Small Country With High Ambition," AIAA-91-5005, *AIAA 3rd International Aerospace Plane Conference*, 3–5 December 1991, Orlando, Florida, pp. 2–4. Permission Per Nobinder.
30. "SÄNGER News," April 1991.
31. Discussion with Italian and German representatives.
32. "SÄNGER News," DASA-MBB, München, No. 3, April 1991, p. 1.

# Chapter 9

# Aero-Space Planes Around the World

## Global Efforts

American, British, and German organizations are not the only ones to have recognized that aero-space planes hold the promise of making space easier and less expensive to access. Nor can they be regarded as the only leaders in this arena. In addition to their participation in the Interim-HOTOL, the former Soviets have been pursuing a variety of paths ranging from air-launched vehicles all the way through to NASP-type scramjet designs. The Japanese are planning for the long haul and have embarked on a long-term aero-space plane technology campaign, with most vehicle concepts revolving around a combination of liquified air cycle engines and scramjets. The French have a long heritage of aero-space plane studies and have recently decided to pursue an advanced propulsion program aimed at studying and testing scramjets.

## Aero-Space Planes in the Commonwealth of Independent States

### Flight of the Scramjet

In the period immediately prior to the demise of the Soviet empire, an event occurred in Kazakhstan that took the aero-space plane world by surprise. It was reported that on November 28, 1991, the Soviets demonstrated the feasibility of the air-breathing scramjet *in flight*. Up until this point, scramjets had only been tested in wind tunnels around the world [1].

The late November flight was conducted at the missile test range at the Baikonur Cosmodrome in Kazakhstan and involved the launch of a test vehicle on top of a small two-stage solid rocket motor. The rocket was needed to accelerate the test vehicle to sufficient speed and altitude for the ramjet propulsion to kick in at around Mach 3.5. For 20 seconds or so, the liquid hydrogen-fuelled engine operated in the conventional ramjet mode—where the air is shocked down to subsonic speeds as it is ingested into the engine. Then, as the test vehicle continued to accelerate due to the still attached solid rocket second stage motor, at about Mach 6.6 and an altitude of 30 kilometers it is thought the ramjet converted to a scramjet.

Like many of the former Soviet activities, the first flight of their scramjet vehicle was shrouded in an annoying blanket of security characteristic of all high-technology programs dominated by the military. However, the deepening economic crisis in the former U.S.S.R., coupled with international interest in the scramjet tests, lead to the Russians' opening up their program and offering it for cooperation. One year after the first scramjet flight, a second essentially identical test was conducted but this time it was partly funded by French industry. Other more advanced joint C.I.S.-French scramjet engine tests are under discussion, some of which may involve a free-flying (after boosting by a solid rocket motor) scramjet-propelled vehicle.

Both scramjet flights involved an expendable, axisymmetrical engine test vehicle positioned on top of a two-stage solid rocket. A ground test version of such an engine was displayed publicly at the Aeroengine '90 exhibition in Moscow [2], and this configuration was reminiscent of the earlier U.S. Aerospaceplane research program that was cancelled, even though the engine had been partly completed [3]. Overall, the program is led by the Central Institute of Aviation Motors (TsIAM) and includes significant participation by the Tupolev aircraft design bureau and the Central Aero-Hydrodymanics Institute (TsAGI).

These first flight tests may well be heralded as a watershed event for hypersonic, air-breathing aero-space planes. The capabilities to test such vehicles would also provide valuable information to assist the beleaguered U.S. NASP program. Indeed, the restricted NASP program is proposing a similar test series, called Hyflite, involving the boosting of scramjet engines by Minuteman Missiles (Chapter 6).

It is also reported that the former Soviets have been working for some time on their own version of the U.S. X-30. Designated "TA" this vehicle would weigh around 70 tonnes

**Figure 9.1** MAKS concept.

and be able to fly up to speeds of about Mach 10–12 [4]. According to German Zagainov, "We intend to develop a project for an experimental vehicle similar to the X-30. But with other takeoff weight and other combinations of engines. We don't intend to achieve a Mach number of 20–25—it's not realistic. It seems to me to be impossible to reach Mach 20–25 air-breathing."

### MAKS: The Other End of the Spectrum

The scramjet test vehicle and alleged TA concepts are pursuing one line of activity toward the development of aerospace planes. Potentially more significant work is being undertaken at the other end of the technological spectrum on nearer term pure-rocket concepts. Like Interim-HOTOL, the availability of the AN-225 enabled the Soviets to propose the MAKS vehicle concept (Figure 9.1). Indeed, one of the fundamental mission requirements for the AN-255 was to carry MAKS. MAKS would have a comparable performance to Interim-HOTOL, the principal difference being that MAKS would utilize a discardable external propellant tank [5]. The rationale is that the extra cost of having to build a discardable tank is offset by the advantage of being able to construct a rather modestly sized, and therefore less expensive, orbital vehicle. The U.S. Space Shuttle also uses an external tank, except that this tank forms the structural backbone of the Shuttle. By contrast, the MAKS external tank is more reminiscent of a simple droptank. The major disadvantage is that MAKS has a relatively small payload bay with a cross-section ranging from 2.1 meters up to 3 meters in diameter. The bay length is a more reasonable 8.7 meters. (See Chapter 16 for a discussion of optimal payload bay dimensions.) MAKS is designed primarily for space station logistics support missions and comes in manned and unmanned versions.

Work on MAKS has been underway since 1986, and an estimated 20% of the development program has been completed. A full-scale engineering mockup of MAKS has been built. The concept owes much to similar work started in 1964 by Gleb Lozino-Lozinsky, chief designer of the Artem I. Mikoyan Experimental Design Bureau. The project was named Spiral and the vehicle itself designated "50/50," with the orbiter designated "50" (Figure 9.2). It involved a small lifting body orbiter with movable outboard wings. An expendable rocket stage was to have propelled the 50 orbiter

**Figure 9.2** Spiral and the 50/50 vehicle configuration (*Gleb Lozino-Lozinsky/Molnija*).

into low Earth orbit following launch from the back of a hypersonic jet travelling at a speed of around Mach 5.5–6 [6]. This is essentially the same as the Star-H proposed by Dassault for launching Hermes.

A full-scale subsonic version of the Spiral orbiter was built (Figure 9.3) and its first flight tests occurred in 1965. Initially, this research vehicle was carried to an altitude of 4–6 kilometers by a Tupelov TU-95K aircraft from which it was dropped, rather like the U.S. lifting body drop tests. (See Chapter 5.) The vehicle glided to a landing on a soft surface runway, using a unique four-post ski undercarriage for the touchdown. Later, this test vehicle was equipped with its own turbojet engine enabling horizontal runway takeoffs and landings. The pilot of this research vehicle was Igor Volk—the man who was to later make analogous flights in the Buran test orbiter. Although the project met nearly all of its goals, it was scaled back in 1966 due largely to the burden of the abortive lunar program and the death of Sergei Korolev, a strong supporter of the Spiral effort. Tests with the research vehicle continued into the 1970s, with final program termination in 1978.

Even today, the Spiral vehicle would be considered a challenging concept, and models of it are often displayed at aerospace exhibitions. Although the project died, much of what was learned in the orbiter design was applied to the Buran space shuttle effort. In particular, a small vehicle designated BOR with a remarkable resemblance to the Spiral orbiter was constructed and used for reentry testing (Figure 9.4). The BOR vehicle measured 3.4 meters long and 2.6 meters across, and it made several orbital flights beginning in 1982. After hypersonic reentry and cross-range maneuvers, the BOR vehicle deployed a parachute and was recovered from the ocean. Initially, it had been thought that these flights were in preparation for the introduction of a small winged vehicle to replace the aging Soyuz. It was also thought that such a vehicle was to be launched on modified versions of the Energia strap-on boosters. Such a program apparently did not exist, because it was not economical given the three or four manned flights conducted each year to the orbiting Salyut and Mir space stations. Such problems characterized the cancellation of the U.S. X-20 effort and the European Hermes.

The current MAKS concept has heritage that spans over nearly three decades, and TsAGI hopes that the project can proceed to completion. As German Zagainov explains, "The next intention of our industry is to finish the MAKS project. A big part of this project has been proven during maybe five years we have been developing this project. Now we have good achievement in this project. Not only technology achievement, but the project in total. We have constructions, detailed technologies, some materials, and some investigations in wind tunnels, and other installations. It is very realistic for us to finish this project, to prove it in total. Especially at this time because we have good change in our economic situation. Our government has no money in order to support this project. But some commercial structures in our country have money—much money. Banks, commercial associations, commercial companies, etc. And now, in the situation of inflation with our money, they are interested

**Figure 9.3** Full-scale version of the Spiral orbiter first flown in 1965 (*Gleb Lozino-Lozinsky/Molnija*).

in investing in prestigious and new technology projects. They are proposing to us to invest in our project. This is a new situation for us. It seems to us, we will have the money for this project."

The C.I.S. has been studying other forms of reusable launch systems. Of special interest, various concepts have been developed by the Energia Design Bureau for recovering the Energia core stage and its strap-on boosters. One concept, shown in Figure 9.5, would involve drastically modifying the core stage with wings, allowing it to launch and return about 20 tonnes. Four strap-on boosters would be needed, and they would use a scissor wing design and a turbojet in the nose for a long-range recovery [7].

### Technology and Facilities

Western observers have nothing but admiration for the level of technology developed within the Soviet Union and C.I.S. Materials and propulsion are areas of key strength. For example, Energia contains about a dozen different rapid solidification rate metals [8]. In addition, they have constructed large carbon-carbon and carbon silicon-carbide composite panels and other structural elements, and assembled subscale models of integral fuselage/wing structure concepts. Much of this work is comparable to that done in the NASP and other programs. The Russians complain that even though they know how to make advanced materials, they lack the production means to manufacture the large quantities needed for utilization. Production is obviously an area for possible cooperation with the West.

In the realm of propulsion systems, the former Soviets have pioneered the development of engines which use both hydrogen and kerosene as the fuels and oxygen as the oxidizer. They believe that such engines are critical to the development of air-launched and single stage designs, including MAKS and other vertical takeoff concepts. According to Gleb Lozino-Lozinsky, [9]

> The most important creation condition . . . is from our point of view the development of three-component sustainer engine. Such engine is developed in the USSR and its reality is beyond doubt.

This engine is designated the RD-701 and various components have been tested, although a full engine has not yet been built. NASA has suggested using the RD-701 in a proposed single-stage-to-orbit launch vehicle.

The use of three-component or dual-fuel engines is important during the initial ascent through the lower atmosphere when drag is high and when performance of high-energy liquid hydrogen/oxygen engines is sharply curtailed. For example, in space the specific impulse of kerosene/oxygen engines is about 150 seconds less than an equivalent hydrogen/oxygen engine. However, at sea level the difference is reduced to around 90 seconds. This difference is still high, but it is compensated by the fact that kerosene is so much more dense than hydrogen. As a result, using kerosene allows a smaller and lighter-weight tank design and, therefore, reduced mass. This would also reduce aerodynamic drag due to the lower overall vehicle volume and, therefore, reducing the propellant mass. Thus, the kerosene would be used during the initial ascent through the thick lower atmosphere, and after a sufficient altitude had been reached, the vehicle would switch over to the higher performance hydrogen fuel.

Another area of activity where the C.I.S. exceeds the West is in hypersonic wind tunnel test facilities. For example, the T-117 wind tunnel at TsAGI can sustain speeds ranging from Mach 10 to as high as Mach 20 for up to 2 minutes, with a test cell diameter of about 1 meter and using air as the medium [10]. By contrast, the best continuously operating wind tunnel in the West can only achieve Mach 8 and shock tunnels must be used for higher speeds.

## Aero-Space Planes in Japan

### Planning Today for Tomorrow

As with most things, the Japanese have a long-term plan for their space program—one in which aero-space planes play a central enabling role. The Japanese have recognized—perhaps more clearly than any other nation—that all space activities are severely stifled by reliance on the existing generation of expensive, untestable, throwaway launchers. For example, according to a paper by Y. Ohkami et al., [11]

> To be skeptical to space development, many pro-space scientists in various fields or "space fans" among ordinary people are very much frustrated. This is because the space infrastructure has not improved in these 25 years or so. For example, no systematic experiments in microgravity have performed in orbit, since a few trials on Skylab and its follow-ons. The space station Freedom is not defined in detail yet, just to recall the fact that the year 1992 was originally the target year of the construction completion. Quite a few astronauts in several countries are waiting for their turn aboard the Spacelab to come . . . Space activities in the 21st century must be free from this situation, and the key is an assured, flexible and really operational space transport system, and it is this role that the spaceplane should play.

And, according to a report by the U.S. General Accounting Office, [12]

> In 1986 the [Japanese] Science and Technology Agency's Advisory Committee on Space Plane [sic] was established to review Japan's long-term research and development of a spaceplane. The Committee recommended that:
>
> - space transportation systems, such as spaceplanes should be fundamental elements of Japan's vision of a space infrastructure to promote future space activities;

**Figure 9.4** The BOR reentry test vehicle (*Gleb Lozino-Lozinsky/Molnija & Paul Czysz*).

- development of spaceplanes is indispensable to Japan's autonomous space activities;
- spaceplanes will improve international space launch options by eliminating the current reliance on limited launch means; . . .

Aero-space planes are fundamental to Japan's ultimate goal to exploit and industrialize space to the fullest extent. This is a logical and, perhaps, obvious conclusion. Yet, at the same time, Japan also recognizes that it has a long way to go before it can build an aero-space plane or, more likely, contribute to an international aero-space plane consortium. (See Chapter 13.) Observers often point out that Japan's aerospace industry is relatively modest compared with those of the West and the C.I.S., and it has little experience in building high-speed aircraft. Japan is simply not in a position to commit itself to a formal aero-space plane program. Instead it has embarked on a long-term effort to develop many of the key technologies and to invest in the needed test facilities. From this base an informed decision can be made early in the next century.

Precise estimates of how much is being spent on Japanese aero-space plane activities are difficult to make. Between 1982 and 1990 around $128 million was spent, augmented by about $22 million from industry. From 1988 to 1999, Japan hopes to spend $600–700 million directly and indirectly on aero-space plane R&D technology work, plus about $40 million from Japanese industry. Of this money, some $155 million is planned for the construction of facilities. It is estimated that in Fiscal Year 1991 the Japanese government spent around $48 million on hypersonic research [13].

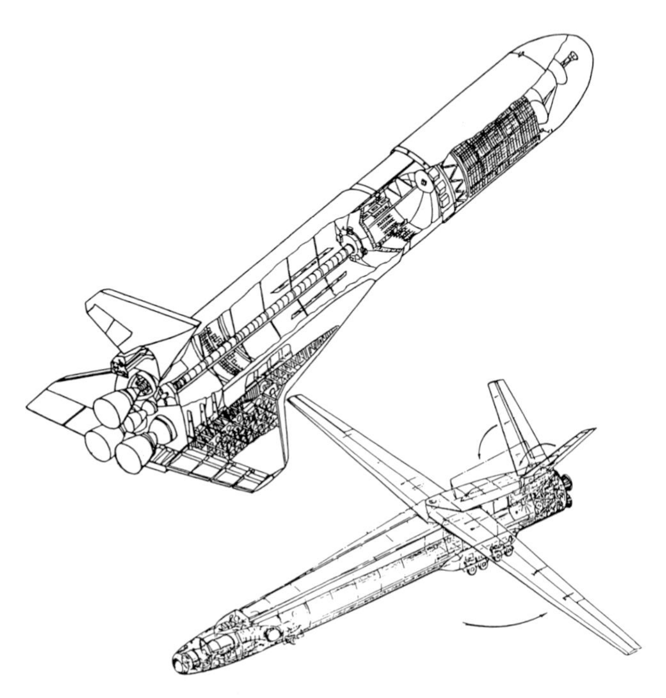

**Figure 9.5** Fully reusable Energia concept (*NPO Energia*).

## Coordination of Activities

The Space Activity Commission (SAC) is the policy-making body that advises the Office of the Prime Minister. SAC makes specific recommendations on the direction and strategy of Japan's space efforts, and coordinates and promotes the activities of the various government agencies chartered to carry out the space program. In 1987, a report entitled *Towards a New Era of Space Science and Technology*, written by the Consultative Committee on Long-Term Policy Under Space Activities Commission, specifically highlighted the development of aero-space planes as a necessary part of Japan's future space infrastructure [14].

Prior to SAC's 1987 report, interest in aero-space plane R&D activities by industry and government agencies led the Research and Development Bureau of the Science and Technology Agency to establish the Advisory Committee on Space Plane in 1986. The recommendations of this committee led to the establishment of a Liaison Group for Space Plane R&D organized primarily by the principal government agencies involved in aero-space plane research. These are [15]

- The National Aerospace Laboratory (NAL)
- National Space Development Agency (NASDA)
- Institute for Space and Astronautical Sciences (ISAS)
- Various universities and research institutes

Japan's aero-space plane activities were initiated by NAL in 1963, which is performing the majority of work directly applicable to future aero-space planes. This work is in the area of system studies, aerodynamics, CFD analysis, wind tunnel testing, hypersonic air-breathing engines (including scramjets), materials development, structural concepts, and simulators. NASDA's primary contribution is supporting development of HOPE and undertaking liquefied air cycle engine (LACE) technology demonstrations. ISAS is pursuing the study of a small suborbital vehicle to be used as a fully reusable technology test-bed. This vehicle is designated HIMES for Highly Maneuverable Experimental Space vehicle.

There are a number of industries involved in aero-space plane research in Japan. Ishikwaijima-Harmia, Kawaski, Mitsubishi Heavy Industries, Sumitomo, and Fuji formed the Space Plane Committee of the Society of Japanese Aerospace Companies (SJAC). SJAC is a liaison body that reports to the Ministry of International Trade and Industry (MITI). SJAC coordinated research activities within this group. For example, in 1989 SJAC established an 8-year program under the generic title of R&D of High-Performance Materials for Severe Environments [16]. This program includes materials research applicable to aero-space planes and is functionally similar to the NASP Materials Consortium.

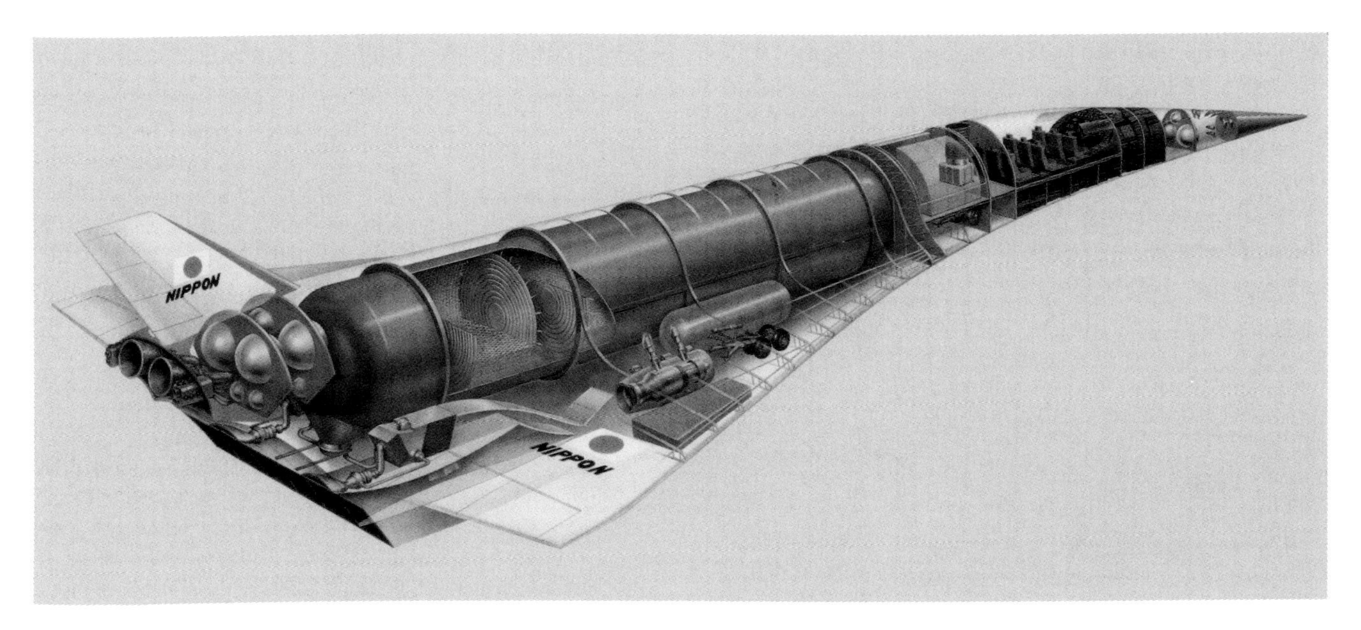


**Figure 9.6** SSTO scramjet/LACE aero-space plane concept (*NAL*).

## Vehicle Concepts and Technology Status

A NAL concept for an operational single-stage-to-orbit aero-space plane is shown in Figure 9.6. This is one of many concepts being studied, and it has been defined purely to help focus basic technology R&D. This SSTO vehicle is designed to place a crew of two plus eight passengers into a space station low Earth orbit. Total estimated takeoff mass is 350 tonnes, with a wingspan of about 29 meters and a length of about 100 meters. The vehicle is powered by an integrated LACE/scramjet propulsion system. The LACE engines are used from a standing start to accelerate the vehicle to around Mach 5. At this point, the scramjet propulsion system is engaged to accelerate the vehicle to about Mach 20. Using onboard oxygen, the rocket motor of the LACE engine is ignited to provide the final push into orbit [17].

As with NASP, NAL believes that scramjet propulsion is a key technology for horizontal takeoff aero-space planes. As a result, in 1993 NAL began testing of a subscale ramjet/scramjet engine. This engine will be fuelled by gaseous hydrogen, and a water jacket is the coolant. It will be tested at speeds ranging from Mach 4 to 8 in a new ramjet engine test facility being built at the Kakuda Research Center [18]. This facility will be capable of sustaining an airstream up to Mach 8 for 30 seconds, at a simulated altitude of 35 kilometers. Its completion and the testing of the subscale scramjet signal Japan's growing competence in critical aero-space plane technology.

Japanese aerospace officials consider that a LACE approach offers the potential for a more efficient and lighter-weight alternative to using a separate turboramjet for takeoff and a separate rocket for the final ascent. NASDA and Mitsubishi Heavy Industries began ground testing in 1992 of a complete LACE demonstration engine. This engine takes advantage of the existing LE-5 cryogenic rocket motor, originally developed for the H-1 launcher, but modified with a heat exchanger capable of liquefying incoming air. Japan is currently the only nation pursuing hardware demonstrations of LACE-type engine cycles for aero-space planes. It also hopes to eventually perform flight tests of LACE engines strapped to the side of an expendable launcher [19].

NASDA also wants to spend an additional $2.73 billion on the H-2 Orbiting Plane or HOPE. HOPE could provide a focus for R&D efforts as well as an independent means for Japan to deliver small payloads to Space Station Freedom and the attached JEM module. In early 1992, though, the Japanese government decided not to give final approval for development of this unmanned glider, for reasons mirroring those that led to the cancellation of the European Hermes program. For example, even in its unmanned configuration, the mass of HOPE has ballooned to about 20 tonnes—twice the payload capability of the baseline H-2. This mandates an upgraded version of the H-2 requiring an additional four solid rocket boosters and the capability to relight the LE-7 cryogenic engine [20]. Interestingly, however, the HOPE project was reintstated and funded in 1993, despite the fact that Hermes had just been cancelled.

Some Japanese officials have suggested that they would prefer to build a much smaller, purely experimental version of HOPE, one similar to the U.S. ASSET/PRIME reentry vehicles (Chapter 5) and the former Soviet BOR test vehicles. Such a vehicle could also fit under the shroud of the

H-2. Further, this vehicle would not be compromised by the need to fulfil a vague operational requirement, nor would it be anywhere near as expensive. It would simply provide the necessary reentry data to verify CFD predictions and determine the performance of new materials. Building HOPE would delay the data return and divert funds from aero-space plane R&D activities and other space program efforts.

As a precursor to the HOPE program, NASDA is taking advantage of spare capacity on the first flight of the new H-2 rocket in 1994 to conduct the Orbital Re-entry EXperiment or OREX [21]. OREX is essentially the nose cap of HOPE, and it will be de-orbited to characterize the performance of high temperature carbon-carbon materials and silicon-based coatings.

ISAS proposed the development of the HIMES vehicle in 1985 as a reusable sounding rocket and a technology testbed for aero-space plane research. If it were built, HIMES would fulfil many of the objectives planned for HOPE. For example, it could test the performance of new thermal protection system materials and possibly also test air-breathing propulsion systems. HIMES would be built from relatively conventional materials and use liquid hydrogen and liquid oxygen as the propellants. Its takeoff mass is estimated at about 14 tonnes and its length is about 14 meters. The vehicle would be launched vertically, fly a suborbital trajectory to about 250 kilometers, and then land horizontally back at the launch site. Eventually, it could be launched horizontally to test the feasibility of linear motor launch-assist systems [22].

Development of HIMES has not been approved. However, ISAS has performed a number of tests with 500 kilogram subscale models of HIMES dropped from helicopters and balloons. In 1988, one of these models was to have been dropped from a balloon and a small rocket then used to boost it to an altitude of about 80 kilometers. The subscale HIMES was then to reenter at speeds of about Mach 4. Unfortunately, the first test ended in failure when the helium balloon tore at 18 kilometers. The second attempt in 1992 was a complete success [23].

ISAS has more recently defined a simpler and less expensive alternative to HIMES. This conical-shaped vehicle would be around 9–12 meters long and have a takeoff mass of 8–12 tonnes. It is very similar in concept to the DC-X built by McDonnell Douglas (Chapter 10). However, as opposed to the DC-X, the ISAS vehicle would be able to achieve speeds up to Mach 7–15, exposing materials to much higher temperatures [24]. Also, the ISAS vehicle would be recovered from the ocean following a parachute descent, unlike the DC-X which lands vertically. Definition of this new demonstrator could also herald increasing interest by the Japanese in vertical takeoff and landing aero-space plane concepts whereas the previous focus was on air-breathing, winged designs.

## *Aero-Space Planes in France*

### Late Into the Act

France has a vested interest in expensive, throwaway launch vehicles. The Ariane 1–4 series of boosters has been an enormous commercial success for the French who have a 57% percent stake in this European venture. France is providing nearly half the funds to develop the European Ariane 5 rocket, a program it initiated and has dominated. Ariane 5 is intended to supersede the Ariane 4 in the late 1990s. It was also mandatory for launching the Hermes spacecraft (although now cancelled)—another program that France initiated and dominated.

France has a lot at stake. It comes as little surprise that its interest in aero-space planes has until recently been relatively lackluster. France builds rockets that are demonstratively successful on the world stage. In the early 1980s, it took on the U.S. Shuttle and won, or so it appeared politically. As a result, the French have some justification in believing that preserving this status quo would be a good thing for France and its rocket business.

Still, even when business is good, it is difficult to ignore the tides of change. In 1992—more than 5 years after the United States, Germany, and Japan—France finally jumped on the aero-space plane bandwagon and initiated the Program of Research on Advanced Hypersonic Propulsion Technology or PREPHA. (See also Chapter 4.) This very significant program originally was funded over a 4 year period at about $90 million (500 million FF), with some $18 million (100 million FF) of this coming from French industry. This funding puts France fourth in order of investment in aero-space plane technology, although economics alone can be a poor method of judging a program [25].

### The Politics of Aero-Space Planes in France

The rationale for France's interest in joining the aero-space plane fray is explained by SEP Vice President Pierre Betin speaking in October 1990, [26]

> Busy with the major medium-term programmes Ariane 5 and Hermes, which it is running on behalf of Europe, France devotes only meagre resources to hypersonics. Working together under the aegis of CNES and in close cooperation with manufacturers Aerospatiale and Dassault, however, Snecma, SEP and Onera have, in the last few years, carried out a study on paper of combined engines integrating successive methods of propulsion in front of a single exhaust . . . In 1989 interest in this preliminary work, the important progress being seen in the USA [with NASP], and the succession of foreign initiatives, caused French industrialists to draw the authorities' attention to the risk involved in neglecting hypersonics . . . It is vital to undertake this [PREPHA] effort so that the

technological gap which already separates the USA from France in this field, and which has become worryingly wide over the last five years, does not become irreparable . . . This may appear nationalistic but it is not. The effort is aimed at preparing us, as a major competent partner, for a concerted European—and indeed intercontinental—action.

France has a hypersonic program but does not appear to be in any hurry to develop an operational aero-space plane. Officially, France does not believe such a payload-carrying vehicle can be built before 2025–30 [27], the reason being that a hypersonic demonstrator program on the scale of the proposed German Hytex cannot be financed before 1995. As such a modest program would require a 10 year development effort, this would put first flight around 2005. Also, by 2005 funding for the current European efforts will be tailing off (Chapter 13). In theory, money could be freed to enable construction of an experimental single- or two-stage-to-orbit vehicle. This could be an international effort and would take another 10 years. Thus, by 2015, France believes a decision could then be made to build an operational vehicle, putting its first flight no earlier than 2025, after yet another 10 year development effort.

For space activities that would benefit now from a cheaper and easier method of accessing space, France's proposed three-decade-long schedule appears painfully slow and out of step with the needs of the users. This is perhaps surprising given that France is enthusiastic about its space technology and is the second largest space spender (not including the C.I.S.). Why wait so long?

One important reason for this ponderous pace is that France believes it will take that long to mature the technology needed for aero-space planes, especially if the scramjet route is taken. This is quite possibly true, although arguably significantly less time would be needed for pure rocket or non-scramjet powered vehicles. Another reason for the long timeframe hinges around justifying France's current space efforts. According to one French aerospace industry official speaking in 1991, "France is working with ESA for implementing large programs—Ariane 5, Hermes, Columbus—and that is the first priority for France within the reasonable future. [However,] no country can invest in space if they're not preparing for the future. But, the amount of money it costs to prepare the future is much less. Also, another little fear is that if you put too much emphasis on the long-term programs, you can destroy the credibility of the short-term programs. Why do something today that is very expensive, when you can do something better that costs you now less money? . . . France is investing in the future, but not putting too much emphasis on it."

Understandably, France is interested in ensuring the successful introduction of Ariane 5 and, prior to late 1992, construction of Hermes. Regardless of their true utility and cost-effectiveness, these programs are important for French and European space industries. Therefore, a 2025–30 introduction date for aero-space planes provides a 20 year breathing space for these programs to exist, thereby justifying their high cost. The only problem with this approach is that the anticipated high costs and limited capability of Ariane 5 and Hermes will preclude all but the highest priority missions. It definitely would change if another organization were to embark on the construction of an aero-space plane. Losing their present 50% plus share of the commercial communications satellite market would be intolerable to France (See Chapter 13).

### PREPHA and its Origins

France has been interested in high-speed air-breathing flight for as long as any other nation. France broke the sound barrier in 1952 and achieved Mach 2 in 1957. The Griffen II aircraft, although partly funded by the United States, achieved Mach 2.19 in 1959 and held a number of aircraft speed records. In 1969, France, in combination with the U.K., built the Concorde Mach 2 airliner—the only successful civilian supersonic aircraft [28].

The Griffen II was unique among France's early achievements because it used a ramjet engine. Ramjet technology was pursued aggressively in France primarily for missile propulsion. In 1964, ONERA's experimental Stataltex demonstrated Mach 5 flight and was for many years the fastest air-breathing vehicle. France is the only nation to have in its armory a large air-to-ground missile, designated the ASMP, which uses a ramjet propulsion system.

In 1971, with continued interest in ramjets and spurred by the U.S. Hypersonic Research Engine (HRE) efforts, France built and tested its first and so far only scramjet engine. The ESOPE engine, as it was called, demonstrated supersonic combustion at Mach 6–7. France was the second nation to demonstrate the feasibility of scramjet systems. Yet, these efforts came to a halt because of a lack of clear rationale for scramjet systems, technology limitations, changing military priorities, and because the United States had pulled out of scramjet research altogether. Further, France had its hands full with its independent ICBM program and startup of the Ariane launcher effort.

For 15 years or so, and like most other countries, France's interest in future aero-space plane launch systems was confined to low-level paper studies. Then in 1986, CNES awarded modest study contracts to a number of companies to explore various combined cycle single- and two-stage vehicle designs. Most notable of these were Aerospatiale's STS-2000 and Dassault Aviation's Star-H concepts. These and other studies are ongoing. Aerospatiale has developed new concepts designated Oriflamme and Radiance, and is pursuing wind tunnel work as shown in Figure 9.7 [29].

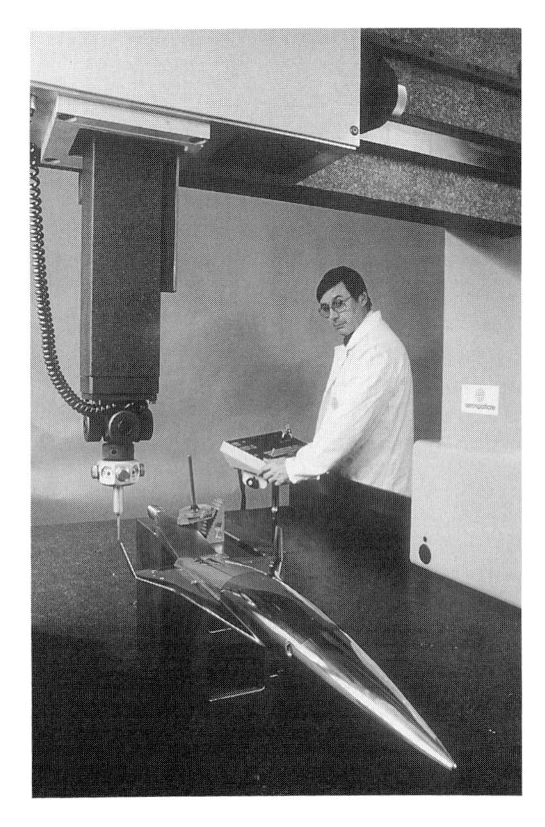

**Figure 9.7** Wind tunnel testing of an Aerospatiale STS-2000 concept (*Aerospatiale*).

In November 1989, Hubert Curion, the minister for science and technology, awarded a $2.5 million (14 million FF) contract to a team composed of the major French aerospace organizations with the aim to propose a French national hypersonic technology program. The team was composed of Aerospatiale, Dassault Aviation, SEP, Snecma, and ONERA. Shortly thereafter, in January 1990, SEP and Snecma formed a company called Hyperspace. Hyperspace merges the air-breathing engine capability of Snecma with the rocket capability of SEP, and is the single point of contact for the two companies with the French government.

This 7 month study effort realized the following conclusions, according to Bruno Debout, [30]

Main conclusions of these studies were:

- ultra high Mach airbreathing flight is a very challenging matter for long term applications,
- supersonic combustion ramjet is the most interesting engine for optimized future airbreathing aerospace planes,
- materials, light structures, CFD are crucial points.

Ultimately this led to the French government launching the PREPHA program, and individual contracts were signed with each of the major team members in January 1992. PREPHA is funded jointly by the Ministry of Defense (DGA or Délégation Générale de l'Armement), the Ministry of Research and Technology (MRT), and CNES. The DGA is providing about 60% of the funds, and CNES and the MRT are splitting the remaining 40%. Overall, the program is directed by a steering committee made up of representatives of the DGA, MRT, and CNES. The DGA is the executive agency, managing the program and supervising the progress of the work. In this respect, it is interesting to note that, like the German Hypersonic Technology Program, the bulk of the funding does not come from the space departments, reflecting the relatively lukewarm attitude of space authorities to aero-space planes.

In addition to the four major aerospace companies and ONERA, a number of other companies are in the team, including Bertin and Air Liquide. Laboratories of the French National Center for Science Research (CNRS) are also supporting the work.

As noted earlier, the principal objective of PREPRA is to develop a scramjet engine and test it in existing (but modified) wind tunnels up to Mach 8—the limit of Western wind tunnels. This engine is not being built for flight testing, but is intended to be fully representative of a scramjet engine flow path, including the forebody compression and aft body expansion. After the 4 year PREPHA effort, France hopes to build a flight demonstrator. A possible configuration is a relatively simple Mach 8 missile.

Although the main emphasis is on scramjet propulsion, PREPHA will engage in four other work areas. These are (1) CFD numerical techniques including development of complete nose-to-tail codes; (2) materials including ceramic, carbon, glass, and metallic matrix composites; (3) vehicles and propulsion systems including low-speed acceleration, slush hydrogen, advanced lightweight structure concepts, and active cooling concepts; and (4) modification of existing test facilities.

Much of this work will benefit significantly from the Hermes program, where materials advances are regarded by officials as being only marginally behind the U.S. NASP program (Figure 9.8). As Paul Donguy notes, " . . . the Hermes vehicle [has] nearly everything except the propulsion system. You have the life-support, the hypersonic aerodynamics, the hot structure, you have a lot of things. It's really an important step. We are not very far from the NASP except we do not yet have the propulsion system. With the studies we have everywhere in Europe with propulsion and with Hermes, we may react very quickly to an American breakthrough."

Some of the carbon-carbon materials for NASP are based originally on French technology. (See Chapter 13.) In fact, Europe might be better positioned technologically than the United States to build a pure rocket Delta Clipper-like ve-

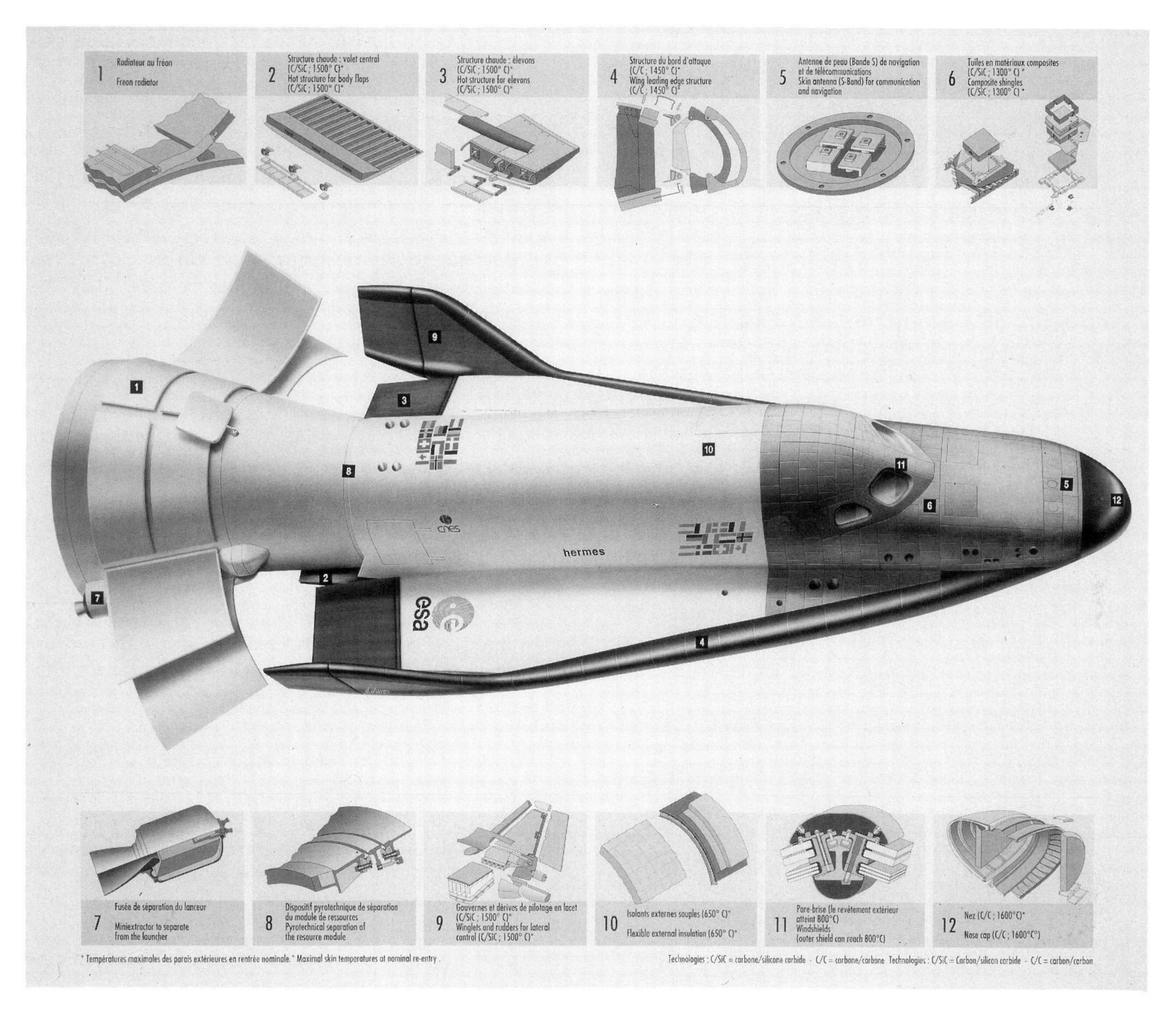


**Figure 9.8** Many Hermes technologies applicable to a European aero-space plane (*ESA/CNES*).

hicle. In addition to materials, the Ariane 5 main engine, built by SEP, is approximately the same size and has nearly the same performance as the engine proposed for the Delta Clipper, as will be detailed in Chapter 10. Modification to this engine, including throttleability and the addition of an extendable nozzle to improve its performance at altitude, would be needed. Unfortunately, tradeoff analysis of pure rocket solutions does not seem to be part of the PREPHA activities.

Another area not currently included in PREPHA is analysis of the applications of future aero-space planes, and the reasons for this are largely tied to attempts to justify the Ariane 5 program. As a result, PREPHA might encounter similar problems afflicting NASP where the rationale and ultimate reasons for the investment are less than obvious. (See Chapter 6.)

Despite this, the potential of aero-space planes is appreciated within industry. For example, according to Jacques Hauvette, "If you can make such a [scramjet] engine, it will change completely the economic balance of spaceflight. [This] will reduce cost, and cost is really the main driver for finding new applications in the future. There are many programs for the space market, however, the current market will remain quite stable. If you want to change the market, you have to change the scale of the price. To change the scale you have three main factors: The first one is air-breathing for reducing mass. The second one is [that you need a] 'reusable car' able to fly and fly and fly. When the Americans introduced the Boeing 707, nobody thought that the price could be reduced to a reasonable fare per passenger. But with such an aircraft flying many hours a day, they could amortize the costs and provide a reasonable fare to the passenger. The third one is are we able to make something reliable enough [so that] if there is a problem we can find an escape trajectory and come back safely to the ground. Because, for the moment it's always a 'fire and forget' strategy. We take insurance at 20%, we have a satellite with an incredible cost because it has to work the first time and work for a long period, otherwise everything is lost. So, if in the future we want to change the market, we have to find something to change the philosophy and have something cheaper in space. Maybe you can send people to repair, to have an escape trajectory—what I call reliability—and in each of these fields we have to make very large improvements to reduce the costs."

## Summary—Overlapping Interests?

Research efforts in the Commonwealth of Independent States, Japan, and France are extensive and comparable with those in other countries. The C.I.S., appears to have a capability unmatched anywhere else in the world, demonstrated by the scramjet flight tests in November 1991 and 1992. Clearly the C.I.S. will not be in a position for some tune to take advantage of its scramjet technology. As current Japanese, French, and American programs rely heavily on developing scramjet systems, if these countries had access to information obtained during the scramjet flight tests, it would speed up their national efforts and avoid unnecessary overlap and duplication. For example, at the present time the French do not plan to build an equivalent test vehicle until after their PREPHA effort is finished in 1995. Sharing data under a cooperative agreement might be one way to help the C.I.S. secure continuation of national support for its efforts.

Both Japanese and French efforts are focussed on long-range R&D objectives, with rather distant dates for introducing an operational system. This is understandable for Japan which does not have a tradition in high-speed flight. It is more difficult to comprehend in France which has both a distinguished tradition in high-speed aircraft and space launchers. Further, whereas Japan has already accepted that aero-space planes are fundamental to its future space ambitions, France seems, on the surface at least, to be less enthusiastic and, perhaps, more skeptical. Nevertheless, France is very confident that it could respond quickly if other countries start to build aero-space planes.

Like Germany and the U.K., the C.I.S., Japan and France believe that international cooperation is essential to their future efforts. This is certainly true if the high-technology approach using scramjets is pursued, although cooperation is arguably less critical for pure rocket solutions. Cooperation for Japan will probably be essential in the next two decades as it has neither the technology nor the trained individuals to build aero-space planes.

## References and Footnotes

1. Reported in an article in *Izvestiya* entitled "Hypersonic Ramjet Engine," (Translated by Technology Detail), February 12, 1992. See also, Covault, C., "Rocket-Launched Engines Sharpen Hypersonic Face-off," *Aviation Week & Space Technology*, February 3, 1992, pp. 18–19, and de Selding, P., "Russia Claims First Scramjet Flight," *Space News*, February 3–9, 1992, p. 3.
2. Reported in the *Aviation Week & Space Technology* article, April 16, 1990, p. 18.
3. Heppenheimer, T. A., *Hypersonic Technologies and the Natiotial Aero-Space Plane*, (Pasha Market Intelligence, Arlington, Virginia), 1990.
4. From discussions with Paul Czysz and also in the *Aviation Week & Space Technology* article of February 3, 1992.
5. Lozino-Lozinsky, G. E., et al., "Reusable aerospace system with horizontal take-off," IAF-90-176 *41st Congress of the International Astronautical Federation*, Dresden, Germany, October 6–12, 1990.
6. From a letter from Gleb Lozino-Lozinsky to the author April 1992. Further information can be found in the *Interavia Space Directory 1991–92*, edited by Andrew Wilson, (Jane's Information Group, Coulsdon, Surrey, England, 1991), pp. 133–134.
7. Gubanov, B., "The Immediate Prospect for the Reusable Space Transport Systems," Soviet Technical Paper No. SSI BIG-1, published by the *Space Studies Institute*, 1990.
8. According to Paul Czysz.
9. Lozino-Lozinsky, G. E., and V. P. Plokhikh, "Prospect for the Development of Reusable Aerospace Systems," IAF-91-210 *42nd Congress of the International Astronautical Federation*, Montreal, Canada, October 6–12, 1990.
10. From viewgraphs produced by British Aerospace (Space Systems) Ltd., and the Central Institute of Aero- & Hydrodynamics (TsAGI) for the AN-225/Interim-HOTOL presentation to ESA, Paris, June 21, 1991.
11. Ohkami, Y., et al., "Space Activities In The 21st Century—Expectation for Spaceplane," AIAA-91-5085, *Third International Aerospace Planes Conference*, Orlando, Florida, December 3–5, 1991, p. 1. Copyright American Institute of Aeronautics and Astronautics © 1991. Used with permission.
12. "Aerospace Plane Technology: Research and Development Efforts in Japan and Australia," GAO/NSIAD-92-5, written by U.S. General Accounting Office, Washington, D.C., October 1991, p. 28.
13. Ibid., p. 90.
14. "Towards a New Era of Space Science and Technology," written by the Consultative Committee on Long-Term Policy Under Space Activities Commission, May 26, 1987. (Unofficial English translation.)
15. Derived from Yamanaka, T., "The Japanese Space Planes R&D Overview," AIAA-90-5202 *Second International Aerospace Planes Conference*, Orlando, Florida, October 29–31, 1990.
16. Sunakawa, M., "Current Status of R&D in Japan on Materials for Space Planes," *Third International Aerospace Planes Conference*, Orlando, Florida, December 3–5, 1991, p. 1.
17. Yamanaka, T., " 'Spaceplanes' R&D Status of Japan," *Third International Aerospace Planes Conference*, Orlando, Florida, December 3–5, 1991, p. 6.
18. Miyajima, H., "Scramjet Research at the National Aerospace Laboratory," AIAA-91-5076 *Third International Aerospace Planes Conference*, Orlando, Florida, December 3–5, 1991, p. 1.
19. Information given during the presentation entitled "A Concept of

LACE for SSTO Space Plane," (AIAA-91-5011) during the *Third International Aerospace Planes Conference*, Orlando, Florida, December 3–5, 1991.
20. " 'Spaceplanes' R&D Status of Japan," p. 3.
21. Ibid., p. 1.
22. "Aerospace Plane Technology: Research and Development Efforts in Japan and Australia," pp. 41–46.
23. Ibid., pp. 47–49. The successful second test was announced at the December 1992 Fourth Aero-Space Planes Conference in Orlando, Florida.
24. " 'Spaceplanes' R&D Status of Japan," p. 2.
25. Kandebo, S., "Japanese Making Rapid Strides in Hypersonic Technologies," *Aviation Week & Space Technology*, December 16–23, 1991, p. 61.
26. Betin, P., "Reflection on the Spaceplane," *Space Policy*, May 1991, p. 144. (Copyright *Space Policy* © 1991. Used with permission.)
27. Ibid.
28. An overview of past French aero-space plane R&D efforts and the PREPHA program can be found in Debout, B., "French Research and Technology Program on Advanced Hypersonic Propulsion," AIAA-91-5003, *Third International Aerospace Planes Conference*, Orlando, Florida, December 3–5, 1991.
29. Laruelle, G., "Potential Hypersonic Vehicles Applications," AIAA-91-5086, *Third International Aerospace Planes Conference*, Orlando, Florida, December 3–5, 1991, p. 6.
30. "French Research and Technology Program on Advanced Hypersonic Propulsion," p. 2. Copyright American Institute of Aeronautics and Astronautics © 1991. Used with permission.

# Chapter 10

# Delta Clipper: The DC-3 of Space

## Seeing Is Believing

Just over 2 years after construction began, the McDonnell Douglas Delta Clipper-Experimental—DC-X—made its first historic flight on August 18, 1993. Powered by four RL-10-A5 engines, the DC-X slowly rose from its launch pedestal in the New Mexico desert. Almost effortlessly, it seemed, it ascended to a high of about 30 meters (100 feet) then, as if from a science fiction movie, the DC-X stopped, and moved laterally another 30 meters. The DC-X started to descend, deploying its four landing legs shortly before gently setting down back on Earth. The first flight of the DC-X was complete to the delight of all those involved. Two more flights occurred a few weeks later, one to 60 meters and the other to 160 meters. More flights were planned, but the tests were halted in late 1993 pending a U.S. government funding decision. Whatever the outcome, these first flights demonstrated that operating a reusable rocket vehicle like an aircraft is feasible and, moreover, cost-effective, giving supporters of aero-space planes around the world hope for the future.

On August 16, 1991, the Strategic Defense Initiative Office (SDIO) selected a team led by McDonnell Douglas Space Systems Co. as the winner of the Single-Stage-To-Orbit (SSTO) program, now officially retitled the Single-Stage Rocket Technology (SSRT) program. Although the focus for SDIO (now Ballistic Missile Defense Organization or BMDO) is on a suborbital vehicle for near-term use, the ultimate goal of the McDonnell Douglas SSTO effort is to enable the development of a fully reusable, single-stage-to-orbit launcher that will significantly reduce the cost and fundamentally improve the ease of accessing space. The orbital vehicle will lift off vertically, ascend to space using rocket propulsion, and then return for a rocket-assisted vertical landing a few hundred meters from the launch point. While the SSTO vehicle doesn't have wings, the shape of the airframe permits the vehicle to fly during reentry and achieve a 3,000 km cross-range—*more* than the Space Shuttle Orbiter.

The experimental suborbital vehicle is designated the DC-X, and the prototype orbital vehicle is designated DC-Y.

**Figure 10.1** The Delta Clipper ascends to space (*McDonnell Douglas*).

McDonnell Douglas has christened it the Delta Clipper (Figure 10.1). Delta refers to the company's Delta expendable launcher, and Clipper refers to the Yankee Clipper ships. McDonnell Douglas hopes the Delta Clipper will open the trade routes of space as the clipper ships opened the trade routes of the seaways. Further, the name evokes memories of a renowned 1940s airplane. As the DC-3 revolutionized air travel, so may the Delta Clipper revolutionize spaceflight. As David Webb observes, [1]

> The extent of the anticipation that it arouses can be seen by the fact that the operational Delta Clipper, as the third vehicle to be built, can and will carry the most famous initials ever to grace a transportation device: DC-3. Should the space version even halfway emulate its famous aviation predecessor, we need have no further concerns regarding the commerialization of space. Nearly 12,000 DC-3 aircraft were built in this country [U.S.], and a further 6,500 were built under license in the Soviet Union. This revolutionized air transportation throughout the world. May their successors do the same in the new milieu of space.

For the moment, the initial contract of just $58.9 million will only cover the cost of the experimental DC-X vehicle, ignoring the contribution from McDonnell Douglas [2]. Even though this money seems relatively small in comparison with other programs—*equivalent to launching less than 3 tonnes on the Space Shuttle*—it may well be the most cost-effective investment ever made in space transportation. Indeed, from this perspective, the SSRT program might be one of the most important projects in the space industry today.

## History and Background

It is not a new idea to build a wingless but fully reusable space transportation system that takes off and lands vertically using pure rocket propulsion. Tsiolkovski and Valier were early proponents of such concepts (Chapter 5), and these types of vehicles were often described by science fiction writers including Heinlein (*The Man Who Sold the Moon*, 1941), and in popular culture by Hergé's *Tintin* rocket (*Objectif Lune*, 1953) which can often be seen adorning shop windows in Paris.

Like so many good ideas in space transportation, the early promise of the vertical riser concept succumbed to overwhelming technological constraints. In the early days, for example, many organizations believed that controlled nuclear fission might be the solution they needed. One of the first serious attempts to design a technically feasible nuclear powered Delta Clipper-type vehicle was that of the Douglas Aircraft Co. According to Maxwell Hunter, who was chief design engineer of the Douglas Missiles Engineering Department at this time, and the person who ultimately would be behind the birth of the SSRT program, [3]

> A few months after Sputnik startled us, Douglas, like most companies, formed a "Space Committee" to basically try and figure out "Now what had we gotten into?" I was Chairman of that committee . . . It wasn't long until we looked into nuclear rockets as well as everything else. We were highly polarized to looking for something that if it were implemented would constitute a major leapfrog of the Russians and wrap up the whole of space, at least as far as the Solar System was concerned. Nuclear rockets seemed an answer.

The vehicle they came up with was designated RITA which stood for Reusable Interplanetary Transport Approach. The vehicle was about the size of the current Delta Clipper vehicle, but different in mission capabilities. Because the specific impulse of such a nuclear engine can be about double that of the best cryogenic fuelled engine at about 900 seconds, RITA would, in theory, have not only been able to get into low Earth orbit, but also would have been able to take a full payload from the surface of the Earth directly to the surface of the Moon, and then back to the Earth again for a vertical landing. According to Maxwell Hunter, RITA had a number of technical problems that put a question mark over its feasibility. Perhaps the most significant were the structural and thermal protection system design problems. In any case, RITA was always going to be politically infeasible because of its nuclear propulsion system.

After RITA in the early 1960's, Philip Bono [4], also of Douglas, began to explore the possibilities for a variety of chemically fuelled vertical riser concepts (Figure 10.2), including monster designs such as the Rhombus (1964) concept for a semi-reusable vehicle that was intended to place between 400–500 tonnes into low Earth orbit. Another Bono design that attracted some popular attention was the Hyperion (1969) vehicle that was intended to be launched from a sled travelling along a rail system built on the side of a mountain. Perhaps somewhat surprisingly, one reason for pursuing these massive concepts was that they were considered more technically feasible than smaller-sized SSTOs. For example, the electronic components for a very large vehicle need not be significantly heavier than for a smaller vehicle. Also, a doubling in the propellant volume requires somewhat less than a doubling in the amount of structural mass needed to support it, due to the laws governing the scaling of volumes. Phil Bono did, however, later make some of the first studies of much smaller Delta Clipper-like vehicles, including derivatives of the Saturn 5 second stage.

Douglas was not the only company exploring vertical takeoff and landing launch systems. One of the most interesting efforts was Chrysler's SERV [5] vehicle which it proposed as its Space Shuttle concept in 1970/71. The SERV concept was actually a two stage design. It utilized a very large SSTO-type blunt first stage that would take off and land vertically, and a lifting body for the orbiter that would land horizontally. A similar concept had been proposed by NASA and Douglas Aircraft during the same timeframe.

Boeing (1976) and Rockwell (1983) also got into the act with proposals for enormous SSTO vehicles for large-

1960 Rita (not shown)
1964 Rhombus - Douglas (13)
1965 S-IVB - Douglas (5)
1967 SASSTO - Douglas (6)
1969 Hyperion - Douglas (9)
1969 Pegasus - Douglas (10)
1969 Beta - MBB(Germany) (8)
???? Nexus - Ehricke (14)
1970 Unnamed - NASA/OART/MAD (11)
1971 Serv Chrysler (12)
1972 ATV - NASA Marshall (7)
1974 Phoenix L - Hudson (4)
1976 Big Onion - Boeing (17)
1976 Phoenix L' - Hudson (3)
1977 Unnammed - NASA-JSC (16)
1982 Phoenix C - PacAm (2)
1983 Unnamed - Rockwell (15)
1985 Phoenix E - PacAm (1)

**Figure 10.2** Comparison of vertical riser concepts (*Max Hunter*).

scale space activities including the deployment of future solar power satellites. (See Chapter 5.) SSTO-type work was not solely restricted to the United States. MBB (now DASA) in Germany had performed extensive work on the Beta I and II concepts (Figure 10.3) for a modestly sized SSTO vehicle during late 1960's and then later in the mid-1980's [6]. Alan Bond also spent many years pursuing SSTO-type designs before eventually becoming convinced that the winged approach was more suitable.

Apart from an occasional study, much of the enthusiasm for vertical risers faded in the 1970's because the development of the Space Shuttle ruled out the prospects for funds for some time. The resurgence of interest in Delta Clipper-like vehicles began slowly in the early 1980's, primarily as a result of the efforts of Gary Hudson and his family of Phoenix vehicles, the first of which, the Phoenix L, he had actually proposed in 1976. Later concepts were proposed under his company's name Pacific American (PacAm) in 1982, the Phoenix C, and in 1985, Phoenix E [7]. PacAm's aim was to obtain commercial funding to develop these vehicles. In particular, the Phoenix E was dedicated to exploiting the potentially enormous, untapped market for space tourism that many people consider exists once the costs can be brought down [8]. PacAm signed up Society Expeditions to market the Phoenix E, and many people sent hundreds of dollars as earnest money to reserve a seat for the first flights.

The private funding approach and space tourism bent to PacAm's efforts probably did little for the credibility of the vertical riser SSTO concepts. Indeed, when it became clear that raising the money for Phoenix E was going to be significantly more difficult than PacAm first thought, the earnest money was returned to the would-be space tourists. It is also important to remember that by the early to mid-1980's it was becoming increasingly clear that the Space Shuttle was a far cry from what many had been led to believe it would be. Therefore, proposals to develop economically operated Delta Clipper-type vehicles fell on ears deafened by reports of the staggering financial burden of the Shuttle. Since the relatively conventional Shuttle was so far off the mark, what possible chance would there be for a brand new unconventional vehicle whose feasibility wasn't even proven? or so the arguments went. Even though Hudson's Phoenix ideas did much to rekindle interest in smaller Delta Clipper-type vehicles, the private funding approach was not likely to be the best answer when it came to turning ideas into hardware. Another approach had to be found.

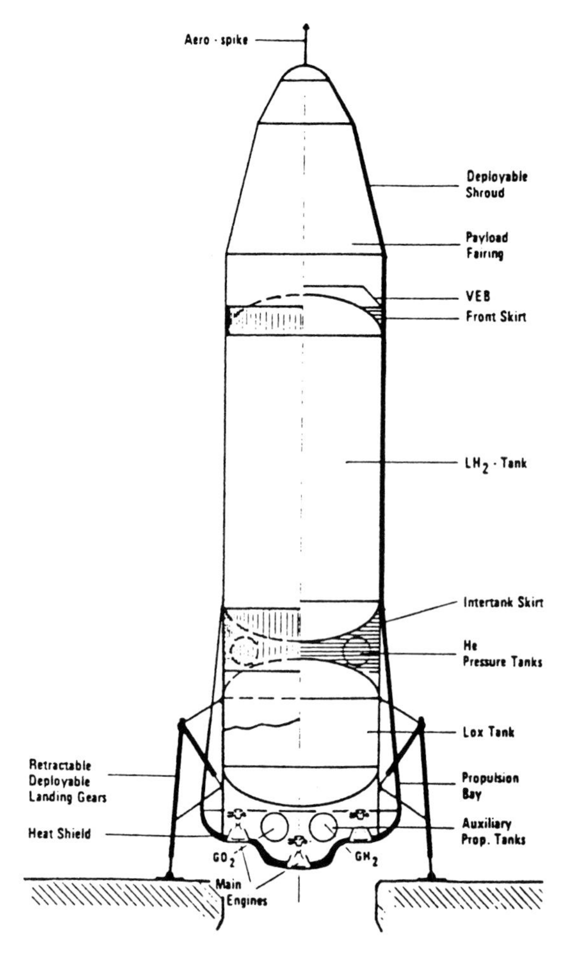


**Figure 10.3** Beta 2 concept (*DASA/MBB*).

## *From the X-Rocket and SSX to SSTO*

In early 1985 Maxwell Hunter, now at Lockheed, picked up from where he'd left off in the early 1960's with RITA and began the eXperimental Operational Program or XOP, which also went by the name X-rocket analogous to the X-planes. As Maxwell Hunter notes, [9]

Along about 1985 . . . , I felt that the Shuttle had occurred long enough ago that a new effort might be tolerated . . . It seemed like the time might be right to take a serious shot at a single stage. I was convinced it should be a reasonably small vehicle.

Like Robert Williams's pursuit of NASP, Hunter's enthusiasm for the X-rocket program was total, as he explains further,

Up until the time I "retired" from Lockheed, this story was extensively peddled to anyone who would listen. There were two major interactions with the Space Division, talks at AFAL and Albuquerque, some discussion with USAF types in DC, briefings to Bill Graham, the President's Science Advisor and Tom Rona of that office, and many talks with Charlie Cook of the Air Force Secretary's Office. In March of 1987, I spent a whole day (eight briefings) in Colorado Springs. Col. Charlie Heimach spent about 2 hours on it before heading to Space Division. Cook arranged in the spring of 1987 for a briefing to Pete Aldridge, then Secretary of the Air Force . . . Ivan Bekey of NASA (now detailed to the Space Council Staff) became interested and arranged briefings to Dick Truly in December of 1986 and Dale Myers in February of 1987 . . . One of the heaviest interactions of all was with SDIO and started in late 1986 while I was still at Lockheed. I was asked to work with the Innovative Architecture Blue Team under Frank Hammil. He asked me to talk about space transportation at the first meeting. So I unloaded the whole show on them, even with such potential competitors as Martin there. This caused an escalating series of presentations which continued after I left Lockheed and eventually reached Abrahamson [then director of the SDI Program].

After a 2 year effort, Lockheed decided to discontinue support for the X-rocket. However, when Maxwell Hunter retired he decided to take it with him and initially renamed it Single Stage eXperimental or SSX. Maxwell Hunter's "general pattern of harassing the system at any opportunity," as he put it, continued at an increased pace. A major stimulus to the SSX concept came from the meetings of the Citizens Advisory Group for Space organized by science fiction writer Jerry Pournelle. In late 1988 Jerry Pournelle organized a meeting of a subset of the Citizens Advisory Group in which both Hunter and Hudson would discuss their ideas. Attending that meeting was General Dan Graham, director of the High Frontier, the Washington lobbying group originally established to support SDI. During this meeting, General Graham decided that the High Frontier should support the SSX ideas and said he would use his offices for this purpose. Also, the name of the SSX was changed. SSX now stood for Space Ship eXperimental [10] (Figure 10.4).

General J. Milnor-Roberts takes up the story of the High Frontiers involvement in the SSX, "Early on it became obvious that the cost of putting up a space-based [defense] system was very sizable simply because of transportation expenses—i.e., $8,000/pound for the Shuttle to $5,000/

**Figure 10.4** Space Ship Experimental (*Max Hunter*).

pound for the Titan. Thinking along that line we were quite open to a long-experienced engineer in space from California [Max Hunter] who came in and said, over two years ago, 'because of the work done on NASP there is now a material available to build the body of a space vehicle that can be a single-stage-to-orbit.' Of course, the reason why we hadn't done it before is because we haven't had the materials strong enough to hold the load and the fuel. That's why we've been throwing parts away like a totem pole, also it's the reason why it is so damn risky. For about another year, we started to develop this idea and we came up with just an outline of what we were looking for and it was called SSX. In February 1989 we got Mr. Quayle, who needs a few brownie points in his quarter, and we also got the head of the SDIO office interested in it."

Maxwell Hunter recalls his meeting in 1989 with the vice president, [11]

In mid-February, Graham, Pournelle and I briefed Dan Quayle for about 25 minutes . . . The SSX story, coming out of the blue, is a rough show for an experienced rocket engineer to absorb, so I have to believe the Vice President missed a few details. He did state, virtually at the beginning, that they were forming the [National] Space Council then and he would ask it to look into this as soon as it was in place. To me, that was all we needed from him at that point.

The meeting with the vice president cleared the way for the National Space Council to assess the SSX concept. Hunter had known Mark Albrecht, the then head of the NSC, for 10 years and explains, "Before, Mark Albrecht in his speeches would mention NASP-type transportation systems, but always within the context of a heavy-lift booster. Now he mentions NASP and SSTO together, and I think this startled some people."

A few months, and several press conferences and briefings later, SDIO finally decided to pursue an SSX-type program under the generic title Single-Stage-To-Orbit (SSTO). The "Statement of Work for the SSTO Technology Demonstration," released to industry in July 1990, stated, [12]

The Department of Defence and the National Aeronautics and Space Administration have sponsored programs in the Strategic Defence Initiative (SDI) and the National Aerospace Plane (NASP) that have delivered impressive advances in propulsion, avionics, structures, and materials technologies. Recent Government and private industry assessments indicated that a significant payoff may be available in applying these technologies to a reusable, single stage to orbit (SSTO), rocket propelled vehicle with significantly reduced servicing and integration requirements and costs.

SDIO was giving the "good words" of Maxwell Hunter and others a run for their money.

## The SSTO Rationale and Design Drivers

The SSRT program is not a research program like NASP, but is intended to take advantage of recent technical advancements from the NASP, ALS, SDI, and other programs, in order to develop an operational and fully reusable space transportation system capable of low-cost, affordable, reliable, and available access to and from space. Achieving these capabilities dictates a vehicle design with essentially the same characteristics as an aircraft. Specifically, this means a robust configuration affording wide operating margins, including a continuous abort spectrum, engine-out capability, and all weather operability. Such a vehicle also demands a high level of supportability to minimize the turnaround time between flights. This is achieved through building a vehicle that facilitates routine servicing and maintenance operations, allowing rapid changeout of components as necessary. These and other design goals are listed in order of priority by the SDIO in the original statement of work, [13]

- *Continuous Abort Spectrum*: Fail-safe operation, engine-out capability during all portions of powered flight, and flight crew escape capability during ascent and re-entry,
- *Aircraft-Like Operations*: Simplified vehicle design with cost-effective integrated logistic support for servicing requirements to reduce turnaround between flights of the same vehicle to a maximum of 7 calendar days and 350 man-days.
- *On-Orbit $\Delta V$*: On-orbit maneuvering velocity change (delta V) of 600 ft/sec [185 meters/sec] in addition to the re-entry delta V.
- *Manned Flights*: 14.7 psia cabin pressure during manned flight.
- *Surge Capability*: Launch rate surge to double the routine launch rate, to be maintained for a 30 day period.

The SDIO rationale was to define the required launch services, and it left it to the contractors to decide on the type of vehicle needed to best meet this need. SDIO was looking for a cost-effective service, not a specific vehicle. For example, SDIO did not define the payload requirements directly in the statement of work. Instead, it gave each of the contractor teams a classified mission model and left it to them to decide on the optimum payload capability. Not surprisingly, this eventually resulted in a medium payload size to and from low Earth orbit up to 12 tonnes, depending on the orbital altitude and inclination. This range is precisely that of all other aero-space plane programs under consideration today. The current nominal payload capability of the Delta Clipper is about 11.4 tonnes to a 400 kilometer orbit inclined at 28.5°.

Achieving these requirements should enable launch costs to be reduced drastically compared with existing launch systems. Already, many people in the program believe that a dedicated launch cost as low as $5 million is realistic, with weekly flights of individual vehicles. But, as the former Program Director Col. Pat Ladner explains, "suppose we're an order of magnitude out and we only achieve a 70 day turnaround instead 7 days . . . then with a fleet of 4 or 5 vehicles this still provides an availability of one flight every couple of weeks." This, he suggests, would still be far superior to all other launch systems where availability is usually measured in months or years. And if the launch cost goal of about $200 per pound ($440 per kilogram) were missed by an order of magnitude, bringing it to $2,000 per pound this would still be significantly better than current launch systems, especially if the other user-friendly operational features, such as payload return to Earth, of the vehicle are taken into consideration.

## *The Uses of the Delta Clipper*

The program is driven by a recognition that current launch systems are too expensive, relatively unreliable and unavailable, and that this is inhibiting the utilization of space, both for military and civilian space activities. In 1991, the SDIO rationale for the program was as follows, [14]

- U.S. launch systems are capacity limited with little capability to sustain any surge,
- Our manned system suffers from launch delays and on-orbit difficulties due to its complexity,
- Expendable launch vehicles are one-shot affairs that either succeed or fail totally,
- Foreign competitors have seized the bulk of the free world launch market challenging our once unchallenged position,
- Worldwide competition for future space launch markets is heating up with [for example] the introduction of the British/Soviet joint venture to produce a reusable air-launched space vehicle, . . .
- The U.S., like airlines, needs to offer a family of vehicles covering multiple payloads and user needs,
- The potential demands of emerging programs such as space station [Freedom] and the Moon/Mars initiative demand increased and diversified lift capacity, personnel transport, and spacecrew rescue.

Col. Pat Ladner and others involved in the program cite one of the most prominent missions for Delta Clipper as being the launch and support of constellations of satellites including GPALS (Global Protection Against Limited Strikes), Navstar 2, and Iridium. (See Chapter 18.) Indeed, the SDIO office believes that if the prototype vehicle meets the hoped-for flight rates and launch costs, then on the successful completion of the test program, its total development, production, and operational costs could be fully recovered just by launching the new Navstar 2 Block 3 satellites. In other words, it might be cheaper to develop the DC-Y and use it to launch the Navstar 2 satellites, than it would be to use existing launch systems. Further, these replacement satellites would be in place far more rapidly than is possible using expendables. For example, the current series of 24 Navstar satellites are being launched by individual Delta 2 boosters at a rate of about six per year over 4 or 5 years. Potentially, a flight-proven Delta Clipper could do the same in under a year, if the payloads were ready. This scenario is, perhaps, somewhat optimistic and would certainly necessitate at least two prototype vehicles. It at least glimpses the magnitude of the potential such vehicles might realize.

The constellation approach to space operations in low Earth orbit would benefit from "the ability to inspect and service spacecraft in orbit," according to Pat Ladner. "If a failure occurred in Sector 23, for example, then we could launch an SSTO vehicle immediately to inspect, repair, or retrieve the failed satellites." Certainly, even if the Shuttle could launch frequently enough to support such missions, it is still far cheaper today to launch a new satellite each time instead of retrieving it from orbit.

It is important to stress that although SDIO's motivation for pursuing the SSRT program stems originally from military launch requirements (i.e., Brilliant Pebbles and GPALS), the Delta Clipper is not intended for exclusive military use, as is apparent from the requirements listed earlier. Nevertheless, this has led some to believe that the future prospects for SSRT are intimately tied to a decision to deploy a Brilliant Pebbles-type antiballistic missile defense system. Thus, as such a deployment is not at all certain and, in any case, is illegal under the terms of the ABM Treaty, some organizations have concluded that the future of the Delta Clipper must also be equally uncertain [15]. According to program officials, however, SSRT is not justified by GPALS, as this mission can be completely supported by the existing Delta and Atlas boosters. It is more a matter of the SDIO Test Offices being the best place to coordinate and support the high technology needs of the program. It is also reasonable to suggest that SDIO was the only organization willing to take on this new program, given that NASA and the Air Force have their hands full with NASP, NLS, and existing launchers.

Certainly, the cost of lifting and supporting the requirements of a space defense system has provided the SDIO with an acute awareness of the need for an inexpensive, available, and reliable launch capability. The ALS program was initiated by SDIO for these same reasons, back in the days when options being contemplated for SDI included the launch of an enormous space laser and particle beam battle stations (See Chapter 3). The launch needs of SDI today are fundamentally no different than any other space activity. The issue is purely one of cost-effective transportation—just as a truck can carry a piece of military hardware and that same truck can also carry a refrigerator.

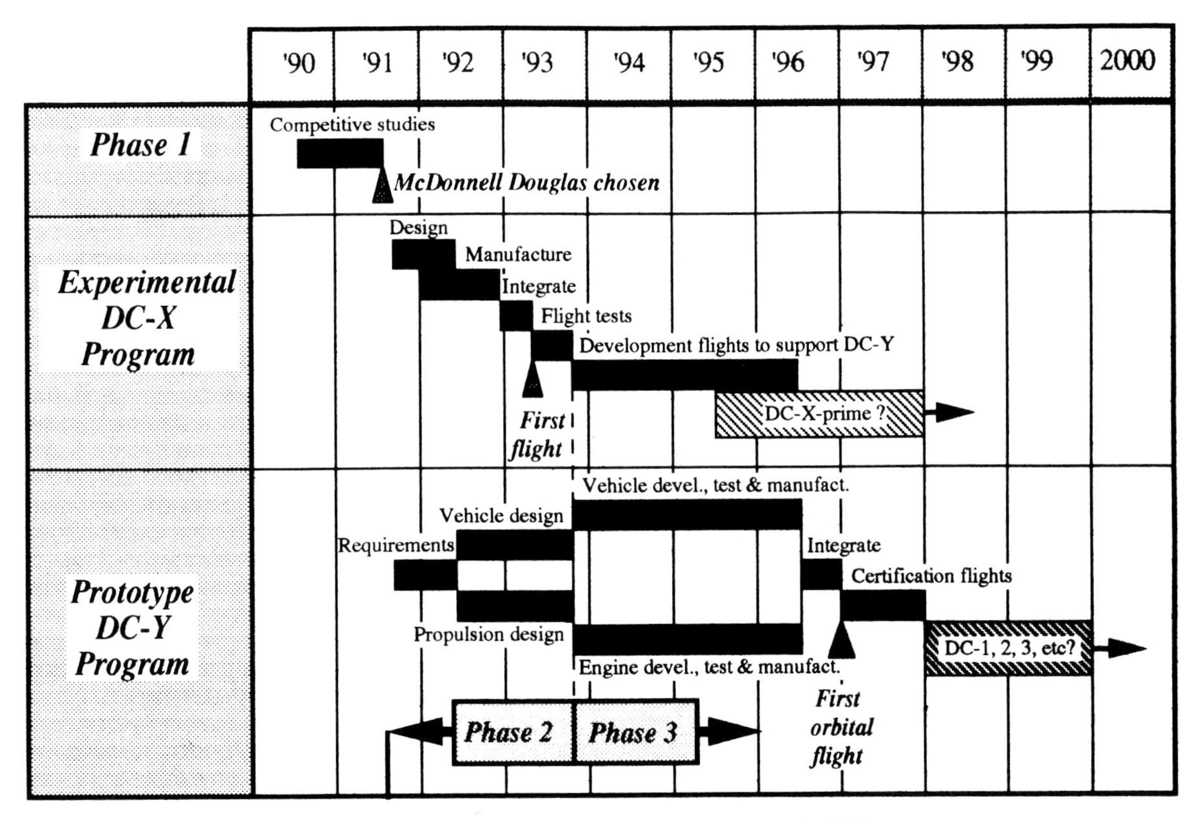


**Figure 10.5** Program schedule for the DC-X and DC-Y.

## Fast Track to Orbit

The Delta Clipper would provide a launch capability much needed today. Therefore, at the start of the effort SDIO had no intention of dawdling through the SSRT program. This was clear from the beginning of the program as noted in the Statement of Work, [16]

- Suborbital technology demonstrator (DC-X)
  First flight in 1992
- Orbital prototype (DC-Y)
  First flight in 1994
- Operational fleet (DC-1, 2, 3 . . . )
  Availability in 1996/97

Four years into the program, these dates have been revised somewhat to reflect budget realities and program redirections. The DC-X made its first demonstration flight in 1993. The first flight of the Y vehicle would be achieved by 1998 or 1999, according to McDonnell Douglas. Since 1992 the U.S. Congress has stopped the SDIO from funding work on the orbital DC-Y vehicle, McDonnell Douglas is studying the DC-Y with its own funds.

Even by modern aircraft standards, this appears a fast-track program if it were achieved. In all likelihood, technical problems will enter into the equation and delay development activities further, but this should come as no surprise. Building in a management process that aims to achieve an orbital capability as soon and as effectively as possible is what is important. SSRT is "lean and mean."

The schedule of the program, as it was in early 1992, is represented more concisely in Figure 10.5 [18]. The initial Phase 1 activity lasted 10 months beginning in August 1990 and completed in June 1991 at a cost of $15 million [19]. Four teams were involved in the Phase 1 effort which focussed on the definition of a number of concepts and was also used to develop plans and requirements for the Phase 2 work. On August 16, 1991, the team led by McDonnell Douglas was ultimately selected. The McDonnell Douglas program manager for the Delta Clipper is Dr. William Gaubatz.

Phase 2 has begun and the DC-X has been built and performed three test flights by end 1993. This was a 2 year effort funded at a total of $58.9 million and involved two primary activities. The first was the construction of the DC-X experimental vehicle. McDonnell Douglas delivered the DC-X to White Sands Test Facility in mid-April 1993, only 21 months after the contract award. The second major activity was to have involved detailed design

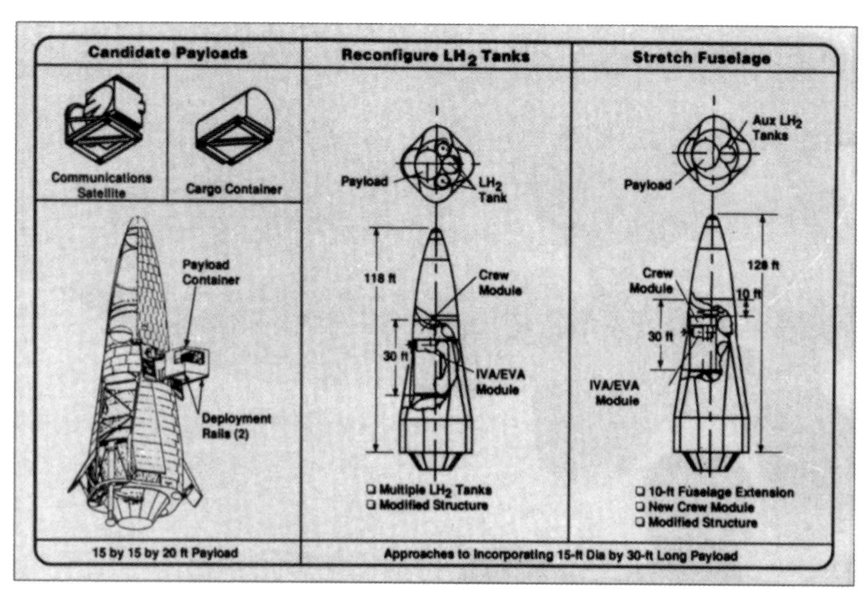


**Figure 10.6** Options for stretching the Delta Clipper (Note the containerized payloads) (*McDonnell Douglas*).

for the DC-Y prototype. The intention was to have the DC-Y almost fully designed prior to the final phase of the program.

Phase 3, currently planned for late 1993, was to have seen continuing flights of the DC-X vehicle to support the final design activities and construction of the DC-Y airframe and propulsion system. If funding could be found, the prototype could be integrated as early as 1998 or 1999, although the exact timing is contingent on early funding for the Delta Clipper's engine development program. If the early flight tests are successful, the program expects to gradually transition into an airliner-like certification mode to qualify the vehicles for the follow-on construction of an operational fleet of Delta Clipper vehicles.

Even though this schedule might appear remarkably quick relative to today's space programs, there is no fundamental reason—perhaps apart from the propulsion system —why it shouldn't be otherwise. The McDonnell Douglas structural design is a relatively simple in comparison with air-breathing approaches. Where the shape of a winged vehicle must be carefully sculpted to provide optimal lift and minimal drag at all speed ranges, a vertical riser can tolerate a much more blunt and symmetrical shape that is obviously more structurally efficient.

As another example, because of the longer time spent in the atmosphere, ramjet/scramjet air-breathers will be exposed to dynamics pressures as high as 70–140 kNewtons per square meter over their complex fuselage/wing shapes. By constrast, a Delta Clipper-like vehicle will experience a maximum of about 30 kN per square meter over its relatively simple symmetrical shape. As a result, the structural demands of a Delta Clipper-like vehicle are significantly less severe than for a high speed air-breather [17].

The problems with having to develop and test brand-new, high-speed air-breathing engines are also avoided. One of the principal reasons for the experimental X-30, is that its scramjet engines cannot be ground-tested beyond Mach 8. The propulsion system development for the Delta Clipper is still a major activity, but is confined to known rocket engine cycles which can be fully ground-tested and characterized long before installation on the vehicle. Further, the relationship between the engine and the vehicle's airframe is far less severe compared with a scramjet-powered design where the airframe is as much a part of the engine as the engine itself. The Delta Clipper engine/airframe integration problems are minor by comparison. This is essentially the same situation for Interim-HOTOL (See Chapter 7).

These differences are important for other reasons. If, for example, during the DC-Y development, unforeseen problems were to arise that resulted in a lower payload performance than expected, such as through weight growth or propulsion system performance shortfalls, then the later operational vehicles could be readily stretched, analogous to the way jet airliners are often stretched by adding new sections to the fuselage (Figure 10.6). With most air-breathing aero-space plane concepts, such simple stretching is close to impossible, which means that the configuration cannot be readily modified into a payload-carrying vehicle unless it is oversized from the very beginning. This inherent "stretchability" allows the construction of a prototype orbital vehicle that can simultaneously double as an orbital experimental vehicle. Of course, the SSRT program still needs a dedicated experimental vehicle (DC-X) of sorts, but only a relatively simple subscale vehicle to demonstrate various basic operational and supportability capabilities, including vertical rocket assisted landings and turnaround of a cryogenically fuelled rocket. By contrast, scramjet-powered aero-space plane programs require a large-scale and very costly experimental vehicle first.

## The First NASP-Derived Vehicle

The SSRT program is not without risk. Still, relatively few show-stopping problems appear to stand in the way of building the prototype vehicle. In fact, the Phase 1 effort lasted only 10 months and cost the U.S. government just $15 million. Further, at the end of this period it was possible for the SDIO to down-select one concept from the four proposed. Therefore, the question is "if it is so cheap and easy, why wasn't it done a long time ago?"

Looking at the SSRT program from this limited point of view ignores the massive technology development effort in other fields that had to occur first. It is only because of NASP that the Delta Clipper is now feasible, because NASP spent the money on the enabling technologies in the first place. If the NASP program had not existed, then the "hundreds of millions of dollars" spent to develop the lightweight/high temperature/high strength materials needed for the X-30 would have had to have been spent by SSRT. NASP paid for much of the required technology which, because NASP is a research program, is precisely what it was supposed to do anyway. In a very real sense, the Delta Clipper is the first NASP-Derived Vehicle.

Still, more than NASP was required, and many other programs have contributed to the required technology base needed. SDI contributed to the development of advanced materials and structures. The advanced health diagnostics monitoring system developed by the original ALS program is being applied, as are the advanced avionics systems developed for the current generation of military aircraft such as the B-2 and the Advanced Tactical Fighter, although as one program official pointed out, "the avionics requirements for the B-2 are an order of a magnitude more complex than those required for the SSTO."

## Management Philosophy: Lessons From the Past

Now that most of the technology appears to be in place to design the Delta Clipper, can building an experimental demonstrator and a fully functional prototype orbital vehicle also be rapid? Maxwell Hunter provides what he feels are good reasons to believe it can, [20]

> Under no circumstance will such a program develop technology for technologies sake. That is someone else's job. On the other hand, if advanced technology really solves the problem beautifully, the sharp program will not hesitate to use it. The question of being *wise* about technology is what mitigates against the use of independent review committees. Such committees once consisted of design engineers looking to get on with it. But in our modern world, they consist of researchers and academics who have a vested interest in seeing more laboratory work performed. The natural recommendation is to delay, and spend more money on basic information *to be sure you're right!*

These are hard-hitting words and how true they really are can, of course, be strongly debated. Advanced programs can be undertaken rapidly if there is the will to do so. As Kelly Johnson explained in reference to the Lockheed Skunk Works, [21]

> . . . the XP-80 jet fighter, was built with just 120 people in 143 days. There were only 23 engineers on the project. There were 37 engineers on the JetStar corporate transport. The U-2 many years later employed a total of 50 people on both experimental and production engineering. On the enormously more difficult SR-71, there were only 135 engineers.

Kelly Johnson's SR-71 is one well-known example that is frequently used. According to an article by Jess Sponable and Gary Payton in *Aerospace America*, [22]

> Despite its use of unproven structural material (titanium) and Mach 3.3 engines, the earliest variant of the SR-71 actually flew for the first time only 30 months after the program was started, and the SSTO vehicle designs use only selective new technologies and are potentially easier to build [than the SR-71].

The SSRT team sees no reason why its vehicle cannot be built quickly and relatively inexpensively because the vast majority of required technology is already in place, and precedents for minimizing the management overhead are available. The one major exception is the propulsion system. Even though the SR-71 was built in 30 months, the basic engine already existed. The same is not true for the Delta Clipper, as will be discussed shortly.

SDIO adopted a hands-off managment approach for SSRT. At the beginning of the program each contractor was allowed to examine various options for the first 4 months with the minimum of interference from SDIO in terms of meetings or reviews. Then, after the concept review, the contractors were given a further 6 months to refine their preferred options, after which the Phase 1 presentations were made. On this basis SDIO was able to select a winner. For Phase 2, SDIO will maintain a rapid management approach that includes streamlined management communications, a small government team, nonrestrictive contractor practices, incentives to encourage innovation, and a focus on producing hardware—not paper [23]. As Pat Ladner explained, it is important to quickly and continuously "demonstrate technology every year" in order to maintain the momentum of the program.

The management approach of encouraging innovation and taking high, but measured, risks was common practice in the days of Kelly Johnson and Ed Heinemann for the development of the so-called black programs. Programs are black primarily to protect technology and military sensitive activities; the Stealth Fighter is a case in point. Being black also provides the opportunity to develop high-risk systems without the constant scrutiny of the media or government accounting organizations. As William Scott notes in *Aviation Week & Space Technology*, [24]

> The SR-71/A-12 program suffered any number of setbacks, including several crashes, that might never have been tolerated in the "white world." But Kelly Johnson and his Skunk Works team were pushing the outer envelope of technology, and risks had to be taken. The spectacular results are now well-known—an aircraft that routinely can cruise at more than Mach 3 and above 80,000 ft [25 kilometers]. The Lockheed F-117 fighter program is another example. In order to achieve a breakthrough in stealth technology, risks were necessary. Several F-117s and their protypes crashed, but the development program continued and the aircraft joined the Air Force arsenal. Can anyone imagine the B-2 program surviving a couple of aircraft crashes and years of fly-fix-fly development? Not a chance.

Turning black is not a panacea for high-risk technology programs, and it is clearly not an option for the Delta Clipper. For the present, the management strategy seems to be working. But as the program completes its first flights of the DC-X and goes in search of more money for the DC-Y, it will be interesting to see how well the situation can be maintained. An accident or setback with the DC-X and DC-Y will cause bad press, even though accidents are inherent in this type of activity. The ability to ride out such failures will be a testimony to the stature of the program and, above all, the determination of program managers and budget appropriators to pursue the ultimate goal of building a true space transportation system.

The actual government management team headquartered at the Pentagon (as of late 1993) is staffed by the equivalent of just "two and a half" SDIO personnel, plus several industry consultants [25]. The management team is supported and advised by personnel from NASA Langley Research Center, USAF Space System Division, NASP Joint Program Office, Phillips Laboratory, Air Force Flight Test Center, and the Wright-Patterson Flight Dynamics Laboratory.

The prime contractor, McDonnell Douglas Space Systems Company, is supported by a limited number of other companies including the McDonnell Douglas Aircraft and Electronics Divisions, as well as McDonnell Douglas Research Laboratories and Douglas Aircraft Co. Also part of the team are Aerojet Propulsion Division, Eagle Engineering, Harris Corp., Honeywell, Pratt & Whitney, and Scaled Composites. The only non-U.S. company is DASA, although both British Aerospace and Aerospatiale were a part of the original McDonnell Douglas team [26].

## *Phase 1 Competition*

Four teams took part in the Phase 1 effort prior to the selection of McDonnell Douglas. Before discussing the Delta Clipper, it is worthwhile to review the other three SSTO designs. This will demonstrate how concepts reflected the strengths of each prime contractor as well as what each considered to be the most important technical issues to overcome. The designs of Rockwell (vertical takeoff and horizontal landing), Boeing (horizontal takeoff and horizontal landing), and General Dynamics (vertical takeoff and vertical landing) provide an interesting discussion of the evolution of the SSTO configuration and illustrate that there may be more than one way to get to orbit with a rocket propelled single stage.

### Rockwell International, North American Aviation

Perhaps the single greatest concern with the feasibility of Hunter's original SSX is the rocket-assisted vertical landing. Of particular concern is whether cryogenic rocket motors exposed to space for a number of days can be safely reignited during the high speed atmospheric plummet. Further problems relate to the storage and conditioning of several tonnes of liquid hydrogen and liquid oxygen within the vehicle during several days in space, as well as ensuring a proper flow of propellant into the engines during the free fall back to Earth.

Rockwell saw these uncertainties as unacceptable and proposed a design that used a small delta wing to allow it to perform an unpowered return to Earth, landing horizontally [27]. Effectively, the mass of the wing replaced the mass of propellant carried by a vertical landing vehicle. The design, would still utilize a vertical takeoff. Not surprisingly perhaps, this concept holds many similarities with the Rockwell-built Space Shuttle Orbiter. Indeed, the configuration of its SSTO, nicknamed the Platypus because of its flat nose (Figure 10.7), does bear an uncanny resemblance to the Orbiter, except for the split V tail configuration and the aero-

**Figure 10.7** Rockwell's Platypus (*Rockwell International*).

**Figure 10.8** Boeing's horizontal takeoff SSTO based on its earlier RASV design (*Boeing*).

spike main propulsion system. This concept was able to draw heavily on Rockwell's TAV studies in the early 1980's. (See Chapter 5.)

### Boeing Aerospace Company

Another concern was the development of the high-performance propulsion system needed to boost the SSTO vehicle into space. Such engines are notoriously expensive and tend to take a number of years to develop, criteria directly in conflict with the low-cost, fast-track approach of the SSTO program. Taking a leaf out of the SR-71 book, Boeing suggested that the SSTO vehicle should use a version of the Space Shuttle main engine modified for greater throttleability, maintainability, and life. Boeing was also able to capitalize heavily on the work it performed during the early 1980's on the Reusable Aero-Space Vehicle (RASV) (Figure 10.8). As Dana Andrews, the Boeing RASV and SSTO study manager, explains, [28]

In 1990 under the SDIO SSTO contract, I re-analyzed the SSTO scene with a new team and got essentially RASV as the best near term affordable SSTO (affordable because it didn't require a new engine).

Thus, the Boeing SSTO was seen as a resurrection of RASV, but updated to take advantage of new materials and structures not available in the early 1980's.

One major problem with the earlier RASV and the SSTO derivative was its mass, about 50% larger than an equivalent vertical riser. The size of Boeing's SSTO was partly due to the large wing area needed to lift the 600 plus tonne vehicle off the ground. The wing on Rockwell's SSTO was relatively lightweight because it was sized for the reentry case when the vehicle weighs about one-tenth its takeoff mass. Thus, because costs are largely a function of the dry mass, the airframe of Boeing's SSTO was likely to be more expensive to develop than the potentially simpler Rockwell SSTO concept. In addition, horizontal takeoff is very inefficient for a rocket-powered vehicle. It results in more propellant and a heavier takeoff mass compared with a vertical riser which is able to head out of the atmosphere as quickly as possible. Alleviating this problem somewhat required Boeing's SSTO to use a powered sled to assist takeoff. The sled not only increased costs, but obviously reduced the operational flexibility of the SSTO by tying it to a particular base. Nevertheless, Boeing felt that the larger airframe and powered sled were small prices to pay compared with the cost of developing a new propulsion system.

After the first half of the Phase 1 effort, it was rapidly becoming clear to Boeing that the horizontal takeoff approach was not winning over the support of SDIO. As a result, Boeing pulled out of the competition. It later agreed to join the Rockwell International team for the last few months of the Phase 1 competition. As Dana Andrews notes, [29]

. . . SDIO was determined to find a new solution and made it quite clear that horizontal take-off was something NASP did. Hence, we decided not to pursue the phase II contract. After our announcement of that fact, Rockwell International contacted us and asked

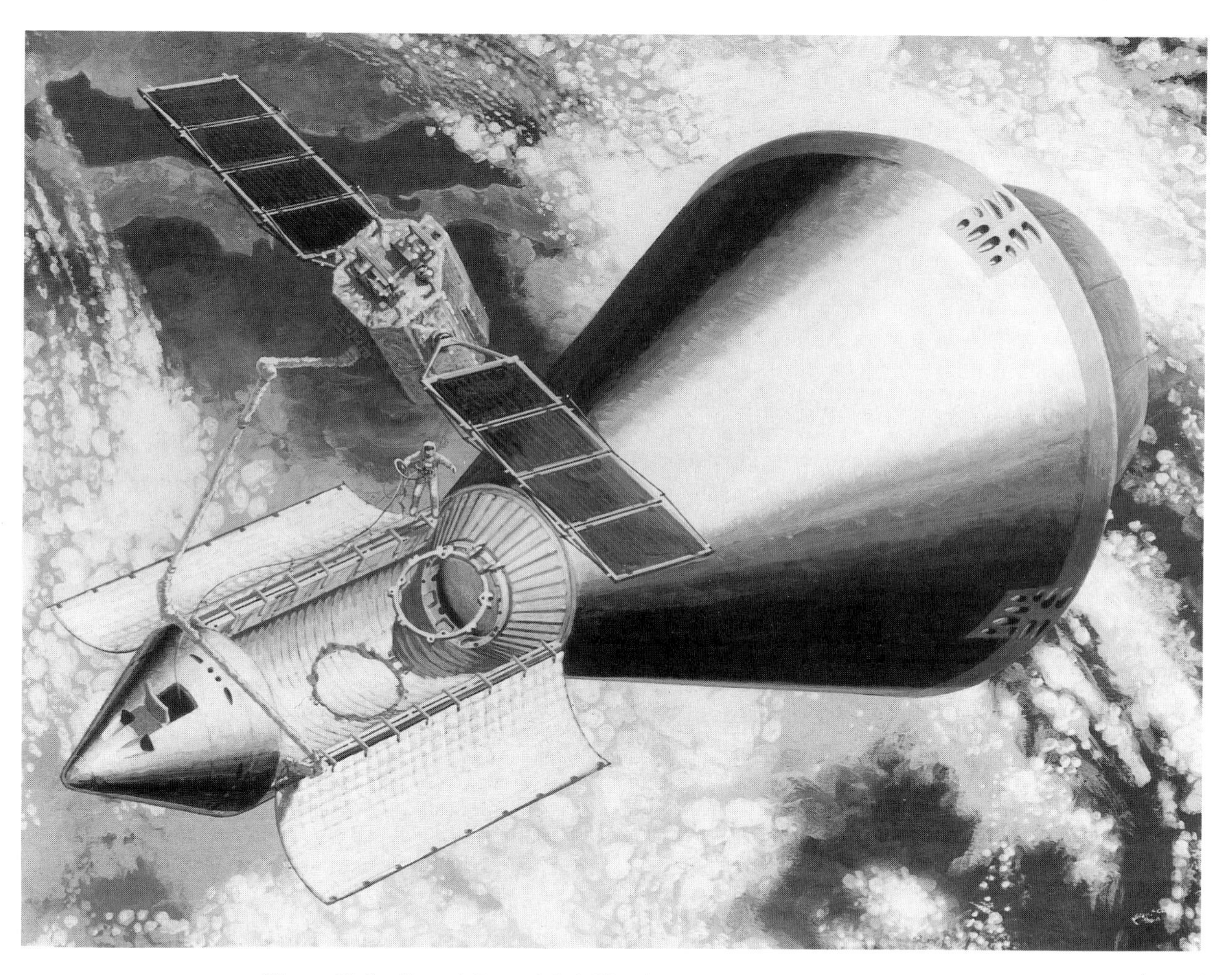

**Figure 10.9** General Dynamic's Millennium Express (*General Dynamics*).

us to join their team. We compared notes, saw it was indeed a good capabilities fit and that we could support their proposed design, so we signed on. Unfortunately, horizontal landing was not a new enough solution for the SDIO and the team lost.

Despite this, Boeing remains confident that single-stage-to-orbit is practical and the most economic means of accessing space. As Dana Andrews states, [30]

We as a company still believe SSTO is doable and will provide the lowest cost to orbit, period.

## General Dynamics

Ideally, a true space transportation system should be as flexible as possible, in terms of where it is launched from, which orbits it flies into, and where it lands. One way to increase operational flexibility is to provide the SSTO vehicle with a lifting aerodynamic shape so that it may fly large distances either side of the ballistic ground track during the reentry phase. In other words, the SSTO does not need to wait until it is in an orbit which passes directly over the landing site, but can choose from a number of orbits that pass within hundreds or even thousands of kilometers of the nominal ground track.

Although SDIO did not specify a cross-range capability in its original Statement of Work, it is clearly useful. However, General Dynamics felt that it was only a desirable capability because its technical and economic impacts outweighed its usefulness [31]. Therefore, General Dynamics pursued a very simple design, called the Millenium Express, that descended from orbit more-or-less ballistically, like Apollo (Figure 10.9). This allowed a simpler shape to be used and reduced the complexity of atmospheric flight, although there were concerns about its stability during reentry. Millenium Express could still achieve a modest cross-range of a few hundred kilometers.

An important technical reason why the General Dynamics concept was not chosen was because of its almost exclusive use of an aerospike engine. This did not tie in well with one of the principal objectives of the SSRT program, which is to have one or more solutions waiting in the wings should the primary options prove inadequate. Evidently, General Dynamics was unable to propose a convincing backup option for the propulsion system, such as one using conventional bell nozzle rocket engines. A related concern was the high heating on the aerospike propulsion system during reentry.

## Designing a Robust Vehicle

As of early 1992, the total liftoff mass of the Delta Clipper DC-Y is estimated at around 590 tonnes (1.3 million pounds), of which the payload consumes less than 2% or about 11.4 tonnes (25,000 pounds) [32]. Given the high risk associated with constructing a vehicle unlike any other, at first glance it might be considered that growth factors will creep into the design that will add to the mass and rapidly eat away at this 2% of payload. However, assessing the feasibility of the Delta Clipper based on the fraction the payload mass consumes of the gross liftoff mass can be very misleading. Even though the current Delta Clipper design has liftoff mass of 590 tonnes at launch, its unfuelled or dry vehicle mass (minus the payload) is only about 48 tonnes. Therefore, although the payload consumes about 2% of the gross takeoff mass, it also consumes 23% of the vehicle's dry mass. Importantly, it is the dry weight where the design uncertainty lies. By contrast, there is no uncertainty over the other 531 tonnes because it is simply propellant, whose mass is purely a functon of the vehicle's dry mass and engine performance. Another way to look at this is that the propellant load cannot grow of its own accord because its design is known precisely.

What all this means is if the dry mass of the Delta Clipper increases by 10% due to the normal type of growth problems most programs experience, then the dry mass of the vehicle increases by just 5 tonnes. The total liftoff mass will increase by around 50–60 tonnes or so, but only because more propellant is required to lift this additional 5 tonnes of dry mass. The current Delta Clipper configuration includes a built-in 15% design margin on the dry mass, a 20% margin on engine mass, and a 2% margin on engine specific impulse, plus oversized propellant tanks in case more propellant is required to account for any dry mass growth. If the dry mass were to grow by 15% between design and manufacture, the propellant tanks will be large enough to allow additional propellant to be loaded to preserve the 11.4 tonne payload capability.

If this design margin is not fully consumed, conceiveably more payload could be carried. However, an overriding objective of the SSRT program is to build highly operable aircraft-like vehicles and not to try to squeeze every last kilokilogram of performance out of the system. This is in stark contrast to expendable rockets and the Shuttle which operate at the limits of their performance. (See Chapter 4.) This is tolerable for today's rockets because they are used just once. But for the Delta Clipper to become a true space transportation system capable of many routine and low-cost flights to orbit, it must have a robust design. This means operating margin. Hence, any extra dry mass margin available will be ploughed back into the margins of other critical systems, such as the engine thrust and performance. *Robustness is a mandatory prerequisite to the success of all reusable transportation systems.*

It is important to note that the dry mass almost exclusively drives the Delta Clipper costs because this is the part that has to be developed, tested, and manufactured. The other 531 tonnes of propellant is made up of liquid hydrogen, which only costs about $3 per kilogram, and liquid oxygen, which costs just $0.10 per kilogram. In this sense, 92% of the vehicle is expendable, but has a recurring cost of just $250,000. By constrast, today's 100% expendable launchers have recurring costs on the order of tens of millions of dollars primarily for the procurement and integration of complex hardware.

## The Overall Configuration

Overall, the Delta Clipper is currently envisaged as about 40 meters tall and 13 meters across the base. The payload bay is 4.61 meters square by 6.15 meters long and situated between the two main propellant tanks (Figure 10.10). Just above the payload bay is a space for a replaceable crew module that can be plugged into the vehicle when crews are required. This module, as presently envisaged, would provide space for a crew complement of two who, during the testing and certification phases at least, would be provided with ejection seats. (See Chapter 16.)

Above the payload bay is the liquid oxygen tank which also acts as the primary load-bearing structure. This tank is currently baselined to be made from an aluminium-lithium alloy. Around and below the payload bay the primary load-bearing structure is composed of graphite-epoxy struts assembled in a lattice-like arrangement. Below the payload bay is the liquid hydrogen tank suspended within this lattice. However, unlike the oxygen tank, the liquid hydrogen tank is nonintegral so that it doesn't take any significant structural loads. The same tank arrangement is also planned for the X-30. The hydrogen tank is currenlty baselined to be made from carbon polyimide ether-ether ketone (C-PEEK), basically the same material as proposed for HOTOL and the X-30 hydrogen tanks. Concern has been expressed over the relative robustness of C-PEEK tanks in operational launch vehicles. Should this be the case, the program will use the same aluminium-lithium material as the oxygen tanks. Another concern was the possibility that C-PEEK might pose a possible leak hazard, so a subscale C-PEEK tank was built during the NASP program. It has already been successfully filled with liquid hydrogen a number of times without experiencing any leaks. (See Chapter 6.) Final results from these tests will help determine whether C-PEEK can be used for the Delta Clipper's liquid hydgogen tanks.

The Delta Clipper is enclosed by an aeroshell made from graphite-epoxy. Over this is the thermal protection system whose precise configuration is dependent on the heating experienced during reentry. Even though the Delta Clipper appears symmetrical, it has a definite surface that takes the brunt of the reentry heating, as does the Shuttle. Nearly all of this reentry surface is covered with a multiwall metallic superalloy. Inconel is a candidate material here. On the cooler leeward side of the vehicle, thermal protection will be provided by titanium honeycomb panels.

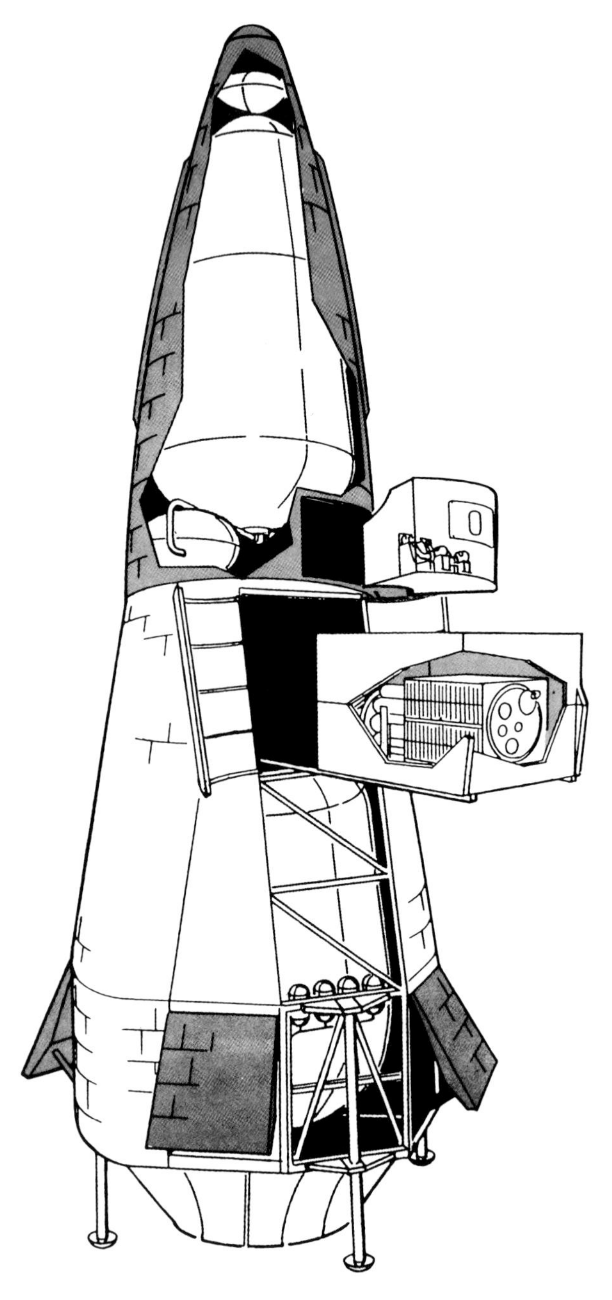

**Figure 10.10** Cutaway view of the Delta Clipper in the original aerospike option (*McDonnell Douglas*).

The nose cap and surrounding region, where the highest heating occurs, are to be covered with a thermal protection system made from carbon silica carbide composite panels. The body flaps at the base are also baselined to use carbon silica carbide. The lighter reinforced carbon-carbon material, as used on the Shuttle Orbiter, has been proposed as an alternative, but it is probably not robust enough for the Delta Clipper's all weather operational design goal. All of these thermal protection panels are isolated from the primary graphite-epoxy aeroshell by means of small stand-off struts, and further protection is provided by multilayer insulation felt blankets. A material called TABI (tailored advanced blanket insulation) will be used to protect the systems under the body flaps [33].

## *The Propulsion Problem*

The original concept for the Delta Clipper's propulsion system, as with the General Dynamics and Rockwell proposals, was to use a linear motor or aerospike. The aerospike consists of a large number of small rocket motors arranged in a ring around the base of the vehicle. Thrust from each individual group or segment of motors is pointed slightly inwards and along a curved surface that effectively takes the place of a conventional rocket's bell nozzle. In this way the individual thrust of each motor reinforces the thrust of the others to form a concentrated spike-like exhaust plume that pushes on the base of the vehicle causing thrust. One advantage of the aerospike technique is that it allows the development of very small engines that, individually, are cheaper to develop and test than the larger and more conventional bell rocket engines. Another important advantage is that the exhaust is automatically adaptable to any ambient pressure.

The aerospike engine was originally thought to offer the potential for a lighter, higher performance, and altitude-adaptable propulsion system compared with conventional bell engines. Because it has never been used on a launch system before, however, its development represents considerable risk and cost uncertainty. In addition, other operational and performance concerns emerged during the McDonnell Douglas analysis. A particular concern focussed on the behavior of the aerospike if one of the segments were to shut down in flight. Exhaust from adjacent segments might spill over into the gap of a failed segment and disrupt the engine's performance. Another concern was whether an aerospike would provide sufficient thrust vector control to affect the delicate rocket-assisted vertical landing. CFD analysis also revealed that the performance of aerospikes might not be as high as first hoped. According to one official, this analysis correlated closely to the measured performance of the aerospike engine tests conducted by Rocketdyne in the early 1970s (Figure 10.11).

Although the aerospike was originally baselined for the Delta Clipper, its operational uncertainties and possible performance shortfalls led the program to discard it in favor of conventional bell engines. The McDonnell Douglas team explored a number of possible options for this engine, including the use of an uprated Space Shuttle main engine [34], but the SSME was not designed for the kind of routine servicing and maintenance needed for the Delta Clipper to achieve rapid turnarounds. Replacing a high pressure turbopump takes several days including the neccessary helium signature testing to verify the seal. The use of high pressure turbopumps also adds a considerable reliability and safety concern.

By late 1991, it was decided that the best strategy would be to develop an essentially brand new engine for the Delta Clipper mission [35]. This engine is likely to use a simple, low-pressure expander cycle similar in size to the Ariane 5 Vulcain engine. (As a backup, a slightly different engine design would use relatively low-temperature and pressure pre-burner turbopumps for improved performance.) The engine is being designed by Aerojet and Pratt & Whitney and is reported to be basically a scaled-up version of the latter company's RL-10 engine, except it is about ten times larger. In addition, four out of the eight engines on the Delta Clipper will be equipped with an extendable nozzle for improved performance at low ambient pressures (i.e., at altitude). This engine is designed for a performance of about 440 seconds, and program officials anticipate that it will cost around $400–700 million to develop and qualify.

A specific impulse of only 440 seconds might appear rather modest for the single-stage-to-orbit mission. Early in the program, it was reported that a specific impulse as high as 470–480 seconds might be required [36], although achieving such a high performance pushes propulsion technology to the limit of the state of the art. The SSME is an appropriate example of the problems inherent in any new technology development. As a result, the McDonnell Douglas team traded vehicle mass for lower performance, but also for a simpler and less expensive engine.

Normally the development of a new large cryogenic engine is a time-consuming process. The Ariane 5 Vulcain engine won't be qualified until 1995, 10 years after the program began [37]. The Delta Clipper team is confident that its engine could be ready within just 4 years given optimal funding. The Pratt & Whitney RL-10 took 3 years to develop and the Rocketdyne J-2, which is similar in size to the Delta Clipper engine, required 4 years, so a 4 year timetable is ambitious, but not unreasonable or precedent setting. Pratt & Whitney's 30-plus years of experience with the RL-10 also provides necessary confidence.

**Figure 10.11** Aerospike engine testing (*Rocketdyne*).

## The Mission of the Delta Clipper

Either manned or unmanned, the Delta Clipper lifts off from its pedestal and begins to accelerate at a modest 1.3 times Earth gravity with the engine operating at 90% thrust to ensure a safe operating margin [38] (Figure 10.12). The vehicle is designed to take off in the same maximum wind conditions as airliners certified under FAA regulations. Should a problem occur at this early point, the option exists to curtail the mission and return back to the launch site. Before landing, the vehicle has to burn off a good fraction of its propellant because, like HOTOL, the undercarriage is designed to support the end-of-mission weight. Although this approach might seem awkward, most large airliners also have to dump much of their fuel if an abort is called shortly after takeoff.

Assuming a safe liftoff, the Delta Clipper continues upward on a ballistic trajectory, although abort options back to the launch site or to orbit are always possible. After the main engines have shut down, auxiliary tanks provide propellant to circularize the orbit at the required altitude. There, the vehicle performs various missions, such as deploying communications satellites and supporting space station activities. During launch, payloads are held within containers, as proposed for the NASP-Derived Vehicle, in order to avoid complexities of integrating payloads to the launcher. Deployment in space is achieved by sliding the container out along rails extending from the payload bay.

**Figure 10.12** A Delta Clipper being prepared for launch while a second vehicle heads for orbit (*McDonnell Douglas*).

After the mission, the Delta Clipper fires its main engines to change the shape of the orbit so that it intersects with the atmosphere. The Delta Clipper actually reenters nose first, compared with most other vertical takeoff and landing launcher concepts that have always reentered base first. While this may seem unique, it is not the first time it has been proposed, as Maxwell Hunter explains, [39]

Although the Rita design first was assumed to enter base-first, it wasn't long till the question of "Why not sideways?" arose. We concluded that it didn't make much difference heating wise and ran a number of wind tunnel tests of both methods to prove the points. Remember, this was 1960!

There are a number of advantages to reentering nose-first for the Delta Clipper compared to those foreseen in 1960. In particular, it allows a long and thin vehicle shape to be used, unlike the General Dynamics concept that required a large base area to evenly distribute the heat of reentry. This shape, coupled to the low vehicle density, enables a semi-ballistic lifting reentry, allowing a cross-range of about 3,000 kilometers to be achieved. The curved nose and flat sides of the Delta Clipper act like an airfoil, and four body flaps around the base of the vehicle control the angle and direction of its glidepath. At hypersonic speeds the Delta Clipper is a very effective glider, and McDonnell Douglas has been able to draw on its considerable experience with boost-glide reentry vehicles to prove the practicality of this reentry technique (See Chapter 5). In addition, a gliding reentry reduces decelleration loads compared to a base-first reentry, allowing a lighter structure and thermal protection system to be used. It also allows conventional bell-nozzle engines to be used as these are more easily protected within the aft-structure of the vehicle. This is not readily achievable with base reentering vehicles.

During descent when the vehicle is about 6 kilometers above Earth, all eight rocket engines are started but initally set at idle. Shortly thereafter, four of these engine are throttled up to about 20% maximum thrust and, at about 3 kilometers, a pitch-up using a vectored engine thrust maneuver will be performed to put the vehicle in its vertical landing configuration. At any point during the powered descent, the Delta Clipper will be able to tolerate as many as two or three engines failing because the other engines could be quickly throttled up from their standby idle setting. Differential thrusting between engines provides control against wind gusting, and also enables the vehicle to be flown to a designated landing point with an onboard GPS-based guidance system. If a crew is on board, they can take over the landing if necessary, just as the Apollo astronauts did for the lunar landings.

After landing, the vehicle is inspected and the crew—if there is one—deplanes. Deplaning can occur relatively quickly after landing because the Delta Clipper does not use toxic hypergolic propellants like the Shuttle. Shortly after landing, the vehicle's onboard diagnostic system transmits a preliminary health status report. This report, coupled with a visual walk-around inspection, is used by ground servicing personnel to prepare for initial maintenance activities. The vehicle is then placed on bogies and towed to its launch pedestal. There, the returned payload can be unloaded and a new payload installed. Full servicing of the vehicle is also performed, with much of the avionics and other systems located under the body flaps and near to ground level for easy maintenance access.

Seven days after landing, according to current plans, the Delta Clipper would be ready for another mission.

## *Certified for Spaceflight*

Building a vehicle that can routinely lift off, ascend to space, and return again for a vertical landing is technically difficult and poses considerable risk. Minimizing this risk is the purpose of the experimental DC-X vehicle shown in Figure 10.13 during its April 3, 1993 rollout. The DC-X was used in the same way experimental aircraft have been used in the past to gradually explore and expand the flight envelope. More specifically, the DC-X demonstrated the basic feasibility of supporting and operating a reusable cryogenic rocket. The DC-X has a fuelled mass of about 20 tonnes, with an empty mass of about 10 tonnes, and stands 12 meters in height with its landing legs deployed.

After completion of initial ground static tests, a series of three flight test programs were planned for the DC-X. The first program started with ground tests where the vehicle is tied to the ground, as has been typical with the first flights of helicopters and even the Harrier AV-8B fighters. As confidence was gained, series one included low-altitude hover testing in windless conditions, vehicle landing and ground effects, and overall system performance, including demonstration of supportability. The second series of flights were planned to include high-altitude (up to 2,000 meters) hover tests to achieve high vehicle dynamic pressure, base drag measurements, simulate the vertical return to launch site abort mode, and control performance in the presence of winds and ground effects. The third and final series was to have included flights up to about 6 kilometers, to achieve higher vehicle dynamic pressures, combined aero/propulsion control, and most critically, the rotation maneuver for transitioning from nose-forward to base-forward flight. All flights involve continuous firing of the DC-X's engines. Importantly, data gathered from each flight was used to modify the overall system supportability, procedures, and schedules [40].

Some of these later test flights may include simulated engine failures to demonstrate various safe abort modes.

**Figure 10.13** The DC-X (*McDonnell Douglas*).

Throughout this testing the DC-X was modified as necessary in a fly-fix-fly type strategy. If finally completed, the SSRT program aims to be able to fly the DC-X as frequently as once every 3 days, thereby demonstrating that a rocket can be turned around like an airplane.

The ground system developed for the DC-X will evaluate all the software and hardware functions required for suborbital and future orbital vehicles. Ground and flight operations will be controlled from the flight operational control center (FOCC). The FOCC is a mobile facility and will be operated by a prime crew of three people. The flight manager is the former Geminii, Apollo, and Skylab astronaut Pete Conrad [41].

The DC-X is not built to the same standard as the orbital vehicle. The propellant tanks are aluminium, and the aeroshell is off-the-shelf graphite epoxy material. Scaled Composites, which made the Voyager aircraft that flew around the world nonstop, is building the DC-X aeroshell. Because the vehicle won't travel very high or even supersonically, reentry can't be demonstrated. According to Maxwell Hunter, "the experimental vehicle has the proper avionics and cryogenic propulsion system, but only about two-thirds of the problems can be solved. The materials and thermal protection system are not covered by the tests because the demonstrator is so cheap. To fill in the gaps, these other things you can do elsewhere, but it is important that you do not hit the heat shield battle unprepared."

At the time of writing, three flight tests were performed in 1993. The SSRT team hopes to complete about 10 flights in 1994, funding permitting. If the program continues into Phase 3, the DC-X will continue to be flown periodically to support the development activities of the DC-Y development and construction effort. In this way, real flight experience can be used to assist the evolution of the DC-Y while it is being put together.

The DC-X flight test program will pave the way for the prototype DC-Y vehicle which, should it be built, would follow basically the same incremental test program, except that the DC-Y will end up in orbit. Once single-stage-to-orbit, abortability, aircraft-like turnarounds, and other capabilities have been demonstrated, McDonnell Douglas envisages the possibility of certifying the Delta Clipper for spaceflight in the same way new airliners are certified to carry cargo and passengers. It might be necessary, for example, for federal aviation authorities to write the procedures that list the type of test campaigns and safety criteria that have to be met before the Delta Clipper is certified for spaceflight. (See Chapter 14.) How adequate the first aero-space plane certification procedures would be is, of course, difficult to judge. At least they would lay the foundations for the standards that future space transportation system would need to be built and tested to before being allowed to routinely carry paying customer payloads to and from orbit. No space transportation system has ever been certified in the way an airplane or helicopter is certified, partly because of the high cost of testing such vehicles, but mostly because they are only used once. Both these constraints disappear with the successful development of an aero-space plane.

## *Funding—Where Does It Leave NASP?*

By late 1993, the BMDO (formerly SDIO) was the sole government funding source for the SSRT program, and officials recognized that if the orbital DC-Y is to be built, funding from other agencies or organizations will be necessary. For these reasons, SSRT was transfered to ARPA in late 1993. ARPA hopes to fund development of a larger version of the DC-X—the SX-2—capable of suborbital flights up to altitudes of 400–500 kilometers [42]. Essentially, this enhanced DC-X would provide ARPA with a reusable sounding rocket capability for short duration technology development and demonstrations in space. This suborbital vehicle would clearly benefit the DC-Y, especially as it would fly hypersonically and be equipped with a reusable thermal protection system for reentry. It would also allow considerable experience to be accumulated in supporting and operating a reusable rocket.

The Delta Clipper is only considered technically feasible today because of advancements made in other programs and, most critically, the developments spurred by the NASP program. Given that the original impetus for NASP was its future promise for building operational aero-space planes with basically the same launch cost and performance characteristics as the Delta Clipper, some have asked "Is it a competitor or a complement to NASP?" At the moment, the technical answer to this question is that it is neither. Both programs have to prove they can achieve single-stage-to-orbit and that they (or their technology) can be used in an operational and economically effective vehicle. Ideally, only then should a decision be made, if one has to be.

Politically, this question may have to be viewed from a nearer term perspective. Funds are notably very tight for all new initiatives today, no matter how worthy they may be. In this light, it is possible to visualize two extreme scenarios. At one extreme, spending $10 billion plus on the experimental and high risk X-30, and then at least that much again to develop prototype and operational launchers, might seem difficult to rationalize compared to the $5–10 billion or so that might be needed to build a pre-operational prototype of the Delta Clipper. In addition, the first air-breathing NDV is unlikely to be buildable much before 2010, whereas the Delta Clipper could conceivably be available by 2000, assuming it too is feasible. At the other extreme, as NASP is justified politically on the basis of pushing technology to the limits—perhaps even more than it is for laying the foundations for true space transportation systems—it may win out over an SSRT program that has no intention of using anything other than existing and proven technology unless it absolutely has to. Also, NASP has the advantage of a far broader national industrial and governmental support base compared with SSRT, which is concentrated primarily on one contractor and ARPA.

If NASP versus SSRT becomes a political funding issue, then the two are likely to become competitors, as indeed, they must compete with all other launcher programs such as the proposed Spacelifter expendable concept. Unless commercial companies are prepared to take over the bulk of the financing, not considered realistic until the feasibility of single-stage-to-orbit has been demonstrated, then a political choice between the two might have to be made. This would be unfortunate as both vehicle concepts have considerable merit. Further, it would seem a pity to shut off one option before the other can be proven as the right way to go. If the Delta Clipper were chosen, this could be interpreted simply as the NASP program transitioning from a long-range *experimental* effort to a nearer-term *operational* space transportation program. This would be ironic, as originally NASP replaced the pure rocket TAV program.

## Summary—The DC-3 of Space?

The SSRT program is not a pure research effort, nor is it being pursued "just for the sake of it." The reason for its existence is to attempt to build a true space transportation system offering the potential to significantly reduce the cost of transporting payloads to and from orbit, while simultaneously making space easier and safer to access. The program also aims to achieve these objectives in a relatively short period of time by evoking the same type of fast-track management techniques used by Kelly Johnson to build and fly the equally demanding SR-71 in just 30 months. In this respect, the goals of the SSRT program are commendable and seem founded on solid rationale.

Achieving these objectives will not be easy. To help minimize risks, the former SDIO and the prime contractor McDonnell Douglas have come up with a phased program that should allow many of the key issues to be addressed early on and at minimum cost. The DC-X is of particular importance because, if the test program is finally completed not only will it demonstrate that rocket-powered vehicles can take off and land vertically and reliably, but that such vehicles can be reused, supported, and turned around in just a few days, like an aircraft.

The DC-3's ability to provide an affordable, reliable, available, and user friendly transportation *service* made it the success it was. The Delta Clipper has the potential to do the same for space. If the publicity the program has already generated is anything to go by, the Delta Clipper stands a fighting chance of realizing that potential.

## References and Footnotes

1. An excellent overview of the SSTO program is provided by David Webb in his *Space News* Commentary, "Building the DC-3 for Space," January 6–12, 1992, p. 15.
2. McDonnell Douglas press release, August 16, 1991.
3. Hunter, Maxwell, "Eve's Name Was Rita," *SpaceGuild*, August 13, 1990, p. 1.
4. Phil Bono's concepts are detailed in his remarkably comprehensive book *Frontiers of Space*, 2nd (revised) edition, co-author Kenneth Gatland, (The Blandford Press, Poole, Dorset, U.K., 1976).
5. Payton, Col. Gary, and Major Jess M. Sponable, "Designing the SSTO Rocket," *Aerospace America*, April 1991, p. 38.
6. Koelle, D. E., and W. Kleinau, "The Single-Stage Reusable Ballistic Launcher Concept for Economic Cargo Transportation," IAF-86-122, *38th IAF Congress*, Brighton, England, October 1987.
7. The Phoenix concepts are described in *Aviation Week & Space Technology* in an article by Richard O'Lone entitled, "Bay Area Firms Pursue Booster Designs," June 25, 1984, p. 169. Hudson was quoted as saying "We are trying to get the price down to outrageously low levels . . . we have the potential of reaching $10 per pound."
8. Woodcock, G., "Economics on the Space Frontier: Can We Afford It?" Space Studies Institute Update, Vol. XIII, Issue 3, May/June 1987, p. 4.
9. Hunter, Max, "Some History of the SSTO Program," memo to Pat Ladner, September 13, 1990, p. 1.
10. Ibid, p. 2.
11. Ibid, p. 3.
12. "Statement of Work for the SSTO Technology Demonstration," SDIO Office, July 1990, p. 1.
13. Ibid., p. 3.
14. "SSTO Program Facts Sheet," SDIO Office, January 1991, pp. 1–2.
15. This appears to be changing, according to recent statements (January 31, 1992, at the U.N. Security Council Meeting), by the Russian President Boris Yeltsin who is in favor of a global antimissile protection system. President Reagan originally proposed sharing the U.S. SDI capability with the U.S.S.R. in the mid-1980's.
16. "Statement of Work for the SSTO Technology Demonstration," SDIO Office, July 1990, p. 2.
17. According to Maj. Jess Sponable of SDIO.
18. Based on the McDonnell Douglas Delta Clipper brochure and press lease pack and discussions with program officials.
19. As stated by Ron Schena (consultant to SDIO) during the Third International Aerospace Planes Conference, Orlando, Florida, December 3–5, 1991.
20. Hunter, Maxwell, "The SSX, SpaceShip Experimental," Draft II, March 11, 1989, p. 30.
21. Johnson, Clarence L., and Maggie Smith, *Kelly—More Than My Share of It All*, (Smithsonian Institution Press, Washington, D.C., London), 1985, pp. 161–162.
22. Payton, Col. Gary, and Major Jess M. Sponable, "Designing the SSTO Rocket," *Aerospace America*, April 1991, p. 45. Copyright American Institute of Aeronautics and Astronautics © 1991. Used with permission.
23. "SSTO Program Facts Sheet," SDIO Office, January 1991, pp. 2–3.
24. Scott, W. B., "Killing the Spark of Risk Douses the Fires of Aerospace Innovation," *Aviation Week & Space Technology*, January 13, 1992, p. 57. Courtesy *Aviation Week & Space Technology*, Copyright 1991 McGraw-Hill, Inc. All rights reserved.
25. As stated by Ron Schena (consultant to SDIO) during the Third In-

ternational Aerospace Planes Conference, Orlando, Florida, December 3–5, 1991.
26. McDonnell Douglas press release, August 16, 1991.
27. Discussion with Rockwell and SDIO officials.
28. Letter from Dana Andrews of Boeing Defense & Space Group, January 14, 1992.
29. Ibid.
30. Ibid.
31. Based on discussions with industry representatives and program officials.
32. Delta Clipper masses (dry, gross, payload) and margin data derived from discussions with Maj. Jess Sponable of SDIO.
33. Most of this technical information was obtained from discussions with SDIO and industry representatives. An overview description of the Delta Clipper can be found in an article by Frank Colucci, "Launching the Delta Clipper," *Space*, November–December 1991, p. 17.
34. As stated by Ron Schena (consultant to SDIO) during the Third International Aerospace Planes Conference, Orlando, Florida, December 3–5, 1991.
35. Discussions with program and industry officials.
36. Discussions with Col. Ladner in March 1991.
37. "Ariane's Vulcain Engine—1991 A Vintage Year," *ESA Bulletin*, November 1991, p. 67.
38. Technical aspects of the Delta Clipper's mission are derived from the McDonnell Douglas Delta Clipper brochure and discussions with program officials.
39. Hunter, Maxwell, "Eve's Name Was Rita," *SpaceGuild*, August 13, 1990, p. 2.
40. Gaubatz, W. A., P. L. Klevatt, J. A. Copper, "Single Stage Rocket Technology," IAF–92–0854, 43rd Congress of the International Astronautical Federation, Washington, D.C., August 28–September 5, 1992, p. 6.
41. Ibid.
42. Ibid.

# SECTION 3

# IMPLEMENTING AERO-SPACE PLANES

*Making predictions about the future is a hazardous occupation, and is more likely to be incorrect in such estimates than not . . . Any 30-year period of transportation history shows that extensive changes have occurred that were not anticipated by those who were making transportation decisions for the future. For example, the Erie Canal was completed in 1825 and sparked a building boom in which 3,000 mi of these waterways were built by 1840. Yet, 30 years after the Erie, which is 1855, canals were all but extinct, having been replaced by railroads. Similarly, in this century that past 30 years has witnessed technology changes in air transportation, computers, highways, and rail transit that could not have been anticipated 30 years earlier.*

—Lester Hoel, *A Look Ahead: Year 2020*
(U.S. National Research Council, 1989)

*In the last 30 years the airline industry has undergone an expansion unrivalled by an other form of public transport. Its rate of technological change has been exceptional. This has resulted in falling costs and fares which have stimulated a very rapid growth in demand . . . Any other industry faced with such high growth of demand for its products while cushioned from competition would be heady with the thought of present and future profits. But not the airline industry. It is an exception to the rule. High growth has for the most part spelt low profits. Increased demand has not resulted in financial success. While some airlines have consistently managed to stay well in the black, the industry as a whole has been only marginally profitable.*

—Rigas Doganis, *Flying Off Course*
(George Allen & Unwin, London, 1985)

*Accelerated rates of capital formation which took place in transport sectors throughout Europe in the nineteenth century also gave rise to a range of externalities, or spinoffs . . . For example, railways (but also canals) trained labor—engineers, foremen and managers—whose skills (initially acquired in transportation) contributed to the development of other industries.*

—Patrick O'Brien, *Transport and Economic Development in Europe*,
(St. Antony's College, Oxford, 1983)

*The difference between the condition of the country if governmental efforts alone had been made, and that now existing, may be more readily imagined than described. The wisest thing governments have ever done in connection with this subject was to authorize companies to make and operate the steel rivers of internal commerce; and the wisest thing they can do in the future is to judiciously blend an avoidance of officious intermeddling with an honest discharge of their legitimate duties to railway investors, travelers, and transporters.*

—J.Ringwalt (Editor), *Development of Transportation Systems in the United States*,
(Railway World 1888)

# Chapter 11

# The Lessons of the Space Shuttle

## A Remarkable Flying Machine

On April 28, 1991, the Space Shuttle *Discovery* thundered into orbit on a mission for the Strategic Defence Initiative Organization. It was the 40th launch of a Space Shuttle in just over 10 years of service, and the 12th for *Discovery*. Although ascent to orbit was as nominal as everyone had come to expect from the Shuttle program, unknown to NASA, the three main engines of *Discovery* were fitted with cracked temperature probes. These probes are about the size of a pencil and are designed to measure the temperature of the liquid hydrogen and oxygen propellants just before they enter the main engine high-pressure turbopumps. Despite the cracks, the probes held together. Had they not, pieces of the probes could have been ingested into the turbines of the turbopumps, the consequences of which might have been the catastrophic shutdown of the engine and the loss of *Discovery*. It was later revealed that five other Shuttle missions had flown with cracked temperature probes. Critically, even though the probes all retained their integrity, a report had been issued *8 months* earlier warning of potential problems with them. This report only surfaced 2 days before the planned June 1, 1991, launch of *Columbia* [1]. The question this episode raises is: if the Shuttle was allowed to fly with an unknown fault—and one that could cause a catostrophic failure—are there other faults lurking in the system? Like a game of Russian roulette, the U.S. space program has owed much to the random fortunes of luck.

This was not the first time the Space Shuttle has been held hostage to random events, or even faulty temperature probes. On July 29, 1985, faulty temperature probes on the main engines of the Orbiter *Challenger* caused the premature shutdown of the center engine 5 minutes into the ascent. Sensors indicated that the engine's temperature was running dangerously high. In fact, the engine was functioning properly. However, other temperature probes began to show a temperature rise in a second engine. If left unchecked, this second engine would have also been automatically shut down, the consequence of which would have been a risky attempt to land *Challenger* on the other side of the Atlantic. This did not occur because of the diligent analysis of the situation by Jenny Stein (née Howard), a Mission Control officer who quickly recognized that the sensors were the cause of the problems and not the engines [2]. *Challenger* went on to complete a successful abort-to-orbit using its remaining two engines.

On January 18, 1986 only hours after the *second* attempt to land the Space Shuttle *Columbia* was called off due to bad weather at the Kennedy Space Center, ground controllers ordered *Columbia's* crew to abandon an experiment in which they were to have fired the reaction control thrusters (RCS) on the nose of the vehicle during reentry. These tests were intended primarily to evalute propellant dumping techniques; normally only the aft thrusters are fired to control the attitude of the Orbiter during the fiery reentry phase. Analysis had revealed that two of RCS thrusters on *Columbia* were of an older variety and firing them during reentry could, in a worst case, have caused the forward RCS pod to explode, leading to the probable loss of *Columbia*. This potentially hazardous situation was uncovered by chance from analysis of another problem when *Columbia* was in orbit. More importantly, the problem surfaced only *after* the previous two attempts to land *Columbia* had been cancelled due to bad weather at the Florida landing site. As former Congressman Bill Nelson, who was a payload specialist on the mission, said, [3]

> If the test had been conducted as planned, the old nose thrusters would have exploded from the atmospheric pressure of re-entry and *Columbia* would likely have become a giant fireball, evaporating into nothing.

The unpredictable nature of Florida's weather may have just saved *Columbia* and the crew.

These are just three examples where the Space Shuttle might be considered to have been very fortunate. In each case the probability of a catastrophic failure might have been less than 1 in 10. Yet, it doesn't take many 1 in 10

**Figure 11.1** The U.S. National Space Transportation System—The Space Shuttle (*NASA*).

probable failures for one catastrophic event to occur, especially in a system as complex as the Shuttle. This is precisely what happened on January 28, 1986. The reasons for that failure and the ensuing repercussions were profound. Much has already been written about this event. Therefore, a detailed discussion will not be presented here.

The Space Shuttle is a remarkable flying machine (Figure 11.1). In itself, it is an outstanding technical achievement, and this success cannot be taken away from the Shuttle. The Shuttle is the first partially reusable space transportation system. It has demonstrated a combined ability to launch, repair, and recover satellites—a capability that never existed before. The Shuttle has enabled a wide range of experiments to be conducted, and has provided the U.S. with considerable experience in manned spaceflight. However, as a critical *operational* element in the U.S. space transportation infrastructure, the Shuttle can only be regarded as a failure when compared with what it *should* have been.

## *Something It Shouldn't Have Been*

By the end of 1993, why is it that the Space Shuttle continues to be plagued by the types of problems that seem to compromise its safety almost every time it flies? And why does it fly so infrequently, experience significant delays before nearly every launch, and cost so much to operate? The answers to these questions are simple, although the reasons behind them are not so immediately obvious. Today's Space Shuttle *is precisely what it was not supposed to be*. It is not the dependable, available, and affordable means of routinely accessing space and servicing and returning spacecraft as originally hoped. Instead, in the opinion of John Logsdon, "the Shuttle is the biggest policy mistake in the history of the space program, and the U.S. will pay for it for many years" [4].

The experience of the Space Shuttle has important ramifications for the development of aero-space planes, both positive and negative depending on what perspective is taken. Positive, if the lessons learned can be applied to the benefit of aero-space planes. Negative, if the failure of the Space Shuttle to meet its objectives leads people to believe that future reusable space transportation systems will all end up like the Shuttle.

A point of view commonly encountered throughout the space industry is, "If the Space Shuttle failed to meet its planned operational objectives, then why should we expect the aero-space plane to fare any better? Wouldn't aero-space planes suffer the same kind of problems as the Shuttle, such as cracked temperature probes and faulty engine controllers?" On the surface at least, this may appear a reasonable argument and is certainly one that haunts advocates of aero-space planes. Engineers working on the design of NASP-Derived Vehicles (NDV) often state that the launch rates, launch costs, and mission types they are predicting for the NDV are almost identical to those given by the Space Shuttle program throughout the 1970's and even up to the first flight [5]. According to the *Space Shuttle 1980 Status Report*, [6]

. . . a $10.5 million [1971 dollars] per flight operations cost could be achieved when a flight rate of 60 missions per year would be reached.

This launch cost of $10.5 million was an estimate made in 1971 dollars, and by January 1980 had been revised to $15.2 million, also in 1971 dollars, and is equivalent to about $53 million in 1993 dollars. Simply put, this would have meant an annual operating cost of about $3.2 billion. This is not that far away from the current $5 billion budget [7]. The critical difference is that today's Shuttle will never be able to fly more than 8 missions annually—not 60. As George Keyworth notes, [8]

The ability to carry out 24 launches a year looked highly dubious even before the *Challenger* tragedy. If we could manage even half that number now we'd consider ourselves to be doing well.

For comparison, by May 1991, after 10 years in service, the Shuttle flew a *total* of 40 missions. It has continued to suffer many prolonged delays, and the average cost per launch is anywhere from $400 million to $1,670 million, depending on the launch rate, if Shuttle capability enhancement work is included, and whether or not development and production costs are amortized [9].

An understanding of the political decisions that gave rise to the birth of the Shuttle and which keep it alive today is necessary in attempting to assess the future of aero-space planes. After all, the decisions on whether and how to build the first aero-space plane are likely to be political.

## Planning for the Post-Apollo Era

The reasons for the Shuttle's failure to meet its hoped-for operational objectives go back to the political environment that gave rise to it in the late 1960's. At that time, the United States was trying to put together a future space strategy for the post-Apollo era, as it was becoming increasingly clear that the Apollo program, and other programs based on Apollo technology, could not be sustained indefinitely due to their high cost. America was going to win the race to the Moon, and thereby achieve President Kennedy's mandate. But Apollo was an end in itself, and once the race was won it would become increasingly difficult to justify a sustained series of further sorties over the following years.

In February 1967, the President's Science Advisory Committee (PSAC) issued a report *The Space Program in the Post-Apollo Period*. This report was concerned with the institutional issues associated with the planning for a post-Apollo program, and it did not make recommendations on particular hardware or technology development. PSAC did recognize that strategies which revolved around a space station, or other activities needing a frequent and sustained manned presence in space, would require some means to lower the cost of putting cargo and people into orbit. As the report said, [10]

For the longer range, studies should be made of more economical ferrying systems, presumably involving partial or total recovery and reuse.

Shortly after the PSAC study was issued, a much more important and comprehensive strategy report was produced by the presidentially appointed Space Task Group (STG) under the leadership of Vice President Spiro Agnew. Its report was published in September 1969, entitled *Report to the President: The Post-Apollo Space Program: Directions for the Future*. Unlike PSAC, the STG report did recommend specific missions, technologies, and architectures. It was rich in ideas including a permanently staffed space station, a lunar base, nuclear propulsion systems, and as an extreme option, a manned Mars mission before 1981. Indeed, Spiro Agnew was one of only a handful of high ranking politicians who publicly pushed the Mars expedition and called for, "a manned flight to Mars by the end of the century."

Although NASA was enthusiastic, the same could not be said of the U.S. public, congress, and the administration. Certainly, the increasing burden with the Vietnam War and a growing range of other domestic problems made solid justifications for increased spending on space activities more difficult to sell to the American people. The late 1960's and early 1970's were a time of great austerity throughout the United States, and the space program was no exception.

While the STG report was seen as out of step with current national priorities, it did include a recommendation to develop a reusable space transportation system as the means to service a manned space station. When Vice President Spiro Agnew asked Olin "Tiger" Teague, chairman of the former House Manned Spaceflight Committee, in 1969 for his opinions on the most critical areas for the U.S. future space programs, Teague replied, [11]

I know of one major contribution that can be made. That is the development of space vehicles that can be used repeatedly, with basic characteristics in common with transport aircraft. In view of the potential in this area, I believe the reusable space transport should stand very high on our list of priorities.

Tiger Teague was one of the strongest supporters of the U.S. space program in the 1960's and 1970's and did much to help guide the Space Shuttle through its difficult political development. His words are as relevant today as they were then.

The PSAC and STG recommendations were all part of the process that eventually would lead to the Space Shuttle. Indeed, the first time public reference was made to "the" Space Shuttle as opposed to "a" space shuttle, was in a speech by George Mueller, NASA's associate administrator for manned space flight, at a meeting of the British

Interplanetary Society, London, in August 1968. Mueller said, [12]

[their] studies show that using today's [ELV] hardware, the resupply costs for a single three-man orbital space station for a year equals the original cost of the space station. This type of cost analysis had led us to carefully evaluate concepts for more efficient resupply systems.

Mueller believed that the *fully* reusable Space Shuttle he had envisaged should be able to reduce the cost per pound into orbit to around $50 (1968), or about $200 (1993). This is the level the Delta Clipper hopes to achieve.

## The Whittling-Down of the Space Shuttle

By 1969, four parallel industrial contracts for the Integrated Launch and Reentry Vehicle (ILRV) were well under way. The objective was to develop concepts for a space transportation system that would significantly lower the cost of reaching orbit. Many classes of vehicles were studied, although all shared a reusable orbital element that would glide back to Earth. The rationale for the Space Shuttle had quickly spread to parts of the Congress, as evidenced in the following statement by George P. Miller, chairman of the former House Committee on Science and Astronautics, on April 23, 1970, [13]

The key to the success of this Nation's future space efforts lies in the development of a low cost, recoverable, and reusable space transportation system. The reusable Space Shuttle will drastically reduce the cost of putting people and cargo into space. In particular, the Shuttle will facilitate construction of a manned orbiting Space Station that will open up new areas of scientific and technological activity in the near neighbourhood of earth.

The Phase B Space Shuttle study effort began in 1970 and was to continue through to the end of 1971—2 years in which the path of the Shuttle was to become irrevocably changed. NASA's initial life cycle cost analysis had indicated that the most effective solution was a *fully reusable*, vertically launched, two-stage-to-orbit vehicle (Figure 11.2). The manned first stage would carry the orbiter to an altitude of 80 kilometers before separating and flying back to the launch site. The manned second stage would continue on into low Earth orbit, after which it would retum to Earth as the Space Shuttle does today.

This two-stage-to-orbit vehicle, if technically feasible, could have reduced launch costs compared with existing expendable boosters because a brand new vehicle would not have to be constructed, stacked, and checked out before every launch. Higher reliability (confidence that the booster and orbiter could be successfully returned in most contingency situations) would have been possible to demonstrate because the booster stage could be test flown a number of times, and with increasing propellant loads, before the orbiter was placed on top. In addition, both the booster stage and the orbiter had multiple engines, allowing a one or more engine-out capability any time during ascent. Also, unlike the solid rocket boosters (SRBs), these liquid engines could be constantly monitored and shut down as necessary. Availability (flight rate) would have been greater because of the reduction in activities required to turn around the vehicles, and because of the higher confidence in not losing them.

Unfortunately during the Phase B study, two serious problems arose with the fully reusable configuration. The first was a technological concern over whether such a vehicle could actually be built. In particular, a major problem was associated with integrating the liquid hydrogen and liquid oxygen propellant tanks inside the orbiter. For example, if the orbiter mass was heavier than expected, stretching it would require a total redesign or "reoptimization" of both the orbiter *and* the first stage—such are the problems with fully reusable launch systems. The second problem turned out to be far more crippling. As the Phase B studies progressed, the first realistic development cost estimates were established, and these costs were far higher than the Nixon administration and Congress were willing to pay.

The *optimal* operational configuration for the Space Shuttle was gradually whittled down partially in an effort to reduce the technical risk, but, more importantly, to reduce development costs. This was done at the expense of increased operational costs because the configuration was now sub-optimized for the operational capabilities it was meant to provide. Whereas before, reducing launch costs and achieving high launch rates were the primary drivers, the design, development, testing, and engineering (DDT&E) costs now became the most significant requirement. For there to be a Shuttle, the DDT&E would have to be drastically reduced. The Nixon administration was strapped for cash and was in no mood to invest up front in a system that would realize operational cost savings many years into the future, when a new administration would be in charge.

Restricting it to a mandated development cost ceiling was to doom the Shuttle's propects of ever lowering the cost to orbit and flying frequently, safely, and reliably. At first, the orbiter liquid hydrogen tanks were taken out of the orbiter and placed in two external tanks on either side of the orbiter. This was later reduced to one large tank for both liquid hydrogen and liquid oxygen, as it is today (Figure 11.3). This initial reconfiguration helped to minimize the problem of tank integration and insulation. It also meant that the operational cost would increase because a new tank would have to be built and attached to the Shuttle for every mission. Further, this tank would not only have to carry the propellant, but would also have to carry the 100-tonne orbiter and provide a path for the 3,000-tonne first stage thrust loads. This tank became the backbone of the Shuttle in every sense

**Figure 11.2** Original Shuttle concept of 1971 (*NASA*).

of the word, making it more expensive than just a discardable one.

If the orbiter fared badly, the large winged first stage fared far worse. Initially, some of the options examined included a much simpler unmanned booster that would be able to fly back to the launch site for reuse. The costs were still too high, so the study teams had to turn to alternatives for a simpler manned and, later, unmanned fly-back first stage, including proposals by Boeing to modify the first stage of the Saturn 5 [14]. All of these options were discarded as unacceptable due to high development costs. Although a pair of liquid boosters was initially thought to be an acceptable solution, this was also discarded in favor of the much simpler and cheaper solid propellant boosters (SRBs). Thus, the first stage was reduced from a manned, reusable, and incrementally testable vehicle that could be safely recovered after an in-flight launch failure, to a pair of unmanned, partially reusable, unstoppable, and untestable solid rocket boosters that must always work.

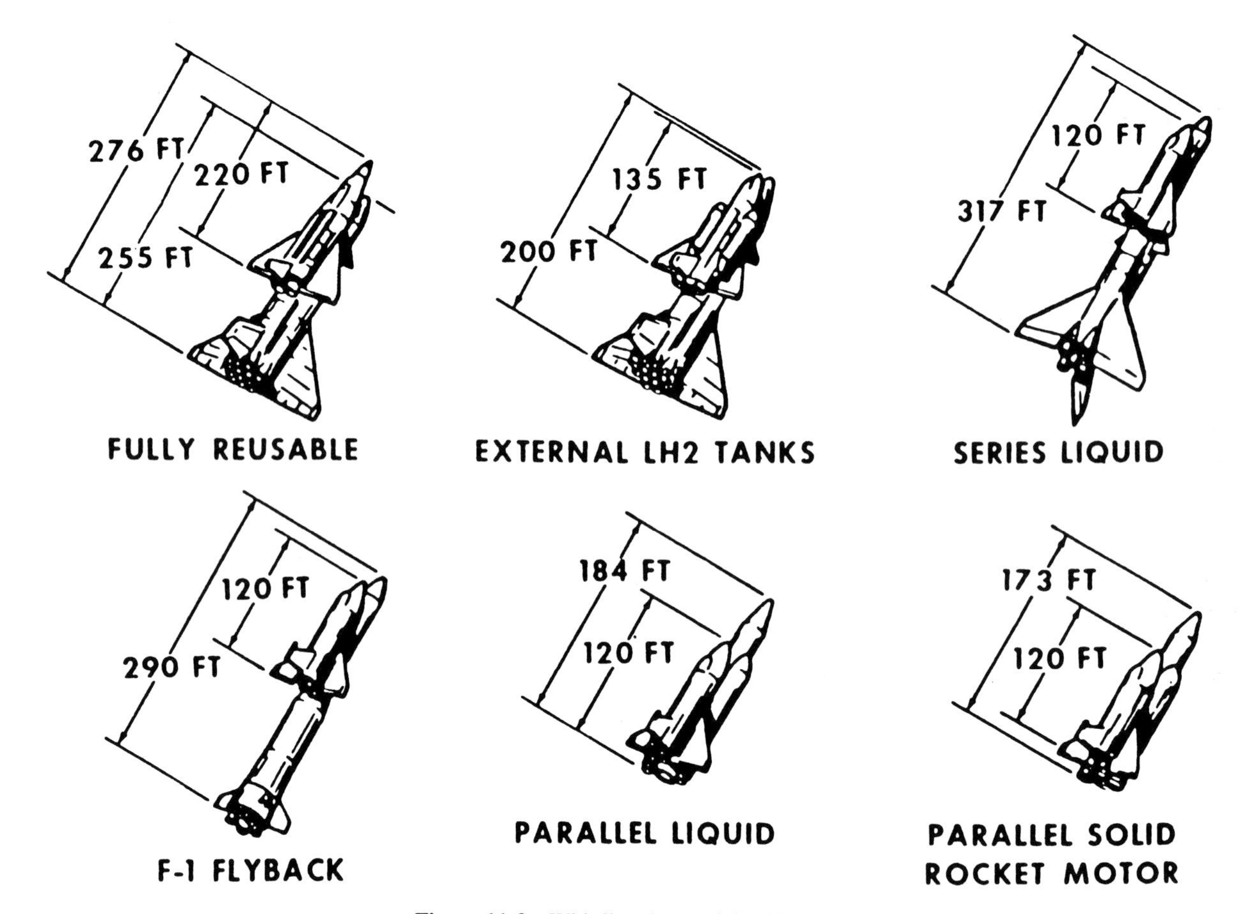


**Figure 11.3** Whittling down of the Shuttle.

One advantage of the SRBs compared with the liquid boosters was their greater robustness, allowing the potential for water recovery. Politically, at least, reusing the SRBs was seen as attractive. In practicality, they are actually more salvaged than reused. More significantly, the selection of SRBs meant that incremental flight testing would be difficult, if not impossible. The SRBs forced the Shuttle to go to orbit on its *very first mission* simply because they cannot be switched off. Indeed, this could have spelled disaster for the Shuttle when, on its first mission, higher than anticipated shock waves were reflected off the launch pad, subjecting it to loads in excess of what it had been designed to withstand in normal operations. This caused some structural damage to the orbiter's wings, and the Shuttle cleared the intertank access arm on the launch tower by a mere 2 meters [15].

Although it may not seem obvious, the original two-stage-to-orbit Shuttle design could have been incrementally tested. With the first stage, NASA could have built and tested it many times before the orbiter was ever placed on top. The first test flight could have involved loading the first stage with only enough propellant for a subsonic ascent to an altitude from where an unpowered landing could be safely performed. Further, the takeoff mass of the first stage for such a test could have been a fraction—perhaps less than 25%—of the complete, fully fuelled load. As a result, the stress placed on the main engines for these early flights would have been considerably less than on later operational flights. This would also have provided wide thrust margins in the event of one or more engines failing. Gradually, over a series of test flights, the first stage could be fuelled with greater propellant loads as confidence in operating it grew.

NASA could have learned *how* to fly this earlier Shuttle concept and demonstrate its performance long before it went into service and before Shuttle-only launch policy decisions need have been enacted.

## The Selling of the Shuttle

While the Shuttle configuration was being compromised, it was still believed that launch costs would be close to those proposed for the original version. At least politically this is how the Shuttle was sold. In reality, it was probably not the case, according to John Logsdon " . . . the economic analysis didn't have anything to do with the Shuttle decision, but was used to justify it." NASA may genuinely have felt that launch costs could be reduced, although not to the levels predicted for the fully reusable version. Analysis performed by the NASA Space Shuttle Task Group in July 1969 indicated that a fully reusable Shuttle would have achieved around $50 per pound, and a semireusable vehicle would have pushed this to about $200 per pound (1969 dollars) or $800 per pound (1993 dollars) [16]. Whether or not NASA really believed that such low launch costs could be achieved is beyond the scope of this discussion. Suffice it to say that the U.S. Congress methodically ripped to shreds NASA's vision of a true space transportation system, to the point where NASA became more concerned about preserving its manned space program—perhaps an even higher priority than reducing cost to orbit.

The Shuttle redesign in 1971 reduced the estimated development costs from in excess of $10 billion to a more politically acceptable $5 billion. The Shuttle was also the only manned element remaining from the Space Task Group recommendations. Originally (1969), NASA wanted both a space station and a shuttle. That idea died almost as quickly as it began when NASA realized in 1970 that it was going to have problem enough just selling the Shuttle. The Station would have to wait for another administration. (In fact, it was not until 1984 that President Reagan gave the Space Station the go-ahead.) For NASA, the choice was clear; either it could take whatever it could get as a manned program (i.e., the less-than-optimal Shuttle design) or not have a manned space program at all. NASA chose the redesigned Shuttle, and on January 5, 1972, President Nixon made the following announcement, [17]

> I have decided today that the United States should proceed at once with the development of an entirely new type of space transportation system designed to help transform the space frontier of the 1970's into familiar territory, easily accessible for human endeavor in the 1980's and 1990's.

The Space Shuttle program was launched, and it went a long way toward revitalizing the U.S. space industry following the Apollo decline. Even at this stage, though, concern was expressed over the change in the launcher's configuration and its impacts on operational costs. Louis Frey, Jr. (R-FL) said at the House Science and Astronautics Committee hearing in February 8, 1972, [18]

> One of our sales pitches on the recoverable craft was the reduction in cost per pound, but now that you're going back to this concept, it seems to me you're increasing the costs per pound, and it almost in a sense destroys the sales pitch that we originally conceived for the craft.

Further, Tiger Teague said at a hearing on February 17, 1972 that in 4 or 5 years, [19]

> we will . . . look back and be very sorry we did not have a more aggressive program.

These haunting words seem to have been proven true. Some observers today have stated that NASA should have been more responsible and not allowed the United States to accept a clearly inferior Shuttle design purely to win funding approval. Instead, NASA bowed to the realization that it had to compromize its vision in order to win some level of support. For example Charles A. Mosher (R-OH), another member of the House Science an Astronautics Committee, and previously an opponent of the Shuttle, said during a 1972 debate, [20]

> So, I stand before you today, confessing that I once was very much a skeptic concerning the Shuttle plans. Now, I have changed and I believe it is for entirely valid reason . . . The Space Shuttle program as now proposed is considerably different from that first recommended. The total development costs for the Shuttle have been reduced from $13 billion to $5 billion, to be spread over some 6 years.

Previously, political opposition to the Shuttle stemmed from the high cost of the program, but by 1972 the drastic reduction in the development cost won over their support. Unfortunately, what the Congress got for $5 billion dollars was a pale reflection on what it, *potentially*, could have had for $10 billion. The U.S. government decided to throw away the goal of "routinizing space," in the words of President Nixon, just to have a manned space program and something interesting for the NASA institution to build.

The slashing of Shuttle development costs also severely restricted the amount of testing that could be performed. As a result, thorough testing of individual components of some major Shuttle elements (e.g., SSMEs, auxiliary power units) were delayed until later into the program when the Shuttle was actually being manufactured. Therefore, when these components were finally tested and failures occurred, retroactive measures were expensive. In other cases, some elements were assembled from essentially untested or unproven components. An example of the impacts such an approach had on the SSMEs is provided by Wilson, [21]

> The obvious weak points [of the SSMEs] are the high-speed turbines as the blades are subjected to enormous stresses, but NASA adopted a policy of testing them in engines which themselves were under test. Standard procedure is usually to test components separately and then integrate them step by step. The result was that from 24 March 1977 to 4 November 1979 there were 14 engine test failures, eight of them resulting in fires that damaged the engines.

During the austerity budget measures of the 1970's, NASA was forced to try and stay within 10% of the original Shuttle program budget of $5.15 billion (1971 dollars.) The funding profile NASA was given did not provide the high *initial* investment needed for many of the critical technology developments, pushing many of them into the manufacturing phase, as noted above. But the Shuttle was a unique vehicle—nothing like it had been built before—and it pushed the early 1970's technology very hard. According to Vice Admiral Levering Smith, a consultant to the Space Shuttle program in 1979, in a letter to the NASA Administrator Robert Frosch, [22]

> I strongly support the observation that the component and subcomponent problems cited by various reviewers are symptoms of greatly inadequate funding in the early part of the development program and that these problems were generally unpredicted because of unfounded optimism of what could be accomplished with the funds made available. I am painfully familiar with the budgetary problems of providing adequate funding for the early part of a development program of this magnitude at the forefront of technology . . . Nevertheless, had NASA successfully resisted these budget pressures, i.e. without causing the program to be cancelled (which has been a continuing fear in many minds), today's program plans would undoubtedly be on a much firmer foundation.

To expect NASA to keep to the initial funding limitations was a tall order. As it turned out, the development cost of the Shuttle up to its first flight was within 30% of the original estimate, or $6.65 billion (1972 dollars). The Shuttle that resulted was not the same Shuttle planned, even in the pruned-down version. For example, the SSMEs were designed for 55 missions before overhaul but, because of cutbacks in funds for qualification tests, even today the SSMEs are partially stripped down, and critical bearings and turbines in the high-pressure oxidizer and hydrogen turbopumps are inspected and replaced after almost every mission.

Problems such as these, and others associated with a lack of spare parts and serviceability and maintainability provisions, all contribute to making the Space Shuttle as expensive, unreliable, and unavailable as it is today. Even if these aspects of the program had been properly funded, however, it is unlikely that the Shuttle would have been much different from what it is today. *The problems the Shuttle currently faces were cast in concrete the moment the decision was made to build the configuration that flies now.*

## The Shuttle-Only Debacle

DOD support for the Shuttle in the 1969–71 period was tepid at best. At the same time, it was clear to NASA that justifying even the pruned-down $5 billion Shuttle solely for the civilian space program was not going to be sufficient to win funding approval from Congress. The space agency needed the military's support, or at least not its opposition. After intensive bargaining campaigns across the DOD, national security communities, the White House, and Congress, NASA finally got what it needed—a military commitment to use the Shuttle. According to John Logsdon, [23]

> almost to the last day of the Shuttle decision process [NASA] insisted that the only Shuttle worth developing was one that would meet all DOD requirements. While the military did not offer to bear any significant share of Shuttle development costs, except those associated with creating a facility for launches into polar orbit at Vandenberg Air Force Base, the Air Force did agree in 1971 not to develop any new launch vehicles of its own. Leaders of the national security community communicated their support of the Shuttle programme to both the White House and Congress. Its military potential was a key factor in Richard Nixon's decision to approve the Shuttle NASA wanted to build.

This was the single most important factor that finally unlocked appropriations for the Shuttle, but it also proved to be a serious space launch policy mistake, the repercussions of which are still being felt today. The administration never actually issued a national launch policy stating that the Shuttle was to be the sole source of space access for the United States. It was more a matter of an agreement between the DOD and NASA leaders. Although as John Logsdon notes further, [24]

> The number of Air Force and other DOD personnel directly involved in the Shuttle decision was 'quite small.' Military support for the Shuttle was thus not based on wide or deep DOD exposure to the concept and many in the Air Force opposed dependence on the Shuttle.

The price of military support was that the Shuttle had to be significantly upgraded to carry Titan 3-class payloads and, more significantly, new payloads the DOD had originally planned to launch on stretched versions of the Titan booster. The DOD demanded and got a higher performance launch system than NASA originally wanted. Specifically, the DOD demanded the Shuttle payload performance be upgraded so that it could carry at least 18.1 tonnes (40,000 lb) into a polar orbit, and 29.5 tonnes (65,000 lb) to a 28.5° inclination orbit. NASA wanted a Shuttle with a payload capability of about 11 tonnes (25,000 lb) to a 500 kilometer low Earth orbit inclined at 55°—a little larger than the payload specifications of most aero-space plane concepts [25].

The military also demanded that the Shuttle be given a much higher cross-range capability than NASA intended, specifically up to 1,500 nautical miles either side of the orbital ground track. One of the prinicipal requirements for this was to enable the Shuttle to return back to the West Coast launch site after one polar orbit revolution in the event of a failure. This is called the abort-once-around (AOA) maneuver [26]. Such cross-range is needed because the Earth rotates about 1,350 nautical miles in 90 minutes underneath the Shuttle's polar orbit. By contrast, when the Shuttle launches into a low inclination orbit from Kennedy Space Center, the cross-range needed for an AOA is much less because the Shuttle is orbiting roughly in the direction of the Earth's rotation. Building a cross-range capability into the Shuttle, necessitated by the AOA and polar launch requirements, led to a larger delta-wing design and a heavier thermal protection system than previously planned.

The combined impacts of a higher payload capacity, the polar orbit mission, and large cross-range capability resulted in a significant growth in the size of the Shuttle vehicle, the net effect of which was to increase the technical risk of the program and push development costs higher. Both effects were in direct conflict with the drive to reduce develoment costs. According to David Webb "The military were as much to blame as NASA, and the Shuttle was almost unusable because of the changes the military demanded. They were hoping that the vehicle would become so impossible to use that it would all go away. The worst part [for the military] was that NASA was able to bring it off."

With the support of the reluctant DOD secured, the National Space Transportation System—as the Shuttle became known—switched gears and was able to move into its expensive development phase. Unfortunately, the military changes and the NASA/DOD decision that the Shuttle was to be "all things to all people" were to have far-reaching impacts on the U.S. space program once the Shuttle began to fly. According to George Keyworth, [27]

> in the early 1980's . . . our national policy placed nearly total reliance on the Shuttle as our means of access to space. It was a policy doomed to fail.

The impacts were felt by both the military and civilian space programs. During its first years in operation, the frequency of Shuttle launches was nothing like that promised by NASA. After the eighth mission in September 1983, NASA planned a further 25 flights by the end of 1985. Only 15 were flown [28]. Flights were being delayed, and the backlog of payloads began to grow ever larger. By early 1985, the DOD had become concerned about the Shuttle's ability to replenish its military space assests—a situation made more acute by the rapidly diminishing stockpile of Titan 34D boosters. After heated debates between NASA and the USAF, President Reagan issued a directive allowing the USAF to keep open the Titan production, but only to purchase ten *Complementary* Expendable Launch Vehicles (CELVs), essentially an upgraded version of the Titan 34D, designated the Titan 34D-7. This increased payload-to-orbit capability of the CELV was necessary so it could carry payloads designed for the Shuttle. At the same time, Reagan's directive would also ensure the DOD would have to use, on average, one-third of the Shuttle capacity [29].

It was fortunate the decision was made at that time. Shortly thereafter, the *Challenger* accident grounded the Shuttle fleet for over 2 1/2 years. The CELVs would still not be ready until 1988 or 1989. Until that time, the U.S. DOD had to rely on the few Titan 34D launchers remaining in its inventory and the hope that its aging military satellites would stay the course. To make matters worse, a few months after the *Challenger* accident, a Titan 34D exploded, eliminating the DOD's ability to place critical intelligence-gathering systems in space for over a year. These events seriously compromised the U.S. military space capability, although exactly how much is not known. It would be interesting to know, for example, whether the outcome of the Gulf War would have been as decisive as it was for the allies had it occurred in the period between 1986 and 1989.

The commercial community was hit equally as hard. The high cost of the program, coupled with the Shuttle's national, all-purpose flavor and promises of dramatic reductions in launch costs, meant that more conventional boosters were gradually phased out. According to Logsdon and Williamson, [30]

> . . . during the 1970's grossly unrealistic expectations regarding the shuttle's cost and capability led to a national program that depended solely on the shuttle for access to space. Hence from 1972 until recently [early 1989] virtually no Government funds were invested in other launch-related technologies or in new launch facilities . . . To make matters worse, in the early 1980's, when the European Ariane ELV's began to be marketed successfully as alternatives to US systems . . . , the Government responded by actively promoting the shuttle as the American entry in that global commerical competition. By 1985 marketeers from the National Aeronautics and Space Administration were roaming the world with glossy brochures stating that the shuttle was "the most reliable, flexible and cost-effective launch system in the world" and asserting to prospective customers that "you can't get a better price." . . . The Government's mercantilist response effectively blocked attempts by the U.S. private companies to commercialize their ELVs and forced them to suspend production of the vehicles altogether. Hence, when shuttle flights ceased [in 1986], the U.S. had no supply of alternative launchers immediately available.

In the early 1980's, NASA was commercially marketing the Shuttle. The test orbiter *Enterprise* was displayed at the Paris Air Show in 1983 for just such a purpose. It was clear that the launch *price* offered by NASA was considerably lower than that needed to fully recover the actual launch costs. This was an issue which Arianespace fiercely contested as an unfair U.S. government subsidy of the commercial launch industry, although NASA responded by saying that most flights included development testing of the Shuttle itself and U.S.-sponsored experiments, making it difficult to differentiate between costs to the commercial user and to the government.

With its lower price, the Shuttle was winning business for commercial payloads, and because of the national launch system policy, production of commercial launchers like the Delta and Atlas/Centaur was brought to a halt. No new money was to be spent on these boosters. Government deregulation did allow companies like Transpace Carriers, Inc., to take over the marketing of the handful of Delta rockets remaining, but they never won any orders because they were uncompetitive with the Shuttle.

In the early 1980's, commercial users could effectively

only choose between the Shuttle or Ariane. After the *Challenger* accident, President Reagan reversed the national launch policy and asked NASA to remove all commercial payloads from the Shuttle's manifest, except those which required the Shuttle's unique capabilities. As a result, nearly all commercial payloads were transferred to Ariane, thus giving the Europeans a considerable headstart. Ariane itself also suffered a 16 month standdown from May 1986 to September 1987. Nevertheless, it was the only launch vehicle in service capable of supporting commercial satellite needs. Not until the Delta 2 was ready in 1989 and the Atlas 2 in 1991 could the United States launch industry handle commercial payloads again.

## Implications for Aero-Space Planes

Understanding the story of the world's first attempt to build a reusable space transportation system is very instructive for the would-be developers of aero-space planes. Initially, a well-thought-out strategy was laid that recognized the need to significantly reduce the cost of accessing space. This strategy eventually became the centerpiece of U.S. space policy. Unfortunately, to realize this objective required a massive initial investment, far more than the Nixon administration and Congress were prepared to make. Thus, to preserve NASA's manned space program the Shuttle design had to be compromised, pruned down to an inferior configuration. Undoubtably, the chosen design was less expensive and risky to develop, but as a consequence, it was inherently far more expensive to operate, less reliable, and less available.

The strategy that could have led to the production of the first true space transportation system went out the window along with the dreams of making space cheaper and easier to reach. So far was the final configuration of the Space Shuttle away from its original objectives that even if it flew twice as often as it does today (about 16 flights per year), and cost five or even ten times less to operate (about $100 million a launch), it still would have been considered a failure. The strategy was, to all intents and purposes, ignored. Unfortunately, every time the Shuttle is launched, it holds much of the U.S. space program hostage because the economic and schedule impacts of another failure would be enormous. What happens, for example, if a Shuttle failure occurs midway through Space Station Freedom's assembly sequence?

Is it reasonable to say that just because the Shuttle failed to meet its objectives, aero-space planes will inevitably go the same way? There are many technical reasons to believe that aero-space planes will not. As already discussed in Chapter 4, the inherent ability of aero-space planes to be fully reused and incrementally tested is the key to learning how to fly these vehicles cost-effectively. The Shuttle is capable of neither. Viewed from a technical perspective, to suggest aero-space planes will fail just *because* the Space Shuttle failed is obviously an argument made from a position of considerable ignorance. The Shuttle is *not* an aero-space plane.

The technical arguments for aero-space planes stand to reason. But do the political arguments? What is to stop political appropriators from demanding development cost reductions of the type that stripped the Shuttle of any semblance of aircraft-like characteristics? Nothing. The basic configuration of the aero-space plane itself may protect it however. If the United States decides to replace the Shuttle with a fully reusable, single-stage-to-orbit vehicle, then a certain minimum investment will be needed to realize that objective. Without this investment, the aero-space plane will probably not be able to reach orbit, such is the sensitivity of the vehicle's performance to technology.

This, of course, doesn't stop appropriators from pruning back on the infrastructure expenditure required to support the vehicles. Nor does it stop funding caps being put on the number of vehicles produced. Such cutbacks can be tolerated because the basic configuration and mission of the aero-space plane cannot be compromised for the reasons given above. Inadequate infrastructure spending might make it difficult to fly aero-space planes frequently. Yet, at least the *right type* of vehicle will have been built. By contrast, the Shuttle's operational performance characteristics were determined at the beginning. No amount of money spent on the support infrastructure—or, for that matter, upgrades to the Shuttle itself—would allow the Shuttle to become significantly cheaper, safer, or more available to use. Not so with aero-space planes.

## Summary—The Fundamental Lesson

It cannot be emphasized strongly enough that the Shuttle is an outstanding technological achievement, and the United States should rightly be proud of it. But it is not, *and it will never be*, a railroad to space.

It costs what it costs to achieve a particular objective, and not what someone mandates it should cost. In an ideal world, NASA probably should never have accepted an inferior design simply because $5 billion was all it could get. But building a true space transportation system was only part of NASA's agenda—the other part was to preserve the prized manned space program and the institution it supports. The Shuttle was under threat of cancellation from the beginning. Thus, in NASA's eyes, having any Shuttle was perhaps better than having no Shuttle at all, even if its operational capabilities had to be radically changed. The drive to reduce launch costs and make space easier to reach had to take a back seat in the flurry of activity to reduce the Shuttle's development costs to politically acceptable levels. From this perspective, it could be argued that the Shuttle did meet its objectives, *if* it is assumed that its objectives were really to preserve the U.S. manned space program and meet politically mandated cost ceilings.

Operationally, however, the Shuttle's high launch costs, low flight rates, and undemonstrated reliability can only lead to the conclusion that it was a failure compared to the original vision. But, the history of terrestrial transportation systems is littered with failures: the Comet, the "Spruce Goose," the *Hindenburg*, the *Titanic*, the Edsel. Yet, from these failures, the invaluable experience gained has led to the enormous success of the Boeing 707 and 747, the Airbus A-300 series aircraft, the QE2, and the Chrysler minivan. To give up on the promise of building an affordable, reliable, and available true space transportation system just because of the Shuttle would seem irrational.

The aero-space plane provides a second chance for policy makers, budget appropriators, and the technical community to build a true space transportation system. One lesson clearly stands apart from the rest, and it is one that the would-be developers of aero-space planes must abide by if they are to succeed. The bottom line is: *you get what you pay for*.

## References and Footnotes

1. *Space News*, May 25-June 2, 1991, p. 2
2. Letter to the editor of *Ad Astra*, July/August 1991, p. 4.
3. Quote taken from Bill Nelson's and Jamie Buckingham's book *Mission: An American Congressman's Voyage to Space*, (Harcourt-Brace-Javanovich), p. 164. Also, the problem was reported in *Aviation Week & Space Technology* under the title, "Ground Controllers Cancelled Mission 61-C Thruster Test Because of Explosion Danger," March 17, 1986, p. 24. This article explained that toward the latter part of the *Columbia* mission, ground controllers noticed that the temperatures of the nose thruster propellants were lower than normal and that these low temperatures could affect the planned thruster tests. The ensuing analysis required a review of *Columbia's* documentation which subsequently revealed that two of the nose thrusters were of an older variety and, if fired below 25 kilometers, could potentially explode.
4. According to one European space official, the U.S. Shuttle decision, "also tricked the U.S.S.R. into Buran and ruined the U.S.S.R. space program."
5. Based on discussions with various NASP industrial representatives. Rockwell International provided a particularly useful appreciation of the problems.
6. "Space Shuttle 1980," Status Report for the Committee on Science and Technology, U.S. House of Representatives, Ninety Sixth Congress, (U.S. Government Printing Office, Washington D.C., January 1980), pp. 4–5.
7. Ibid., p. XI. It is actually more complex than this, of course, since 60 flights per year would require 60 externals and 3 new sets of solid rocket boosters every year.
8. Keyworth, G., "Launch Vehicles of the Future: Earth to Near-Earth Space," presented to the IAU Colloquium *Observatories in Earth Orbit and Beyond*, Goddard Space Flight Center, April 26, 1990, p. 2.
9. In a report entitled "Each Shuttle Flight: $1.67 Billion," by G. Koprowski in *Washington Technology*, (September 26, 1991, p. 1), it was explained that a Congressional Research Committee calculated each Shuttle launch so far has cost the United States $1.67 billion, including development and production costs but excluding NASA staff salaries. This analysis is perhaps a little harsh because the development and production costs should be amortized over the total Shuttle lifetime. For example, if the Shuttle were to complete another 100 missions over the next 15 years, the amortized cost per launch would be about $1 billion. NASA, by contrast, claims each Shuttle flight costs "just" $400 million. This is calculated by taking the annual Shuttle operating costs—as defined in NASA's budget—and dividing it by the number of flights. So in 1991, NASA spent about $2.8 billion on pure operations, giving about $470 million per flight. However, these cost calculations should also include costs associated with other NASA salaries and what is termed "Shuttle Production & Capability Development," which is funded at about $1.3 billion. Most of this money must be spent to safely fly the Shuttle. Therefore, it is by definition operational cost. As a result, in 1991 the *minimum* NASA spent on Shuttle operations was about $4.1 billion, giving a cost per flight of about $680 million. If the Shuttle achieves 8 flights per year, the current target, the launch cost will be about $500 million per flight. Other estimates are reported to put NASA's real Shuttle expenditure at nearer $6 billion per year.
10. NASA, "Technology Influences on the Space Shuttle Development," Johnson Space Center, June 8, 1986, pp. 3–4
11. Hechler, K., *The Endless Space Frontier: A History of the House Committee on Science and Astronautics, 1959–1978*, AAS History Series, Volume 4 (American Astronautical Society by Univelt, California, 1982), p. 242.
12. "Technology Influences on the Space Shuttle Development," pp. 3–5.
13. *The Endless Space Frontier*, p. 249.
14. Baker, D., *Space Shuttle*, (New Cavendish Books, London, 1979), p. 9.
15. Derived, in part, from an interview with STS-1 Commander John Young during the British Broadcasting Corporation program *Horizon*, "The Ultimate Explorer" in 1982.
16. "Technology Influences on the Space Shuttle Development," pp. 4–33.
17. Allaway, H., *The Space Shuttle at Work*, SP-432, EP-156, (NASA, Washington D.C., 1979), p. 33.
18. *The Endless Space Frontier*, p. 260.
19. Ibid., p. 261.
20. Ibid., p. 265.
21. Wilson, A., *Space Shuttle Story*, (Crescent Books, Hamlyn Publishing Group Ltd., London, 1986), p. 29.
22. "Space Shuttle 1980," p. 579.
23. Logsdon, J., "The Decision to Develop the Space Shuttle," *Space Policy* (Butterworth & Co., London), pp. 109–110. With permission of *Space Policy*.
24. Ibid., p. 109.
25. *Space Shuttle*, p. 7. Also, according to William Haynes, the Shuttle's payload capability into polar orbit was only 8.2 tonnes (18,000 pounds) before the Vandenberg launch site was shut down.
26. National Space Transportation System: Overview, NASA Kennedy Space Center, September 1988.
27. "Launch Vehicles of the Future: Earth to Near-Earth Space," p. 1.
28. *Space Shuttle Payloads*, Morton Thiokol, August 1983. Also, Wilson, A., ed., *Interavia Space Directory 1991–92*, (Jane's Information Group, Coulsdon, Surrey, 1991), p. 33.
29. Covault, C., "Presidential Directive Expands U.S. Space Launcher," *Aviation Week & Space Technology*, March 4, 1985, p. 18.
30. Logsdon, J., and R. Williamson, "U.S. Access to Space," *Scientific American*, March 1989, Vol. 260, No. 3, p. 34.

# Chapter 12

# Investing in Leading-Edge Technology

## The Problem With Aero-Space Planes

The technology needed to build any aero-space plane is extremely demanding, and until the late 1980's, many of the key technologies just did not exist. Others would argue that many of the critical technologies still are not in place. In addition, aero-space planes will need to be quite large just to be able to reach orbit. This is an important point because, unlike developing aircraft, which can start off small, the aero-space plane must be large from the beginning of its development. For example, the current NASP experimental X-30 calls for a vehicle requiring a total takeoff mass in the region of 200 tonnes to get itself (i.e., zero payload) into orbit, which is about the same takeoff mass as Concorde or a Boeing 767. For operational vehicles like HOTOL, Sänger, and the Delta Clipper, masses in the region of 300–600 tonnes are probably more realistic, about the same takeoff mass as a Boeing 747 or the Antonov AN-225. From the point of view of size, it is not possible to build the equivalent of the Wright Flyer if the objective is to put the aero-space plane in orbit.

Using advanced, and essentially unproven, technology to build a very large vehicle inherently means a high degree of risk associated with its technical feasibility. Generally speaking, the net effect is that such vehicles will be difficult and time-consuming to build. Above all, they will be expensive. If the go-ahead had been given in 1993, the simplest possible vehicles, such as the Interim-HOTOL or the Delta Clipper, could take around 7 to 10 years to build and qualify for operational space launch missions at a cost of up to $10 billion. More technically challenging systems such as the proposed scramjet-powered NASP-Derived Vehicle or complex vehicles like Sänger (with two high-performance vehicles instead of one), may take 15 to 20 years to produce at costs in excess of $20 billion [1].

Thus, the problems with aero-space planes are (1) they take a long time to produce, (2) their initial development costs are high, and most critically, (3) they present very high levels of risk.

None of this is unexpected. Any new development in transportation, whether it is building a transcontinental railroad, an airline, or even an automobile, carries with it significant cost, schedule, and risk. The conventional Boeing 767 design and definition process required 6 years to complete [2]. The length of time was more a function, though, of the need to optimize each airliner's configuration to best suit the market needs. Thus, the risk was not entirely technical but stemmed from the need to ensure commercial competitiveness, a factor which can present at least as great a risk as technology, especially in the highly competitive market for air transport services.

At the other extreme, Howard Hughes's "Spruce Goose," the de Havilland Comet, and the Tupolev TU-144 "Concordski" are all good examples where, for one reason or another, the technical risk prevailed and each failed to meet its intended operational objectives. This was in spite of the fact that each was supported by a vast amount of engineering data and flight experience from a variety of other multi-engined turboprop, turbojet, and supersonic aircraft that had already flown for years. The Comet could have been the aircraft remembered today for ushering in the era of low-cost, long-haul jet transportation. Indeed, it was the first such airliner introduced commercially. Then a relatively simple structural design flaw around the windows caused a number of fatal accidents. Before the problems of the Comet could be fixed, the Boeing 707 stole the limelight and the market. Only 87 Comets were built, in contrast to the 972 Boeing 707/720s [3]. As aero-space planes don't have a precedent, there should be no surprise if the first one fails to fully meet its objectives.

This discussion raises three important questions. First, will the payoff, in whatever form it comes, be worth the high initial investment? Second, what can be done to minimize the probability of failure and how much will this cost? Third, what are the politics of the high-technology R&D decision-making process?

## Is It Worth the Investment?

Investing in aero-space planes is worth at least some initial effort for two distinctly different reasons. The first reason—and the one from which the theme of this book was inspired—is the potential economic and operational impacts such vehicles could have on the way space activities would be conducted as a consequence of lower cost and easier access to space. The specifics of this are discussed in detail in Section 4. If an operational fleet of aero-space planes has the potential to save more money than it costs to develop, compared with alternative launch systems, then there might be good reason to believe that such vehicles are worth the investment. As this payoff may take more than a decade to realize, though, long-term funding commitments are more difficult to justify. The Space Shuttle has proven this point precisely, as discussed in the previous chapter.

This situation is further exacerbated by the high technological risks inherent in aero-space plane programs. Ironically, it is the high-risk nature of aero-space planes which provides the second reason for supporting their investment. Aero-space planes stand at the forefront of technological progress in many important fields. As a result, the research phase of an aero-space plane program can act as a stimulus for advanced technology in general. Therefore, during the many years it may take to achieve the intended operational objectives, aero-space planes carry with them the means to pay their way in the first few years as a result of the application of the technology they are responsible for advancing. In general, this is referred to as *technology transfer*, although the terms *spinoff* or *spillover* are sometimes used.

At the very minimum, the chances of an aero-space plane transitioning from a good idea on paper to the initial technology development phase are enhanced if the investment needed to minimize the technical risk of the program can be justified on the basis of the technology transfer benefits. In other words, are the direct and indirect benefits to other sectors of the economy greater than the money that must be spent to show that the critical technologies are available for aero-space planes? This is an extremely important question, and one which drives the prospects of some aero-space plane programs today.

### Return on High-Tech Investment

The Americans, Germans, Japanese, and Russians are investing in high-technology aero-space plane programs more or less for these reasons. One of the main rationales for the Germans funding Sänger is an acute awareness of their limited experience in high-speed aircraft. The former Soviet Union has also recognized the potential payoff of investing in aero-space planes. According to aerospace officials of the Central Aerohydrodynamics Institute, [4]

> It is profitable [to build a single-stage-to-orbit aero-space plane] because of tremendous development in new materials technology, computer technology, automation and so forth. A typical example of this is the American moon flight program . . . It paid for itself twice over, owing to the extensive use of its advanced technological achievements in ordinary terrestrial applications.

The Apollo program is a particularly appropriate example from the perspective of spinoffs. Estimates vary on how much the Apollo program has paid back to the U.S. economy—studies indicate anywhere from $2 to as much as $24 for every dollar spent on the program [5]. Some of this money has come from products spun off directly from technologies first developed for space applications. From Apollo and other NASA space programs have come a wide range of products including an implantable medication system for arthritis, diabetes, and kidney diseases; flame resistant materials for airplanes, ships, and buses; a scratch resistant plastic lens for eyeglasses; an anticorrosion pain additive; and an advanced lightweight wheelchair [6].

Europe has also benefited. A report commissioned by ESA and performed by the Bureau d'Economie Théorique & Appliquée in Strasbourg stated that from 1977 to 1986, every dollar spent on ESA space programs indirectly realized $3.20 [7]. Examples include technologies designed for the hostile environment of space being used on the equally hostile environment of North Sea oil platforms, and a technology developed for satellites to track Earth stations has been applied to television cameras as a way to enhance their focussing ability.

### Wishful Thinking?

Whether these products would have collectively paid for NASA's or ESA's programs is difficult to judge. Indeed, such claims for recent spinoffs from NASA's activities have been strongly refuted by some organizations. A notable analysis is presented by John Pike, [8]

> The most comprehensive recent review of the impact of NASA technology benefits to the commercial sector was conducted for NASA by the Chapman Research Group in 1989 . . . This study evaluated the benefits derived from technologies identified in the annual NASA report *Spinoff* during the period from 1978 through 1986. A total of over $21 billion in sales and savings benefits were identified as resulting from NASA activities. However, the report conceded that only about $5 billion of this total was due to actual spinoff, that is "instances in which a product, process, or even an entire company would *not* have come into existence had it not been for the NASA furnished technology." Most notable among these is the $1.6 billion in medical instruments, frequently cited as a major NASA spinoff. The remaining $16 billion in benefits were in areas where "the NASA technology *contributed* to the sales, but that contribution can vary widely, from a relatively small percentage of the

total sales or savings . . . " And in this area, additional sales of commercial aircraft accounted for over $10 billion.

The significance of these findings is best appreciated in the context of overall NASA spending during this period. The total NASA budget from 1978 through 1986 amounted to over $54 billion, of which about $2.5 billion was devoted to aeronautical research and development. NASA efforts in aeronautical research would seem to have been quite effective, with a $2.5 billion investment yielding $10 billion. Such a 4-to-1 payoff is not too surprising, given that work in this area is specifically targeted to improving commercial products. But the fact that the total NASA investment of $55 billion [sic] yielded a paltry $5 billion in true spinoffs, creating entirely new products or industries, suggests a very poor return of 10 cents on the dollar. Again this should not be surprising, given the highly specialized nature of much of the engineering and development work conducted by NASA.

Despite the apparent large payoff of NASA's aeronautics activities, it is ironic that NASA's contribution to NASP—funding for which comes out of the aeronautics budget—was all but eliminated in the Fiscal 1992 budget. (See Chapter 6.)

Arguments differ about the real issues behind technology investment. Some would say that the space community generates figures like "24 dollars for every 1 dollar" just because it is looking for a reason to support programs that actually have little direct use other than fulfilling their own objectives. The only objective of Apollo was to beat the Soviets to the Moon. Others would argue that investing a fraction of the money spent on these programs directly into each technology sector affected would have been a much cheaper and more productive alternative, assuming that such products were known about beforehand and there had been the political will to invest in such technologies. Further, while the research phase of the Apollo program really did stimulate technology spinoffs, repeatedly launching Saturn 5's—*which consumed the vast majority of the Apollo funds*—did not. The majority of the Apollo investment did not stimulate new technolgies.

Whatever the truth, claims of large spinoff benefits from space programs—especially large-scale programs—probably hold relatively little absolute meaning because it is impossible to know what would have happened had there not been a space program.

## Practical Applications

What the Apollo program actually did was to stimulate the technological capability of the United States by creating a large number of industries with advanced technical capabilities together with a vast army of technically adept people able to do this work. The Apollo program did not invent the microprocessor, but it did put a massive investment infusion into the microelectronic industry because of the demand for lightweight/high-performance computers. Microelectronics without a doubt would have eventually been developed for widespread usage without Apollo. Still, Apollo was in no small way responsible for the rapid pace in which microelectronics were introduced and assimilated into practically every sector of human activity—from toasters to ICBMs.

Likewise, in Europe a large number of small, highly specialized companies have benefited not only directly as a result of the technology spinoff itself but sometimes, according to the ESA technology transfer report, because, [9]

> The required reliability and performance characteristics for space pregams translate into a stamp of quality which is sometimes the deciding factor in a sale.

Evidently, if a particular company can meet the demanding requirements of spaceflight, this is seen as being reflective of the other products that company produces.

## NASP—That Competitive Leading Edge

One of the chief political motivations for the NASP program is the technological benefit to the United States for pushing technology to the limits. Today, the U.S. balance of trade is consistently in the red, averaging around $10–15 billion per month. This deficit would be considerably larger it it weren't for the contribution of the U.S. aerospace industry which, at the moment, accounts for more than half of the total U.S. exports which is estimated to be around $37 billion for 1991 [10]. The threat that Europeans, Japanese, and others are beginning to challenge U.S. dominance of the international aerospace business has given the United States great cause for concern. Whereas before the 1980's, the United States was practically the sole supplier of commercial airliners, today this position has been strongly challenged by the European Airbus Consortium which now accounts for about 30% of total world sales. China and Japan are gradually creeping into the airliner industry, and both already manufacture large sections of some commercial U.S. aircraft (e.g., the Boeing 777).

Likewise, the United States once dominated the original, albeit fledgling, commercial satellite launch industry. Today, Arianespace launches more than 50% of all commercial satellites and has a total turnover of about $1 billion per year. Viewed from an American perspective, the parallels between these examples and the U.S. automobile industry are all too striking. The U.S. automobile industry once dominated the *world* market, but today is struggling to maintain leadership in its own *domestic* market in the face of intense Japanese competition. Congressman Dave McCurdy summarizes these concerns, "I don't know what we're going to export if we give up aerospace. I just don't know where we're going to go. This is probably the best argument we've got for NASP."

The U.S. rationale for funding the NASP Phase 2 program can be directly attributed to the long-term goal of maintaining a strong aerospace export market. Building the X-30s is seen partly as a way of coalescing these efforts and giving the R&D communities a focus. Bob Williams provides an interesting perspective on how this works using the example of materials being developed for the NASP program, "You've got to be focussed. I've been in the technology business for 30 years and I understand this very, very well . . . There's just no question about it. People say 'you're so focussed and you're so narrow that you develop this one material and that's just not going to come to anything.' What they don't understand is when you get American industry focussed on solving this problem, they do internal research and development programs that fill up the technology base. They have to know what the tradeoffs, options and alternatives are, and all the chemistries and the other kinds of characterizations of materials and so on. And then, what happens is that universities come in . . . to consult on the fundamental issues, to perform FME [Failure Mode & Effects] analysis or creep analysis. So it does propagate down into the tech base. Every new material that goes into that airplane [NASP] has got a *huge* tech base."

## NASP Spinoffs

No one could reasonably expect a scramjet propulsion system to be used widely on future aircraft. But during the course of developing that scramjet, new materials and engineering techniques will be produced that will find application in other programs. Attempts have been made to estimate such technology impacts for the NASP research activities. A study performed by DRI/McGraw-Hill for the NASP Joint Program Office [11] analyzed the potential "economic impacts of undertaking basic research, development, production, operations, and maintenance to make the NASP/NDV a viable and productive endeavor for the U.S. economy" between 1990 and 2010. The study took the diverse technologies being pursued by the NASP program and estimated their economic impacts on a large number of industrial sectors ranging from aircraft and spare parts to business services. It was assumed that if the combined NASP and NDV programs cost $23.5 billion over 20 years, this would actually be recouped as early as 2001. By 2010, it was estimated that the total cumulative benefit of the NASP and NDV programs would be nearly $140 billion, of which $30 billion would be a direct economic benefit and $110 billion would be in technology benefit.

The NASP program has also issued a guide for technology transfer entitled: *The National Aero-Space Plane (NASP) Domestic Technology Transfer Program Technology Description Document: A User's Guide to Understanding and Utilizing NASP-Developed Technologies* [12]. This document states that:

The objectives of the NASP technology transfer initiative are to:

1. *Maximize the return on the investment* for the NASP R&D budget through proactive technology transfer
2. *Enhance United States international competitiveness* through rapid and direct transfer of NASP technologies to the U.S. commercial, civil, and federal sectors
3. *Develop, exercise, refine, and rigorize the technology transfer process* to provide the most efficient means of transfer.

The United States has invested modestly in the NASP program but spinoff claims have already been made. Investigators are exploring how advanced materials accelerated by the NASP Materials Consortium—some of which were found to be unsuitable for NASP—can be used in new commercial airliner and military jet engines. According to Congressman Tom Lewis, "because of the new materials, you can make [jet engines] lighter overall, and at the same time build up the temperature threshold so that you can run hotter to get more thrust."

The first genuine commercial spinoff from NASP has already been made. An article in *Aviation Week & Space Technology* stated that, [13]

> Beta 215, a heat-resistant, corrosion-resistant titianium alloy developed under the National Aero-Space Plane program, will make its debut in a commercial role in several areas of the Boeing 777. The material, also called Timetal 215 by its developer, Titanium Metals Corp. (Timet), will be used to fabricate engine nozzles and plugs and possibly aft fairing heat shields on 777s powered by Pratt & Whitney PW4000s and Rolls-Royce Trent engines. Using the material in nozzles and plugs will save designers about 180 lb. [80 kg] per engine installation, or about 360 lb. [160 kg] per aircraft.

Other spinoffs of the Beta 215 material are being studied for medical use, according to Barry Waldman, [14]

> The attractive properties of Beta 21S [a titanium alloy] [are] currently being evaluated by the FDA [Food and Drug Administration] as a material for hip joint implants. Because of its greater corrosion resistance than other alloys, it reduces the body's exposure to toxins and it is more compatible with bone, promoting bone growth and attachment onto the implant.

Another spinoff comes from the computational fluid dynamics (CFD) work. NASP did not create CFD analysis. However, the need to understand the aerodynamic characteristics of the X-30 at untestably high Mach numbers has required a large investment that has spurred CFD work throughout the United States, providing it with an infusion of research funds it might not otherwise have received. Increased fidelity fluid mechanics modelling as a result of the NASP program has already been applied to structural analysis, electromagnetic field analysis, plasma physics, and perhaps most importantly of all, global climate modelling [15]. In addition, as the precision and validation of CFD analysis improve, it is possible that the aerodynamic configuration of most future aircraft will be optimized electronically, rather than through the usual process of iteratively modifying a model and testing it in a wind tunnel. Such a process will be extremely valuable in the design of airliners and other aircraft because it will save time and money, and also yield a more precise aerodynamic shape giving better fuel efficiency.

CFD codes have already been used to design an aerodynamic vehicle. The B-52-launched Pegasus winged small launcher was designed entirely using CFD analysis techniques derived from the NASP program—no wind tunnel testing was used. Indeed, the first Pegasus mission was heavily instrumented in order to compare the actual aerodynamic flight data with the CFD predicted data [16]. The availability of CFD for this application helped to minimize the development costs and risks of this privately developed launcher.

### Worthwhile Endeavor?

The intention of this discussion is not to argue that aero-space planes will lead to widespread technology transfer. Certainly, if the legislative and regulatory mechanisms (including rights of ownership) are not in place, then any arguments for technology transfer are lost—a criticism often levelled at the former Soviet space program. Nevertheless, using history as the best guide to the future, a reasonable answer to the question "are aero-space planes worth the investment?" can be forwarded. Pragmatic reasons suggest that programs in the *research phase* which push technology to the limits merit investment if, at a minimum, what is learned can be applied to other sectors of the economy, either as direct spinoffs or, more significantly, in stimulating work in other high-technology applications.

## What Is the Cost of Minimizing Risk?

Although the cost to build an operational fleet of aero-space planes will be very high, it is important to appreciate that the costs actually associated with proving the technology are not high in the majority o cases. In other words, the size of investment needed to be reasonably confident that an aero-space plane can be built and flown to orbit is a fraction of that needed to actually build a vehicle and complete the program.

During the development of any program, costs come from two principal areas: enabling technology and infrastructure. Bob Parkinson provides an explanation of the difference, "You need technology before you build an aero-space plane, but you *do not* spend infrastructure money on technology—that's the most ineffective way of doing it. If you need CFD to do aero-space planes, then you spend money on CFD, you don't spend it on fixing the life-support system on Hermes. I would argue that you should be spending somewhere between 5 to 10% of your budget on technology, and that technology should not be concerned with flying hardware for any other purposes than to test a bit of technology to see if it works. So if I want to know about high-speed flight and the only way of doing it is to fly a vehicle, I fly the cheapest possible vehicle to test that high-speed flight situation—a robot drone. When my technology is mature, then I engage in an infrastructure program that doesn't push technology at all—because I have what I need—and I would produce my infrastructure for the cheapest possible cost. However, I would spend more money on infrastructure than technology because just building the infrastructure is expensive—building the Channel Tunnel is expensive [compared with proving the technology for drilling the tunnels, for example]."

The enabling technologies work, ideally, should be undertaken well in advance of any infrastructure-type work. Even though the costs to prove these technologies are relatively minor compared with the estimated total program cost, they nevertheless completely drive the requirements and configuration of the infrastructure. If the technology is wrong or unproven, this will impact the infrastructure costs. Bob Parkinson explains further, "All the evidence says that if I'm developing technology at the same time as building my infrastructure, it's going to be more expensive. The Chinese have produced a formal model of the R&D process. Basically what it says is that the further into a program you get, the more your problems will become focused and apparent, but the less room you have for maneuver, so the more work you have to do to get a change at that stage. What it really says is that if you don't do what R&D you can early on, it will be more expensive later on. You have to have something that is pulling everything together, and I'd argue what NASP does is right" [17].

This point has been proven by the Space Shuttle program where the necessary research work could not be properly completed before design solutions were finalized and manufacturing had begun. According to William Anders, a consultant reviewing the Shuttle program in the late 1970's, [18]

> NASA, flush from their outstanding achievement of putting men on the moon and convinced that a shuttle program was vital to our nation, probably had tended to underestimate the degree of some of the technical challenges of the STS and, as problems became more obvious, probably has buckled too easily to budget pressure. The Nixon Administration did not live up to agreements of initial funding and subsequent budget levels nor was the contingency recommended by NASA allowed. Support by subsequent administrations has not been strong. While permitting the program to continue, the emphasis has been to pressure NASA to reduce its annual costs below those required to maintain program schedule and management efficiency. The impact of this approach, inevitably, has been to push NASA towards a higher risk and less efficient program where qualification testing is done concurrently with vehicle manufacture and work performance shortfalls are

pushed into succeeding years—in essence, schedule slip was substituted for adequate funding levels and contingency. This, in turn, has led to a need for continual reprogramming of work (very inefficient) and a stretch in the completion date and overall cost. NASA managers have had to become so caught up in the budget battle each year that their program focus tended to shift toward that of achievement of an annual level rather than the completion of a difficult technical project.

For aero-space plane programs to avoid a repeat of this, it is obviously vital that the critical technologies are fully demonstrated first, perhaps even before a specific type of operational aero-space plane is chosen. For example, it is not entirely true that Germany believes Sänger is the only aero-space plane alternative for Europe—likewise for Britain with HOTOL, or France with STS-2000, and so on. At this early stage in the development of what could become a European aero-space plane program, it is not especially wasteful for each country to study vastly different vehicle designs because the investments are relatively modest. What is of greatest importance is developing the technologies first, especially as the major fraction of the technologies are just as applicable to one configuration as the next. If, for example, Europe were to dedicate itself to Sänger, but several years into the program found that separation of the upper stage at Mach 6.5 was not feasible, Europe would have to go back and reexamine alternative concepts. Not only would this lose several years of effort, but politically it could be very damaging. Although adopting one design may seem politically attractive, technically it might be unnecessarily risky. The SSTO program, by contrast, was able to choose the Delta Clipper configuration with some confidence because of the massive investment in critical technologies made by the NASP program.

It would be valuable for Europe to have a common aero-space plane enabling technologies effort in which the technological strengths and resources of each nation could be pooled. Indeed, this is precisely what should occur in 1994 with the anticipated approval of the ESA Future European Space Transportation Investigation Program (FESTIP) with initial funding at about $30 million for the first 2 years (Figure 12.1). FESTIP will primarily feed aero-space plane technology back into the national vehicle system-level studies in order to provide a roughly equivalent platform from which each can be compared with the others.

The enabling technologies contribution to development costs comes from the fact that certain key capabilities must first be proven through research and experimentation before an informed decision can be made on the infrastructure requirements and what levels of expenditure are likely. It is

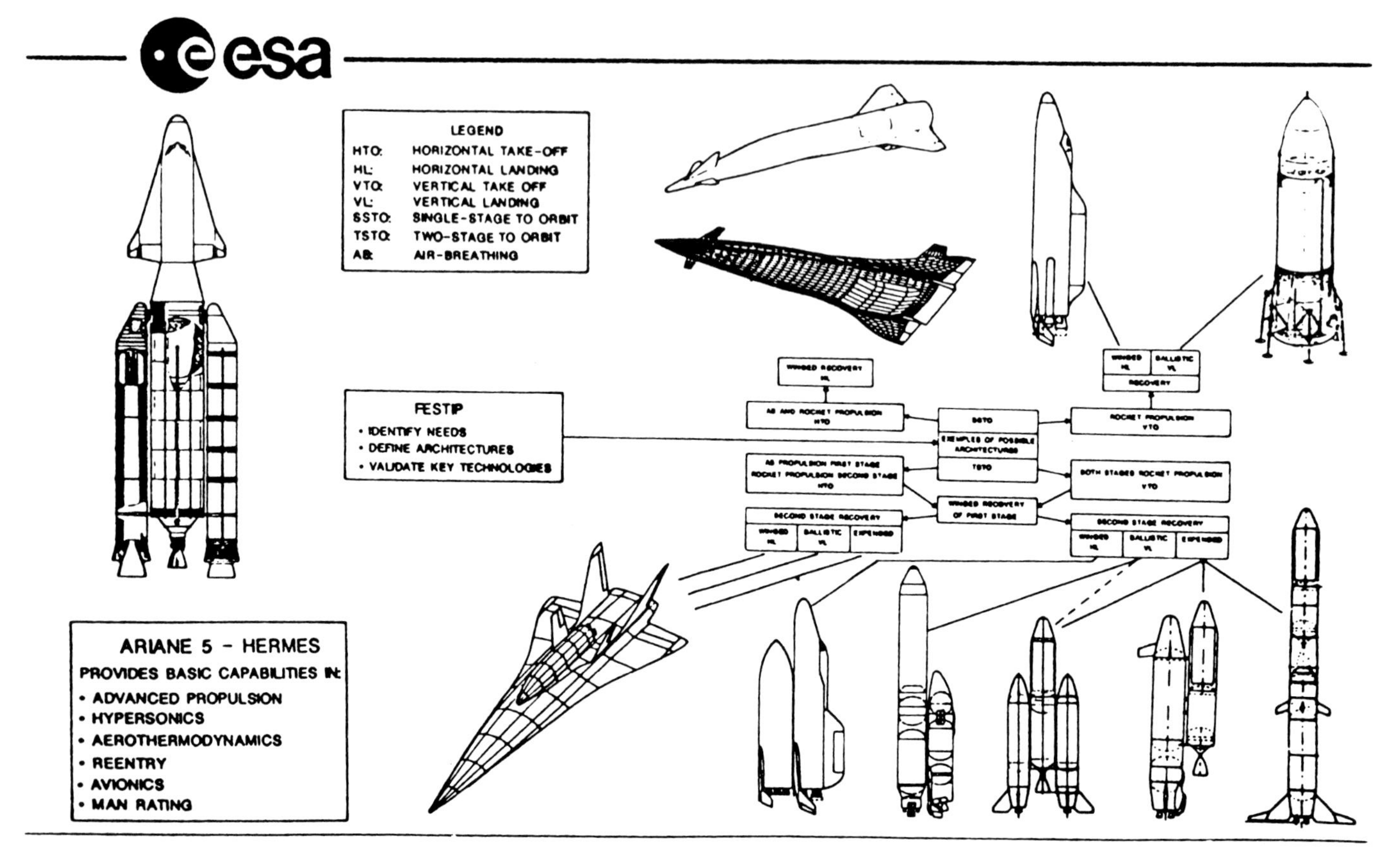


**Figure 12.1** FESTIP overview (*ESA*).

critical to appreciate that enabling technology work only has to be as involved and as expensive as it needs to be to return the required data points. For less complicated vehicles like Interim-HOTOL and the Delta Clipper, where the vast majority of the technology can be tested on the ground before building an orbital vehicle, the required investment to produce the enabling technologies may be less than 10% of the total program costs. NASP will require a greater enabling technology expenditure because it uses much higher technology, some of which can only be researched in flight. For example, if the United States had considered it practical to start from the outset to build an operational aero-space plane launcher, then a lower technology risk solution would probably have been chosen, meaning the enabling technologies expenditure would have been less than that required for NASP. As it is, the NASP Phase 2 effort has already spent $2.8 billion on enabling technology, or about 25% of the total cost of the NASP program through to completion.

Politically, investing in enabling technologies can be attractive for two main reasons. First, this investment pushes the technology and is, therefore, from where the greatest spinoffs will come. Second, it does not commit governments or other organizations to eventually building the vehicle because the expenditure involved is relatively small. Overall, a properly paced technology R&D program is vital to laying the foundation for any operational aero-space plane program because it allows the technical risk to be minimized. On the down side, political support for an aero-space plane R&D effort might give a false impression of the prospects for eventually building vehicles.

## *What Are the Politics of Technology Research and Development?*

Any program that pushes technology in a new direction will probably merit political interest, especially if the project is also able to capture the imagination—and not just of the people involved with the program. To continue this point further, the ability of programs like NASP and Sänger to obtain funds was due, in part, to the "sexy" image they portray. Certainly, Ronald Reagan's reference to NASP as a research program that could lead to a new Orient Express did much to attract attention to this program, albeit for many of the wrong reasons. (See Chapter 6.)

The NASP program is indicative of the way high-technology programs that have a significant civilian content are initiated and funded in the United States. The situation is explained concisely by David Webb, [19]

> The NASP Program is a classic case in point: Initiated by DARPA, directed and funded by Air Force/NASA, supported by Navy, and SDIO, it is administered by the Joint Program Office, which reports to the NASP Steering Group made up of representatives from DOD and NASA. This multi-agency make-up can and does work well in an operational mode; but, when it comes to representing the Program in Congress there is no single body which has overall authority or responsibility to defend, or even explain the Program at the highest levels of government. This is left to DOD and NASA, whose budgets fund NASP; but, neither agency initiated the project and both have differing requirements and reasons why they want, or don't want, an operational vehicle. The result is that the Program finds itself in a defenceless situation.
>
> In Congress the Program's position is made considerably more difficult by the apparent lack of support for NASP at the highest levels in DOD, and the seemingly only tepid support in NASA. This situation first became evident when the Secretary of Detence cancelled the program in his initial week on the job; a shock which has to be followed by the less than enthusiastic attitude displayed by the Secretary of the Air Force, when he stated that the Air Force had no mission, nor need for such a vehicle. NASA said, however, that they would support the NASP as a technology program, but did not indicate support for building a test and demonstration vehicle. Normally, these actions and remarks would have sounded the death knell for any project. It is a tribute to the staying power of the NASP that it has so far refused to go away—proof, perhaps, that it really *is* difficult to keep a good idea down.

Only when key members of the House of Representatives, together with the National Space Council, realized that NASP was in jeopardy through lack of a sponsor, did they act to save it. If it weren't for this action, NASP would probably not have survived into the 1990's. (Chapter 6)

NASP is also typical of the way the United States handles advanced technology R&D efforts. It is a process which can and has worked in the past on military programs, but for civilian programs the process has been more haphazard. The reasons for this are largely due to the changing fortunes of science and technology under postwar administrations. Although science and technology enjoyed strong support during the Kennedy and Johnson administrations, when Nixon came to power the situation changed abruptly, according to Hiskes and Hiskes, [20]

> The tremendous achievement of the Apollo program was possible largely because of an initial and continuing belief in the importance of the project. After the lunar landings, it became rapidly evident that no other science project currently under way or under consideration enjoyed such a presidential mandate. As a result, without an external stimulus comparable to the Soviet launch of Sputnik in 1957, President Richard M. Nixon decided that a politically possible, even desirable, approach would be to lesson the federal government's role in scientific research . . . Nixon did not stop with cutting funds for many R&D projects. In a move of far greater significance and potential damage for the entire scientific community, the president razed the presidential advisory system for science and technology . . . Nixon abolished the Office of Science and Technology, established in 1972, terminated the White House post of science adviser to the president, informally established by Roosevelt, and asked for the resignations of the member of the President's Science Advisory Committee.

Although action by Ford and later Carter helped to restore the belief that technology R&D was important to the country's future, the legacy of decisions like those Nixon made is still present today. For example, before the Clinton administration, no single civilian agency was dedicated to support pure R&D work, unlike the military which has DARPA. Therefore, the responsibility for programs which have a civilian content is usually given to agencies like NASA and DOE who, of course, have their own R&D priorities. Many of the more impressive civilian technology developments have been spinoffs from entirely separate efforts. The commercial airliner industry serves as a good example of a technology derived directly from a military requirement that had little originally to do with building civilian airliners. (See Chapter 16.) The many industries and technologies spawned from the Apollo program were a secondary consequence of the U.S. goal to beat the Soviets to the Moon in the quickest possible time.

Congressman Dave McCurdy sums up the current situation for NASP, "This government deals with crises for the most part and there are a lot a crises in this country: cities, national debt, crime, educational system is going to pot . . . But this [NASP] is one of those continuums that we need to do. We ought to have just a nice area of government that [allows us to] continue to do these long range R&D programs . . . We need a technology policy, and I've always supported a civilian DARPA. There needs to be some advanced research that there is no guarantee that there's going to be a return on or a product from, but the potential payoff is high enough that you should invest in it, perhaps even gamble."

In the late 1970's, attempts were made to establish a formal civilian R&D budget as David Webb explains, "Adlai E. Stevenson was the man in the Senate who pushed the hardest to develop some rational way of dealing with technology, and he spent 12 years beating his head on this. He developed the Stevenson-Wylder act which was to create a huge technology development foundation in which they were going to put '$2 billion' as a start up in to it, and every year they were going to put 'X' amount. This money would all be used for technology development—no second guessing—but just like the European system. Every five years they would have to come back and explain exactly how the money was spent and what they were doing for the next five years."

The Stevenson-Wydler Act in this form never reached a floor vote because of the difference in the length of time between elections of senators and representatives. Most of the senators had little opposition to the act because they are elected every 6 years, meaning they could accommodate the proposed 5 year length of the act. The House members were less favorable because they would have to be elected three times during each 5 year period, meaning that the foundation's activities would be more difficult for them to monitor and control.

One recent initiative in this direction could indicate some action in the near future. A study, *Regaining U.S. Manufacturing Leadership*, produced by the Hudson Institute, a policy organization, proposes the formation of a National Technology Agency (NTA) under the leadership of a technology advisor to the president, heading up the various interagency research activities in the United States. The NTA director would also be in charge of a National Advanced Research Projects Agency (NARPA) that would take the place of DARPA [21]. Following the election of President Clinton, the role of DARPA was changed to include nondefense activities. The "D" was dropped, and the Advanced Research Projects Agency or ARPA was born.

Ideally, a high risk program like NASP would be best served if it did not have to fight each year for its life, as under the current method of appropriating funds. The Stevenson-Wydler Act was intended to provide this type of facilatory mechanism and an organization like ARPA could be a solution. The future of projects like NASP can never be secured unless Congress as a whole supports it, as was the case for Space Station Freedom in its earlier years. Against this background, NASP did remarkably well for a few years.

By contrast, because Europe, Japan, and until recently, the former U.S.S.R. operate under multiyear funding authority, civilian R&D efforts can receive uninterrupted funding over many years. ESA has traditionally operated on a budget profile that can extend over as many as 10 years unchanged and unchallenged—something unheard of in the United States. Even here, as will be discussed in the following chapter, this situation is changing for space programs.

## *Summary—Politics and High Technology*

Aero-space planes are inherently large and complex machines that push technology to the boundaries of the state of the art. As a result, they are expensive in real monetary terms. Perhaps it is not possible to say with total certainty today whether such vehicles can actually be built. For all these reasons, it is unrealistic to expect the private sector to pay for the development of aero-space planes, especially as the payoff will not occur for many years.

Government funding will be necessary. Fortunately, the cost to prove the technologies for these vehicles—and thereby minimize the technical uncertainty and risk—is a fraction of the cost needed to actually build an operational fleet of aero-space planes, although this is a strong function of the type of vehicle being developed. In addition, because aero-space planes are on the leading edge of technology, the technology from such programs may spill over to a large number of other sectors of the economy.

How political decisions are made to support and fund a particular R&D effort may hinder the prospects of promising programs like aero-space planes. Certainly, in the United States programs seem to be selected on a haphazard basis, with little strategic rationale. Part of the problem is that programs like NASP have been sold politically on the basis of technology spinoffs. While this is a good method of getting a program up and running, it may give a false impression of the actual support for the program and the political intent for building vehicles. As the cost of building any aero-space plane will be anywhere from 5 to 20 times the cost of the enabling technologies research, the emphasis placed on the program's technology spinoff benefits must be carefully balanced against the rationale chosen for building the vehicle itself. NASP, in particular, has suffered repeatedly from a generally confused understanding of its ultimate purpose. Without an unambiguous and clearly understood rationale, then NASP will continue to suffer, and its future will remain uncertain because of the inability to win over the support of a dedicated sponsor.

## References and Footnotes


1. These numbers are purely rule-of-thumb guides.
2. Mowery, D. C., and N. Rosenberg, *Technology and the Pursuit of Economic Growth*, (Cambridge University Press, Cambridge, England, 1989), p. 171.
3. Bennett, J., and P. Salin, "The Private Solution to the Space Transportation Crisis," *Space Policy*, August 1987, p. 189.
4. "Soviets Cancel Fall Flight to Mir Station: Move Forces Longer Stay for Cosmonaut; Gorbachev Studying Space Plane Idea," *Space News*, July 29–August 4, 1991, p. 21. With permission of *Space News*.
5. For example, a study performed b the Midwest Research Institute of the United States in 1971 concluded that the $29 billion spent by NASA from 1959 to 1969 would generate a return to the economy worth some $207 billion, giving a 7 to 1 ratio.
6. NASA publishes the book *Spinoff* every year in which it lists the latest technological achievements. The list presented here is a selection of spinoffs detailed in these publications.
7. Details of this report were printed in *Space News* under the title "ESA: Space Industry Spinoffs Exceed Program Investment," April 8–14, 1991, p. 22. ESA has recently published a new catalogue called *Transferable Space Technologies or TEST* (1991) which is intended to be used as a basis for company marketing activities. TEST has been distributed to more than 27,000 companies and printed in four languages.
8. Pike, John, "Testimony before the Committee on the Budget, Task Force on Defense, Foreign Policy and Space of the House of Representatives: Space Station Freedom," *Federation of American Scientists*, July 11, 1991.
9. *Space News*, April 8–14, 1991. With permission of *Space News*.
10. Data from the U.S. Department of Commerce International Trade Commission.
11. "An Analysis of the National Benefit of NASP/NDV Expenditures: 1990 through 2010," DRI/McGraw-Hill & General Dynamics, January 1990.
12. This document was compiled by Science Applications International Corporation, Dayton, Ohio, October 1991, p. 3.
13. "Boeing 777 to Incorporate New Alloy Developed for NASP," *Aviation Week & Space Technology*, May 3, 1993, p. 36. Courtesy *Aviation Week & Space Technology*, Copyright 1993, McGraw-Hill, Inc., All rights reserved.
14. "Joint Hearing before the Subcommittee on Technology and Competitiveness of the Committee on Science, Space and Technology and the Subcommittee on Research and Development of the Committee on Armed Services U.S. House of Representatives," U.S. Government Printing Office, Washington, D.C., March 12, 1991, p. 64.
15. Williams, R., "Forces of Change and the Future of Hypersonic Flight," Keynote address, *First International Conference on Hypersonic Flight in the 21st Century*, University of North Dakota, September 20–23, 1988.
16. Noffz, G., and R. Curry, "Summary of the Aerothermal Test Results From the First Flight of the Pegasus Air-Launched Space Booster," AIAA-91-5046, *AIAA Third International Aerospace Planes Conference*, Orlando, Florida, December 3–5, 1991.
17. For a more detailed discussion of this topic see also: Parkinson, R. C., "Organizational Impediments to the Reduction of Cost of Space Programmes," IAA-91-639, *42nd Congress of the International Astronautical Federation*, Montreal, Canada, October 5–11, 1991, p. 4.
18. Quote taken from a letter to former administrator Robert Forsch which was printed in "Space Shuttle 1980: Status Report for the Committee on Science and Technology, U.S. House of Representatives, Ninety Sixth Congress," U.S. Government Printing Office, Washington, D.C., January 1980, p. 579.
19. Webb, D. C., and R. J. Hannigan, "Economic and Socio-Political Impacts of NASP-Derives Vehicles: A Technical Report," International Hypersonic Research Institute, November 1990, p. 23.
20. Hiskes, A. L., and R. P. Hiskes, *Science, Technology and Policy Decisions*, (Westview Press/Boulder and London, 1986), pp. 46–47.
21. "Washington May Consider Concept of National Agency to Coordinate Government/Industry Science Efforts," *Aviation Week & Space Technology*, January 20, 1992, p. 26.

# Chapter 13

# To Cooperate or Not to Cooperate?

## *Questions to Consider*

Nearly every nation and organization pursuing aero-space plane programs has suggested that such vehicles can only be built under the auspices of an international consortium. Indeed, formal policies have been adopted promoting international cooperation as a result of the realization that projects like HOTOL, Sänger, the Japanese and former Soviet efforts, and others depend on the success of cooperation. In fact the German government will not transition its Hypersonic Technology Program to Phase 2 unless sizable and direct international contributions are forthcoming. The U.K. Government has insisted that significant international participation in the HOTOL program is mandatory before it will consider restarting funding. The exception in all of this is the U.S. NASP program. Although there have been numerous discussions with European, Japanese, and former Soviet representatives, U.S. officials have made it clear that as the NASP program stands today (1993), they have little desire to open up the program to foreign nations.

According to a U.S. GAO report entitled *Aerospace Plane Technology: Research and Development Efforts in Europe*, the advantages and disadvantages of international cooperation are considered to be as follows: [1]

> [The] advantages of international cooperation in the NASP Program include sharing technology data and information, expertise, and approaches; having access to greater resources such as test facilities; sharing costs; reducing or eliminating duplication; and increasing the market size. Disadvantages include inherent difficulties in different program goals and objectives, concepts and size; sharing technology; sharing ownership; difficulties in integrating diverse national bureaucracies; delays in reaching decisions due to differing political and legal systems; complications resulting from different decision processes, priorities, and competencies; increased administrative costs; political inertia which may make projects hard to start and even harder to stop; competition for funding with countries' other national and international commitments; a tendency to undertake low-priority projects only; conflicts between cooperation and competition; and a decreasing market size.

Although these advantages and disadvantages are from the perspective of cooperation in the U.S. NASP program, most of these issues are considered generic. At this point, it is important to differentiate between the various types of international cooperations. Discussion of international cooperation within this context can refer to formal government-to-government agreements for those particular programs funded by governments—the International Space Station is an appropriate example. At the other extreme are contractor-to-contractor relationships which generally, although not exclusively, stem from competitive requirements.

This chapter tries to put into proper context some of the critical issues surrounding international cooperation, both generally and specifically. First, what is the status of international cooperation at the present time, and what are the motivations for pursuing further cooperative agreements? Second, and perhaps more important, what are the principal barriers to international programs? Indeed, is an international, government-funded aero-space plane program seen as the best way to achieve the development of a true space transportation system? Are there alternatives that may lead to competitive programs driven by market needs rather than nationalistic goals?

## *Existing Cooperative Efforts*

There is already a broad range of international cooperation on aero-space plane research at a variety of levels.

### ESA

The European Space Agency is, of course, bound in one very large international cooperative agreement formed by many of the individual sovereign nations that make up the collective notion of Europe. As a result, ESA automatically conducts a number of international studies in line with its charter. The Winged Launcher Configuration (WLC) studies, for example, that started in 1987 were intended to compare various winged, single-stage launchers with other two-stage to orbit concepts. Not surprisingly, this work is being jointly conducted by British and German companies and has been seen as a first attempt to rationalize both HOTOL and Sänger. Therefore, even though ESA has spent about $3 million on six WLC studies by 1993, the work has drawn partly upon the more substantial investments both countries have made during their individual national efforts. It should be noted that certain critical technologies, such as the HOTOL RB-545, could not be taken advantage of for classification reasons. In conjunction with WLC, ESA has hopes to establish the much more significant FESTIP program by mid-1994 to pursue enabling technologies and other systems work on future launch vehicles.

### Interim-HOTOL

The HOTOL program was reborn in 1990 after an invitation by the Soviets to study a modified version of HOTOL launched off the back of their AN-225 aircraft. In September 1990, a protocol was signed for an initial 6-month cooperative study. Positive results from this effort led to further, more detailed studies. As discussed in Chapter 7, this was a substantial effort for the British and former Soviets. The latter constructed various models of the Interim-HOTOL to perform separation tests in their wind tunnels. In addition, detailed engineering design and economic work on the proposed vehicle was performed. British Aerospace is currently trying to attract other European countries to participate in the next phase of the effort, as well as to convince Europe of the importance of cooperating with the C.I.S.

### German Hypersonic Technology Program

The Sänger effort has secured its first direct government-to-government international cooperative agreements for the continuation of Phase 1 efforts with the signing of two Memorandums of Understanding (MOUs). In particular, Sweden (September 1990) and Norway (December 1990) are each contributing about 2%. Italy may agree to contribute 20% to the Sänger Phase 1 effort, although it will probably invest in FESTIP instead. Germany is actively pursuing other possible agreements and hopes to have Spain and Austria sign MOUs in the near future. Prior to this, DASA had an industry level technology exchange agreement with British Aerospace Australia.

### SSRT

The SSRT program has so far included some international cooperation at the contractor level, with the blessing of BMDO at the Pentagon. This may seem somewhat surprising for a DOD-funded program, and especially one which uses many of the technologies developed from the classified NASP program. SDIO has previously funded other international SDI research, and its experience in cooperating in other countries has probably filtered through to the SSRT program. The original (1990/91) McDonnell Douglas team included British Aerospace, Aerospatiale, and DASA, and the original Rockwell team also included British Aerospace. Currently, only DASA is involved in the McDonnell Douglas SSRT efforts, having built the landing legs for the DC-X.

### NASP

Although the classified NASP program does not have or expect to have any government-to-government cooperative agreements, A modicum of work has been performed by foreign countries. In Britain the NASP program is using the University of Sheffield's hypersonic shock tunnel to perform generic work in the area of high Mach number aerodynamics in order to help understand and minimize the scramjet performance losses [2]. NASP contractors have been studying the possibility of using a carbon fiber-reinforced silicon carbide composite material developed by SEP of France and licensed by DuPont of the United States. In addition, a silicon carbide fiber developed by British Petroleum is also being evaluated and, because this material needed a U.S.-developed coating to make it suitable for NASP, this has led to one of only a handful of government-approved transfers of NASP technology outside the United States [3].

## *The Power of International Cooperation*

As government space programs become more complex and involve use of higher standards of technology, program costs invariably increase. The time required to bring them to fruition also increases, which consequently tends to drive costs higher. Naturally, governments wishing to pursue a particular project may need to enlist the support of other nations to spread out these costs, minimize technical risks, and obtain technological capabilities they would otherwise lack.

Enlisting foreign partners can also have negative impacts and risks. In particular, the success of one partner's effort can be held hostage to the possibility of another partner pulling out. These drawbacks must be weighed against the political value of having partners onboard a major program. The survival of the U.S.-supplied Space Station core has been in no small way a consequence of the international political repercussions that almost certainly would ensue if the United States unilaterally cancelled Freedom—an act that would leave Canadian, Japanese, and most of the European hardware with no place to go (Figure 13.1).

Already, Europe, Canada, and Japan have invested about $2–3 billion in various contributions and are planning to invest several billion dollars more over the next decade. On June 4, 1991, Jean-Marie Luton, director general of ESA, and other senior officials of Canada and Japan, testified before the U.S. House Science, Space and Technology Com-

**Figure 13.1** The International Space Station in 1986 (*NASA*).

mittee that oversees the NASA budget about the possible threat of the United States cancelling Freedom. This, in addition to a coordinated lobbying effort that included letters and joint communiques, was the first time NASA had played its international card in support of its Fiscal 1992 Space Station budget request. Two days after these precedent-setting international testimonies, the House of Representatives reversed a decision by the appropriations subcommittee that had previously voted to eliminate funding for Freedom. Whether the House reinstated Freedom funding primarily because of U.S. commitments to its international partners is not certain. It clearly was an important factor. According to an interview with Robert Frietag, former NASA deputy director of advanced programs, in *Space News*, [4]

> We knew if we found ourselves locking in with international agreements, it would be awfully hard [for the U.S. Congress] to say no to the program.

Much can be read into the implications of this situation. If Freedom were a U.S.-only program, would it have survived as long as it already has, and would NASA's budget *in toto* have received the large increases it has over recent years? It may well be that the future funding of NASA's major initiatives, like Freedom, is precariously dependent on its current international cooperative agreements.

## The Single Act: Europe '93

The development of the first aero-space plane, whether it is experimental, prototype, or operational, may need to be an international cooperative program funded by governments. Europe has no other choice, as cooperation is rapidly becoming the norm when it comes to conducting business. Few major European high-technology programs, including military programs, are conducted on a national basis. The Helios optical military reconnaissance satellite is being built by France, Italy, and Spain. The new European fighter aircraft is being developed by Britain, Germany, Spain, and Italy, and the Airbus airliners are built by a European consortium of industries, and so on.

After December 31, 1992, all impediments to free trade within Europe were dismantled with the Single Act of the Commission European Community (CEC), and the concept of international cooperation in Europe may take on new meaning. A primary reason for the Single Act is to ensure that things like mass-produced consumer products, such as VCRs and personal computers, are built to the same technical standards. Prior to 1993, such products were built to the standards set by the country where the products will be sold. Space programs, for the most part, do not fall under such constraints. More significantly for European space programs, the Single Act is intended to open up a large fraction of previously closed government markets—a market which accounts for around 50% of the total GNP of each nation—to equal and fair competition within Europe. For example, it will allow an Italian company to compete fairly for government-sponsored work in Germany, and vice versa. The only exceptions to this are basic R&D activities and, of course, certain military-sensitive programs.

ESA has functioned under a policy of *juste retour*, meaning that if a country contributes 20% of the funds to ESA, its national industries will get around 20% of the contracts. This system has worked well since the birth of ESA, as each European nation has had the right to give government-funded contracts to its own industry. Because ESA was formed outside the CEC, the juste retour policy will be in direct violation with the Single Act. ESA member states are currently grappling with this problem, although a possible resolution might involve classifying the work of ESA as R&D, as, indeed, most of it is. Speaking in 1991, in the opinion of George van Reeth, "I think the danger in there is indeed that companies—not states—go before the European court of justice in Luxembourg . . . And it is there you could run into problems. I realize it could happen. But as long as the Single Act contains a disposition that R&D contracts are not subject to the open competition, we're OK for 90% of what we do. We have about 10% of our activity which you cannot call pure R&D."

Of course, the success of a government-sponsored program like a European aero-space plane will still depend entirely on the priorities of each government. Even though the fair competition rule now applies, each nation may still demand that if it puts $1 billion in, it will expect to receive $1 billion of work. How this will function legally is not at all clear once a European aero-space plane program transitions from the enabling technology R&D efforts under FESTIP to the construction of operational vehicles and their associated infrastructure. One option is suggested by George van Reeth, "We will probably have just as much geographical distribution, but it will be presented in a more palatable way. You can turn the thing around. Today you get a program running, you ask everybody 'you participate for 3%, you participate for 7%, you participate for 35%.' And then you look for work in each country . . . Instead, you can do the opposite. You can conceive a program, see where the work can be done, write all the Request for Quotations, get the package together, and then say 'gentlemen, your participation in view of your industrial capabilities is going to be a certain percentage.' More distinctly, you put the people in who want to be there and say that that's the package and, in order to pay for it, 'you'll have to pay 25%, you'll have to pay 5%, you'll have to pay 3%', etc. Now it's not going to be all that clear. There will be iterations all the time, but you can work it that way."

International cooperation within Europe is a way of life, as the above discussion shows, and once a European cooperative program is initiated, it is generally extremely difficult to stop. Europe is a tightly integrated business community, and one country threatening to renege on its promises can bring swift reprisals from the other partners. For example, in 1987 the U.K. threatened to renege on its promise to increase the ESA space science budget by 5% each year until 1992. The sum of money involved was small, a few million dollars. But because the space science budget is a mandatory budget item, as opposes to an optional budget item such as Columbus or Ariane 5, the threat of the U.K. pulling out effectively would have meant the U.K. intended to veto the planned 5% increase to the ESA science budget *in toto*. In other words, the remaining nations originally agreeing to this budget plan would have been unable to make up the shortfall until after the program expired. The U.K. failing to meet its 5% increase promise would mean that other nations would be unable to also increase their contributions by the agreed 5% per year—thus freezing the budget and slowing down the planned programs.

A concerted diplomatic lobbying effort on behalf of the European nations forced the U.K. to reverse its decision and stick to its original agreement. It is important to note that even though the actual business impacts to the U.K. would have been relatively minor, the more significant impact would have been in the area of trust. If the U.K. could one minute promise to fund a program that was relatively cheap, and then the next minute turn around and renege on a promise that was made in good faith, then how could it be trusted when it came to cooperation on more expensive programs?

The Hermes and Columbus programs also provide another example. It should be appreciated that rarely does ESA itself invent programs, especially those in the optional budget item category, even though technically it is supposed to, according to the ESA Convention [5]. Columbus was initially driven by a bilateral cooperative study effort between Germany and Italy, while Hermes was the brainchild of France. It is a generally held view within Europe that Germany and Italy agreed to support the French-led Hermes program because this was the only way to secure French funding for Columbus—and vice versa, of course. According to Roy Gibson, [6]

> . . . the two programmes [Hermes and Columbus] are intended (at least by some of the ESA member states) to be inextricably linked. Put bluntly, the price of French support for the German-inspired Columbus programme appears to have been German support for the French-inspired Hermes. If this condition is maintained when the reviews are carried out before or just after the end of the first three-year period, it may well mean that the only way to get approval for one of the programmes [at the 1991 Council meeting] will be to approve the other simultaneously—whatever its shortcomings.

If Germany had unilaterally decided to pull out of Hermes at the November 1991 Council meeting in Munich, France would almost certainly have pulled out of Columbus. By the 1992 Ministeral Meeting in Granada, however, Germany decided to withdraw funding for Hermes. Hermes effectively died, taking the Columbus Free-Flyer with it. Only the Columbus Attached Module survived.

## *Technology Transfer: A Temporary Barrier to Cooperation*

As already noted in Chapter 12, aero-space planes will push technology to the limit, and funding for many of today's programs have been justified on this basis alone. As might be expected, if a nation or organization spends money on particular aspects of leading-edge technology, it might be somewhat reluctant to hand it over to other nations and organizations. It is this issue, perhaps above all others, which today hampers the prospects of international cooperation to develop aero-space planes. As technology advances over the coming decades, and as applications for this technology are sought, these barriers should become less of a hindrance.

The HOTOL project was one of the first aero-space plane programs to demonstrate the problems with technology transfer. The HOTOL propulsion system design was classified secret by the U.K. Ministry of Defence in order to stop details of the engine falling into the hands of the Soviets. Reluctance of the U.K. government and industry to remove the classification was one of many factors that made it impossible to attract international partners. As a result, the original concept of an air-breathing, single-stage-to-orbit HOTOL vehicle ground to a halt.

HOTOL was also promoted as a potential program for cooperation with the United States. In September 1988, the NASP and HOTOL teams were given classified briefings on each other's propulsion systems designs and according to a GAO report, [7]

> . . . after the exchange of engine data, British Aerospace officials were reportedly less interested in cooperating with the U.S.

This situation is not surprising, given that the objectives and magnitude of the two projects were completely different, but this was not the only roadblock. After these briefings, some U.S. officials expressed an interest in working with BAe and Rolls Royce to explore the feasibility of incorporating the RB-545 into the X-30 for the low-speed flight regime. After an import license was drawn up to allow the RB-545 data into the United States, U.S. State Department regulations allegedly put a stop to this initiative. The reasoning was because NASP was a classified program, once the RB-545 became a part of the NASP activities, it too would be classified under U.S. technology *export* regulations. Simply put, this meant that BAe and Rolls Royce officials would not have been allowed to talk to U.S. officials about *their own engine* [8]. Therefore, it was not surprising that the British were seen as less interested in cooperating with the United States. It was also reported that the United States was not as interested in genuinely collaborating with the U.K. as it was in having the U.K. contribute to a U.S. program. From a U.S. perspective, this is understandable given the probable large funding disparity between the two nations. From the U.K.'s perspective, it is also understandable that it would not be interested in playing a subservient role in a program where it had little control.

German officials have also complained about what they were actually getting from their contributions to some U.S.-led programs. In the opinion of Heribert Kuczera, "I do not know enough about the X-31, but it's more or less a one-way road. We are happy to be in a position to contribute, but what do we get? A label which says 'MBB,' and the German flag on the X-31?" He elaborates further on the subject of German-U.S. international cooperation from the perspective of technology transfer, "I do not want to exaggerate, but the current status is—and it is somewhat frustrating—that cooperation with the U.S.S.R. is much easier than with the U.S. That's a problem because we would like to cooperate with the U.S. We get a lot of comments from our American friends who say privately of the Sänger concept 'it's easy, it's conservative and it's realistic—and we are concentrating on this X-30 that does not seem to be feasible, at least not to be able to reach orbit.' So, they have problems with their political direction because they have a program that they cannot change easily, and I think that is clear. They cannot say 'up to now we were aiming at a single stage to orbit vehicle, and from now on we think a two stage vehicle is much better.' From my point of view, the Americans would like to cooperate, but it probably would be a one-way road—just 'come into the boat, and then we will see.' They did this several times before, and this is not what we want. And then they come up with restrictions, technology transfer and so on . . . It's a pity."

European nations, Japan, and the former U.S.S.R. have expressed interest in cooperating with the United States. According to Bruce Abell, "The U.S. is not interested in participation by other countries at this point. If that was unclear a year ago, it certainly is much clearer now. The U.S. doesn't feel the other countries have anything to offer in terms of technology right now, which is quite possibly correct if you're trying to push the Mach number for air-breathing as high as you can. They wouldn't mind having the [foreign] money, but you can't get the money without sharing the technology."

Congressman Dave McCurdy provides a perspective on the attitude of the U.S. Congress towards international cooperation on NASP, "If we can shoulder it alone, in the long-term I think American industry benefits, and I'd like to see them benefit. So, from the standpoint of the potential for great creation of wealth in the future, it's a good investment and I think we ought to continue. If it gets to the point where we're not going to shoulder it, then, sure, we'd bring in some of our allies if they're willing to do it."

As the title of this section suggests, technology transfer is probably only a temporary restriction, despite the overwhelming problems it presents today. When this technology can be applied to saleable products these barriers are likely to collapse—as they have for the sale of military fighters and other systems to foreign countries. Once again, as Bruce Abell notes, [9]

> What do we gain, and what do we lose when we invite foreign participation in our NASP program and in what we hope will follow? My belief is that, under the existing circumstances, we gain far more in terms of program vitality than we might lose in terms of transferred technology. This program is at risk, and we have to find a way to generate stronger public appreciation of its importance. Because if we don't, and if the pace of technology development continues to slow, then our leadership position in aerospace technology is threatened. This is no longer 1985; the NASP program's successes have awakened worldwide interest in hypersonics, and we no longer have the luxury of deciding how fast the technology will develop worldwide.

Speaking in 1991, Bruce Abell believes further that "when push comes to shove, when you've got a $10 billion program and it looks like you're not going to be able to swing it yourself, then you might find it is much more interesting to bring in some partners to work with you on it. I know that the Europeans would very much like to become cooperatives in whatever the U.S. is doing. Since money could become an issue, then there may be a need to look afresh at cooperation. If you want to participate in a worldwide market, then you have to have worldwide participation in it. It's going to be very difficult to get launch business from other countries if they have competing systems."

## Multiyear Versus Yearly Funding

Past space cooperation between the U.S. and European nations in space, including the International Solar Polar Mission, Spacelab, and Freedom/Columbus, has not been without problems, and this is likely to continue with any international cooperation with the United States on a future government-sponsored aero-space plane program. One reason for this is that opposite methods are used by the United States and Europe to fund programs. According to Joan Johnson-Freese, [10]

> . . . the U.S. budget process makes it impossible for NASA to guarantee the continuation of an international project beyond a yearly basis.

Specifically, the United States provides budget authority to programs, including NASA's and the DOD's, on a yearly basis so that every year each line item is scrutinized and, depending on the political whims at the time, can either be deleted, cut back, fully funded, or increased. A mechanism does not exist to ensure that once a program is initiated it will continue to completion. The NASP program has been severely disrupted by this process, as was discussed in Chapter 6.

Europe, on the other hand, works on a multiyear funding basis—as indeed do Japan, Canada, and, until recently, the Soviet Union. In Europe, the ESA member states normally will agree on a budget package for a fixed period of time—up to 5 years for example. This timeframe can be less, as was the case with the optional Columbus program. For such programs, the ESA Convention stipulates that, [11]

> . . . if the reassessment shows that there is a cost overrun greater than 20% of the indicative financial envelope . . . any participating State may withdraw from the programme.

Inside the budget margins, the member states are committed to funding the program to completion with little possible recourse, unless cumulative cost overruns escalate beyond 20%. It should be noted that preserving the program within the 20% limitations sometimes is done at the expense of the system's utility, as with Hermes and Columbus, and therefore what the member states get for their money is not what was originally planned. On the positive side, multiyear funding makes it nearly impossible for any nation to withdraw from the program once it begins. On the negative side, it also means that European nations have to be reasonably conservative in their allocations as they are making commitments that must stand for many years, even under a new government.

Although the U.S. system of yearly funding can be disruptive, it does permit higher funding levels to be obtained, especially if a particular program is considered a priority and is well supported. In this respect, yearly funding is responsive, and once large sums of money have been committed, institutional momentum will tend to carry it forward. Conversely, although multiyear funding is less disruptive, it is also less responsive and maintains funding at more modest levels.

Interestingly, the political advantages of the responsive-type nature of yearly funding appropriations have begun to creep into ESA's space efforts. At the ESA ministerial meeting in Munich (November 1991), it was decided to postpone the decision to start full-scale development of Hermes and Columbus by 1 year, instead of committing Europe to the 14-year funding plan originally proposed. Further, European ministers resolved to meet every 2 years to review the progress of the optional programs over the coming years [12]. As a result, Europe is now likely to experience some of the problems the United States faces.

This brief overview of the different budgetary processes at work in the United States and Europe helps highlight the potential problems standing in the way of aero-space plane development. What conclusions can be drawn? If the U.S. administration and Congress say they want to develop an aero-space plane launcher, then their budget process is set up to permit this. If Europe wants to develop an aero-space plane within ESA, then technically it may have to wait until after the commitments to Ariane 5 and Columbus expire early in the next century. Because the operational costs of these European programs are expected to be high, it is likely that any available money, which could otherwise be spent on a new program such as an aero-space plane, will be needed to keep Ariane 5 and Columbus alive.

## *The Potential Threat to U.S. Dominance in Aero-Space Planes*

Given this fundamental difference in the way the United States and the rest of the world fund technology programs, combined with U.S. reluctance to open the NASP program to international cooperation for technology transfer reasons, what threat is there that a non-U.S. aero-space plane will emerge at some stage in the future?

Currently, Europe is collectively spending around $50 million directly on aero-space plane research. Europe's expediture is spread out over a range of separate, and basically, isolated efforts. Although the United States may not feel particularly threatened by Europe at this time, over the next few years the situation could change profoundly. By 1995, the amount of money Europe will be spending on aero-space planes may rise with the buildup of the French PREPHA program, the start of the ESA FESTIP Step 2 program, and the various other contributions being made to other national endeavors. In total, Europe may be spending closer to $100 million by 1995. The situation will become more interesting if the European states agree to focus on a single vehicle configuration. At such a point, the inertia such an international program creates might make it difficult to stop, especially if each nation agrees that building an aero-space plane is good for Europe.

This may not be an unrealistic expectation because if European nations ever intend to build an aero-space plane, then they *will* have to cooperate and, more than likely, with each other first. Such a move could come more quickly than is currently planned, especially if France begins to recognize the potential of aero-space planes and believes the technology might be closer than previously thought. Europe may have lost Hermes, but it has not lost the *technology* Hermes was responsible for developing. For the most part, this technology can be applied to aero-space planes.

The European Community has also started to express an interest in aero-space plane-related technologies. In a major policy report entitled *The European Community: Crossroads in Space*, by an advisory panel chaired by Roy Gibson, it was stated that, [13]

> It is already time to be considering the post-Ariane 5 period and to decide how Europe should proceed. Several EC countries are already funding programmes investigating single or dual stage-to-orbit recoverable launch systems. It is impossible to predict an entry into service date for such advanced systems, but the tendency is very clear. This sector will become the technological spearhead of the launcher industry, and could also have a marked influence on the aviation industry . . .
>
> The minimum ambition for Europe should be to establish as early as possible a flexible strategic framework that has been agreed at the European level, and within which over the next few years national and European programmes can work with the least practical danger of future conflict between exploratory concepts. Europe reveals a significant lack of basic R and D in this whole area. Long lead times are involved and there exists a wide range of disciplines and communities (e.g. in hypersonics, computational fluid dynamics, propulsion and control, materials technology, avionics) so far insufficiently connected to the European space programmes. The widest possible involvement of the relevant existing facilities and expertise in both public and private sectors is needed. European cooperation could clearly be beneficial especially in basic technology, and here the EC could assist through the Framework Program for Research and Development which already included an aeronautical element.

A general consensus in Europe that aero-space planes can be built and can reduce the cost of space operations might be all it takes to get such a program moving.

## *Cooperating With the Commonwealth of Independent States and Japan*

The C.I.S. has capabilities in materials, structures, and propulsion systems that rival those of the United States and Europe. Therefore, it is not inconceivable to imagine a Euro-C.I.S. aero-space plane program. Indeed, it seems ludicrous for Europe to ignore these vast resources in hypersonics and launch systems when the investment to obtain this knowledge would pale in comparison with the benefits to Europe. The recommendations of the Interim-HOTOL joint presentation made to ESA in June 1991 included the following, [14]

> - Soviet facilities and capability represent a valuable resource which Europe should tap into,
> - Soviet experimental capabilities are world class (eg. in hypersonic aerodynamics)
> - Soviet facilities and capabilities need some Western support to be sustained
> - Europe will be investing about 92 MAU [$110 million] per year in hypersonic research. If 5% went to the USSR facilities for relevant inputs, the immediate and long term rewards would be disproportionately high.

Part of the problem about cooperating with the C.I.S. is that Europe (and the United States, for that matter,) finds it difficult to believe that the C.I.S. could have a comparable, if not superior capability. Bob Parkinson, speaking about his experience cooperating with the former Soviets, explains, "It's very difficult to get some people, even those people who ought to know better, in Europe [to understand this] because they think we're all getting all excitable or something. And I think that the Lüst era was responsible for a lot of this which was 'we have this set program, we have to go down this set program, and we have Hermes and, therefore, it's got to be justified that we have it, and when we've done that, then we can pay attention to something else.' It's an insecurity thing; they believe if you rock the boat you won't get funds for anything. When I come back from the Soviet Union and say 'I have seen panel-sized lumps of titanium aluminide fabricated with optical welding ready for a vehicle of this sort, I've touched them, I've handled them,' they say 'you can't make titanium aluminide, we don't believe you.' Their attitude was that we were doing all right. They didn't believe that it is possible to be totally outclassed."

Cooperation might become more compelling if Europe decides that a cooperative aero-space plane program might be one way to help stave off a total collapse of the C.I.S., especially with the former Soviets on Europe's doorstep. This situation has been broadly recognized by the European nations. At the November 1991 ministerial meeting, ESA was directed to start exploring avenues for limited cooperation with the former Soviets.

Japan's desire to build a single-stage-to-orbit vehicle may inevitably lead it to Europe, as it is not in a position to fund its own program past the enabling technology phase. Japan, which already spends tens of millions of dollars on aerospace plane research, might therefore become a strategic partner in a joint Euro-C.I.S.-Japanese aero-space plane program. Such an alliance could clearly more than afford the costs to build an aero-space plane fleet. This scenario is supported in the conclusion of the GAO report, [15]

> Individually, European countries do not pose a serious challenge to U.S. preeminence in hypersonic aerospace plane technologies. However, the combined convergence of national interests, expertise, approaches, funding and sharing of test facilities in a major international collaborative effort in hypersonics under the European Space Agency or among European countries, Japan and/or the Soviet Union, could, in the long term, prove to be competitive with the NASP Program.

Viewed from an American perspective, much of this might be considered as wild speculation—cultural differences alone might preclude it. Yet it must be appreciated that if Europe ultimately intends to build an aero-space plane it *will have to cooperate*, within Europe and potentially with other countries. Certainly, as technology relentlessly continuous to advance, the feasibility of aero-space planes will become much clearer. It is probably unwise to second-guess the forces of change at work in today's highly internationalized world. The next few years will be very interesting for the future of aero-space planes.

## Competition: An Alternative Scenario

Consider a scenario where the United States is successful in demonstrating that aero-space planes can be built and can launch payloads more frequently, more reliably, and, above all, at a fraction of the cost of other launchers. A large number of successful test flights of the Delta Clipper prototype vehicle or other vehicles may be required to demonstrate such a capability. In this scenario, the United States might subsequently decide to build an operational aero-space plane. If this went unchallenged, the United States would almost certainly win back the commercial launch market in its entirety from Europe. (See Chapter 15.) Clearly, if Europe didn't want to give up this market, it would have to act quickly to remain commercially competitive.

Competition radically changes the boundaries of international cooperation. Investments are made with a view to producing an operationally effective system that can generate revenue. (See also the discussion in Chapter 15.) Now the SSRT program has examined possible alternatives for future operational versions of the Delta Clipper vehicle following a successful prototype development. Potentially, if the Delta Clipper prototype is able to prove conclusively that such vehicles can be built, then a commercial consortium, or quasi-government-commercial consortium, could take over the production and operations of a Delta Clipper fleet for the launch of commercial, civilian, and military payloads. In which case, finding parties to invest in such a fleet is likely to become more important than *where* that investment actually comes from. This view is shared by David Webb, [16]

> If U.S. corporations and financial houses are unwilling to risk the necessary capital (as they have been on numerous occasions in recent years—particularly in the computer and information sectors) it should be allowable for the initiator of the project to seek partners in other nations . . . Although such a possibility may appear too visionary for our generation, we should not underestimate the power of change. We only need to refer to the last few years of the eighties to see that.

It seems obvious that competition can transcend the normal barriers to international cooperation and issues like multiyear versus yearly funding or technology transfer will become redundant because the real issue at stake will be what aero-space planes *do*, in terms of operational capability, rather then what aero-space planes *are*, in terms of what they represent technologically or symbolically. Potentially, the realization that aero-space planes are possible may push some countries to suggest the setting up of an international aero-space plane group or groups that either supplies launch services to all countries or regulates the supply of aero-space plane vehicles to individual states. Certainly, if the cost to develop aero-space planes, together with the cost to put the necessary support infrastructure in place, turns out to be higher than currently anticipated, then a single development entity might become a possible solution. As David Webb indicated, this may appear visionary when compared with current space practices, but the proposed capability of aero-space planes, assuming they are feasible, may well make such a situation worth considering. Indeed, it may be inevitable.

## Summary—Roadmap for Aero-Space Plane Cooperation?

If governments are responsible for the development and operations of the first aero-space planes, international cooperation will be an intrinsic part of that activity in Europe, and almost certainly in the C.I.S. and Japan. The fusion of each nation's technical specializations, the political robustness cooperation gives, and the ability to spread out costs, all work together to support this notion. On the other hand, the United States may decide to go it alone because of concerns over technology transfer to other nations, and also because the United States is probably the only nation in a position to fund such a program in its entirety.

From a European and possibly also a Japanese perspective, these nations may be reluctant to play what they may feel to be a subservient role to the United States in a cooperative venture. Europe would probably only participate in any U.S. program if it could contribute to the definition of the vehicles and have unrestricted access to these vehicles for its own purposes. If past international cooperative experience is any indicator, together with the conflict between yearly and multiyear funding, Europe is unlikely to be enthusiastic about building part of an American vehicle that is controlled by American authorities. From an American perspective, however, if the United States allowed cooperation, that contribution would have to be very significant to be meaningful. The United States might be quite justified in demanding control of an aero-space plane program if all the international partners were prepared to spend was on the order of 10–20%, as they currently plan to do on the International Space Station.

This situation changes radically if the motivation for building aero-space planes is driven by commercial competition, rather than strictly for nationalistic or regionalistic purposes. If a U.S. government-funded development program were to prove that aero-space planes are feasible, then almost certainly Europe would have to respond if it intended to keep some of the commercial launch market. The reverse is also true. Technology transfer worries become redundant in this scenario because what is now important is what aero-space planes do rather than what they're made from. Potentially, it could be that commercial organizations (although probably government insured) would take on the role of building and operating the first aero-space plane fleet. In which case, obtaining the investment will be more important than where that investment comes from.

## References and Footnotes

1. *Aerospace Plane Technology: Research and Development Efforts in Europe*, (GAO/NSIAD-91-194, U.S. General Accounting Office, July 1991), p. 114.
2. Ibid.
3. "Bulk Manufacture of X-30 Materials in Offing as NASP Design Matures," *Aviation Week & Space Technology*, April 15, 1991, p. 69.
4. Lawler, A., "Station Partners Rally U.S. Government for NASA," *Space News*, June 17–23, 1991, p. 1. With permission of *Space News*.
5. *Convention of the European Space Agency & Rules of Procedure of ESA Council*, (The Scientific and Technical Publications Branch of the European Space Agency), June 1984.
6. Gibson, R., "Europe's Next Move," *Space Policy*, August 1989, p. 187. With permission of *Space Policy*.
7. GAO/NSIAD-91-194, p. 116.
8. Based on discussions with U.S. aero-space officials.
9. Abel, B., "NASP Policy Lessons from its Perilous Transition from Laboratory to Runway," Talk at the *AIAA Second International Aerospace Planes Conference*, Orlando, Florida, October 31, 1990.
10. Johnson-Freese, J., *Changing Patterns of International Cooperation in Space*, (Orbit Book Company, Malabar, Florida, 1990), p. 35.
11. *Convention of the European Space Agency & Rules of Procedure of ESA Council*, Annex II, Article III, (The Scientific and Technical Publications Branch of the European Space Agency), June 1984, p. 78.
12. A full report of the Ministerial Conference in Munich is presented in the ESA Bulletin, No. 68, Nov. 1991, pp. 9–40.
13. Report by an advisory panel on the European Community and Space, *The European Community Crossroads in Space*, (Office for Official Publications of the European Communities, Brussels, 1991), pp. 10–11.
14. Viewgraph presentation pack produced by British Aerospace (Space Systems) Ltd., and the Central Institute of Aero- & Hydrodynamics (TsAGI) for the An-225/Interim-HOTOL presentation to ESA, Paris, June 21, 1991.
15. GAO/NSIAD-91-194, p. 122.
16. Webb, D. C., and R. J. Hannigan, "Economic and Socio-Political Impacts of NASP-Derived Vehicles: A Technical Report," The International Hypersonic Research Institute, November 1990, p. 31.

# Chapter 14

# Spacecraft or Aircraft?

## Thinking Ahead

Sometime early in the 21st century, an aero-space plane sits on a runway, and within its cargo bay is its first commercial payload. The research and development test flights are finished; so are the government-subsidized demonstration flights. Now this aero-space plane can prove its worth as an operational space transportation system. The combined effects of engine thrust and wing lift cause it to take off from the runway to begin the initial ascent. A small banking maneuver heads it out over open water. Here, the aero-space plane cruises for a time until the required latitude is reached. With engines throttled up, and its nose pointing skyward, the aero-space plane accelerates toward space. At some almost unperceptible and undefined point during the ascent, the aero-space plane transitions from an aircraft to a spacecraft and goes into orbit, circling the Earth at 27,000 kilometers per hour.

Only hours after the aero-space plane has deployed this first payload, the engines on the vehicle are restarted to change the orbit parameters to a new orbit that intercepts the atmosphere. Gradually, the vehicle again experiences the slowing and frictional heating effects of the atmosphere, although it is a while before its elevators and ailerons regain their grip on the thin upper atmosphere and the vehicle can be said to be flying. In one scenario, the aero-space plane cannot descend directly to Earth. Instead, it skips off the atmosphere and ascends briefly back into space, before reentering again for further skips. Eventually, the aero-space plane slows to the point where the aerosurfaces can completely take over from the vehicle's thrusters for the remainder of the descent, and it can carve out its final trajectory to focus the sonic booms away from highly populated areas. At subsonic speed the aero-space plane turns toward the runway and lands.

Constructing a vehicle that can routinely fly to and from space is an immensely difficult technical, financial, and political challenge. Other issues have the potential to be just as challenging, even though they may appear to be secondary concerns at first glance. The flight of aero-space planes calls into play a number of legal, regulatory, insurance, and environmental issues:

- Will aero-space planes be allowed to overfly foreign territory?
- What is the extent of the operators' liability in the event of accidents?
- What are the environmental impacts of aero-space planes?
- Will aero-space planes be required to meet certification standards?
- Which regulatory body will oversee the certification and operation of aero-space planes?

It could be argued that many of these issues are not relevant today because no one has built an aero-space plane or, indeed, proven conclusively that it can be built. "Let's build an aero-space plane first, and we'll cross the legal bridge when we come to it," is an attitude often expressed. Unfortunately, it may not be as simple as that. The question "Are aero-space planes spacecraft or aircraft?" is a convenient way of summarizing the fact that the laws governing the flight of aircraft through the air do not neatly dovetail into those governing a spacecraft orbiting the Earth. Air law and space law are not just different, they are *profoundly* different. If a vehicle flies in both air and space, therein lies the basis for a conflict (Figure 14.1).

## International Air Law Versus Space Law

To begin with, consider air law. At the present time, all nations have sovereignty over the airspace immediately above their territory. Thus, airplanes must have prior permission to overfly the airspace of another nation. In addition, the type of aircraft and its mission and payload have to be defined and registered and, if necessary, even inspected. (Military reconnaissance flights are another matter, of course.) If such permission isn't granted, then incidents like the Soviet downing of the Korean Airlines Flight 007 in 1983 can arise. The subject of national sovereignty over airspace is governed by major conventions including the Convention of Paris (1919), the Warsaw Convention (1929), and the Chicago Convention (1944) [1].

Now consider space law. Objects launched into Earth orbit have little choice but to travel around the Earth in an elliptical path that has the center of the Earth as roughly the center of the orbit. Thus, if a spacecraft is launched in an easterly direction from the Kennedy Space Center located 28.5° north of the equator, the plane or inclination of its orbit will be such that the spacecraft will, to all intents and purposes, always travel 28.5° north *and* south of the equator. Further, because the Earth spins once every 24 hours and the plane of the spacecraft's orbit remains more or less fixed relative to the Sun, the net effect is that a spacecraft launched from KSC will overfly all territory between the latitudes of 28.5° north and south after only a few days in space. If the spacecraft is launched into an orbit whose plane is perpendicular to the equator, it will have no other choice but to overfly every single nation within a few days of launch. There are some orbit exceptions, but these are few. *Orbital mechanics, therefore, have dictated space law.* This was recognized by the United Nations Treaty on Outer Space (1967) which was signed and ratified by the majority of the world's spacefaring nations including the United States, the former U.S.S.R., and the countries of Europe. Importantly, this treaty deals with the critical issue of property rights by declaring that no property may be owned in outer space.

Space law obeys the physical laws of the universe. A satellite must be allowed to orbit the Earth, as it cannot change its path just because some country has forbidden "by law" its territorial space to be transgressed. It is easy for an airplane, on the other hand, to skirt around a country if the law will not allow passage.

Despite this absolute, there have been attempts to modify space law to make it more in line with air law. Some South American nations have suggested that sovereignty over airspace should extend all the way to geosynchronous orbit, to the disadvantage of nations north and south of the equator because geosynchronous orbit is the most lucrative satellite position. Others insist that a spacecraft travelling or returning from space must request permission before traversing a nation's borders. Such attempts have been resisted because they would severely constrain the nature of space activities, especially military operations.

In a situation where a space activity and an air activity are essentially the same, it is ironic that different laws govern them. This conflict is best exemplified in military air and space activities. Air law allows an airplane invading the airspace of a sovereign nation to be shot down. On the other hand, space law allows satellites to pass over all territories.

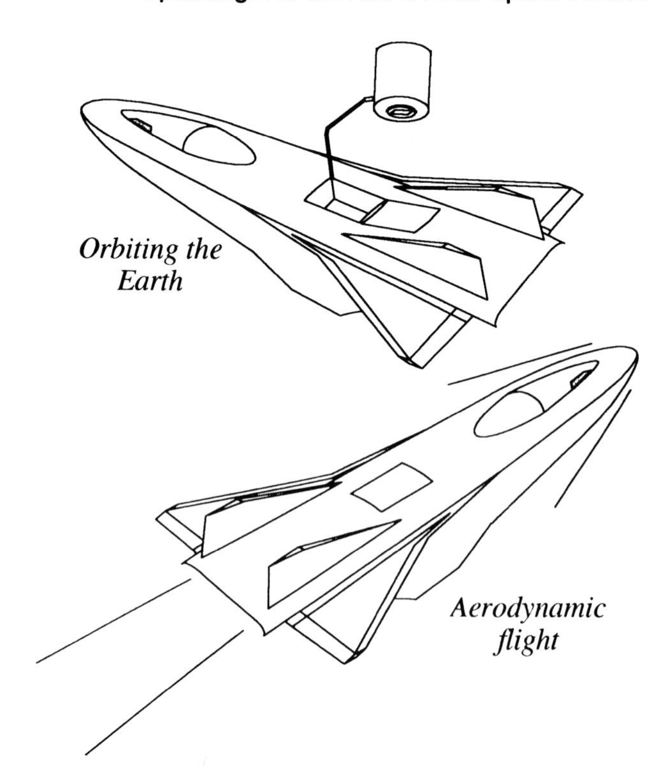


**Figure 14.1** Spacecraft or Aircraft?

A military spy satellite performs essentially an identical function to a spy plane. Why, therefore, is it permitted to shoot down a manned airplane, such as Gary Powers's U-2 in 1962, and it is not permitted to shoot down an inanimate satellite?

The reason seems to be linked to technical feasibility rather than logical reality. It has always been possible to shoot down airplanes; the same cannot be said for spacecraft. In recent years, it has become technically possible for some classes of reconnaissance satellites to be destroyed in orbit. It has also been recognized that the risk of losing a nation's military space "eyes and ears" has prevented the escalation of antisatellite weapons. It is considered better for opposing sides to have an orbital spy capability, than for either side to have none at all.

The U.N.'s Treaty on Outer Space gives any nation the right of "innocent passage" over every other nation's territory "for the benefit of all mankind." Clearly the terms *innocent* and *benefit* have different meanings in space than they do in the air.

## Current Launchers and the Space Shuttle Precedent

Most launch systems today are oblivious to the legal definition of where space begins and therefore the obvious conflicts between space law and air law have not applied. A typical expendable booster is usually launched over oceans or, in the case of the C.I.S. and China, over relatively uninhabited regions. Thus, the booster stages can be disposed of and if a problem occurs, the range safety officer can destroy the vehicle and ensure that it has no chance of swerving off course and falling on populated areas. Once a booster has successfully placed its payload in space, most of the vehicle has either crashed back to Earth or will eventually destructively reenter.

If a booster or orbiting spacecraft were to inadvertently crash into an inhabited area, an aircraft in flight, or a ship on the ocean, then, under the U.N. Outer Space Treaty, the launching nation must pay for any damages caused. Specifically,

> Each state which launches or procures the launching of an object into outer space, and each state from whose territory or facility an object is launched, is internationally liable for damage to a foreign state or to its component parts on the earth, in air space, or in outer space.

And, according to the Convention on International Liability for Damage caused by Space Objects (1972),

> A Launching State shall be absolutely liable to pay compensation caused by its space object on the surface of the earth, or to aircraft in flight.

Such a situation occurred in 1978 when the Soviet Cosmos 954 spilled radioactive material over Canada, and the Soviet Union was forced to pay for the cleanup [2]. In the minority of cases where a spacecraft is intentionally returned to Earth, that reentry usually occurs over water or the nation's own territory. If not, the limited number of occasions permits governments to inform other nations of their plans for an overflight.

The Space Shuttle is the first space transportation system that "routinely" flies to and from space. It is wired with explosive charges in order to destroy it during launch, but only on the solid rocket boosters and the external tank, and not on the orbiter itself. The reason being that the SRBs cannot be switched off like a jet engine, meaning that the Shuttle could, in theory, turn back to Florida after launch if, for example, the SRB nozzles became locked in one position. The SRBs of the *Challenger* accident had to be destroyed for this reason.

Even though the Shuttle orbiter is capable of flying back to Earth like an airplane, it is considered by most countries as a spacecraft rather than an aircraft throughout its flight regime, as David Webb explains, [3]

> The first flights of the Space Shuttle initially caused considerable controversy within the international legal community. Fortunately, however, this quickly died down because of two circumstances: First, NASA sensibly aware that there might be a problem, had in 1978, amended the National Aeronautics and Space Act of 1958 to include the Shuttle as a "spacecraft" within the meaning of the Act; second, some eminent international jurists pointed out that the Shuttle was quite evidently more spacecraft than airplane, as, with self-contained power-plants (i.e., rocket engines) it did not require the atmosphere to attain space, and it had no powered capability in the atmosphere when it returned from space. This argument was accepted by most nations and the Shuttle became a spacecraft.

The Shuttle is also considered more of a spacecraft because it is launched over water, and its reentry usually begins over the Pacific Ocean. Unlike an aircraft, the Shuttle has only been launched from KSC, and landings either occur at the Dryden Flight Research Center or KSC, although White Sands was used on one occasion. Backup transatlantic abort landing sites are available in Ben Guerir, Morocco; Moron, Spain; and Banjul, The Gambia [4]. The U.S. government has negotiated separate agreements with each of these countries to cover the eventuality that an orbiter may have to set down there.

The Shuttle can be a spacecraft—or at least not an aircraft—because its operations are so precisely constrained. Specifically,

- It can only be launched and landed at fixed sites.
- Aerodynamic flying in the atmosphere almost always occurs over water or over a limited number of predeterminable countries.
- The abort landing sites are predetermined and agreements with other nations are made well in advance.
- The Shuttle is currently the only vehicle of its type to fly operationally.
- Most importantly of all, because it doesn't fly often enough to justify a new "Shuttle Law," ad hoc intergovernmental agreements suffice.

The possible continuation of the C.I.S. Buran and even the future introduction of the European Hermes would not change this situation, for many of the same reasons. The nominal flight rate of Hermes (before it was cancelled) is one to two flights per year, while Buran's is believed to be around two or three flights per year. If several different aero-space plane fleets flew a hundred times per year to space using various launch and landing sites, then it might be a different story, as will be discussed later.

## Implications for Government-Operated Aero-Space Planes

Given the precedent set by the Shuttle, will there be a need for an "Aero-Space Plane Law," or can aero-space planes fit within the confines of already existing laws and treaties? Since the objective of aero-space planes is to reduce the cost of launching payloads and to make it easier to access space, it might be perfectly acceptable to base the first U.S.

aero-space plane fleet at the Kennedy Space Center, the first European fleet in Kourou, and the first Japanese fleet at Tanegashima. The reason is that the cost of transporting the payload to these facilities will still be minor compared with cost of the launch. It costs on the order of hundreds of thousands of dollars to transport a typical satellite to KSC from the U.S. West Coast. Thus, if it costs $10 million to launch a payload on an aero-space plane compared with the hundreds of millions of dollars to launch a payload on the Shuttle, then the issue of where it is launched from is not too important.

In the above scenario, a government owned and operated aero-space plane could fly a trajectory little different from the Shuttle. That is, it could take off from runway 15 or 33 and accelerate to orbit long before reaching Africa or Europe, thereby avoiding the problems of overflying another nation's territory as airplanes do. Similarly, the reentry could begin over the Pacific or western United States with an unpowered landing back to KSC. The sonic boom pattern and intensity could probably be kept within the limits of the Shuttle, as will be discussed later. Similar arrangements could be made for launches from Kourou, and the agreements made to cover the Hermes program would probably suffice.

This type of trajectory would be conceivable for air-breathing or pure rocket-powered, winged one- and two-stage-to-orbit aero-space planes and would enable the vast majority of space activities to be conducted on a more effective basis than possible today. The one exception could be vehicles that require powered landings, such as the Delta Clipper. This vehicle as currently conceived will either abort back to the launch site or abort to orbit, and it is questionable whether an abort down-range facility will be required. If it is, then a few large and uninhabited areas could probably be designated as abort sites.

In the aero-space plane era, it will still be necessary to obtain agreements with individual countries for emergency landing sites, as it is for the Shuttle. If aborts remain infrequent—one every few years—the need for a new aero-space plane legal regime to cover these eventualities seems unnecessary. Conceivably, aero-space planes could also inadvertently and uncontrollably fall out of orbit, as indeed can any large spacecraft (e.g., Skylab and Salyut 7). In such an instance, any injured or damaged parties are covered under the terms of the U.N. Outer Space and other treaties. As David Webb points out, [5]

> Air and space law are subsets of international law. All nations are bound together inextricably by a series of treaties and conventions that have evolved over the centuries to avoid international anarchy. Each nation is dependent on the others recognizing a common code of conduct that has grown out of these laws (e.g. the recognition of diplomatic immunity). If one nation unilaterally transgresses these laws, it is likely that many of the others will band together in a common action to persuade that a nation return to the original status quo.

The U.N.'s uncompromising resolution to extricate Iraq from Kuwait is an obvious example of this theory in action.

To summarize, it is possible to foresee a situation where a future government owned and operated aero-space plane fleet could, if it had to, function properly and effectively within the type of international agreements that currently suffice for the Space Shuttle. Nothing substantial would need to change from a legal perspective. For commercial operators, though, the situation is radically different.

## *Implications for Commercially Operated Aero-Space Planes*

Aero-space planes have the potential to be more than just government owned and operated replacements for the Shuttle and other expendable launchers. A government-developed aero-space plane fleet could be handed over to a commercial entity (Chapter 16). Alternatively, an international consortium—either government or privately financed —could sell aero-space plane vehicles to organizations and countries requiring them. These and other options are perhaps realistic propositions assuming an elastic market response to significantly reduced launch costs.

Given such scenarios, aero-space planes will have to be built and operated within the type of practices and regulatory measures all other businesses must adhere to today. Principal among these are the issues of insurance, liability, the environment, and airworthiness, most or all of which will impact the prospects of commercial aero-space planes.

## *Insured Access to Space*

It is extremely unlikely that the commercial operator of an aero-space plane fleet will be allowed to fly its vehicles unless appropriate measures have been taken to insure against failures and litigation brought by damage caused to third parties (see following subsection). The issues of insurance and third party liability coverage have been a significant concern already in the commercial launcher field and were a principal reason why the Reagan administration dropped any plans to hand over the Shuttle fleet to the private sector, despite moves by companies like Astrotech, and its subsidiary Space Shuttle of America Corporation, to seriously consider purchasing orbiters. According to the report by the Shuttle Operations Strategic Planning Group released in April 1985, [6]

> The liability for systems failure must rest with the government, particularly in the case of a catastrophic failure. A private operator

may be able to get first-layer insurance coverage for his actions, but the group believes the government would have to retain the responsibility for major accidents. If the responsibility still largely rests with the government, it does not seem sensible to turn the system over to private enterprise, which may or may not operate in the government-approved manner.

The commercial development and operations of aero-space plane fleets are likely to come up against many of these problems. Consider a situation where a private organization attempts to build and operate the first aero-space plane fleet shortly after the government had already paid for and flown an experimental vehicle like the X-30 or a prototype vehicle like the Delta Clipper. Even though the government may have been able to demonstrate that aero-space planes can function as reliably and economically as hoped for, maintaining an operational fleet that is capable of flying a large number of missions year after year (e.g., greater than 20 or 30) will be a different matter entirely. Further, if problems are going to occur, they are most likely to happen during the initial years when experience is low on the learning curve. Unfortunately, this coincides with the period of time when the users are still trying to adjust to this fundamentally different means of accessing space. The ability of the fleet operator to attract existing customers, and stimulate new users (Section 4), will be very sensitive to the early success of the aero-space plane. A catastrophic failure in the first few years may have devastating effects on the economic viability of the commercial endeavor. At that point, the market will not have had sufficient time to develop, meaning revenues will be relatively modest. Additionally, the cost of the failure and the failure recovery activity is likely to be enormous, representing a large fraction of the original fleet investment cost.

In this scenario, it seems reasonable to conclude that the commercial insurance industry would be unable to provide adequate capacity to cover this immature and highly risky business venture against the potential billion dollar claims. As Ann Deering explains, "since the start of the commercial 'space' insurance industry the maximum amount of insurance available to cover a single launch event against physical damage to the satellite and launch vehicle is $300–$325 million. The available amount of liability insurance had a maximum world capacity of $1 billion per launch event. It is not likely that any more capacity will be available if poor vehicle performance and potential large losses are expected to occur."

This situation contrasts sharply with the development of the aviation industry. Today's large airliners and other aircraft can all be traced back to a multitude of early aircraft programs. The first airplanes were small and relatively inexpensive compared with today's, and each could only carry small payloads over short distances. Most of these aircraft were developed and proven as the result of military requirements (Chapter 16), and different military requirements demanded various types of planes, such as fighters, bombers, and transporters. Therefore, if a particular type of aircraft crashed, such failures were tolerated because the cost impact, *as a fraction of the total investment in the aviation industry*, was relatively modest. Even though there were many failures in the early years of aviation, these did not seriously jeapardize the development of the industry. This was because a large number of aircraft development programs were already underway around the world. It didn't take long before the potential advantages of powered flight were realized. The market for air transportation was ready-made because of the precedents set by other forms of fully reusable transportation, such as boats, trains, cars, and horses.

The same cannot be said of the aero-space plane business. For an aero-space plane to carry a significant payload to orbit, very large vehicles have to be built from the start—*small* aero-space planes cannot be built. As a result, such large, high-technology machines are going to be expensive and, initially at least, only a few organizations are likely to be in a position to pay the development costs (Chapters 12 and 16). Therefore, the potential failure of the first commercial aero-space plane fleet is likely to have profound repercussions on other programs—just as the Shuttle has done, albeit for different reasons. This situation is further complicated by the fact that the use of space is so underdeveloped today, and it will take a successful aero-space plane to determine whether space can be more widely exploited.

Should a commercial company obtain the requisite insurance coverage, one positive outcome might be a significant reduction in the cost of launch insurance for the customer. The cost of a catastrophic aero-space plane failure is likely to be many times greater than the cost of the payload—roughly $1 billion for the aero-space plane and $50 million for the payload. Therefore, the cost the aero-space plane operator charges for a launch could include a small contribution to the insurance, with this being proportionate to the ratio of the payload cost to the aero-space plane cost [7]. To take a simplistic example, if the cost of a dedicated launch is $10 million, perhaps only 5% or $500,000 of this would be launch insurance (i.e, the satellite costs 5% of the value of the aero-space plane it is being launched on). The user would still have to obtain insurance for the other activities of the spacecraft, such as raising the orbit to geosynchronous altitude, but at least the riskiest part of the mission would be covered. For comparison, current insurance costs have skyrocketed to about 16–18% due to launcher failures in recent years. Because insurance must also cover high launch costs, the average premium today is around *$25 million* for a $150 million launch event.

## *Who Is Liable?*

Unlike government-operated transportation systems, commercial airliners, ships, trains, and cars are insured against third party actions, as will be aero-space planes, which will need some form of insurance against damage or injury caused to third parties. The case of the current U.S. commercial expendable launch vehicle industry is explained by Ann Deering, [8]

In the mid-1980's, the commercial ELV industry experienced many obstacles in developing a regulatory environment which responded to the third party liability issues and also met the requirements of the Air Force, NASA, and ELV operators in 1984. The Department of Transportation Office of Commercial Space Transportation (DOT OCST) was established to regulate the ELV industry through licensing procedures to protect the interests of public safety.

After a great deal of inter-agency dispute, the administrator of the Department of Transportation was given paramount authority over all other agencies to settle third party liability and government indemnification issues. The DOT OCST sets third party liability requirements for commercial ELV operators for damage to persons and property or government launch facilities. The amount of mandatory third party liability insurance differs for each operator. DOT OCST conducts risk assessments which evaluate the maximum probable loss which could occur at launch. This estimates the maximum possible injury to persons or property.

The ELV operator is required to purchase insurance in that amount which could be as low as $20 million or as high as $80 million. The U.S. Government then indemnifies the operator should an accident occur and if damages exceed the set insurance limit. In other words, ELV operators have a cap on the maximum amount of their exposure for any single launch event. The reasoning behind this is to prevent a commercial operator from exposing the company to unlimited liability. Arianespace also provides third party liability insurance and indemnification.

To the extent that an aero-space plane is not a pure aircraft, intended to carry passengers from one point of the Earth to another, its status is currently unregulated. Even if such regulations are not in place before the first commercially operated vehicle takes off, this doesn't mean that there won't be any liability issues to contend with, but merely that liability will be governed by the domestic law principals of the state. Eventually, organizational structures will need to be established to regulate aero-space planes. Existing air law can be used as a starting point to see what form such regulatory structures might take and how they might affect litigation concerns.

The draftsmen of the Warsaw Convention (1929), and those who amended it at The Hague (1955), were originally concerned with formulating legal constraints to save operators from large litigation claims, thereby minimizing damaging effects on the aviation industry as a whole. Their eventual solution was to agree to set limits on the operator's liability. At the same time, in the interest of fairness, the draftsmen also ensured that passengers and cargo owners had a way to claim damages without the need to prove fault. In addition, they dictated where the plaintiff had to sue, that is, the forum. The combination of these solutions was eventually to lead to a number of problems that hang over the airline business even today, as Robert Webb QC explains, [9]

The effect of this cocktail mixture of a limit of liability and a compulsory forum in which to sue was, like so many cocktails, to give geat initial satisfaction in the short term, but then to give way, as it is now doing, to an icy hangover. Firstly the limits, too low for rich countries and too high for poor ones, were passed unilaterally by governments; they made it a term of landing permissions that limits be increased or Article 20 defences waived; they were passed by Courts—who strained the language of the Convention to achieve "just" results in given, tragic, cases but above all they were passed by litigants who chose, instead of suing the operator/carrier, to take advantage of the developing law of product liability to sue other Defendants: manufacturer, licenser, Air Traffic Control, airport owner and so on. These decisions were not based on perceptions of fault, or blame; they were either simply an attempt to circumvent the limitation of the Convention, or they were an attempt to find "deep pocket" companies who might be willing to pay large sums to protect their product's name . . . The truth today is that *where* you can sue someone is often more important than how much they are to *blame*, or even if they are to blame at all.

Robert Webb QC goes on to provide an example of how this situation tends to work today,

. . . a legal claim has no universal value, unlike a gold nugget, a French Franc or a U.S. Dollar. Its value depends on whose Court computes it. [Consider] the case of an American who was killed on an internal Bangladesh-Biman flight. In Bangladesh his claim was worth £900 maximum by government decree. In England, where Bangladesh could be sued because they had an office, it was worth £85,000. In America, it was worth $3 million—same death, same man, same widow, but different jurisdiction. With these disparities it is not surprising perhaps that the legal question has turned from "Who is to blame?" to "Where can we sue?"

Certainly, if an international convention for the aero-space plane is derived directly from the existing Warsaw Convention, then would-be commercial aero-space plane operators, manufacturers, launch site owners, and so on will be in a constant state of trying to defend themselves from being sued left, right, and center when damage or injury happens. In the United States, litigation has been controlled to some extent by mandatory signing of cross-waiver and inter-party waivers of liability by all contractors and subcontractors involved in commercial launch vehicle programs. Because aero-space planes have more in common legally with airliners than they do expendable rockets, it seems clear that a *commercial* aero-space plane industry will similarly require its own convention and legal infrastructure to set limits of liability. Otherwise, astronomical claims could conceivably afflict so much damage as to cripple this immature and unproven commercial industry. This important issue will arise at some stage with aero-space planes, unless appropriate regulations can be agreed upon, if not internationally, at least under the laws of the states owning and operating aero-space plane fleets.

## *Sonic Booms and Takeoff Noise*

One area where litigation could run rife is alleged damage or injury caused by sonic booms of reentering aero-space planes. Even though an operational aero-space plane is likely to accelerate to orbit directly over water, thereby minimizing or eliminating sonic booms on the land, there are likely to be instances where the sonic boom of a returning vehicle will impinge on populated areas. In addition, the test phase of an aero-space plane program will involve flying the vehicle over land areas in order to allow a premature abort landing to be made. (See Chapter 16.) In this instance, sonic booms will be almost unavoidable. Unfortunately, for the United States at least, under the present FAA regulations (FAR 91:51) no testing of sonic booms is permitted and no amount of overpressure is considered acceptable which will expose the public to this noise.

How loud the sonic boom of aero-space planes will actually be is difficult to answer and depends largely on the vehicle configuration. An aero-space plane like HOTOL or the experimental X-30 is expected to produce sonic booms around that of Concorde at low altitiudes, while at higher altitudes aero-space plane sonic booms will not be as loud on the ground. According to Maglieri et al. for the X-30, [10]

At low Mach-altitude conditions, a boom level of just over 2.0 lb/ft$^2$ is projected while at the high Mach-altitude conditions, which involve a major portion of the flight envelope, boom levels are less than 1.0 lb/ft$^2$.

At a given altitude and during the reentry phase, aero-space plane sonic booms will also be less loud than those of the Shuttle orbiter because the lighter weight and superior flying characteristics of aero-space planes mean that they will decelerate to subsonic speeds more quickly and at higher altitudes.

Indeed, it may be possible for aero-space planes to decelerate to subsonic speeds over water before turning toward land for an unpowered landing. The HOTOL project has looked at this issue in some detail. For a landing at Kourou, HOTOL would be able to carve out a trajectory that would take it around the northern coast of South America (Figure 14.2). The trajectory would require flight over Central American territories, but at high altitudes where the sonic boom noise would be minimal. The one exception in this discussion is the Delta Clipper which, because it follows a ballistic trajectory beneath hypersonic speeds, may well expose land areas to sonic booms.

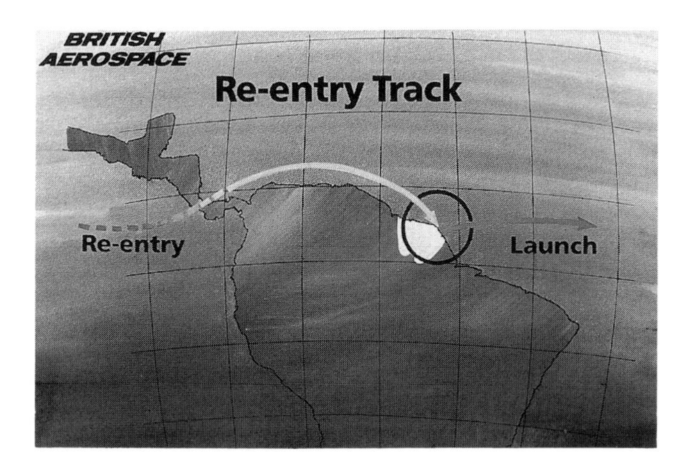


**Figure 14.2** HOTOL proposed re-entry trajectory (*BAe*).

While it is possible to minimize the sonic boom overpressures on the ground, it will be all but impossible to eliminate them completely. How loud is too loud, though? At the time of their development, there was concern that sonic booms from Concorde and the U.S. Super Sonic Transport (SST) would cause structural damage to buildings and would disturb the public in general. Most of these claims have proven fallacious, as John H. Wiggins says, [11]

At the nominal sonic boom strengths of 1–2 psf [pounds per square foot] (nominal cruise conditions of the SST) there should "essentially" be no sonic boom damage. The word "essentially" is put into quotations because there is a statistical probability that highly sensitive objects which are about to fail under natural background conditions anyway may be damaged proximately by a boom . . . It is estimated from our test program [over Oklahoma City in 1964] experience that about one in 100 million window panes exposed will crack under a 2 psf nominal boom. Sensitive or precariously situated bric-a-brac sitting on shelves or hanging on walls may fall at the same statistical frequency or lower, possibly one per billion objects per 2 psf nominal boom. No structural damage or plaster cracks are expected.

It seems unlikely that sonic booms from returning aero-space planes will cause damage to property. Despite this, litigation will still be inevitable for other reasons. Ann Deering provides an example of the type of case that could be expected to arise, "What happens if the sonic boom from an aero-space plane returning to California causes a dentist to jolt in a way that injures a patient? How do you handle such an situation?"

Another environmental issue is the noise made by aircraft during takeoff. Whether the aero-space plane uses air-breathing or pure rocket propulsion systems, the demands on the engines will make these vehicles some of the noisiest machines around, as Maxwell Hunter explains, [12]

It would be very desirable, of course, to be able to launch from normal airports . . . The biggest launch constraint, however, is likely to be the sound level. Rockets have higher velocity exhausts than jet engines and higher jet velocity means more noise. Almost no work has been done on the problem of quieting rocket engines. Yet once all the other safety features of the SSX rockets are understood, sound level is likely to be the constraining factor in the ability to use normal airports or build launch terminals convenient to cities.

Only one of the current generation of aero-space plane concepts is capable of meeting the FAA's noise control requirements. This is the Interim-HOTOL because takeoff is achieved using the AN-225 high bypass jet engines. Apart from this, it is for the takeoff noise reason alone that no one in the aero-space plane community anticipates launching the vehicles from Heathrow or Dulles airports—despite media claims otherwise.

The takeoff noise question has its origins in the "success" of U.S. environmental lobbies in their efforts to kill the SST and their failed attempts to stop Concorde from coming to the United States. Concorde, in fact, was found to be generally less noisy during takeoff and landing than many other aircraft such as the Boeing 727 and 747. Between 1977 and 1986 the average number of noise violations made by Concorde per 100 departures from JFK International Airport was 1.43 compared to 1.67 for the B747, 2.12 for the B727, and 4.08 for the B707 [13]. According to Ann Deering in a letter to the FAA regarding a proposal by the company SuperSonic Express or SSX (no relationship to Maxwell Hunter's SSX) for a one-time flight of Concorde over the United States in 1984, [14]

SSX has done extensive research which indicates that several complex factors are involved in *informing* the public of the *true facts* about supersonic flight. The public will hopefully assess the facts, rather than the *misinformation* about the harmful environmental effects of SST's, which never came to fruition. Every allegation made in the late 1960's and early 1970's regarding the negative impacts of SST flight by so-called experts in the field has now been factually proven to be wrong.

Sometimes facts alone are insufficient to put to rest issues like sonic booms and takeoff noise. As many aero-space planes do, unfortunately, bear a resemblance to the SST and Concorde, a repeat of the SST saga could loom on the horizon. Therefore, it would be in the interest of the potential commercial aero-space plane operators to try to obtain an early ruling or, at least, a consensus on the sonic boom and takeoff noise issues in order to minimize the threat of similar action being taken against their vehicles. These issues should not be underestimated and—whether pertinent or not—they could turn out to be real showstoppers, as they were with the SST and Concorde. The noise issue alone might dictate the type of vehicle configuration chosen and/or where it can be operated from. *This is especially the case in the distant future when aero-space planes may fly to and from space as regularly as airliners fly the Atlantic today.*

Certainly, it would be a tragedy for spaceflight if the first aero-space plane program was stopped in midflight because of a misunderstanding over environmental noise impacts.

## Atmospheric Environmental Issues

Noise is not the only environmental issue to be addressed. Another, and potentially more serious issue, is the possible damage aero-space planes could have on the atmosphere and, in particular, the ozone layer. All air-breathing aero-space planes are likely to use liquid oxygen and liquid hydrogen as propellants, with an exhaust "pollutant" that is basically water, although even ice crystals formed by water vapor can cause some ozone damage. If water were the only pollutant, there might be little to argue about. But like most aspects of aero-space planes, it is not so simple. Water is not the only byproduct of air-breathers. At the high combustion temperatures, and especially in supersonic combustion ramjets, the usually inert atmospheric nitrogen molecules combine with oxygen to form highly reactive nitrous oxides. The various species of nitrous oxides react with almost anything, including the highly unstable tri-oxygen atom ozone molecule.

It is important to note that all jet-powered aircraft produce some level of nitrous oxides. This has not been a particular problem as far as the ozone layer is concerned, because most of these aircraft fly far below the ozone layer (i.e., 10 km versus 30 km). At these heights, the normal mixing effects due to winds and turbulence quickly disperse nitrous oxide concentrations. Some aero-space plane configurations, however, will come much closer to or even pass directly through the ozone layer, injecting their nitrous oxide byproducts directly into the area where they have the potential to react with ozone. Vehicles like Sänger and the original HOTOL breathe air to an altitude of about 25 kilometers, after which pure rocket propulsion systems are used. The X-30 is intended to breathe air up to around 50 km, all the way through the densest regions of ozone layer. Concerns have been expressed over the effects on the ozone layer.

Studies on this issue are being conducted at various institutes around the world, and the general conclusion is that the damage cause by nitrous oxides will be minimal. The reasons are relatively straightforward to understand. First, the *time* most aero-space planes spend near or in the ozone is relatively short, on the order of a few minutes. The exception here is Sänger-like vehicles which are intended to cruise close to the ozone layer for several hours before releasing the second stage. Second, the number of aero-space plane flights per year is low, meaning the total annual exposure time will be small. Studies, such as those conducted by NASA [15], have shown that any effects nitrous oxides could have on the ozone will be very localized and the normal atmospheric transport mechanism will quickly repair this damage. Engineering solutions may also allow nitrous oxide emissions to be reduced in the same way the noise and pollution emissions of jet engines have been steadily reduced compared with their earlier predecessors.

The purpose of this discussion is not to argue that ozone damage will be minimal, but rather to show that it is an issue of considerable concern and that there appear to be good reasons to believe that any damage will be slight when the first aero-space plane takes flight. In the meantime, careful and rigorous work and testing will have to be done. This work must include a well-planned public relations campaign that focusses entirely on the facts and always compares the possible environmental impacts of aero-space planes with other launchers like the Shuttle and Ariane 5. A goal is to avoid the emergence of the type of technical misunderstandings that characterized the plight of Concorde and the SST. Once again, the visual similarities of aero-space planes to supersonic or proposed hypersonic airliners may lead many people to believe they are one and the same. Referring to the NASP program, David Webb provides an interesting perspective, [16]

> The real facts are, however, not the point. It is the general perception of these facts that counts . . . It appears almost inevitable that there will be same reaction from many people in the environmental community to the HSCT [High Speed Civil Transport] concept as there was in the seventies to the SST. It is too much to expect that they will attempt to differentiate between the NASP/NDVs and the SST/HSCTs when many of them are fighting the entire concept of a technological society. Assuming that to be the case, it is incumbent upon the NASP Program and its supporters to publicize the differences *now*; using every opportunity to try and fix in the mind of the media and the public that the NASP and subsequent NDVs, are first and foremost, space launch vehicles. This is another and, perhaps, the primary reason why it is essential to press for the NASP to be classed as a space vehicle, not an airplane.

The issue of ozone depletion is, quite rightly, a subject of great concern politically and for the general public and, more likely than not, an issue of even greater prominence at the time of the first aero-space plane flights a decade or so from now. Some may argue that "the less the public knows the better," and that environmental impact data should be withheld. First, this approach would be an insult to the intelligence of the public. Second, and potentially far worse, it could release a Pandora's box of negative publicity, especially if the public felt that this information was being withheld because it proved that aero-space planes would cause unacceptable damage to the ozone layer.

It would be unfortunate for the future of commercially operated aero-space planes if their activities were constantly being interrupted by legal injunctions placed by misinformed groups. On the other hand, if certain aero-space plane configurations really are capable of causing irreparable damage to the ozone, it is clearly important to known which configurations these are at the earliest possible opportunity. If it does become a problem, then efforts may have to focus on pure rocket systems like the Delta Clipper or the Interim-HOTOL which only emit water.

## *Certified for Spaceflight*

Aero-space planes have the potential to be more than just straight replacements to the Shuttle, as stated previously. Assuming that adequate solutions are found for many of the above issues, there might be operational advantages to launching an aero-space plane from within a nation and allowing it to fly like an aircraft over land and across foreign territories. Noise constraints are likely to preclude ascent to orbit over land because of the sonic booms, nevertheless an overland cruise at subsonic speeds might be useful for test flights and delivery of aero-space planes to the launch site. For two-stage systems like Sänger and the Interim-HOTOL, overland flight might enable the launch base to be located at more convenient inland sites or allow the required launch latitude to be reached more directly rather than having to fly around the land and over water.

Whatever the reason, allowing a commercially operated aero-space plane to fly over populated areas will necessitate that the vehicle be registered as an aircraft under air law and certified by an internationally recognized certification process. For a civilian aircraft to fly over another nation, that aircraft must meet a list of internationally recognized standards. In Europe, these are know as the Joint Airworthiness Requirements, while in the United States they are known as the Federal Airworthiness Requirements. Airworthiness requirements force aircraft to undergo a rigorous series of tests before they can be used operationally to carry people and cargo. Typically, airworthiness requirements define the total number of flying hours that have to be accumulated, the ascent and descent performance, the maximum runway roll, under what environmental conditions the aircraft must be flown, the ability to shut down or restart an engine in flight, and the capability to abort a takeoff as well as perform various other aborts. Airworthiness requirements dictate the aircraft design in terms of structural strength, design margins, ground testing, materials that can be used, the number of engine tests, and so on. All of these have to be met before a new airliner can be given a Certificate of Airworthiness.

In the case of military aircraft, and other government

owned and operated aircraft, the situation is obviously different. Military aircraft are only allowed to fly over another nation's territory if it is in the strategic interest of that nation, and prior government agreements assure that accidents can be satisfactorily cleaned up.

The format and precise details of the airworthiness requirements for an aero-space plane will need to be specifically tailored to the vehicle configuration, the level of technology used, and operating modes. For example, the type of structural margins airliners must meet are likely to be too severe for an aero-space plane. Such margins, if enforced, would mean that most aero-space plane concepts would be too heavy to reach orbit. Therefore, allowances must be made.

While the writing of an aero-space plane certification document may take time, it is precisely the same procedure as for a new type of civil airliner. The requirements for Concorde—an aircraft which, like the aero-space plane, had no predecessor to base the requirements on—were established by the joint airworthiness authorities of Britain and France and written in the Transport SuperSonique (TSS) document. One important aspect, the TSS forced Concorde to fly a total 5,000 hours throughout its test program before it could carry its first paying passenger [17]. The aero-space plane is a greater leap than Concorde was from other aircraft. Therefore, it would be prudent to begin the compilation of certification procedures many years before the first aero-space plane is built or even designed. Whether this will be the job of new branches of the FAA or CAA, a new government office such as the U.S. Office of Space Transportation, or by the U.N. itself is to be determined. The point here is that if commercially owned and operated aero-space planes are to be allowed to behave with the flexibility of aircraft, they will be classified as aircraft under air law. As a result, they will have no other choice but to be certified by law.

No space transportation system has ever been certified like an aircraft, as it is all but impossible and incredibly expensive to certify something that is thrown away after every flight test. Even though it would be an expensive and time-consuming process to certify an aero-space plane, the end result would be a *proven* capability that the user could rightfully trust and have confidence in. Certification is one reason why airliners have been as successful as they are and may be a key reason to the success of the aero-space plane. At the end of the day, *it is the user that counts*.

## Summary—Planning Now for the Future

Developing aero-space planes imposes a multitude of technical, economic, and political challenges. The legal, liability, environmental, and regulatory issues have the potential to pose equally great problems under certain circumstances. These circumstances can be split into aero-space plane fleets owned and operated by either the government or the private sector. In the case of a government fleet, aero-space planes could probably fit within precedents set by the Shuttle and still achieve a significantly improved means of accessing space. New agreements would probably have to be reached with those few countries where the aero-space plane might overfly or perform emergency landings.

Since aero-space planes have the potential to be more than Shuttle replacements—should the hoped-for performance and launch cost objectives be achieved—it is realistic to consider private ownership and operations. Under such circumstances the situation changes completely. Problems with obtaining insurance for a highly risky investment and to cover third party litigation are likely to be a major constraint; the only solution may be government-provided insurance. Limits on liability will probably have to be determined by international consensus.

Concerns about the noise generated by sonic booms and takeoffs impose regulatory and public acceptance barriers that, if the Concorde and SST experiences are anything to go by, may be difficult to dismantle without proper public education. The perception that aero-space planes could damage the ozone layer may also pose constraints that could limit the choice of vehicle configuration to rocket-only solutions.

Finally, if commercial aero-space planes are to ever operate with aircraft-like flexibility and be allowed to regularly fly over nations, then internationally agreed standards of certification will have to be developed. Imposing airworthiness standards would, as a byproduct, provide the user with confidence that the aero-space plane will function reliably.

All of these problems are significant and, in combination, could severely restrict the type of vehicle chosen and how it is operated. Most critically of all, they cannot and must not be ignored once an aero-space plane program transitions from the paper design stage to the hardware stage. If organizations are serious about building effective aero-space plane launchers, these issues need to be attacked with as much diligence as any other problem.

## References and Footnotes

1. Webb, D. C., and R. J. Hannigan, "Economic and Socio-Political Impacts of NASP-Derived Vehicles: A Technical Report," The International Hypersonic Research Institute, November 1990, p. 33.
2. Wilson, A., ed., *Interavia Space Directory*, 1992.
3. "Economic and Socio-Political Impacts of NASP-Derived Vehicles: A Technical Report," pp. 33–34.
4. *National Space Transportation System: Overview*, NASA Kennedy Space Center, September 1988.

5. "Economic and Socio-Political Impacts of NASP-Derived Vehicles: A Technical Report," p. 33.
6. Details of this report were printed in *Aviation Week & Space Technology* under the title, "NASA, Defense Dept. Drop Idea Of Private Shuttle Management," April 29, 1985, p. 42.
7. A similar idea was proposed originally by British Aerospace for HOTOL.
8. Letter from Ann Deering.
9. Webb, R., "The Extent of the Liability of the Operator With Regard to the Flight Personnel, Passengers and Third Parties," presented at *The Spaceplane and the Law Symposium*, May 14–15, 1991.
10. Maglieri, D. J., and V. Sothcott, "Influence of Vehicle Configuration and Flight Profile on X-30 Sonic Booms," AIAA-90-5224, *Second International Aerospace Planes Conference*, Orlando, Florida, Oct. 29–31, 1991.
11. Letter from Dr. John H. Wiggins to Ann Deering, August 23, 1983. Dr. Wiggins is the former technical director of the White Sands Missile Range Sonic Boom Test, and an expert U.S. government witness in many litigation proceedings.
12. Hunter, M. W., "The SSX: Designing for Flight Safety," July 1988, p. 5.
13. Data obtained from an article by William Good entitled "Concorde Diamond Aniversary—A Sparkling Success (Draft)," June 23, 1987, p. 4.
14. Letter from Ann Deering to J. Lynn Hems, administrator of the Federal Aviation Administration, September 10, 1983.
15. Kinnison, D., D. Wuebbles, and H. Johnston, "A Study of the Sensitivity of Stratospheric Ozone to Hypersonic Aircraft Emissions," *Proceedings of the First International Conference on Hypersonic Flight in the 21st Century*, University of North Dakota, September 20–23, 1988, p. 389.
16. Ibid., p. 38.
17. Clavert, B., *Flying Concorde*, (Airlife Publishing, Shewsbury, England, 1989), p. 169.

# Chapter 15

# Accepting Failures: Backing Up Aero-Space Planes

## Aero-Space Planes Will Fail

At some point in the operational life of an aero-space plane, something unexpected is going to happen on the way to orbit that will force the mission to be abandoned. The event could be a result of an overheating engine, failure of a particular appendage to retract properly, a bird strike in the intakes of an air-breathing vehicle, an incorrectly secured access panel that comes loose. Whatever the cause of the failure, the aero-space plane may be unable to achieve the required orbit and will have to abort the mission and return to base for repairs. On even rarer occasions, the aero-space plane may fail *catastrophically*.

While such failure events are unexpected when they occur, they can nevertheless be planned for. The aero-space plane should be able to detect something wrong and then take appropriate action that should lead to a safe recovery of the vehicle, its valuable payloads, and/or the crew. This is in stark contrast with most current launch systems as, *even* if they can detect a failure, there is little that can actually be done about it. Current launchers either achieve an orbit or fail catastrophically. (See Chapter 3.) Because of the experience of an extensive test program, the aero-space plane should, in the vast majority of cases, be able to abort and return safely to Earth at any stage in the mission. Still, a few situations will remain where, unfortunately, a failure will lead to the total loss of the aero-space plane.

Although at first glance this is a gloomy scenario, it does not conflict with reality. Terrestrial transportation systems, whether airliners, cars, trains, or ships, suffer unexpected failures at some time during their operational lives. Failure is just as much a part of life as success—**failure defines success**. Of critical importance to this discussion is what can be done to minimize the negative impacts of failures on space operations. Even though aero-space planes are fundamentally different from the current generation of launchers, might they be expected to suffer long standdowns after a catastrophic failure? Would such failures be as highly disruptive to space operations as the failure of an Ariane or Shuttle? Would aero-space plane failures be tolerable?

## Backup: The Way to Minimize the Impacts of Failures

If the Space Shuttle experience has shown one thing, it is that the success of future space activities depends intimately on the ability to always be able to access space. Fundamentally, this means that if the primary launch system fails, there must be an alternative available—a backup. Yet, this lesson appears to be going unheeded today and for future space activities: as of 1993 the Shuttle is the only vehicle currently funded to support Space Station Freedom with crews and logistics. As Logsdon and Williamson explain, [1]

> Some observers are also disturbed by NASA's decision to rely on the shuttle as the sole launch system for the space station. What happens, they ask, if another shuttle failure grounds the entire fleet while the station is under assembly or in operations? Even if the Shuttle is 99 percent reliable, they point out that a simple statistical analysis shows there is an even chance of one major failure in the next 75 missions.

Although numerous discussions and recommendations have called for an alternative crew launch system, designated the Personnel Launch System (PLS) [2], it seems unlikely to be built in this time of budget austerity. Recommendations for an Assured Crew Return Vehicle (probably based on the PLS) have also been made (Figure 15.1). It is not funded either and in view of NASA's commitments over the next decade, funding for this program also seems very unlikely.

It is difficult to understand how Freedom can be funded at many tens of billions of dollars, but at the same time funds cannot be found for a relatively simple alternative means of launching crews and logistics in the event of another Shuttle standdown. Nor is it easy to understand why funds cannot be found for a crew escape capability particularly in a time-critical emergency—such as a collision with a large piece

**Figure 15.1** Assured crew return from Freedom (*NASA*).

of orbital debris. Fortunately, the collapse of the U.S.S.R. has provided a cost-effective ACRV in the form of modified Soyuz capsules.

The Space Exploration Initiative is another example. Today the majority mindset has most SEI planners focusing on using a single type of heavy-lift launch vehicle (HLLV) to place large components and structures in orbit for crewed lunar and Mars missions. The NASA *Report of the 90-Day Study on Human Exploration of the Moon and Mars* described scenarios that require the use of three HLLVs launched in rapid succession to assemble the space transfer vehicles needed to propel crews and cargo to the Moon [3]. What happens if one of those HLLVs fails or is delayed due to a problem with the booster or payload? This question is seldom answered satisfactorily. The Synthesis Group has stated that the first stage of their proposed launcher needs to be designed to be 100% reliable [4]. Given that the Synthesis Group's scenarios would require the launch of several expendable HLLVs every year just to support a modest permanently staffed Lunar Base—let alone a Mars Base—and must continue to do this over many decades, is this a realistic expectation?

Importantly, the threat of such failures would drive the HLLV operator to spend considerable time and money on *every single* booster before each is launched. While it is conceivable that such an HLLV-based SEI program could successfully launch five or more vehicles every year for three or four decades, the cost of achieving it would be unavoidably enormous. This is considered one of the primary reasons why an SEI program centering around heavy-lift expendable boosters would cost hundreds of billions of dollars. (See also the discussion in Chapter 19.)

The failure of a launch vehicle is, in the West at least, extremely disruptive compared with the failure of a terrestrial transportation system. When an airliner crashes, although tragic, such failures seldom lead to a significant standdown of that particular make of airplane. Even if a standdown is ordered, as with the DC-10 in the mid-1970s, it is seldom for very long. The reason is because of the experience gathered over many thousands of hours of flying an airliner. If the airliner has performed successfully many times in the past, then there is no reason to presume that all other aircraft of the same make are going to start falling out of the sky. Further, airworthiness certification regulations ensure that the operating characteristics of any new airliner are fully understood long before it enters service.

Most critically of all, the impacts of standdowns on the *user* of air transportation services are minimal simply because there is an array of alternative airlines to choose from. This is certainly not the case with space transportation, especially in the area of standard payload interfaces as discussed in Chapter 4 and as Gale et al. note, [5]

A fleet of different vehicles with overlapping capabilities provides advantages for resiliency of launch capability. This is a fact of life in most transportation industries: when DC-10's were grounded, Grandma could still come to visit on a 747, because she was a payload with standarized interfaces for both vehicles. But when commercial satellites were essentially barred from using the Space Shuttle, they had to endure years of delays and millions of dollars of modifications before being launched on Deltas or Arianes. For these customers, launch capability resiliency is not yet a reality.

Another significant difference is that when an airliner crashes, the economic impact, *as a fraction of the total airliner business*, is relatively small. Although an airliner loss may lead to hundred million dollar insurance claims, because there are thousands of airliners each requiring insurance, the insurance pool is large and, therefore, robust. One airliner crash does not ruin the insurance market for everyone else.

When a Space Shuttle fails, the investment lost is enormous compared with the total cost of the program. As Bob Parkinson explains, "We don't have two-and-a-half year hangups when an airliner kills 200 people. We do have a hangup, but we still fly the things. In the 1930s they discovered gold in the centre of New Guinea and it was totally inaccessible apart from by air. The aircraft used were as 'experimental' as the Shuttle, but they built a whole gold mine in the centre of New Guinea, all ferried in by air, everything. They lost people, they lost aircraft, but the difference was that the failures were relatively inexpensive, and that's critical. If I could fly the Shuttle 40 times per year and if it cost no more than it costs at the moment—in total annual costs—I wouldn't feel bad about allocating a Shuttle flight to investigate a new part of the envelope with no payload on-board. If the risk of losing a vehicle meant that I was going to wipe off a 'hundred *million* dollars' rather than 'three *billion* dollars' I would take more risks and learn faster. I know this because B-1B bombers cost about $100 million and yet they fly a lot more flights [than the Shuttle]."

It seems clear that for a rational and expanding space program, an ability to back up the primary launcher is a prerequisite. In this light, what is the proper mix of alternative vehicles in a future launch architecture that centers around aero-space planes?

## The First Aero-Space Plane Fleet

If unrealistic and politically motivated demands are put on the aero-space plane to launch all payloads—just as with the Space Shuttle—many of the potential advantages provided by the unique capabilities of aero-space planes will not be fully realized. This would especially be the case for the first aero-space plane fleet in its first few years of service—a time when aero-space planes will be at their most vulnerable. For example, aero-space planes are perfect vehicles for resupplying space stations with logistics as well as launching new components (Figure 15.2). They have an inherent potential to perform this type of mission more cheaply and more frequently than the Shuttle and Soyuz/Progress. If, however, despite a highly successful test program, an aero-space plane is seriously damaged or lost during its first years of operation, then a standdown of some

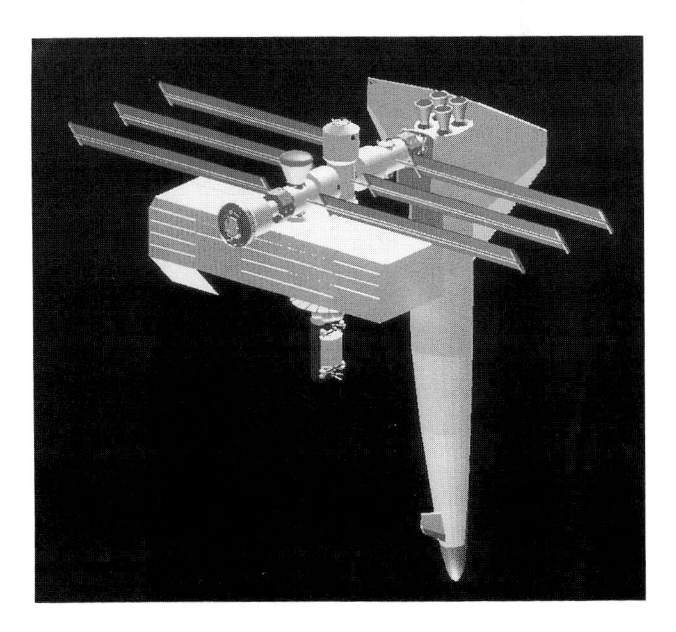

**Figure 15.2** Even in the aero-space plane era a backup will be needed to access space stations (*BAe*).

period will be inevitable. Potentially, the duration of this standdown will not need to be as long as for expendable rockets because of the ability to test another vehicle up to the point where the failure occurred, and because of the large data base available from the extensive test program.

Nevertheless, delays on the order of several weeks or months might be expected simply because of the high cost of the few vehicles available. Demand for aero-space planes will be relatively low in the first years, and the number of vehicles built will be small. Therefore, the cost of the loss of one vehicle as a fraction of the total number of vehicles will be large. Unduly and unnecessarily risking a large fraction of an expensive new investment like this might be considered unacceptable, especially if the first aero-space plane fleet is being operated commercially.

From the issues raised in the above discussion, it seems obvious that the first aero-space plane fleet will require a backup for time-critical payloads such as space station logistics, military satellites, and planetary payloads. This does not pose a significant problem. Although, in principal, a suitably sized aero-space plane could launch all types of payloads (Chapter 17), it is likely that there will be a minority of existing or near-term planned payloads that are either too heavy or too voluminous for an aero-space plane. The need for a vehicle to launch these payloads could automatically provide the backup capability, as will be explained shortly.

## Reliving the Shuttle Experience

Even if the aero-space plane can launch all payloads, a number of years may pass before users will be confident in switching their payloads from the known quantity of an expendable rocket to the unknown quantity of the aero-space plane, due to a fear of reliving the Shuttle experience. If this is true, expendable rockets will continue to be fired for many years after the aero-space plane becomes operational, therefore automatically providing the necessary backup capability.

This fear of following in the wake of the Shuttle may, in actuality, be groundless. Critically, aero-space planes *can be, will be*, and *must be* tested thoroughly long before they enter service. To put this in perspective, the first aero-space plane may fly as many missions (including suborbital flights, and repeated takeoffs and landings) in a 2 or 3 year test program as the Ariane 4, Atlas 2, or any other rocket will fly operationally over one or two decades. Indeed, the number of test flights might be an order of magnitude greater than the meager four test flights of the Shuttle. *To suggest that aero-space planes will go the way of the Shuttle is to ignore the very precept on which they are based.*

Confidence is always experience-based. Because their test and certification flight experience will rapidly surpass that of existing launchers, aero-space planes could be expected to perform at least as well as expendable rockets from the beginning. And remember, current launchers are very expensive to use, they are frequently delayed, they do not fly very often, they fail with uncomfortable regularity, and they suffer prolonged standdown periods after failures. In this regard, aero-space planes do not need to be vastly superior to expendable rockets in order to utterly dominate the existing market.

Even if the first aero-space plane costs as much to use as existing expendables, its ability to safely recover the payload after an in-flight failure will be an overwhelming selling point. If this is true, then commercial satellite customers might switch to aero-space planes as soon as possible. As payloads are taken away from expendable rockets, their flight rates drop and, therefore, the cost of each launch increases. For economic reasons alone, expendable launchers would not survive commercially very long after the aero-space plane is introduced, much as the turbine engine made the piston-driven engine obsolete almost overnight for military and long-haul aircraft.

Clearly, the impacts would be profound for Arianespace, General Dynamics and the other suppliers of commercial expendable launch services, unless, of course, they were also operating the first aero-space plane fleet. The Delta Clipper, for example, is being built by the same company that operates the Delta 2 booster.

## Forming a Prudent Launch Architecture Policy

Only about a third of the western world's payloads can be classified as commercial, with the rest being government-funded payloads serving military, scientific, and applications requirements. Therefore, maintaining a backup capability for the aero-space plane will depend almost entirely on the shape of government space launch policies. In forming a launch architecture policy that includes an aero-space plane fleet, it is absolutely vital that the decision makers understand the critical operational factors. The notion that aero-space planes are only for launching people, and expendable rockets are only for launching cargo is simplistic and fallacious. It is, nonetheless, a perception that currently seems to characterize some thinking. Aero-space planes are as much cargo carriers as they are people carriers—if not more so.

It is possible to marry the individual operational strengths and weaknesses of two vastly different types of launcher in a manner that allows each to simultaneously complement *and* back up the launch of time-critical payloads. Consider the following:

- Expendable Rockets
  *Strengths*: Large payload mass and volume possible
  Ability to be stockpiled
  Well-understood operability
  *Weaknesses*: Inability to be launched very frequently
  Unreliable
  Inability to be launched on demand
- Aero-space Planes
  *Strengths*: Low dedicated cost to orbit
  Launch on demand
  High flight rate
  Modest payload size and mass (cargo bay always nearly full)
  *Weaknesses*: Inability to launch very heavy and voluminous payloads

Based on the features of each vehicle, it is possible to envisage a nominal scenario where all but one or two of the current expendable rockets are phased out of service over a relatively short period after the aero-space plane fleet is introduced. The policy of phasing out these rockets is *only* to be undertaken once the aero-space planes have demonstrated, through an extensive test program, a superior performance and the ability to meet the flight rate demands. This was not the case with the Shuttle where the decision to start phasing out expendable vehicles was taken long before its performance could be properly assessed. A handful of flights is not enough.

Even though the remaining expendable launcher may eventually have little, if anything, to launch, it is not prudent to phase it out of service completely. Instead, the rate at which it is manufactured is steadily cut back to save costs so that, over a period of time, a large number of vehicles are stockpiled. The rate at which this backup rocket is launched is also severely cut back—perhaps to only one or two every year—just enough to ensure that the ground team is always ready to support a launch. It is important to understand why this might be a desirable decision to make. Although expendable rockets have a high failure rate, the possibility of a failure occurring is minimal if only one or two launches are performed each year. Therefore, if the expendable rocket had to be called upon to back up the aero-space plane, it wound be possible to sustain a short surge through launching some of the stockpiled boosters. Provided this surge does not need to be sustained for any significant length of time, the probability of a double failure will, theoretically at least, be similarly minimized.

The payloads for these rockets could include relatively simple and inexpensive technology demonstration satellites, or any surplus payloads where the cost to modify them for an aero-space plane launch would be greater than an expendable launch. Potentially, if this rocket was of the heavy-lift variety, such as the proposed Spacelifter, then it could be used to launch large outsized structures, such as space station modules, or bulky items, such as propellant for future solar system exploration missions. (See Chapter 19.)

Variations on this theme are also possible depending on the requirements of the launching state. Specifically, if the United States develops its own independent aero-space plans, then it might decide to keep a version of the Titan 4 stockpiled and in service, while a second medium-lift launcher like the Delta 2 could be stockpiled, but not launched, to act as a further insurance policy. Europe might decide to keep the Ariane 5 stockpiled and in service, while stockpiling the Ariane 4. The C.I.S. might decide to keep the Soyuz booster in service, while stockpiling the Proton.

## A Robust Launch Architecture

*If the expendable launcher fails*, the economic and schedule impact would be minor when the aero-space plane is flying normally. This is because the payloads being carried would be noncritical and relatively inexpensive. The impacts on the schedule of a recertification program would also be minor because there wouldn't be any particularly urgent need to return the launcher to flight status as the next planned "proof-of-manufacturing" flight might be a year or so away. Also, and not insignificantly, there probably would be no insurance claims to worry about as these launches would likely be government-funded.

*If an aero-space plane failed*, and was out of service for several months, then these stockpiled launchers would be called into service to launch a limited number of time-critical payloads such as military satellites and space station logistics. Commercial satellites and other scientific and applications missions would probably have to wait until the aero-space plane reentered service. This is not considered a fundamental concern because these one-shot spacecraft types are often subjected to delays of months or years today. It must be emphasized that this scenario only deals with missions similar to those conducted presently.

Having a space station in orbit demands a PLS-type capsule placed on ground standby to cover an aero-space plane loss, thereby allowing continued crewed access to the station. This capsule could be equipped with an escape tower, similar to that employed by the Soviets on the Soyuz capsule, and as was used two decades ago on Apollo. (Hermes had planned to use ejector seats similar to those used during early Space Shuttle flights.) Using this capsule will be more expensive and risky than using an aero-space plane, but this is not the issue. If the aero-space plane downtime is long, it will be vital to have continuous access to the space station in order to supply it with basic logistics until normal operations can resume.

A space station the size of Freedom is too great an investment to leave untended in orbit. Further, because the bulk of work on Freedom might be expected to be associated with maintenance, abandoning the Station for any length of time could mean that, when a crew finally returns, the cost and risk to the crew of repairing the failed elements might be considerable—as it was when Salyut 7 was reactivated after being left abandoned for 8 months [6].

Within the constraints of the current spacecraft launch requirements, it is reasonable to suggest that the first aero-space plane fleet can be effectively backed up by one or two existing types of expendable launchers and thus ensure continuous access to space for a rational space program. The cost of having these expendable rockets effectively on standby will be significant, and even higher when surged to back up the grounded aero-space plane. However, the cost of not having a backup would be far higher, as the Shuttle experience has aptly demonstrated.

## The Second Aero-Space Plane Fleet

If aero-space planes eventually live up to expectations, then ideally it might be desirable to eliminate the role of the expendable launcher as a backup to the aero-space plane. Indeed, for economic reasons it might be desirable to eliminate expendable boosters completely, but only if users had access to more than one type of aero-space plane. Vehicles like the proposed Delta Clipper would operate in parallel and competition with a vehicle more reminiscent of the winged HOTOL-type vehicle, for example. Appropriate analogies can, once again, be made with jet airliners used for long-haul flights. Originally, the first successful airliner was the Boeing 707. Although some turboprop powered airliners remained in service for a while after the first 707 fleets, the introduction of the DC-8, and other long-haul jet airliners, quickly sealed the fate of turboprop airliners.

Using the type of failure scenarios discussed earlier, if one aero-space plane was grounded for an extended period of time, priority payloads could be switched to the alternative vehicle, assuming there was space available or that the backup aero-space plane could be surged. Whether this is practical will depend on the launch policy regulations of the launching state as well as under what auspices the fleets are operated. For example, the U.S. military may decide it is in its best interest to have its own independent and secure aero-space plane fleet. Thus, either the military would need to keep its Titan 4 in service as a backup to its own aero-space plane, or it could phase out the Titan 4 completely but be in a position to commandeer the civilian fleet. The latter case is more or less analogous to how the U.S. Civil Reserve Air Fleet operates during crisis periods. Civilian airliners were extensively used throughout the Persian Gulf crisis, with relatively minor impacts on normal air service [7].

The threat of the military commandeering the civilian aero-space plane might be considered very damaging to a competitive aero-space plane business. In the early 1980s a similar threat constantly hung over the Shuttle. Although it is irrelevant now, there were concerns that military requirements could, without warning, take over the Shuttle and, as a result, seriously jeopardize its competitive commercial standing. The Ariane rockets did not have to live with such concerns. In hindsight, the fears of negative impacts from the military commandeering the Shuttle were probably unfounded given the small number of commercial payloads launched each year, and the fact that most of these experienced long delays anyway.

## Cooperation or Competition?: Benefits and Impacts

If it is perceived that no single nation or subgroup of nations can afford to develop its own indigenous aero-space plane, then an international consortium of government agencies may be the only way such vehicles can be developed. The obvious concern then is that all nations will be dependent on a single *type* of aero-space plane. Thus, if one aero-space plane were lost early in its operational lifetime, all aero-space planes would probably be grounded, as with the de Havilland Comet (Chapter 12).

While an international or global cooperative program may be attractive politically, *operationally* it could be undesirable with respect to attaining assured access to space. (See Chapter 13.) The connotations of such a decision are profound. If the objective of an aero-space plane program is truly to build a space transportation system that makes space access simpler and more affordable, then a global cooperative program might be unacceptable on these grounds alone. If the objective is *just* to build *an* aero-space plane, a global program might be acceptable. If this is the case, however, aero-space planes shouldn't be built at all.

It could be argued that a sufficiently thorough test program of such a global aero-space plane would be able to catch all significant problems long before it enters service. Therefore, when failures occur, they will be a result of vehicle-unique problems rather than a generic problem affecting the entire fleet, meaning standdowns will be short or not at all, as with most airliners.

While the above may be true, it ignores a number of fundamental influences. First, it would be naive to believe all the necessary funds will be supplied for a complete test program. Even after a fully funded enabling technologies program, technically demanding programs like aero-space planes will be prone to problems that may inevitably increase costs. As costs increase, there will be a tendency to stretch out the program at the expense of capability. For example, a program that originally planned 100 qualification test flights might seem rather "luxurious" in the eyes of political appropriators if the program is already over budget. While politically attractive, operationally such a cost-cutting exercise would be a disaster. (See Chapter 11.)

Second, when a new airliner is developed, it must be cost-effective—otherwise no one will buy it. Therefore, its specifications are driven entirely by commercial pressures. Some of the cheaper aero-space plane concepts could follow this approach, but a single global program to develop an expensive aero-space plane would clearly not benefit from competitive pressures to keep it cost-effective. Thus, the drive to maintain a cost-conscious edge in the development of such a global aero-space plane might be lost among stronger politically driven concerns. Most notably, these could include political directives to keep within budgets set long before the start of the program, and micro-management from some of the more dominant partners in the venture. Demands by some partners that their funding be contingent on the use of certain technologies, regardless of whether they are cost-effective, is one example. Another might be other partners insisting that their participation is dependent on a larger than optimal payload capability, rather like the Shuttle experience.

Political bargaining, and the requisite endless stream of committees and working groups could also align to make an international effort less optimal, at the risk of compromising the end result. Again, the experience with the Shuttle provides an example. According to John Logsdon, [8]

> . . . the decision to develop the Shuttle was the result of bargaining among multiple participants in the policy-making process, and the compromises, negotiations, and coalition-building that accompanies such bargaining. This kind of process, in the absence of a more clearly 'rational' approach to policy choice, can often produce outcomes that approximate rationality in a particular situation. It can also lead to decisions that take insufficient account of crucial factors that are not powerfully represented in the bargaining process, e.g., the future. Such a dysfunctional result seems to have been the case with respect to the Shuttle decision.

Domestic politicking could certainly have an enormous impact on an international aero-space plane consortium, and the policy-making process described above would also have a strong influence among the international partners. Such influences would probably result in a compromised vehicle, inadequately tested and, therefore, susceptible to failures and global standdowns.

Competition is the key to the cost-effective operational success of any high-investment and unproven transportation system. According to German Zagainov, "The world must have two systems. You know the history about American Shuttle—America had no possibility to launch. It seems to me after some time we should decide to join our technology—BAe, Germany, etc.—and combine with other countries' technologies. So, we agree to start two aero-space plane programs. It is very important for Europe to have two aero-space systems, not single systems—it's a mistake. For example, the first system may be Interim-HOTOL. And, after that, the Sänger."

Competition holds the potential to significantly reduce launch costs, as Bennett and Salin describe, [9]

> *Competition, which generates rapid, small innovative steps, results in a quick spread of innovations through imitation of successful products.* No single organization or development effort comes up with all the important cost-reducing innovations. No organization can generate all the right answers at one time. Without intense competition, major cost-lowering strategies can go undiscovered for years. With competition, not only are valuable innovations discovered more quickly, but they also spread more quickly. It is wrong to think that a crash programme to build one big reusable vehicle [like the Shuttle] will teach us how to design and operate cost-effective reusable vehicles faster than would the development of several smaller, competing reusable vehicles, which can quickly be modified to take into account internal discoveries (a firm's own R&D) and successful innovations of competitors (another firm's R&D).

The stimulant for competitive aero-space plane launchers could be the commercial satellite market, even though it might appear too small. As suggested above and in Chapter 13, if the United States—or a U.S.-led consortium that included foreign companies—were to demonstrate that aero-space planes were feasible and embarked on a development program, it would win the commercial satellite market in its entirety. Europe—or a European-led consortium that included Japan, the C.I.S., and other foreign companies—may feel that this market is something it does not particularly want to give up and, in response, may embark on its own program. The end result would be operationally effective aero-space planes that benefit not only the commercial market, but also dramatically reduce the amount of money governments spend on space transportation. If launch costs can be reduced low enough, such vehicles may stimulate new commercial markets that were previously impractical (See Chapter 20).

One obvious beneficiary would be international space station programs like Freedom, because the United States, Europe, and perhaps Japan would have a comparable means of supporting the space station on a regular basis. Even if one country insisted its aero-space plane should be the primary means of accessing the space station, the availability of another country's vehicle would ensure that access to and from these valuable facilities is not significantly jeopardized in the event of a grounding. This contrasts sharply with the current Freedom program which, as presently configured, is completely held hostage to the continued success of the Space Shuttle.

## Summary—Backups Are Mandatory for Rational Space Programs

Undertaking rational space missions—commercial, scientific, crewed, military, or others—will demand the need for a backup launcher to ensure access to space. Aero-space planes can be thoroughly tested long before they enter service and therefore can demonstrate a higher reliability than expendable rockets. Yet they must also have a back-up, especially in the first few years of operation when they are at their most vulnerable. Certainly, each individual aero-space plane represents a significant fraction of the total investment of the fleet, therefore, after a failure there will be no hurry to restart operations until the problems have been resolved. Existing expendable rockets can provide such a backup for time-critical payloads, but only if these rockets are stockpiled and launched as infrequently as possible so that they will be available to support an aero-space plane standdown. These operational issues must be considered by policy makers before aero-space plane launch policies can be written.

Importantly, from the perspective of requiring a backup, the prospect of an international government-funded program to develop a single type of aero-space plane for all users is probably doomed to failure from the start, especially as the requirements for such a vehicle would not necessarily be driven by commercial factors. Competition is fundamental to reducing launch costs and easing access to space, and it is an important factor in the future success of aero-space planes. It is also fundamental to the ultimate objective of eliminating expendable rockets, except those required to launch unavoidably heavy or voluminous payloads. Phasing out expendable boosters will only be possible if the user has unrestricted access to a second, independently developed aero-space plane fleet. Overall, aero-space planes seem to carry with them the potential for having assured, robust, and continuous access to and from space—something notably lacking from today's space efforts.

## References and Footnotes

1. Logsdon, J., and R. Williamson, "U.S. Access to Space," *Scientific American*, March 1989, Vol. 260, No. 3, p. 37.
2. Phillips, E. H., "Langely Refines Design, Begins Human Factors Test of Personnel Launch System," *Aviation Week & Space Technology*, July 15, 1991, p. 52.
3. NASA, *Report of the 90-Day Study on Human Exploration of the Moon and Mars*, November 1989, pp. 3–16.
4. Statement made during the Synthesis Group's presentation at the Le Bourget Air Show in June 1991.
5. Gale, A., R. Simberg, and R. Koenig, "Space Transportation and Payload Integration," AIAA-90-5272, *AAIA Second International Aerospace Planes Conference*, 29–31 October 1990, Orlando, Florida, p. 3. Copyright American Institute of Aeronautics and Astronautics © 1990. Used with permission.
6. Wilson, A., ed., *Interavia Space Directory 1991–92*, (Jane's Information Group, Coulsdon, Surrey, England, 1991), p. 141.
7. See, for example, an article by James Ott entitled "Foreign Ownership of U.S. Carriers Feared as Limit to Future Military Airlift," *Aviation Week & Space Technology*, April 22, 1991, p. 96.
8. Logsdon, J., "The Decision to Develop the Space Shuttle," *Space Policy*, May 1986, p. 105. With permission of *Space Policy*.
9. Bennett, J., and P. Salin, "The Private Solution to the Space Transportation Crisis," *Space Policy*, August 1987, p. 186. With permission of Butterworth-Heinemann, Oxford, U.K.

# Chapter 16

# Aero-Space Plane Strategies

## Defining Strategies

The previous chapters have demonstrated that the implementation of aero-space plane programs is complex and fraught with critical constraints. These issues must be dealt with by every aero-space plane program that hopes to see its vehicle regularly shuttle to and from space, picking up or dropping off payloads on the way. Important concerns and tradeoffs underpin the process of choosing a particular approach. What are the requirements for aero-space planes and how are they defined by the user's needs? What are the support infrastructure requirements and cost? What are the alternatives for building and operating the fleet? What impacts would various launch prices have on the utilization of aero-space planes?

Different aero-space plane programs will adopt strategies which best fit their particular needs and configurations. Key strategic issues, and how they apply to current aero-space plane program thinking, are discussed here.

## Payload Size

If the ultimate goal of an aero-space plane program is to build an economical operational launch system, it is obviously important to first understand what size payload the vehicle should carry. This is the starting point for all aero-space plane strategies. If the payload capability is too small, the aero-space plane might not be able to capture enough of the payload market. If too large, then this may lead to a bigger, more expensive vehicle that is underutilized. At either extreme, the aero-space plane could end up being less than optimally economical.

Determining the ideal payload capability is the most basic aspect of aero-space plane strategies. Of the limited number of satellites launched today, about three-quarters of all payloads have launch masses of less than 8 tonnes (Figure 16.1). The other quarter are composed of very heavy and, importantly, unique payloads of varying shapes and sizes. In order to avoid the trap that the Space Shuttle fell into of trying to be the one launcher for everyone, aero-space planes should be limited to launching the *majority* of payloads. A backup booster could then be used to launch the remaining unique and outsized payloads as necessary (See Chapter 15). Therefore, the launch mass performance of the first aero-space plane fleet should be restricted to about 8 tonnes for low Earth orbit. As Lozino-Lozinsky et al. observe, [1]

> A profound additional analysis is necessary of the advanced SS (space system) efficient load mass characteristics, but, as it seems, mass characteristics of the advanced systems . . . will survive and, consequently, the nearest task is a creation of reusable winged horizontally take-off systems with 7–8 [tonne] class efficient load mass as loads comprising the main part (up to 70–90%) of the general cargo launched to the near space. This is confirmed with a selection of loads almost of all orbital plane variants currently considered . . .

The lower payload mass limit of all aero-space planes should, at the very minimum, be defined by the ability to launch the majority of the current generation of communications satellites complete with either a simple solid rocket or storable liquid propellant upper stage. Commercial communications satellites are, after all, the one sure market for economically operated aero-space planes.

The configurations of today's commercial communications satellites, space stations, and polar orbiting payloads are designed for the current generation of rockets. The introduction of a user-friendly launch system has the potential to alter this equation fundamentally, as will be discussed in Section 4. Nevertheless, in order to maintain a convincing economic case, the developer of the first aero-space plane fleet has little other choice than to match the vehicle's payload capability to the existing market.

As has been seen before, political mandates could try to force the aero-space plane to carry most of the heavy payloads as well. After all, the Shuttle originally was to carry 11 tonnes into low Earth orbit, but was later "upgraded" to carry 29 tonnes to ensure military use and, therefore, politi-

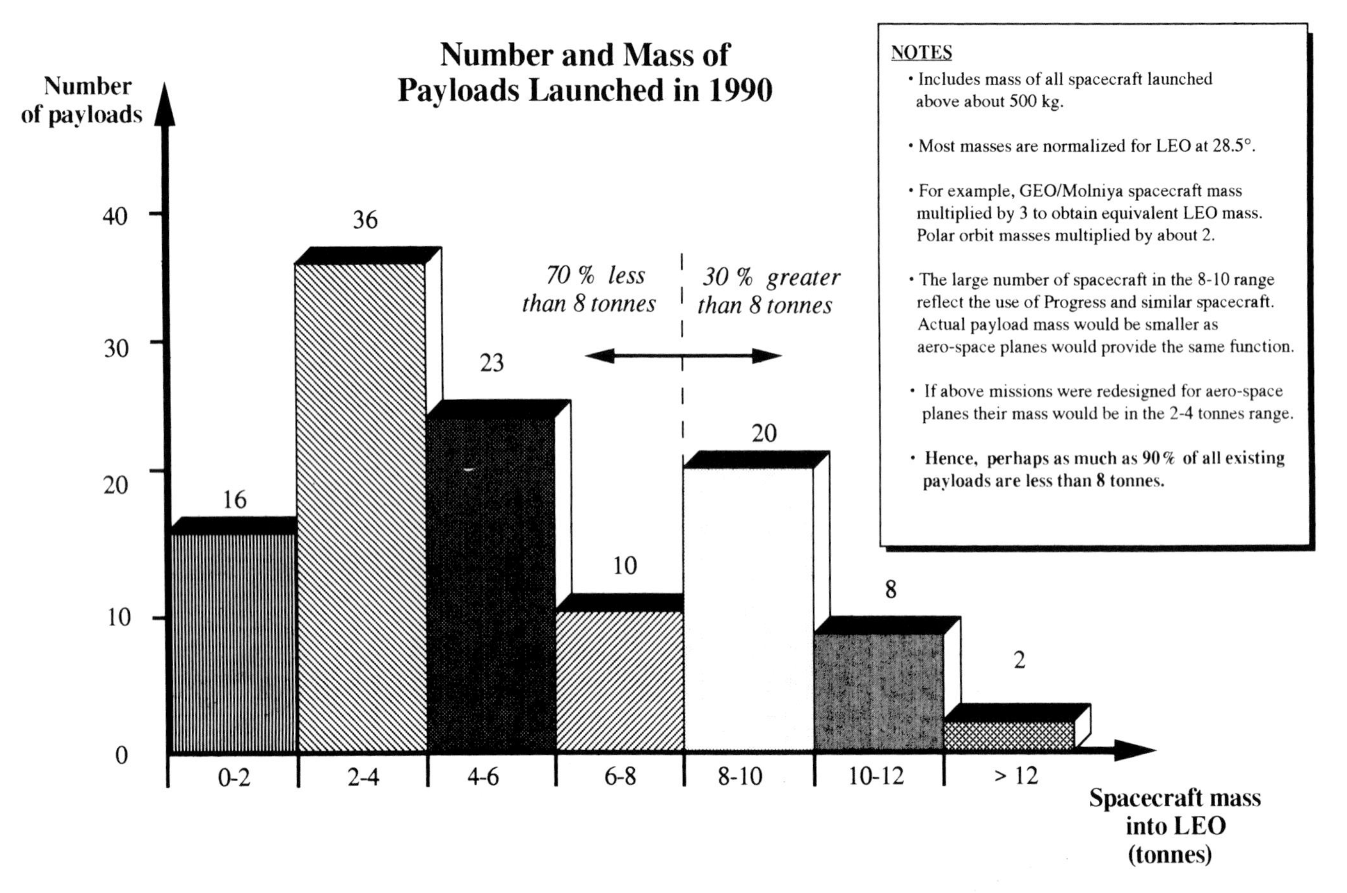


**Figure 16.1** Number of payloads launched as a function of their mass.

cal support. Could this story be repeated for a government-funded aero-space plane program? To find a possible answer it is important to first ask another question, What is the largest size aero-space plane that can *sensibly* be built?

As mentioned, aero-space planes must be large in order to carry *any* payload mass. Although the very high technology X-30 will not carry a payload, if it were built it would have a takeoff mass of about 200 tonnes just to get itself and a crew of two into orbit. For lower technology vehicles such as the Interim-HOTOL or Delta Clipper, equivalent zero-payload takeoff masses would be about 200 and 300 tonnes, respectively. To obtain a payload capability, a given type of aero-space plane will need to be significantly larger than 200 or 300 tonnes.

How much larger will an aero-space plane have to be, though? As a simple rule-of-thumb, the first aero-space plane should probably not be significantly heavier or longer than the largest aircraft flying today. Aircraft such as the Boeing 747 have a gross takeoff mass of about 350 tonnes, while the AN-225 can have a takeoff weight of up to 600 tonnes. (In the case of the Interim-HOTOL, the maximum size of this combination at takeoff is, by definition, the maximum takeoff mass of the AN-225.) Vehicle dimensions are also important from the viewpoint of ground handling, inspection, and support facilities. The maximum length of an aero-space plane probably shouldn't be much beyond aircraft like the Boeing 747 (Figure 16.2).

It can be argued that because airplanes must be able to take off and land at various airports equipped with standard facilities, their size has to be restricted to ensure the maximum flexibility in operations. An airplane requiring an 8 kilometer runway would find few airports able to fly from. In addition, takeoff noise regulations can limit the maximum thrust of an airplane's engines and, consequently, constrain the aircraft takeoff mass. By contrast, the first aero-space planes could quite satisfactorily be based at only one or two specially outfitted sites located in remote parts of the world. Still, because building an aero-space plane is an immensely difficult technical challenge in itself, coupling this with the burden of having to build a vehicle which is also significantly larger than any aircraft that has ever flown would seem unwise.

The purpose of the above discussion is to clearly define the payload range. Based on current aero-space plane programs, if the takeoff mass is large enough to allow a payload, then it will only account for about 2–3% of this mass. Thus, an aero-space plane weighing some 350 tonnes at takeoff implies a payload capability of around 7–10 tonnes. As the inclination of the orbit increases away from the equator, less assistance is provided by the Earth's rotation, leading to a reduction in performance. Therefore, for a mission to Space Station Freedom at 28.5°, the payload capability will be reduced by around 25% to roughly 5–7 tonnes. For polar orbiting missions at 98.5°, this reduction will be around 60% to roughly 3–4 tonnes.

This payload performance is typical of vehicles like the original version of HOTOL with a takeoff mass of about 275 tonnes or Sänger with a takeoff mass of about 370 tonnes. Higher payload fractions are possible, but require more advanced systems. Because liquid oxygen is the primary contributor to the takeoff mass, breathing air to the highest possible speed has the potential to reduce the amount of liquid oxygen that has to be carried and, therefore, increases the payload fraction compared with lower technology alternatives. The NASP program aims to demonstrate just this, and it has proposed that payload fractions of 5% or more might eventually be possible with second-generation NASP technology, assuming air-breathing nearly all the way to orbital speed (e.g., Mach 18–20). Achieving such a capability—and one which can also be applied to a future operational system—will be technically and economically challenging, assuming it can be done at all in the near future.

The Delta Clipper is something of an anomaly in this discussion because it doesn't look or function like an airplane. Making direct comparisons with winged aircraft could be misleading, as the discussion in Chapter 10 explained. Although its nearly 600 tonne gross takeoff weight is heavier than all aircraft except the AN-225, speed over wings is not used to lift it from a runway. However, because the Delta Clipper's physical size (i.e., 40 meter length by 13 meter diameter) is well within the range of many large aircraft, and as size is an important parameter in servicing and turning around the vehicle, more or less the same rules can be considered to apply.

Overall, it is considered that Shuttle-like political requirements forcing an aero-space plane to carry all payloads are unlikely because they would lead to enormous and, potentially, impractical vehicle designs. To carry 29 tonnes into low Earth orbit would require a vehicle like HOTOL to have a gross takeoff mass of more than 1,200 tonnes—twice the size of the largest aircraft ever flown. Even if such a vehicle could be built, its costs would be similarly enormous because of the proportional relationship between vehicle dry mass and cost. For all of these reasons, the first operational aero-space plane will have a payload mass capability on the order of a few tonnes simply because there is no other choice.

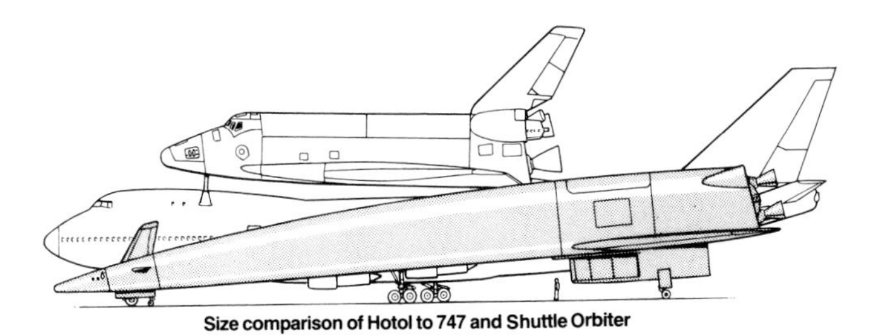


**Figure 16.2** HOTOL comparison with a Boeing 747 (*Flight International*).

## Payload Bay Cross Section and Volume

Payload bay cross section and volume are another important factor in determining the market capture. Here, as with the payload mass, vehicle constraints play a profound role in the size of the payload bay. For most current aero-space plane concepts, large hydrogen tank volumes are required to contain the low-density liquid and/or slush hydrogen fuel. Thus, depending on the fuselage shape, the large hydrogen tank volume generally facilitates the efficient incorporation of a large payload bay cross section, although perhaps a more modest bay length. The actual size of the payload bay isn't critical, although there is an advantage to making it as large as sensible. Specifically, its size should not significantly compromise an optimum vehicle configuration or force a higher technology solution to be adopted just for the goal of obtaining a slightly larger cross section or length.

With single-stage vehicles, larger diameter payload bay cross sections are generally easier to incorporate because of the large hydrogen tank requirements. With two-stage vehicles that separate at hypersonic speeds, the smaller hydrogen tank and the need for efficient aerodynamic shapes can be a significant constraint on payload bay size. Heribert Kuczera explains the situation for Sänger, "Is there a need to have a payload bay on the unmanned Sänger upper stage which is compatible to the Shuttle? If you increase the diameter of the upper stage to a payload volume that says you can deploy a space station module, then it would be a significant increase in volume of the upper stage, and this seems not to be very good in terms of integration of the two vehicles."

Ideally, a payload bay cross section of at least that of an Ariane 4-type launcher (3.65 meters in diameter), or in the region of that of the Shuttle (4.62 meters in diameter), with a length dimension about double this, would be good to aim for—if only because most user organizations are familiar with this and know how to use it. The Shuttle bay cross-section size would probably be the most suitable if it could be easily achieved, because many payloads are already designed for it. These include the Eureca, the Multi-Purpose Experiment Support Structure, the Spacelab Pallets, SPAS,

Astro-SPAS, the Multimission Modular Spacecraft, and various cradles for launching comsats. Further, Space Station Freedom, Columbus, the Japanese Experiment Module, together with their multitude of constituent parts are all sized for a 4.62 meter cross section. These types of spacecraft and structures are likely to fly far more frequently than they do today simply because of the aero-space planes' unique capabilities (Chapter 17). At a minimum, it would probably be necessary to modify the payload-to-aero-space-plane interfaces in order to facilitate easier installation in the payload bay (Chapter 4).

Viewed from the perspective of meeting the needs of the *existing payload market*, the aero-space plane payload mass capability and volume size defined above are eminently suitable for the vast majority of space applications. This payload range would allow the largest of the current generation of commercial communications satellites to be deployed in a low equatorial orbit, allowing use of relatively inexpensive storable liquid and solid propellant upper stages needed to place the spacecraft in geosynchronous transfer orbit. Similarly, the 5–7 tonne range to a Space Station Freedom orbit is suitable for rotating crews, handling a large fraction of the regular logistics requirements such as food, water, and propellants, and delivering new expansion modules and other hardware elements. The same is true for missions to Mir at a 50° inclination. For polar orbiting missions, 2–4 tonnes would be suitable for spacecraft like the Spot Earth observation spacecraft and the NOAA weather monitoring spacecraft.

## Manned or Unmanned?

Another important strategic issue in planning for aero-spane plane launch systems is whether they should be piloted or automatic vehicles. According to Maxwell Hunter, who describes the NASA man-rating procedures as "really a very expensive, understandably neurotic, method of creating *man-rated ammunition*," [2]

> A most frequently asked question concerning the SSX idea is, "Is it manned or unmanned?" And usually, this question is usually accompanied by, "Why don't you make it unmanned. It's so much cheaper, and who needs men anyway?" We automatically and intuitively associate unmanned with expendable (and cheap) and manned with recoverable (and expensive).This is all a natural outgrowth of the paradigm conflict between ammunition and transportation.

Whether aero-space planes should be manned or unmanned is a complicated question. Nevertheless, it is possible to highlight some of the principal issues by separately examining the options for the flight test and the operational phases of a particular aero-space plane program.

During the first aero-space plane test flights, probably the greatest concern is being able to safely recover the vehicles when unanticipated failures occur. This is important for vehicles like the X-30 whose suborbital flight test program is planned to occur over the continental United States. According to Wiezbanowski et al. [33]

> The main reason such profiles must be flown is to allow emergency recovery of the X-30 when, not if, problems occur. A pragmatic review of previous X-30 series aircraft flight programs more than justifies this approach. As an example, the X-15 made emergency landings at fields other than EAFB ten times during its 199 flight test program . . . Obviously, there are significant implications if these profiles are to be safely flown with an unmanned X-30. The first orbital flight of the Soviet Buran proved that today's technology can support unmanned vehicles, with known characteristics, flying relatively benign trajectories to a single recovery site. There is considerable doubt, however, as to whether or not the technology is there to develop the true "Artificial Intelligence" system that would be required to safely control and recover the X-30, with all its unknown characteristics, to an undefined multitude of recovery sites.

It is important to remember that the first test flights of aero-space planes will not go to orbit. Indeed, they may not even pull up their undercarriage. Therefore, building an automatic flight control system that can routinely handle landings at various nominal and abort air fields, *and one that must work on the first flight of the aero-space plane*, might be more difficult and expensive than putting a crew compartment within the vehicle. Further, governments might be reluctant to allow an essentially untried, robot guided aircraft the size and mass of an airliner to roam around the country. This reason alone may dictate that aero-space planes are manned during the test phase.

Now, consider the operational phase. It is not necessary to have a manned presence for most space missions, such as the deployment of a communications satellite. Further, a crew compartment, together with the systems required to support several astronauts in space for several days, has a mass of a few tonnes, and this will obviously reduce the amount of payload mass. Thus, it might be appropriate to fly crews only when required (e.g., in the cargo bay), while unmanned cargo missions would use an automatic system to maximize the payload capacity. This is the philosophy adopted by the HOTOL, Sänger, and Delta Clipper proposals.

It is possible to envisage an aero-space plane strategy using a crew compartment just for initial flight tests and an automatic system for the first orbital flight tests. The first crewed test flights would allow the basic handling properties of the vehicle to be fully characterized by experienced pilots. These flights could be limited to takeoff and landing demonstrations and up to a flight speed where safe ejection is always possible. Data gained from the manned flights would be used to teach an automatic system how to fly the vehicle during the critical flying phase of a mission. After many test flights, enough confidence would be gained in the automatic system to allow it to go solo. Gradually, the automatic system would expand the rocket part of the envelope all the way to orbit without endangering human life. Once a number of automatic flights had been successfully performed, then manned orbital missions could begin.

## *Proving It Can Be Done: Who Should Pay for It?*

A question which is often raised is, if aero-space planes are such good ideas then shouldn't the commercial sector pay for them? Indeed, the demise of the original HOTOL concept was partly due to this attitude, as explained by Sir Geoffrey Pattie, "the HOTOL program was an example of how the British deal with this kind of a program which is to try and get the private sector to fund it, which is impossible. That was never going to be a runner because of the long-term, high-risk nature of the program." Conversely, it might be considered unrealistic to expect governments to fund such a program in its entirety.

An aero-space plane's size, and whether it is manned or unmanned, are factors important in determining the vehicle's cost. The heavier an aero-space plane's unfuelled weight, the more expensive it will be. Regardless of the exact strategy choices made, aero-space planes *will* be expensive. Therefore, who should pay for the construction of the first aero-space plane? Although difficult to answer, pointers can be obtained through discussion of analogous high-technology programs. The best example is the commercial airliner industry which, much like aero-space planes, emerged soon after the technology became available from military programs. According to Mowery and Rosenberg, [4]

> [Military] research investment was not intended to support innovation in any but military airframe and propulsion technologies. It has nevertheless yielded indirect but important technological spillovers to the commercial aircraft industry, notably in aircraft engines . . . The development of the first U.S. jet engine was financed entirely by the military. More recently, military-supported research on turbofan engines for the C-5A transport influenced the development of high-bypass engines that power the latest generation of commercial transports, including the Airbus Industrie A300, A310, and A320, as well as the Boeing 737–300, 747, 757, and 767 . . . The development of commercial aircraft also has benefited, however, from military-sponsored research and procurement in airframes . . . The Boeing 707, for example, was based closely on the design of a tanker, the KC-135, developed by Boeing to provide in-flight refuelling for strategic bombers (the B-47 and B-52) developed previously by the firm.

Clearly, the initial success of the U.S. commercial airliner industry was possible because of the spillover from the military. The DC-3 is a prominent example of a highly successful passenger and cargo aircraft which was the success it was because the U.S. military ordered over 10,000 of these aircraft during World War II. After the ending of hostilities, hundreds of these aircraft became war surplus and were modified to serve as short-range passenger and cargo transports. Another example, this time for long-haul flights, was the Lockheed Constellation. This elegant tri-tail aircraft was derived directly from the military C-60 transport, and civilian versions were sold to airline companies like Pan American, TWA, BOAC, and Air France [5]. A similar situation arose outside the United States; the Olympus engines of Concorde were basically the same as used on the British Vulcan long-range bomber [6].

The last few decades have seen a steadily widening gap between military aircraft requirements and civilian airliner requirements. This is not of any particular significance because the airliner is now, and has been for the last two decades, a well-established and self-supporting commercial industry in its own right. Consequently, some of the revenues generated from the commercial volume of sales can be channelled back into private R&D efforts or used to finance upgrades or new airliners. The question raised here is would there have been a commercial airliner industry without this military spillover? Because of the need and market for passenger and cargo transportation, it seems reasonable to presume that 707s, D-10s, and other such airliners would have emerged—*eventually*. It is also fair to say that the development of commercial airliners would have taken far longer. This is not only because of specific technology needs. The development of airworthiness standards used today to certify airliners to safely carry people evolved from the military's need for reliable and available transport services. The commercial airliner industry is indebted to the military.

Making analogies between vastly different systems is always a dangerous game to play. Even though some aero-space plane concepts have a passing resemblance to airplanes, all similarities end there—nothing remotely like an aero-space plane has ever been built before. But this is not the point of choosing airplanes for this analogy; railways, highways and other transportation infrastructures could also have been chosen. Major ventures that are inherently expensive and push the boundaries of technology are unlikely to be pursued by the commercial sector until the risk of failure has been practically eliminated. In other words, the first Aero-Space Planes, Inc. will probably not be established until someone else has spent the money to prove it can be done. In most cases, this means governments. No one would have expected a commercial company to privately finance an airliner like the 707 without, at a minimum, first having a vast amount of applicable experience in airframes and high-performance jet engines to call upon. The risk of failure would have been just too great because of the very high up-front investment and the long length of time required to build and qualify such an aircraft.

It seems reasonable to conclude that the majority of the funding for development must come from governments. In the initial stages of an aero-space plane program, government funding can also be an effective way of pursuing advanced technological developments, as it allows advantage to be taken of many diverse companies, research centers, and other specialized agencies—a possibility that may not be open to a commercial organization. The NASP program, and in particular the Materials Consortium made up of over 100 companies, has aptly demonstrated the value of nationally directed and supported research (See Chapter 6). In addition, because the investment required for the initial enabling technologies work is relatively modest, the private sector is in a position to contribute a substantial *fraction* of the funds for this work, although the specific magnitude of that contribution might be relatively modest. This, in turn, makes government support easier to rationalize. This is the case for programs like Sänger where some 20% of the Phase 1 research is funded privately.

Once governments have proven the technical viability of such advanced systems beyond all reasonable doubt, a point will be reached when the technology is sufficiently mature for the commercial sector to consider its application in a profitable activity. This may mean governments will have to finance the development of an experimental or even pre-operational prototype aero-space plane to demonstrate conclusively that such vehicles can be flown to orbit and back, turned around in a short period of time, and flown again, and again, and again. If the burden is put on the private sector to finance the development of the first aero-space plane, it may be a long time before any are built.

## Investing in the Support Infrastructure

A significant contribution to the costs of aero-space planes will be associated with putting in place the support infrastructure required to enable construction and operation of the vehicles. For development activities, appropriate facilities and toolings are needed to allow the design, manufacture, and testing of the vehicle. In addition, it is important to know whether the materials needed to build the vehicle are available in sufficient quantity, especially in the case of a new alloy that has not been previously used. For operational activities, facilities to handle aero-space planes, including those for maintenance, servicing, fuelling, and cargo handling, are needed, as well as launch facilities (runways, trolleys, or pedestals), and extensive radar tracking and communications systems. Finally, and not insignificantly, further investments are required to educate and train the work force needed to build and fly such a vehicle.

The support infrastructure for the first aero-space plane is going to be expensive to create, as Robert Barthelemy explains in the case of the experimental X-30, "The dollars to conduct the [NASP] Phase 3 program is dependent on some fixed dollar amount plus a variable amount which is proportional to the weight of the vehicle, that is, $y = A + Bx$. The 'A' is a big number. It may be as much as $4 or $5 billion regardless of whether the vehicle weighs '10 pounds' or '500,000 pounds.' The reason the fixed cost is so high is because of the infrastructure you need to create this thing. Then you need an engineering force and a test force, and you need time—['A'] ends up as a big number. Now you can fool with that number, and lower it by really getting in and reducing the cost. But you're buying risk reduction, safety, and completeness of data. It's investment, and sooner or later you're going to put in that investment. Also, what you learn there is directly applicable to what you do on NASP-Derived Vehicles. You don't want to make a shoddy job of the infrastructure and then find out that you got a great airplane but you don't understand how you launch it, how you fuel it, or whatever."

The principal reason why the cost is going to be so high is because no nation or organization has ever attempted to build an aero-space plane. The first aero-space plane will be as unique as Stevenson's "Rocket" [7]. This contrasts sharply with most other forms of transportation which have evolved over time from relatively small and simple vehicles. It is possible, for example, to trace the development of today's ocean-going freighters and cruise liners all the way back through the millennia to the ships of Columbus, the barges of Alexander the Great, the galleys of the Greeks, and the canoes of the Polynesians. Each ship was able to build on the experience of the former. All of these early ships provided the same basic functions and capabilities as today's vessels. They were fully reusable, incrementally testable, and were capable of carrying cargo and people back and forth from one point to another. Thus, the support infrastructures were also able to evolve slowly. The majority of the world's major trading centers developed directly as a consequence of the needs of the evolving seaport infrastructures.

The same cannot be said of aero-space planes. The first aero-space plane will represent the start of its own new evolutionary family tree. It will be its own heritage. Because the first aero-space plane capable of reaching orbit will be large and require the use of advanced technology, the initial support infrastructure costs will be high, although perhaps less than the cost of building the specialized launch gantries and integration and assembly facilities characteristic of expendable rockets and the Shuttle.

In this light, it might be considered unreasonable to expect a commercial organization to pay for the initial investment cost of establishing the aero-space plane support infrastructure, on top of actually building the vehicle itself and meeting all the regulatory requirements. Fortunately, the support infrastructure does not have to be completely reinvented every time a new aero-space plane is built. Once that initial investment has been made for the first aero-space plane fleet, it can be propagated for subsequent fleets, provided the interval between building such vehicles is on the order of a few years. The support infrastructure is an investment for the future.

## *The Operators: Government or Commercial?*

It is reasonable to conclude that governments will have to fund the development of the first aero-space plane together with its support infrastructure. But who should operate the first fleet? Consider the differing strategies of the Shuttle and Ariane.

Because NASA's charter is primarily to perform research and development, it has been criticized in recent years for spending too much on operations. The operational Shuttle consumes between one-third and one-half of NASA's budget. The Shuttle cannot be privatized, because the high cost of its operations will never be commercially attractive. Further, the cost of a failure is enormous and beyond the boundaries of normal insurance practices. Only NASA can operate the Shuttle.

By contrast, Ariane is a commercially viable launcher and is operated as such by Arianespace. This quasi-commercial organization is driven by the underlying need to make a profit. Arianespace does not have to rely on yearly or multiyear appropriations from governments to keep it alive. For example, it has a market base which allows it to make bulk purchases of launchers, taking advantage of economies of scale. ESA fully-funded the development and testing of Ariane and the support infrastructure.

Theoretically, the first aero-space plane fleet could be operated by government or commercial agencies, or a combination thereof. In the United States, vehicles like the Delta Clipper or the air-breathing NASP-Derived Vehicle could be developed, procured, and then operated by NASA or the DOD. In Europe, ESA could develop an equivalent capability and leave it to an organization modelled on Arianespace to operate the vehicles commercially.

The final choice of government versus commercially operated aero-space planes will depend on a large number of factors, as discussed throughout Section 3. On the one hand, a dedicated government controlled fleet might be necessary to support military missions. On the other, as the reason for transitioning to aero-space planes is fundamentally economic, strategies that enable cost-effective operations are clearly desirable. After all, the function of aero-space planes is only to provide an economic transportation *service*.

Government-run space programs are generally not conducive to meeting cost-effective operational objectives. The problem is that organizations like NASA have large additional overheads because of the need to support a nationwide institutional structure. This institutional surcharge will ensure that the real cost per flight will be significantly higher than it needs to be. As one example, even though the Shuttle is launched from Cape Canaveral, Florida, its missions are planned and controlled in Houston, Texas, while the overall program is managed in Washington, D.C. The Shuttle is expensive to operate for the reasons defined in Chapter 11, and the institutional surcharge does little to help.

Economic operations of an aero-space plane fleet would seem to be best met through centralizing the activity and, if at all possible, adopting commercial practices. Can aero-space planes be commercialized? The Shuttle experience might initially lead to a negative conclusion simply because of the high cost of an aero-space plane loss. But this ignores fundamental differences between the Shuttle and aero-space planes. Inherently, aero-space planes are testable. Therefore, it is possible to envisage a situation where a government organization funds the production of just one or two experimental or pre-operational prototype aero-space planes. The government then proceeds to fly them in an incremental test program so as to *demonstrate* basic systems reliability, the ability to safely recover the vehicle after various deliberate in-flight failures, the schedule required to turn around the vehicle for the next flight, the maintenance and servicing needs, launch mass performance, and other characteristics that will determine future operational success or failure.

In this scenario, the government will have spent only enough money to prove the basic concept and feasibility, while minimizing its investment in the support infrastructure and personnel. The commercial community will have a known quantity on which to base business plans. With proven feasibility, the commercial community can procure vehicles and proceed with the operations of the fleet. As with airlines today, they won't have to be afraid of frequent vehicle losses. Government funding will still be required for construction of the support infrastructure, much in the same way airports and railway stations are funded today. In the

**Figure 16.3** Skylon (*Reaction Engines*).

beginning, at least, governments may also have to act as the insurance agents.

Aero-space planes offer another possibility new to space transportation. Commercial organizations might consider selling vehicles to other organizations or countries, rather like airliners. This has obviously not been practical with expendable launchers because new vehicles must be built for every mission. Alan Bond is currently pursuing such a strategy for his proposed Skylon vehicle (Figure 16.3), for which, he believes, there is already a market. As he notes, "if you look at the current launch requirements of individual countries, even at annual launch rates of about one or two satellites per year, it would be cheaper for such countries to procure a single Skylon vehicle and use it over 10 or 20 years because today's launch costs are so high."

## Re-couping the Investment: Approaches and Choices

In a conservative scenario, the prospects for constructing an operational aero-space plane are clearly heightened if the first fleet can be economically justified purely on the basis of the existing *commercial* market demands. The case for an aero-space plane is further enhanced if it can respond to an increase in the market size as it occurs. Bob Parkinson explains, "How often do you have to fly? You're always up against two problems. One is that you build these vehicles to fly often, but at the same time you've got to convince people that if there aren't '64' payloads per year, but there's only '15' payloads per year, it's still worthwhile. What you want is a vehicle that breaks even at about 15 flights per year which you can guarantee, but if you find that you have a market of 60 or 70 you can still fly . . . We've tended to take the same attitude [with Interim-HOTOL]. I'm not talking about a market that might happen, but something that has happened. Ariane 4 launches enough to keep us going now, this year. It doesn't take account of a space station, just to take one example."

"Worthwhile" means that it would cost less to develop, manufacture, and operate the Interim-HOTOL, launching around 15 payloads per year, than to launch those same payloads on existing boosters. For so few payloads, launch costs probably wouldn't be significantly different from current launchers, but the greater reliability and availability of aero-space planes would obviously give them an important competitive edge. Contrary to some popular conceptions, some of the simpler aero-space plane configurations may not have to fly very frequently in order to be more economic than existing boosters.

The attitude of the HOTOL program, as it also appears to be with Sänger and Delta Clipper, has been to try and develop strategies that would allow the full life cycle cost recovery of the program over a typical lifetime of 20 years. They argue, "It is not possible to assume that the development of such a vehicle would be 'written' off" [8]. In the past, the development costs of the majority of launch vehicles were paid for by governments. The Shuttle and Ariane rockets were developed specifically as launch systems and paid for directly by NASA and ESA, respectively.

Alternatively, it can be argued that the life cycle cost recovery philosophy adopted by these programs is excessively burdening. Not only do they attempt to significantly reduce launch costs by developing a technically demanding and unproven vehicle, but they *simultaneously* hope to recover the life cycle costs in the long run. If virtually all other launchers had their development costs written off by governments, it might be reasonable to expect the development costs of aero-space planes to be written off by governments.

There is an important distinction in this discussion. Although the price a customer pays for an Ariane or Delta launch does not go toward paying for the development costs, it does pay for the production costs. Therefore, it might also be reasonable to expect the user to pay a contribution toward the production costs of a government-run or privately operated aero-space plane fleet, or a combination of the two. In this sense, strategies that include the ability to amortize vehicle production costs over the expected lifetime of the vehicle may be essential to the prospects of aero-space planes.

## Pricing Policies: The Commercial Users' Perspective

Forcing the aero-space plane operator to charge commercial users a price that includes amortized development and/or production costs may have overall detrimental economic impacts. Only a drastic reduction in launch costs, perhaps by one or even two orders of magnitude, may be what it takes to stimulate the market in a way that allows new, self-supporting commercial space opportunities to emerge (See Chapter 20). This is also true for new scientific, applications, and military missions. Obtaining such a reduction in costs might only be possible if development and production costs are *not* amortized into the launch price. There probably is a threshold where the benefits of new commercial missions to the nation as a whole outweigh the need to recover production and development costs. By contrast, an aero-space plane fleet that charges a small number of initial users a price comparable to current launchers may do little to encourage new users to enter the market.

The original road and railway infrastructure of many nations might be appropriate analogies here. The U.S. transcontinental railway that started in 1866 was built by two commercial companies, but was heavily subsidized by the government. According to commentator Alastair Cooke, [9]

> Working summer and winter, it took the Central Pacific two years to hurdle the formidable barrier of the High Sierras. A thousand miles or more back East, the Irish gangers frequently fainted from the midsummer heat, but the company officials were revived by the thought that the government had promised a subsidy of $16,000 *per mile* of track. Once they started to climb the Rockies, it went up to $48,000 per mile, with wide stretches of free land bordering the track thrown in.

The government effectively paid for and wrote off the bulk of the railway network procurement cost, while the bulk of the operational costs were paid by the users. Ultimately, this massive government investment in transportation infrastructure was more than recouped by the tax revenues accrued as a result of the new industries and trade that became possible *only* because of the availability of an affordable transportation means. An immediate impact that occurred is described by Cooke, [10]

> . . . while the continental line was creeping across the plains, there came a twenty-nine-year-old livestock trader from Chicago named Joseph McCoy. He looked at the railroad and he looked at the map, and he got an idea. He knew there was lush grassland all the way to southern Texas. What would be simpler than to connect a cow with a railhead and make a fortune and, incidentally, add beef to the diet of millions in the East? . . . McCoy now spent the considerable sum of $5,000 getting out advertising circulars and sending riders down to Texas to promise the cowboys a safe trail and a fair price at the railhead . . . A hundred days after McCoy had posted his offer, the village of Abilene heard a thunderstorm of hooves coming up from the south. It was a start of the cowboy legend, and of McCoy's reign as emperor of the cattle kingdom. He had boasted that he could deliver two hundred thousand cattle in his first decade. He was wrong. In the first four years he shipped over two million out of his stockyards. He was one of the rare American promoters whose production exceeded his propaganda. He was, they liked to say, "the real McCoy."

If the users of the first railroads had had to pay a contribution toward full cost-recovery over a *commercially* short period of time (i.e., 5–10 years), it is reasonable to suggest that today's wide-scale railroad networks would probably not exist, and neither would the new businesses and tax revenues. The private sector has never been required to pay fully for the establishment of an unproven national transportation infrastructure.

The above analogy and the example of Abilene, Texas,

may appear somewhat simplistic, and, for reasons that will be presented in Chapter 17, it is flawed. However, if space is to be more thoroughly exploited, an affordable, available, and reliable space transportation infrastructure is a prerequisite. In this sense, aero-space planes are directly analogous to highways and railroads. Remarkably enough, the United States sometimes views the Space Shuttle in this context, as it is often referred to as national resource. The $5 billion spent annually on the Shuttle is typical of the funding given to building and maintaining national transportation infrastructures, but the low flight rate of the Shuttle makes the expense look ridiculous. If the Shuttle flew 500 regularly scheduled flights per year at *$10 million* per flight, instead of the actual case of around 8 frequently delayed flights per year at *$600 million* per flight, then it could properly be called a national resource that would allow for the opening of the space frontier.

## Summary—Planning for the Unknown

Discussing aero-space plane strategies is like planning for the unknown. Aero-space planes do not exist much beyond drawings, computer analysis, and small-scale test articles. Yet, a number of basic and fundamental issues underlie possible strategies. The payload capability of the first operational aero-space plane will be in the region of 7–10 tonnes. Much below this range, few existing payloads could be carried, especially commercial communications satellites. Above this, aero-space planes rapidly become unwieldy and beyond the range of the current largest aircraft.

The optimum payload capability broadly determines the size of the aero-space plane which, in turn, determines the technology choices. All payload-carrying aero-space planes will have takeoff masses of at least 250 tonnes and require various combinations of advanced and unproven technologies. By the time the first aero-space plane has successfully completed its maiden flight, *someone* will have spent a large sum of money. In spite of their potential, the high risk/high cost nature of aero-space planes is not conducive to conventional commercial investment practices. If the commercial community must bear the brunt of this cost, aero-space planes are unlikely to be built in the foreseeable future. Government funding to some minimum level will almost certainly be mandatory, as it has been for the railroads, highways, and airlines.

The decision to develop aero-space planes will be—or should be—driven by economic imperatives. Therefore, strategies that will eventually enable commercially viable operations should be pursued. Governments might spend only enough money to prove the basic operational feasibility of aero-space planes, such as by building one or two prototypes and flying them through a rigorous incremental test program. *Demonstrating* that aero-space planes are practical and economically competitive with existing launch systems might be all that is necessary before private dollars are invested in building a fully operational fleet.

How the first aero-space plane is eventually priced will depend on many factors. This cursory examination has highlighted the importance of minimizing the price the user pays for an aero-space plane flight. Forcing the user to pay a price that includes development and production cost recovery might ensure that launch costs can never be reduced significantly below the threshold where new commercially self-supporting space missions become practical. Low prices could, potentially, be compensated by revenues generated by new spaceflight activities that were not affordable or even possible before. Again, the first major railroad, highways, and airliner transportation infrastructures were all directly or indirectly subsidized, and justified on the basis of minimizing the user price in order to maximize utilization and, ultimately, widespread economic growth.

Whether aero-space planes will provide opportunities for new commercial space enterprises is difficult to say. Given an appropriately priced user cost and many regularly scheduled flights, aero-space planes might do for spaceflight what the railroads did for steakhouses in New York. How this may happen is the subject of Section 4.

## References and Footnotes

1. Lozino-Lozinsky, G. E., L. M. Shkadov, and V. P. Plokhikh, "Reusable Aerospace System With Horizontal take-off," IAF-90-176, *41st IAF Congress of the International Astronautical Federation*, Dresden, Germany, October 6–12, 1990, p. 2.
2. Hunter, M. W., "The SSX: A True Spaceship," *The Journal of Practical Applications in SPACE*, Vol. 1. No. 1, Fall 1989, p. 51.
3. Wierzbanowski, T., and T. Kasten, "Manned Versus Unmanned: The Implications To NASP," AIAA-90-5265, *AIAA Second International Aerospace Planes Conference*, October 29–31, 1990, Orlando, Florida, pp. 6–8.
4. Mowery, D. C., and N. Rosenberg, *Technology and the Pursuit of Economic Growth*, (Cambridge University Press, Cambridge, England, 1989), p. 185.
5. Blay, R., *Lockheed Horizons*, Issue 27, December 1988, p. 27.
6. Calvert, B., *Flying Concorde*, (Airlife Publishing Ltd., Shrewsbery, England), p. 148.
7. This is not a perfect analogy, as there were several attempts to build steam-power trains before the Rocket. Stevenson's design proved to be the first useful train built and, as a result, allowed the first basic railroad lines to be laid. Interestingly, the name "Rocket" was given because bemused observers at that time commented that they'd sooner sit on a rocket than Stevenson's invention.
8. Parkinson, R. C., "A Total System Approach Towards the Design of Future Cost-Effective Launch Systems," *IAA Symposium on Space Systems Cost Methodologies and Applications*, San Diego, California 10–11 May 1990, p. 1.
9. Cooke, A., *America*, (London), p. 228. Extracts reproduced from *Alistair Cooke's America* with the permission of BBC Enterprises Ltd.
10. Ibid., pp. 229–232.

# SECTION 4

# NEW DIRECTIONS IN SPACEFLIGHT

*A transport infrastructure is a prerequisite—though by no means a guarantee—of economic development.*

—Hans Adler, *Economic Appraisal of Transport Projects*,
(World Bank, Washington D.C., 1987)

*The ability of the railroad to produce transport services at a lower cost has an analogue in lessened resource requirements. Thus overland transport rates of fifteen cents a ton-mile imply the consumption of fifteen cents of labor, capital, and entrepreneurial inputs for the unit output. A railroad rate of three cents, say, means that twelve cents of inputs formerly needed can be applied to other tasks. The difference in costs of transportation due to the introduction of the railroad therefore provides one measure of the increased production it made possible.*

—Albert Fishlow, *American Railroads and the Transformation of the Ante-bellum Economy*
(Harvard University Press, 1965)

*The building of an improved highway is a very small affair to the commerce which is to pass over it. The former is the pioneer of the latter. The expenditure of $1,000,000 upon a railroad is the occasion for the expenditure of ten times that sum to develop what the former has rendered available.*

—*American Railroad Journal*, XXIX (1856)

*The steamboat greatly reduced the cost of transportation by water, but the steam locomotive was nothing less than revolutionary in its effect on transportation by land. The railroad has made cheap transportation possible for vast areas of the earth's surface. This cheap transportation was one of the basic facts on which the economic life of the nineteenth and twentieth centuries was built.*

—Philip Locklin, *Economics of Transportation*
(Business Publications, Inc., 1966)

*Without railroads the social and economic development of the great American West would have been impossible. Nature, indeed, had endowed the area with fabulous riches and boundless opportunities, but in doing so, as if to tease puny but acquisitive Man, had scattered her gifts over an enormous landlocked territory. There they might have remained, isolated and unused indefinitely, had it not been for the coming of the rails.*

—Richard Overton, *The Railroads*
(University of Texas Press, Austin, 1953)

# Chapter 17

# Are There Enough Payloads?

## Will It Take a "Leap of Faith"?

Are there enough payloads to justify the high costs of developing aero-space planes? Consider the extremes of opinions often cited by critics and advocates. *Advocates* expect aero-space planes to herald a "revolution" in the exploitation of space. Analogies with the past are often made comparing the advent of aero-space planes to the railroads that opened up the Western frontier or the laying of national highways. *Critics* have often charged that aero-space plane launch costs can only be reduced significantly if the vehicle were to fly very frequently. They believe that we don't need to launch any more satellites, and that demand for space access is inelastic, regardless of launch costs.

The arguments of the advocates certainly paint a rosy picture for aero-space planes. The trouble is that it is not at all obvious what is going to fuel this revolution. Analogies with terrestrial revolutions might be seen as standing on shaky ground. The massive investments in the laying of transcontinental railroads in many countries came long *after* population centers and trade had already been established. This pre-railroad trade was, of course, extremely limited in extent. The railroad investment was justified because some trade already *existed*, and its limited growth could be seen as a direct consequence of the lack of an efficient transportation infrastructure.

The same cannot be said of space. This is not because there is nothing to do in space. Rather, the cost and operational limitations of current launch systems have made trade, in the form of exchange of products, entirely impractical. The pre-railroad trade was possible because of the existence of other forms of transportation such as the horse and buggy, sailing ships, and canal barges. In this sense, aero-space planes might aptly be described as the equivalent of the horse and buggy [1], with one fundamental difference: cost. A revolution might occur in space, but in order to determine whether it *will* happen will necessitate a much more substantial investment than the first horse and buggy required.

Certainly, a one or two order of magnitude reduction in launch costs, coupled with an equivalent one or two order of magnitude increase in launch opportunities and reliability, could facilitate a revolution in the exploitation of space. However, justifying a new, risky, and expensive program on the basis of faith alone might be seen by outside observers as being naive in this time of tight budgets. There is a credibility problem, and faith alone may not be enough to overcome it.

At the other extreme, the critics' idea that demand for space access is inelastic has damaged—and still does—the case for aero-space planes, much in the same way some people continue to perceive these vehicles as being just some sort of Super-Concorde. The main problem is that most aero-space plane concepts don't need new payloads to justify their development. HOTOL and Delta Clipper are both designed to be economically attractive purely on the existing commercial satellite market, perhaps to the point where life cycle costs can be recovered. Whether this is realistic is difficult to prove. Aero-space planes, of course, are capable of launching more than commercial satellites. In the United States alone, about one-third of payloads launched are commercial. In 1991 the United States launched 19 expendable rockets and Shuttles, only 6 of which were commercial [2].

More than enough payloads seem to exist to justify the development of most of the simpler aero-space plane concepts. While this can be seen as a persuasive argument, it can also be misinterpreted. If an aero-space plane is seen as being justified purely on one existing type of market (e.g., commercial communications satellites), then it may appear that such a vehicle would only be marginally better than existing expendable launchers. This is especially the case if life cycle cost recovery is imposed, as this may increase launch costs to the point where they may become little different from existing expendable rockets. Given the perceived high risk and technical uncertainties with aero-space planes, there is a high probability that problems may arise during a development program that could rapidly erode these marginal benefits. In such a scenario, it could be suggested that it is better to stick with the known launch systems than to risk the unknowns of aero-space planes. It is important to note that this argument fails to take into account the higher demonstratable reliability and availability of aero-space planes.

## Taking the Middle Ground: A Means to Get Started

The "existing market" scenario is probably not much more convincing than the "revolutionary impacts" scenario discussed earlier. The main problem is that the former scenario divorces the manner in which space missions are conducted and the way payloads are designed to best match the capabilities of the aero-space plane. Virtually all of today's space activities are held hostage to the capabilities of the current generation of launchers, as discussed at length in Chapter 2. Therefore, it is reasonable to examine the payload market for aero-space plane launch services in relationship to how those payloads might respond to the potential for a radically improved means of accessing space.

Another reason for choosing this approach is because it is the only realistic method of assessing the probable impacts of aero-space planes in relationship to more conventional launch systems. Usually, when assessments of alternative launch options are performed, the actual reasons for the payload and missions are all but ignored. A perspective on this situation is described by Bruce Abell, [3]

> . . . there's been a self-deceiving tendency to downgrade the seriousness of our constricted access to space by developing economic models showing that demand for space access is pretty well known and can be determined by looking at current and projected demand. Some of those models conclude that not only can we meet it with existing systems, but that we have overcapacity in lift. But those kinds of projections are fundamentally flawed. Why? Because they define the price, then assess the market. But at $5,000, or more, per pound, there are very high barriers to participation. We learned that when we tried to invent economically meaningful industrial roles for the Space Station—so-called space manufacturing—only to find that access costs blew those arguments out of the water . . . There's an assumption at work that presumes demand for space access is inelastic. I think that badly misreads the situation, yet it continues to underlie space access planning processes at work today.

And according to Kirkpatrick et al., [4]

> Numerous studies have been performed in recent years to assess the need and justification for new launch systems . . . all of these studies took a "demand side" approach to determine the utility of new launch systems. They assessed the ability of postulated launchers to meet projected needs as embodied in detailed government compiled mission data bases, such as the Civil Needs Data Base developed by NASA . . . such a "mission model" presupposes the existence of a certain set of launch vehicles—the payloads, in both size and weight, are planned and designed based on existing launch system capabilities. This allows no scope for new types of purposes, or amounts of payloads which might be permitted by the development of a new launch system. It also is a budget constrained model, again based on the assumption that the total funding available for space is limited and known, and that transportation costs are going to absorb a fixed percentage of it, leaving little opportunity for the purchase of additional payloads. Thus, it cannot account for such factors as: 1) more money could be made available for additional programs if the transportation costs were reduced, 2) the costs of the payloads themselves might be reduced due to reduced reliability requirements if space transportation were both more accessible and lower cost or, 3) the possibility of new funding sources from the private sector. In other words, the fixed mission model did not allow for any elasticity of demand based on changes in launch price, responsiveness, availability or reliability.

Most of these fixed mission model-type assessments are based on a mass-by-mass approach that calculates the economic utility of a launcher based on how effectively a particular mass—not a spacecraft—can be launched into a specific orbit. Simply put, if one launcher option can place 1 kilogram in space for less cost than another, it must, by definition, be a better launcher. This argument is used frequently by supporters of expendable, heavy-lift boosters. It assumes that today's missions and payloads are only loosely restricted by launcher performance and, moreover, that this is the way payloads will always be. As discussed in Chapter 2, today's space programs have the following characteristics:

- Space programs are forced to rely on munition-derived expendable launch vehicles (ELVs) that cannot be flight-tested before putting a payload on top of them.
- ELVs cannot safely return to Earth if a problem occurs during ascent, and must either achieve an orbit or fail catastrophically.
- ELVs cannot be launched on demand but are flown very infrequently and are prone to long delays, especially after a failure.
- ELVs cannot collect or service payloads in orbit (except for priority Shuttle missions or the return of materials in a small capsule) when required but must leave them in orbit to function independently.

It is not surprising that today's space programs are often delayed, spacecraft take many years to build, relatively few missions are conducted each year, and each mission is, above all, very, very expensive. As noted in Chapter 3, boosters like the proposed Spacelifter are going to have to launch very large payloads to stand a chance of significantly lowering the cost per kilogram. It may be possible to reduce launch costs to $1,000 or $2,000 per kilogram, but if 50–100 tonnes worth of payloads have to be launched to achieve this, then the practicality of being able to exploit this low cost is questionable. Since the vast majority of payloads, especially commercial and other satellites *plus* those missions requiring a return capability, are already rather modest in size (around a few tonnes), such arguments would appear to be hopelessly flawed. Heavy-lift rockets might, in the long term, be able to save some launch costs, but that's about it.

A simplistic mass-by-mass approach for estimating the impact of aero-space planes would be like justifying the commercial development of the 747 based on the same number of passengers that flew on DC-3s 50 years ago. By way of another analogy, the cost per passenger-mile on a 747 jet-liner is about $0.25, whereas the cost per mile by taxi is about $5. Yet, how many people take 747s to go 5 miles for a business appointment? This type of launcher-payload mismatch will be explored further in the next chapter.

If the first operational aero-space plane—in whatever form it came—were to realize at least some of the capabilities envisaged today, it would improve access to space in a profound and perhaps even radical way. Whereas before, the introduction of each new launch system has provided an incremental improvement in access to space, usually by allowing more mass to be launched, the same is definitely not true of aero-space planes. Therefore, it is reasonable to ask the following questions:

- How would the current type of space activities change in the era of aero-space planes when access to space is dictated by the requirements of user?
- If today's spacecraft and missions are designed to make maximum use of the current generation of launchers, how might that change for an aero-space plane launch?
- Would aero-space planes act as a catalyst for increasing the number of payloads, either because space missions can be conducted in a new way or as a result of the emergence of new missions that just weren't economically or technically practical before?

In the following discussion, emphasis will be placed on the impacts of the aero-space plane on the current types of missions and payloads. This is reasonable because the success of aero-space planes ultimately depends on how useful they actually are in the first few years of operation [5]. If the introduction of aero-space planes is seen to lead to a significant increase in the number of payloads purely as a consequence of the re-optimization of current payloads and missions, and *with no more than current levels of funding*, then the case for the aero-space plane is considerably strengthened.

If governments are already prepared to spend the large sums of money that they already spend on space today, at the very minimum, it is reasonable to expect that they might also be prepared to spend up to that amount on space operations in the aero-space plane era. Assessing the impacts of the aero-space plane on this basis does, if nothing else, allow a normalized comparison to be made between aero-space planes and current launchers. For example, NASA spends $5 billion to fly the Shuttle about 8 times per year. If an aero-space plane fleet could cost just $500 million per year to operate and, for this money, was able to fly 50 times per year at $10 million per flight, could the $4.5 billion savings be used for more space activities, including human exploration of the solar system?

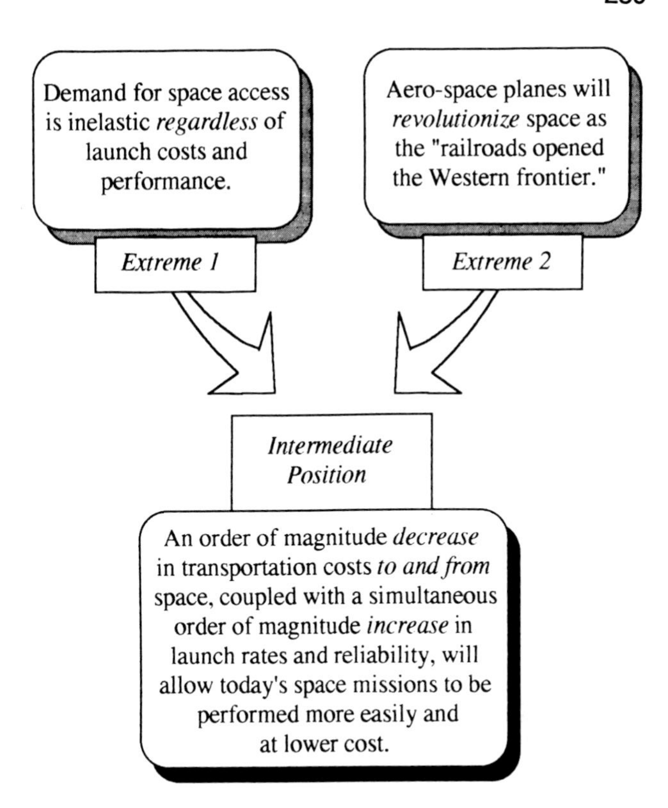


**Figure 17.1** An intermediate approach to assessing the effects of aero-space planes.

This intermediate approach (Figure 17.1) is a compromise between the highly conservative "demand for space access is inelastic" notion and the "aero-space plane will revolutionize space" proclamation. Perhaps an order of magnitude reduction in transportation costs to and from space, coupled with a simultaneous order of magnitude increase in launch rates and reliability, will allow today's space missions to be performed more easily and at lower cost.

## Haven't We Heard All of This Before?

One of the chief selling points of the Space Shuttle was its impacts on space operations and, in particular, on how missions would be performed and payloads designed to suit the capabilities of the Shuttle. As late as 1979, Howard Allaway stated in an official NASA publication, [6]

> Since about 80 percent of the cost of space missions has been going into payloads, and only 20 percent into launch costs [sic], still bigger savings—30 to 40 percent of total payload program costs—will result from the changes in spacecraft design made possible by the Shuttle's great cargo capacity and by what it can do that one-way launch vehicles can't.

At the time the above was written, it was still believed the Shuttle would significantly reduce launch costs compared with expendable launchers, as well as achieve hopelessly optimistic flight rates,

> Based on traffic projections of more than *fifty Shuttle flights a year* when the system comes fully into use, the launch savings alone could be a half a billion dollars a year more, depending on inflation. [emphasis added]

Allaway goes on to explain,

> Thanks to the Shuttle's relatively gentle acceleration [7], designers of the spacecraft it carries may be able to use some off-the-shelf parts rather than creating and testing costly and rugged one-of-a-kind equipment. Because of the Orbiter's large payload bay and great lifting ability—twice that of the biggest expendable vehicle commonly employed—satellites can be simpler: less tightly packed, less limited in weight. Standardized parts and modular components may be used, and virtually the same spacecraft can be employed for different purposes by changing its cameras or other sensors. The Shuttle's ability to check out satellites in space while they are still in the Orbiter and again after they are deployed, to repair them in orbit, and to return them to Earth for overhaul also justifies designers in relaxing some reliability precautions, such as redundant circuits. This too saves money. Studies of how past spacecraft could have been designed differently if the Shuttle had been available to launch and service them showed that their costs could have been reduced substantially.

Most of these statements are, indeed, true. A large number of payloads have been able to take advantage of the unique features of the Shuttle. Almost every time the Shuttle is launched, it carries a standard Spacelab pallet or other standard structures on which a variety of payloads can be bolted. The Deutsche Aerospace Shuttle PAllet Satellite (SPAS) has been used on a number of occasions where payloads have been deployed and retrieved on the same Shuttle missions, such as in 1984 when a SPAS was used to photograph the *Challenger* in space (Figure 17.2) and during the 1991 SDIO mission. Other widely used platforms include the Multi-Purpose Experiment Support Structure (MPESS), the Hitchhiker G&M support structures, the Get-Away Special canisters, and the sounding rocket-derived SPARTAN Flight Support Structure [8]. The Shuttle's unique capabilities have made these payloads and missions possible (Figure 17.3).

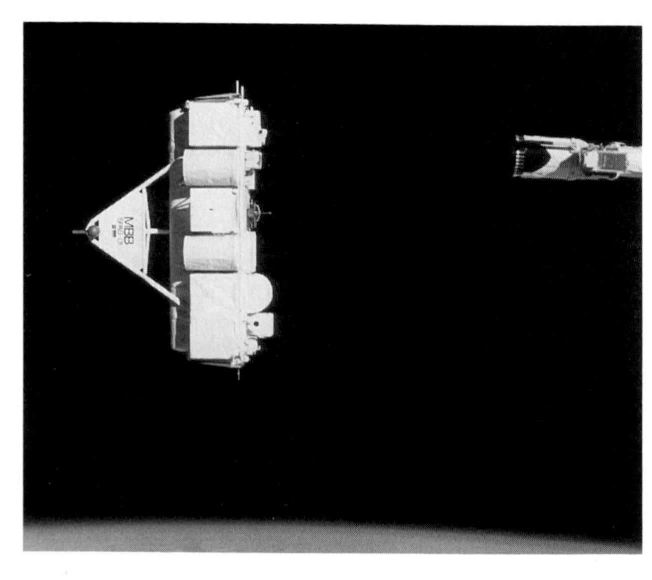


**Figure 17.2** SPAS-01 being deployed by the Shuttle (*NASA*).

The Multimission Modular Spacecraft (MMS) was developed to allow various combinations of modular packages to be easily exchanged on orbit, as occurred with the Solar Maximum satellite in 1984 [9]. The MMS has also been used on Landsat 4 and 5, Topex, and the Gamma Ray Observatory (GRO). The Hubble Space Telescope is designed to be repaired in space and periodically returned to Earth. If Hubble had been designed for a launch on a one-way expendable rocket, then it would be impossible to repair its defective optics, and replace the failed gyros and jittery solar arrays. The Spacelab pressurized modules are another example of a payload designed to be reused many times and in a variety of configurations. As with Hubble, to undertake comparable Spacelab-like missions using an expendable rocket would involve launching a brand new mini-space station for every mission à la Skylab—something probably even more expensive than using the Shuttle/Spacelab combination.

The Shuttle has facilitated simpler and more flexible satellite designs as a means to save costs and has also enabled new mission opportunities that were impossible before. Unfortunately, it has not been possible to capitalize on the cost savings because of the high cost and limited availability of the Shuttle. Considering Shuttle launch costs on the order of $500 million, small savings in spacecraft costs seem almost irrelevant. Worse, perhaps, while cost savings are possible compared with expendable versions, the *actual* costs of many Shuttle spacecraft are very high. For example, the GRO observatory cost around $800 million while Hubble cost $2 billion. They are so expensive because the combined effects of the very low Shuttle flight rate, long lead times, and frequent delays inevitably mean such payloads must wait

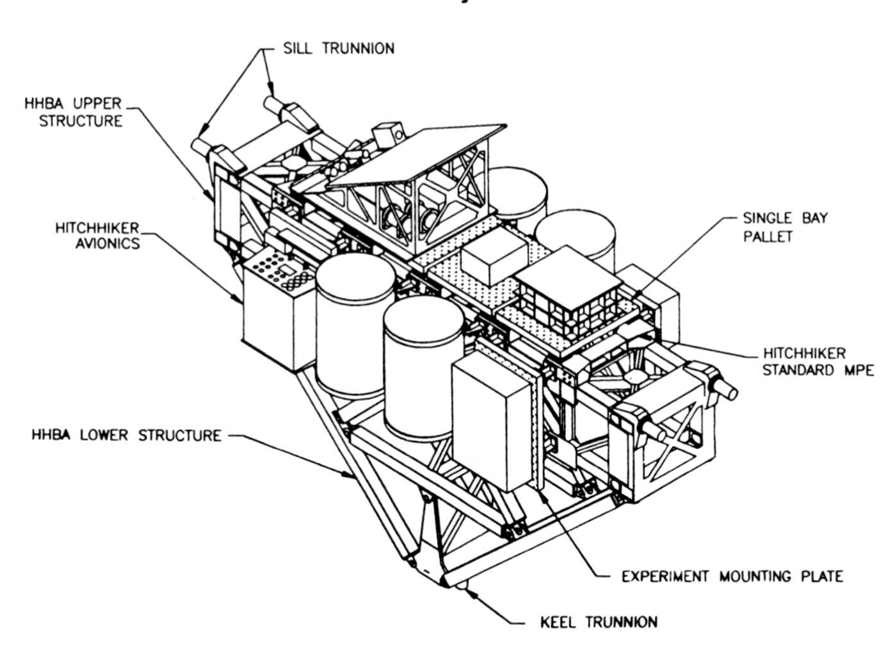


**Figure 17.3** A standard Shuttle structure for carrying diverse experiments (*NASA*).

many years before they are launched. Hubble would have cost less had it been launched in the mid-1980's as planned, rather than in 1990, and if it could be serviced on-demand, rather than after nearly 4 years in orbit.

The Solar Maximum Mission satellite is another reusable spacecraft that was allowed to reenter and burn up simply because of Shuttle unavailability. In addition, and as discussed in Chapter 2, Solar Max is the only satellite designed for repair on orbit, *to have actually been repaired on orbit* up to late 1993 (Figure 17.4). In December 1993, Hubble became the second. The Landsat 4 spacecraft, which also used the MMS platform, was planned for a Shuttle repair, but because the Shuttle Vandenberg facility was closed, it was impossible [10].

**Figure 17.4** Solar Max repair mission in 1984 (*NASA*).

The unique Shuttle capabilities have indeed enabled some simpler payload designs, together with new mission options previously considered impractical. Unfortunately, the benefits of this have been lost among the absurdly high costs of each Shuttle mission. Coupled with its limited flight opportunities, this has had negative impacts on a typical space program schedule. In the era of aero-space planes—where higher flight rates, on-demand availability, full payload recovery, ability to abort, and lower costs to orbit should be possible—the effects on space activities are anticipated to be more obvious—and profound.

## *Defining the Effects of Aero-Space Planes*

The payload market for launches will be enhanced or stimulated in four distinct ways during the first few years after the aero-space plane's introduction. These impacts, introduced here and discussed in more detail later in the chapter, are defined as follows [11].

1. *Redistribution of launch cost savings*. The simplest method of understanding the impacts of aero-space planes

is to determine the financial savings that could be achieved purely as a result of a reduced launch cost, ignoring any possibilities for payload redesign. If the financial savings are significant, then, depending on the type of mission, this money could be reinvested into other missions or enhanced operations. A typical space mission today costs $200 million, of which $100 million is associated with launch costs. In the aero-space plane era when launch costs could be nearer $10 million, the savings of $90 million could be reinvested into other space operations. For government funded missions, this is considered a likely consequence as a greater throughput of orbital activity can be performed for the same funds. For commercial missions, such as communications satellites, it is more likely that a balance would be found between reducing the cost of the transponder lease charge and construction and launch of more satellites.

2. *High availability, on-orbit servicing, and recovery*. For any degree of expanded space operations it is clearly fundamental to have the capability to frequently and routinely return to orbital platforms, rather than having to discard spacecraft at the end of their lives or as a result of simple satellite failure or orbital decay. The aero-space plane capability is ideally suited for this type of mission. In contrast to the Shuttle, the ability to launch frequently, regularly, and on-demand would greatly enhance the existing missions in virtually all areas of space operations, from Earth observation, to microgravity, to crewed space exploration. When the demonstrated ability to launch on time is combined with low launch costs, the potential for servicing and recovery missions as one of the primary activities for the aero-space plane becomes apparent.

3. *Mass constraint relaxation and payload redesign*. With current launch systems, the design tradeoff of a spacecraft almost inevitably favors cramming as much capability into as small a volume and mass as possible in order to achieve the maximum return from the expensive launch—whether financial, scientific, or otherwise. Essentially, it is cheaper overall to spend money in making the spacecraft as lightweight and highly integrated as possible, than to allow the mass and size to grow and pay for the additional launch charges (Chapter 2). In the aero-space plane era, where launch costs could be an order of magnitude lower, this trade could be reversed to the point where much simpler and heavier "sub-optimized" spacecraft could be built for significantly lower costs while maintaining the same mission objectives and performance as current spacecraft.

4. *New mission opportunities*. The reduced cost and the flexible and routine access to space that may emerge from an aero-space plane program are likely to open up additional or new mission possibilities impractical today with existing launchers. Aero-space planes could allow new uses which demand low cost, continuous, and unrestricted access to space. Such missions could include the type of commercial opportunities that have, until now, failed because of current launchers.

Each of these possibilities will be discussed within the context of the basic aero-space plane mission. Each will be applied in sequence in a manner that gradually reveals the scale of the influences that aero-space planes could have on the current payloads and mission types. In reality, of course, all of these changes would probably occur simultaneously.

## To Orbit and Back: The Basic Aero-Space Plane Mission

To begin with, the most basic form of transportation involves going somewhere and then coming back. There would be few terrestrial activities today if cars or airplanes only went in one direction with no return option. Likewise, the most basic type of aero-space plane missions should involve the deployment of an unmanned spacecraft platform in low Earth orbit, and after a period of time ranging from days to years, this platform would then be recovered and returned to Earth on a later aero-space plane mission. In every respect this would appear to be an extremely simple mission, but the reality of the first three-and-a-half decades in space reveals that this type of mission has only been performed three times by late 1993 (Chapter 2), even though, of course, it was one of the chief selling points of the Space Shuttle.

Today, the only space platform available for launch and recovery by the Shuttle on two separate missions is not even American, but is the *Eu*ropean *Re*trievable *Ca*rrier, or Eureca (Figure 17.5). Although the Long Duration Exposure Facility (LDEF) was launched and recovered by the Shuttle, it is not considered a true space platform because, unlike Eureca, it was only a simple, passive structure. Eureca provides not only mounting for various experiments but also electrical power, thermal control, telecommunications, attitude and pointing control, etc. It can be returned to Earth and, after 12–18 months of refurbishment, relaunched with a new set of experiments. The experiments for the first mission are devoted to astrophysics, microgravity, and life-sciences research, as well as technology development activities including testing of the platform itself. The total payload mass is around 1 tonne, while the total spacecraft mass at launch is about 4.5 tonnes [12].

The Eureca program began in 1982 and was originally scheduled for launch in 1987, but Shuttle delays and higher priority payloads pushed this back until mid-1992, with recovery about one year later. It should also be noted that Eureca can only be launched on the Shuttle. Eureca was scheduled to fly a minimum of three times, with the next missions planned for 1995 and 1997 [13]. However, these

**Figure 17.5** The European Eureca space platform (*NASA*).

re-flights were planned to occur around the time Space Station Freedom is scheduled to be launched in 1996. Remembering that Freedom was redesigned in early 1991 and 1993 just to lesson the burden on the Shuttle, it might be seen as being unrealistic for the United States to make the Shuttle available for two flights of this European satellite within the space of 6–9 months in both 1996 and 1999. Furthermore, the inevitable effect of delays experienced by almost every payload launched by the Shuttle is likely to push these launch dates further into the Freedom assembly sequence and logistics support, making opportunities to launch and recover Eureca increasingly unlikely. At best, it is considered that this recoverable, refurbishable and reusable platform—which has already cost Europe about $250 million—will fly *twice* before 2000 [14]. Indeed, as of early 1993 it has been reported that ESA is unlikely to fund a re-flight of Eureca.

Aero-space plane-like capabilities would have profound impacts on Eureca-type missions. The following discussion is presented to ensure that the key issues are properly justified and understood. It is first important to understand why things are done the way they are, in order to understand how they could be done given a change in the fundamental constraint of getting to and from space.

## Redistribution of Launch Cost Savings

The first impact of aero-space planes on Eureca-like missions can be seen immediately by comparing the per-mission operational costs of the current Shuttle-launched Eureca versus an equivalent aero-space plane–launched version. This is shown in Figure 17.6. This calculation assumes that a brand new suite of science experiments are developed for each flight. In reality, some experiment reuse might be expected; Eureca-2 was planned to fly most of the same experiments as Eureca-1 [15].

The $280 million is the amount of money that *has to be spent by someone* for one Shuttle-launched Eureca mission, and it is clearly dominated by the "real" Shuttle launch costs calculated purely on a proportional mass-only basis using a Shuttle operational cost of $500 million (Chapter 11). An aero-space plane launched Eureca, by contrast, would cost around $120 million per mission based on a launch cost of $10 million. Thus, the net savings to European and U.S.

| Fixed Costs Per Mission | |
|---|---|
| Science equipment development/integration | $50 million |
| Flight operational costs (6 months) | $25 million |
| Platform turnaround costs | $25 million |
| *Hence, fixed cost per mission* | *$100 million* |
| **Launch Costs** | |
| Space shuttle ($90 million, up & down) | $180 million |
| Aero-space plane ($10 million, up & down) | $20 million |
| *Hence, launch cost savings* | *$160 million* |
| **Total Mission Costs** | |
| Shuttle-launched Eureca | $280 million |
| Aero-space plane launched Eureca | $120 million |

**Hence, savings of $160 million, roughly equivalent to one additional Eureca flight.**

**Figure 17.6** Comparison of Eureca mission costs for a Shuttle launch and an aero-space plane launch.

taxpayers would be around $160 million. If the aero-space plane launches were twice as expensive ($20 million), it should still lead to savings of over $100 million.

Potentially, therefore, this money could be invested in one or two additional Eureca missions, translating into two to four new aero-space plane launches above and beyond what might be expected if a straight payload-to-payload swap between the Shuttle and an aero-space plane was performed. Is this a realistic expectation? Considering the initial development cost investment of about $250 million for Eureca, it is reasonable to believe that governments would like to see their investment properly exploited. If the purpose of building Eureca was to fly it just once or twice, then it probably should not have been funded in the first place.

## High Availability, On-Orbit Servicing, and Recovery

Looking at the Eureca mission purely on a launch cost savings basis is over-simplistic because the pacing issue for each mission is Shuttle availability. In particular, if the Shuttle had been available in 1987, then almost certainly Eureca would have flown in 1987—and at less expense. The Shuttle restricts the frequency at which Eureca can be launched to every 3–5 years, and it also puts constraints on the time the platform remains in space. Nominally, the Shuttle-launched Eureca is scheduled to remain in orbit for a minimum of 6 months, followed by a 3 month dormant phase in which time the Shuttle is planned to retrieve Eureca. If the Shuttle is severely delayed, Eureca will reboost itself back to its operational altitude until the Shuttle is ready [16].

In the era of the aero-space plane, missions like Eureca could be undertaken when appropriate for the needs of the research, not subject to the whims of an uncertain launcher schedule. Some research activities, such as those associated with microgravity materials processing, might be facilitated by frequent re-flights with on-orbit durations as short as a few weeks or days. Other missions such as those dedicated to astronomy or life sciences might prefer longer stays on orbit.

The elimination of Shuttle-type launch and recovery constraints would mean that Eureca-like platforms could be launched as soon as they were ready, as frequently as necessary, and left in orbit for as long as research activities demanded. When this capability is coupled with the low aero-space plane launch costs, the total program costs are less, as long delays are avoided. More than one mission every 3–5 years can be undertaken because, for more than any other reason, *they can be performed*. The net result is almost certainly an increase in aero-space plane launches.

## Mass Constraint Relaxation and Payload Redesign

The discussion so far has limited the effects of aero-space planes to a spacecraft similar in design to that of Eureca. As noted in Chapter 2, spacecraft are always optimized to best suit the needs of their launchers, and the same is true of Eureca. Therefore, how could a Eureca-type spacecraft be designed to best suit an aero-space plane, and in what way would this affect development and operational aspects of the Eureca mission?

These are complex questions, and to answer them requires a good appreciation of the driving factors that shape a spacecraft design. One of the most interesting factors is that it takes about 12–18 months to refurbish Eureca for a new mission. This may seem excessively long, but because access to the Shuttle limits missions to every 3–5 years, it would be pointless to have a faster turnaround time. Incorporating a faster turnaround time incurs higher total program costs for two principal reasons. First, there is a cost associated with developing the hardware and procedures to allow the spacecraft to be disassembled and reassembled rapidly. It involves dividing the spacecraft up into standard and interchangeable modules. For example, a modular approach would locate the propulsion system (including tanks, thruster, etc.) in an independent and exchangeable unit, whereas the actual Eureca propulsion system is located inside the main spacecraft structure, making access to it complex and time-consuming.

By definition, a modular approach is less mass efficient than a fully integrated design, and this leads to the second and by far the most constraining issue: *Shuttle launch costs*.

The price of a Shuttle launch is based on the size of the payload relative to the Shuttle's capacity. If a payload consumes 25% of the lift weight, then it will be charged 25% of the "artificially" set launch costs of a dedicated Shuttle mission. If that same payload also consumes 50% of the payload bay length, it will actually be charged 50% of the Shuttle launch costs, even though it only consumes 25% of the Shuttle's lift weight. Thus, Eureca, like most Shuttle payloads, was mass and volume optimized to ensure the minimum launch costs, and this is why it has a mass of 4.5 tonnes, and not 10 tonnes, and uses only 2.5 meters of the length of the payload bay, as opposed to 5 meters [17].

Overall, the combined impacts of Shuttle launch costs and the long time between missions meant that it was cheaper overall to build a compact, highly integrated, and lightweight spacecraft than to build a larger, modular, and heavy spacecraft.

How does this equation change for an aero-space plane launched Eureca? First, in order to take advantage of the high aero-space plane flight rates, the spacecraft would have to be designed for rapid refurbishment—modularization. Normally, it might be expected that modularization would lead to more expensive spacecraft and higher launch charges, as described above. In the case of the aero-space plane, this may not be true because the cost of using a *dedicated* aero-space plane is significantly less than that of a *shared* Shuttle launch. Also, as aero-space planes are likely to be able to lift about one-third of the Shuttle's payload mass, a modularized Eureca would probably consume most of the payload capacity anyway. Thus, an aero-space plane launched Eureca would not have to be compromised through having to share a ride, neither would it matter if it weighed 5, 7, or 10 tonnes, as the launch cost would be the same. Indeed, the mass of Eureca could be allowed to grow as large as necessary, all the way up to the limits of the aero-space plane launch capability.

Allowing a spacecraft like Eureca to grow can save development and production costs *provided* minimal additional launch charges are incurred. The greatest savings come as a result of the simpler overall design. Consider the following examples.

### Structural and Thermal Design

The structural design of Eureca is made from a trellis-like network of carbon fiber reinforced plastic (CFRP) struts. Although this allows a very rigid and lightweight structure, it has a number of cascading impacts on other aspects of the platform. One important impact is that the structure has a low thermal mass. As a result, every time Eureca passes from the Earth's shadow to direct sunlight, parts of the spacecraft heat up more rapidly than others. Thus, to ensure that high-power payload experiments do not overheat, Eureca is equipped with a thermal control system that circulates fluid around the spacecraft to even out its overall temperature. Such a fluid loop is expensive because of the difficulty in building a system to function in a microgravity environment and to work reliably and untended for at least 6 months. Further, a fluid loop can induce low level vibrations into the platform that compromise the relative purity of the microgravity environment, although the impacts on Eureca are small.

If a Eureca-like structure were re-optimized for an aero-space plane launch, then it would be possible to use a different construction technique and heavier materials. This would allow a more flexible approach to mounting payload units. In addition, the higher thermal mass of a metallic structure could significantly reduce the extremes in the daily variation in temperature on the spacecraft, providing a more uniformly heated or isothermally stable platform and possibly eliminating the need for an active fluid loop.

It is important to note that the cost of the structure of Eureca is probably about 5% of the total development cost. Although the cost savings of a structural redesign alone would not be significant, the net effect on the spacecraft system could be to reduce the development and production costs and the time to construct and refurbish the spacecraft, because a redesign reduces restrictions on many other subsystems.

### Propulsion System Design

Uncertainty over the ability of the Space Shuttle to rendezvous with Eureca after 6 months has forced the inclusion of a hydrazine propulsion system in the design. After launch, the spacecraft will propel itself from a 300 kilometer Shuttle orbit up to a 525 kilometer orbit, thereby reducing the rate of orbital decay due to atmospheric drag.

Incorporation of a propulsion system was a necessary if not mandatory solution entirely because of the Shuttle. Unfortunately, it has considerable impacts, not only in designing and integrating the propulsion system itself, but also on the other subsystems such as the structure, thermal control, data handling, flight control, mission management, and communications. Use of a propulsion system means that great care has to be exercised in ensuring that it is designed, built, and handled properly, as an in-space failure would almost certainly write off $250 million in development and about $280 million in mission costs.

The high aero-space plane availability and on-demand aspects of the vehicle operations could completely eliminate the need for an orbit-raising capability. All that might be required is a basic hydrazine-based system, or even a pressurized cold gas propulsion subsystem, just to control the attitude of the spacecraft with small thrusters. Indeed, it

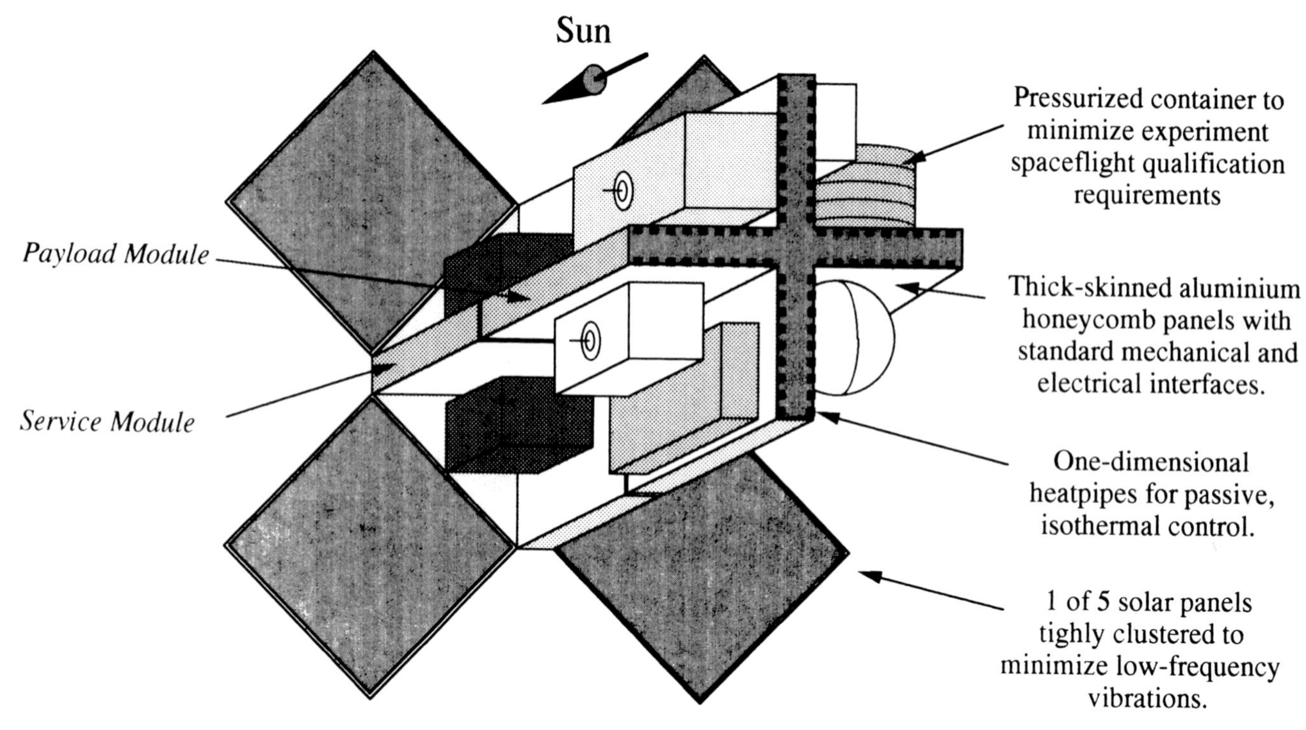


**Figure 17.7** A possible aero-space plane-launched version of a Eureca-type platform.

might be more suitable to eliminate the propulsion system completely through the use of reaction control wheels—devices that if commanded to spin in one direction will cause the spacecraft to rotate slowly in the other direction. An aero-space plane launched Eureca might not need to carry one drop of propellant.

### Other Subsystems

The above exercise can be performed for many other subsystems with similar results. These are summarized in Figure 17.7. It would be possible to develop a range of truly standard payload modules constructed in a similar manner to the Shuttle Get-Away Special canisters but larger. Such containers could contain experiment packages within a pressurized environment cooled by circulating air around the chamber—just like the majority of former Soviet spacecraft in orbit today. Such modules would allow the experiment equipment to be simplified by eliminating the need for lengthy and expensive payload development programs currently used in the West to qualify experiments for spaceflight, although they would still have to be designed to withstand launch loads. This is not a particular concern because the larger mass margins afford more robust construction. Breadboard quality experiments more representative of terrestrial laboratories could be flown.

Solar arrays are another example. On Eureca the long solar array "wings" flex slightly every time the spacecraft passes into sunlight, rather like the solar arrays of the Hubble Space Telescope before it was repaired (Chapter 18). The effect of this is that the microgravity purity will be disturbed a little every 45 minutes or so. An aero-space plane version of Eureca could use relatively thick panels firmly secured to the platform and tightly concentrated around the aft end of the spacecraft to minimize their flexibility.

Although it is difficult to attach hard numbers to the above discussion, intuitively it seems reasonable to conclude that if a Eureca-like platform were optimized for an aero-space plane launch, its development and operational costs could be reduced compared to the Shuttle-optimized version. The consequence is that the cost of the research would decrease while the rate, volume, and quality of research would improve. Importantly, it would also lead to a greater number of aero-space plane launches than otherwise predicted by a simplistic demand-led payload market analysis.

## *New Mission Opportunities*

Eureca, like the majority of Western spacecraft built today, is designed to take maximum advantage of a minimum number of very expensive launches. Even though this means spacecraft are complex and expensive, as a rule, they are nevertheless generally more cost-effective than the alternative of building several simpler and heavier spacecraft because the number of launches are minimized. Today, the tradeoff between building a few complex spacecraft versus building several simpler spacecraft is performed more or less automatically because of the high costs, low availability, and low reliability of the current generation of launchers. Given that the proposed aero-space planes have the potential to radically improve access to space, could this tradeoff be reversed?

In the era of the aero-space planes, the construction of complex, multipurpose spacecraft like Eureca, or even the proposed re-optimized Eureca configuration outlined above, may no longer be the norm. Ideally, rather than building one spacecraft whose utility is compromised by the need to meet the conflicting requirements of each individual experiment, it may become viable to build several simpler spacecraft dedicated to supporting each individual experiment type. For example, instead of building one spacecraft that has to function simultaneously as a platform for astronomical observations and materials processing research, as is the case for Eureca, two platforms would be built, each dedicated to fulfilling the unique objectives of either activity. In such a scenario, the platform dedicated to astronomical observations could be allowed to frequently change its attitude to observe different parts of the sky. Meanwhile, the platform dedicated to microgravity research could remain in a fixed attitude relative to the Sun so as to minimize disturbances resulting from attitude changes and, thus, achieve the lowest possible microgravity conditions. There are many other advantages to using dedicated platforms. Another is the time a platform remains in orbit. While an astronomical platform might require a year or more to achieve certain observations, a microgravity platform may only require a few weeks.

The above discussion shows some benefits of having individual platforms dedicated to one particular mission type. Other examples could equally have been chosen, including Earth observation, upper atmospheric sciences, plasma physics, technology development and demonstration, micrometeoroid measurement and collection, and so on. The point of this discussion is to demonstrate that *if the option exists* to put certain types of experiments on a dedicated platform whose orbital activities and stay time in space are optimized to meet the objectives of the research, then the quality of that research is improved as a function of total costs.

Wouldn't the construction of five platforms each carrying one experiment type be more expensive than one platform carrying five totally different experiment types?' This not a simple question to answer convincingly without a good deal of programmatic and spacecraft design analysis. A number of factors might lead to the intuitive conclusion that five platforms could be cost competitive with one. It is important to recognize that the process to integrate a single platform with a range of experiments with conflicting requirements is difficult and costly in terms of managing the design so one experiment doesn't interfere with another.

- A number of people of different disciplines must be coordinated and made sensitive to the needs of each other's disciplines.
- Design analysis has to be continuously performed to ensure that as changes or modifications to the experiments occur, their impacts on the other parts of the spacecraft can be controlled (e.g., if an instrument requires a cooler environment to function properly).
- Problems are associated with having to wait for the late delivery of some experiments, meaning that those experiments that can be constructed in rapid time must wait on the ground until their co-experiments are ready.
- Full-up, integrated testing of the complete spacecraft is needed to verify that all the individual elements work properly and as a single cohesive unit.

All of this is expensive because building and testing complex and highly integrated spacecraft is a very time-consuming process and time is the single greatest factor driving costs. Obtaining the funds to build an expensive spacecraft is difficult at the best of times, and usually limitations are put on the rate at which funding is appropriated. The impact of such limitations is that this further increases the time and total program costs. The Hermes program is a classic example of this process as, in an attempt to win approval for this program, it was necessary to reduce the yearly contributions. This less than optimal funding has forced a stretching out of the program by 3 or 4 years. Even though the *annual* funding became less than that originally planned, the *total* funding was greater because the time to build this payload became that much longer. The Hermes program is typical of most programs in this respect—the more expensive the program is from the start, the more this will tend to drive the time required to complete it. This in turn further stretches the schedule and increases costs at the expense of utility.

A criticism that is often levelled is that complex payloads often delay launches. Therefore, it could be argued, even with the higher availability of the aero-space plane, launches are still going to be delayed. However, the reason why Eureca and many other Western spacecraft can themselves delay launches is directly because the *launcher* forces most spacecraft to be complex, highly integrated, and mass optimized in the first place. Also, the use of standard payload containers (see Chapter 4) ensures that if one payload is not ready, the next in line can be launched without delaying the aero-space plane schedule. Aero-space planes offer a choice that previously has not existed in the way spacecraft are designed and how missions can be performed.

Building unique spacecraft that must fulfil a myriad of mission requirements is a difficult, time-consuming and expensive process, *even if* such spacecraft were designed for launch on an aero-space plane. Building several simple and identical platforms might be a more economically attractive way of performing Eureca-type experiments. When a particular experiment is ready to fly, it could be bolted to a simple, robust, and truly standard platform. This dedicated platform would then be launched, left on orbit, and recovered as necessary to meet the specific needs of the experiment.

This platform itself might be composed of separate modules that can be assembled as appropriate to provide the required propulsion, communications, power, and other services to the various types of experiments. Such add-on services would be designed to support a maximum load and not to match precisely the needs of the experiment and mission. Therefore, a power module capable of providing 2 kW of electrical power might be flown on a mission that only requires 1 kW simply because it is cheaper to reuse this oversized module than to modify it or build a specifically optimized module for the 1 kW mission needs. In fact, it is the ability to reuse this module that enables it to be economically attractive, despite being oversized.

It is not acceptable to fly oversized components on most spacecraft today because each extra kilogram has to be paid for in launch costs. In the aero-space plane era, the cost to launch a spacecraft could be a factor of 10 or more less than today. The implication once again is: the aero-space plane allows cheaper spacecraft designs because mass is no longer such an overwhelming design constraint.

## The SPARTAN Precedent

The above discussion might seem rather theoretical, but precedents for these types of missions have already flown on the Shuttle. These have included standard platforms that have either remained in the cargo bay or have been deployed and recovered on the same mission. The U.S. Hitchhiker structures and GAS Cans, and German-built SPAS platforms are particular examples, as discussed earlier.

Another is the SPARTAN platforms [18] (Figure 17.8). As the name implies, the philosophy behind these platforms was to keep them as simple as possible by dedicating each to fulfilling a minimum of objectives at the lowest possible cost. The first SPARTAN mission occurred in June 1985 on Shuttle flight 51G. During this mission, the 1,008 kilogram SPARTAN-1 (including 136 kilograms of astronomy instrumentation) was deployed from *Discovery* using the remote manipulator system. It then flew in proximity with the orbiter for 45 hours, performing astronomical observations of the Perseus clusters of galaxies and the core of the Milky Way. After which, SPARTAN-1 was retrieved and returned to Earth. The total cost of the mission was just $3.5 million, excluding launch costs [19].

It wasn't until April 1993 that the second SPARTAN-201 mission could get finally underway—8 years after SPARTAN-1. SPARTAN has had to give way to the higher priority and more expensive payloads competing for the handful of expensive Shuttle launches. The fact that SPARTAN has not flown more frequently does not mean there isn't any research that justifies its use. Only the high cost and limited availability of the Shuttle has restricted use of this inexpensive facility. Currently, about 10 SPARTAN missions are planned for the 1990s.

## Aero-Space Planes: More for Less

Those simpler and cheaper SPARTAN or SPAS-type missions flown on the Shuttle were all carried as secondary payloads because of the availability of extra capacity on certain missions. Seldom are "cheap" payloads launched on expensive rockets—normally payload costs tend to mimic launch costs. This is not a coincidence and is a consequence of the interrelationship between payloads and launchers. Bob Parkinson explains in the case of communications satellites, [20]

> Communications satellites cost approximately the same as the cost of providing the launch services. If launch services were charged strictly on a cost-per-kilogramme basis, and satellites were mass-cost optimized, then we would expect the two parts of the total cost to be the same . . . Satellite costs decrease *for a given mission* as the satellite mass increases, but the cost of launch steadily increases. The actual minimum is never exactly where the two costs are equal for a variety of reasons, but the general impression is that if launch costs per kilogramme decreased, satellite mass per transponder would increase, and total costs would go down. [emphasis added]

The subtleties to this argument are important to understand. Currently, a typical 2 tonne commercial communications satellite costs around $60 million, while a shared Ariane launch costs about $50 million. If an attempt is made to squeeze such a spacecraft into just 1 tonne—while maintaining precisely the same mission—the cost of the spacecraft would go up considerably, assuming it could be done at all (Figure 17.9). However, as launch costs per kilogramme are roughly constant, it will cost half as much to launch a 1 tonne satellite as a 2 tonne satellite. Unfortunately, extremely lightweight/high efficiency electronics, structures, and other components would be needed to build this 1 tonne spacecraft. Such systems would be beyond the state-of-the-art technology. Therefore, building 1 tonne comsats would necessitate a very expensive technology development program. Further, once this new technology was invented, its use would be initially limited, meaning production costs would be high. Consequently, the amount of money spent to build a lightweight spacecraft would *not* be equally offset by the amount of money saved from the lower launch cost. Therefore, the total cost to build and launch a

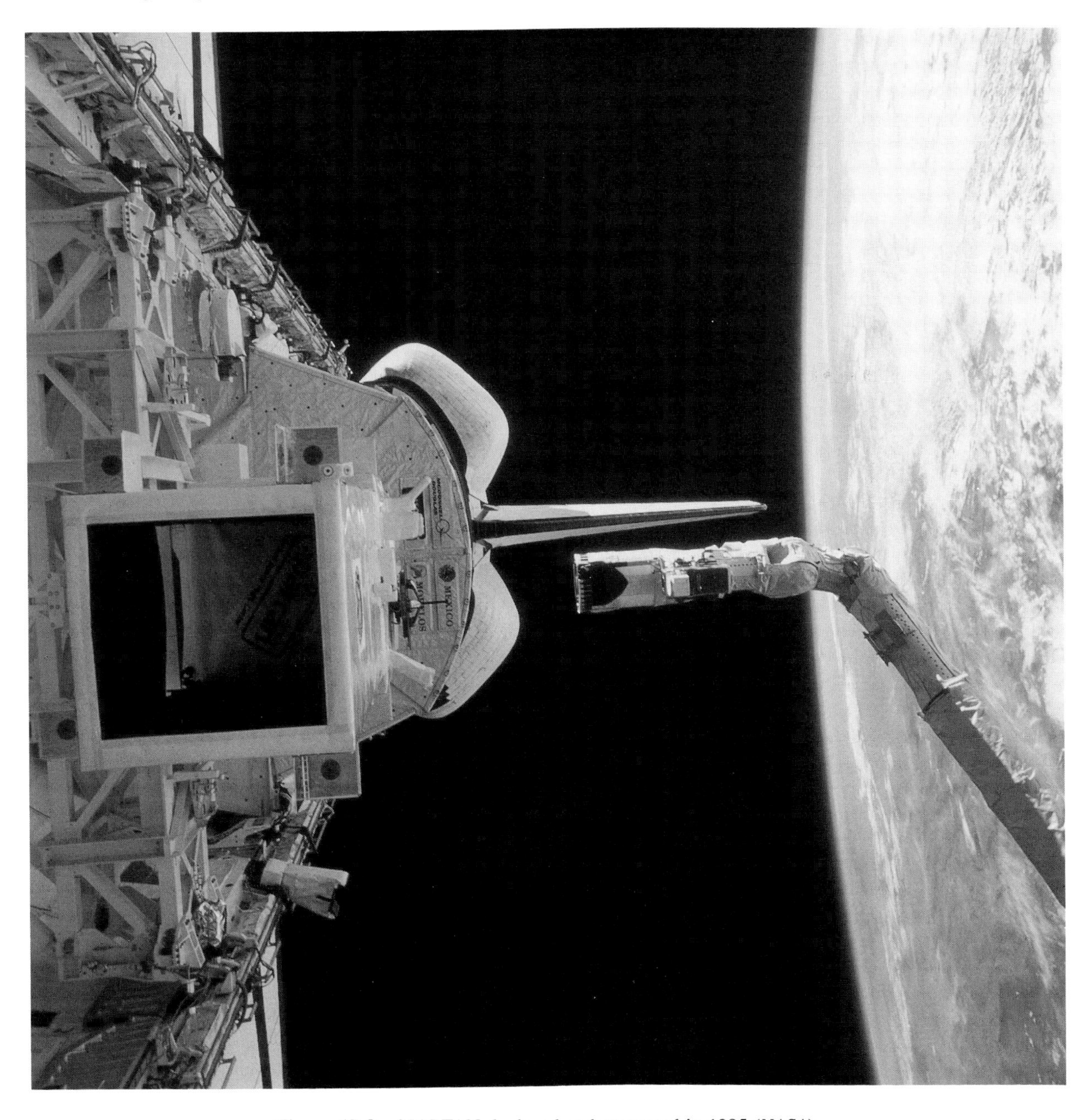

**Figure 17.8** SPARTAN deployed and recovered in 1985 (*NASA*).

1 tonne would be larger than the 2 tonne comsat and its launch.

At the other extreme, allowing the mass of a 2 tonne comsat to increase by a factor of two would reduce spacecraft costs, as noted earlier in this section. The actual magnitude of this reduction becomes less and less significant the more the comsat mass is allowed to grow because the curve of spacecraft mass versus spacecraft cost gradually becomes less steep the heavier the spacecraft, as also shown in Figure 17.9 (i.e., the spacecraft cost will always be greater than zero). At the same time, launch costs increase at a constant rate and, as a result, overwhelm the savings possible through spacecraft cost reductions. For example, increasing a $60 million spacecraft's mass from 2 tonnes to 4 tonnes might reduce its cost by 50% to about $30 million—saving $30 million in spacecraft cost. But where a 2 tonne spacecraft costs $50 million to launch, a 4 tonne spacecraft would cost $100 million to launch. Hence, it would cost $20 million *more* to build and launch a "cheaper" 4 tonne satellite than a mass optimized 2 tonne comsat.

The critical issue here is that the *cost per kilogram* to launch 1 tonne is about the same as for a 2 tonne or 4 tonne spacecraft. However, the cost per kilogram of a spacecraft increases sharply when greater demands are put on the technology needed to build it. Hence, although launch costs change *linearly*, spacecraft costs change *exponentially*. A minimum total cost occurs roughly where the two are equal.

**It is reasonable to conclude, therefore, that if aerospace planes reduce launch costs compared with existing rockets, payload costs are likely to follow suit because of the desire to minimize the total payload and launcher cost. For commercial programs in particular, this relationship is expected to be very pronounced.**

A complementary, though more qualitative, explanation can be obtained if viewed from the perspective of value for money. For example, if people can only afford to spend $500 on airfare for a vacation, they are also likely to spend at least $500 on the vacation itself. People who can afford to spend $5,000 on airfare are likely to spend at least $5,000 on vacation activities. More importantly, people capable of spending $5,000 on airfare are hardly likely to spend just $500 on the vacation simply because they wouldn't be getting much value for their money.

This value for money concept appears true for space programs, except for the important difference that there isn't the equivalent of a $500 airfare. Today, if the U.S. government can afford to pay $5 billion for 8 Shuttle launches, it should likewise be able to pay a similar amount on payloads in order to get the most from these high launch costs. It would be absurd to pay $500 million just to launch payloads whose total value was only $10 million. Shuttle payload costs are usually on the order of the launch costs. Similarly, few space users that can only afford to build a $5 million satellite will be able to pay $90 million for a dedicated Ariane 4.

In the era of the aero-space plane when launch costs have the potential to be on the order of $10 million per flight, cheap payloads like SPAS and SPARTAN will be able to justify their own dedicated launch—and not flown purely on a space-available basis. It may be considered a flagrant waste of tax payers' funds to spend $500 million to launch a $10 million payload. It may not be a waste of money to spend $10 million to launch a $10 million satellite.

What this also means is that as launch costs drop, those cheaper payloads waiting for a "free" ride will be launched much sooner and, presumably, more frequently than otherwise. While the launch market for billion dollar Hubbles

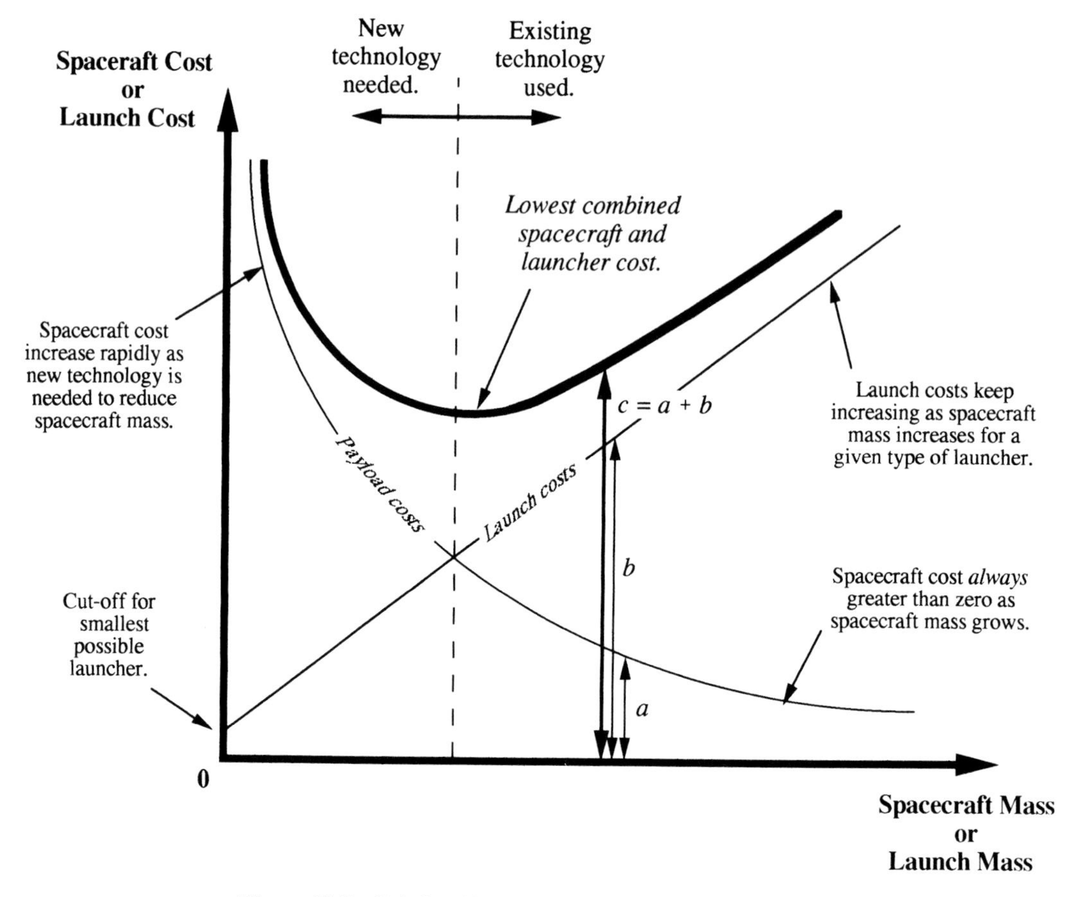


**Figure 17.9** Relationship between payloads and launch costs.

and Galileos is very small, the launch market for million dollar SPARTANs and SPASs will be much larger. Ten SPARTAN missions would cost much less than *one-tenth* of the cost of a single Hubble Space Telescope [21].

This discussion raises an important issue that future launchers must address. Specifically, launch costs are important, but they are only one aspect of space transportation services. A launcher that reduces costs to orbit must *simultaneously* respond to the increased demand for launches. Bob Parkinson notes further, [22]

The optimum cost-per-kilogramme theory for a satellite ought to work as well in low Earth orbit—implying a cost for Space Stations of about $10,000/kg [of hardware]. Actual costs for Freedom appear to be ~$100,000/kg—well away from that optimum. One reason for this is that *availability* of the launch vehicle (in this case the number of Shuttle launches required) is a key constraint. The ability of the *Hotol* system to fly *frequently* may therefore be as important as its reduced operational cost.

It is nearly useless to develop a new launch system only capable of reducing launch costs. Launch opportunities must also be increased. If a new expendable launcher is introduced that is capable of reducing launch costs, but cannot fly more frequently than existing launchers (or return payloads to Earth), this vehicle will be reserved for launching the high-priority—i.e., expensive—spacecraft. In this sense, even though Ariane 5 may reduce the cost per kilogram, it will *not* stimulate new commercial uses of space, because it is designed for a maximum of eight one-way launches per year—little different from Ariane 4.

**Reducing launch cost alone will do nothing to stimulate the use of space as payload costs are intimately a function of launch costs *and* launch opportunities.**

In the aero-space plane era many of today's traditionally expensive spacecraft may be re-optimized so their development and production costs begin to gravitate toward the reduced launch costs. Although this may seem to be wishful thinking, it is the high costs and limited access to space that force *most* spacecraft costs to be as high as they are today. Given the opportunity to remove the demanding requirements to obtain as much as possible from a minimum number of very expensive launches, and adding in new capabilities for frequent launch and payload recovery, the costs of individual spacecraft have the potential to drop substantially.

**This situation has never occurred before because all past and present launch systems offer basically the same performance. There is absolutely no inherent reason why space operations must remain as expensive and limited as they do today. Given a true, user-friendly space transportation system, there will be more than enough real payloads to keep the first aero-space plane fleet busy.**

## Summary—A Look Into the Future?

Even though aero-space planes have the potential to radically improve the means of accessing space, it is not sufficient to bank on the possibility of many new applications arising out of thin air, causing a revolution. Yet, it is nonsensical to suggest that the number of payloads launched by aero-space planes will be the same as launched today. There is probably a middle ground somewhere between these two extremes, where the number of payloads would be expected to increase relative to today, but without an increase in funds to support these "new" payloads. This is because the inherent capabilities of aero-space planes facilitate more cost-effective approaches to conducting space activities, particularly spacecraft design and schedule.

One specific example is used as an introduction to this complex subject—the Eureca recoverable space platform. In the aero-space plane era, reliance on unique, multipurpose, highly integrated, and expensive spacecraft could eventually be discarded in favor of several standard, single-purpose, simple, and inexpensive spacecraft as a more cost-effective means of achieving the same mission objectives. The following chapters will explore other examples of this process in further detail.

Payload costs seem to gravitate toward launch costs as this has the effect of minimizing total costs. As a result, significant reductions in launch costs may automatically stimulate both significant reductions in spacecraft costs and an increase in the number of payloads being flown. The notion that spacecraft costs will always remain as high as they presently are is apparently fallacious. However, this situation can only arise if the launch cost reductions are coupled with an increase in launch opportunities.

Clearly aero-space planes would have a positive impact on how space operations are performed. They herald a future where space programs are conducted in a manner more characteristic of today's successful terrestrial activities. The aero-space plane has the potential to eventually turn space into just another domain for human endeavors. The following chapters will show one vision of how this may indeed be so.

## References and Footnotes

1. An interesting discussion comparing revolutionary impacts of the Model T Ford and the DC-3 with those which might occur with the advent of aero-space planes can be found in Gulcher, R. H., and L. E. Brown, "Prospects for Routine Space Transportation," AIAA-89-2028, *AIAA/AHS/ASEE Aircraft Design, Systems and Operations Conference*, Seattle, Washington, July 31–August 2, 1989.
2. Data extracted from an article by Vincent Kiernan, "Soviet Launches Decrease in '91," *Space News*, January 13–26, 1992, p. 29.
3. Abell, B., *Second International Aero-Space Planes Conference*, Orlando, Florida, October 1990.

4. Kirkpatrick, J., R. Simberg, and J. Bryan, "Spaceplane: The Key to a New Era of Space Utilization," AIAA-91-5083, *AIAA Third International Aerospace Planes Conference*, Orlando, Florida, December 3–5, 1991, p. 2.
5. This approach was used previously in a report by Webb, D. C., and R. J. Hannigan for SAIC entitled "Economic and Socio-Political Impacts of NASP-Derived Vehicles: A Technical Report," *The International Hypersonic Research Institute*, November 1990. In addition, an IAF paper published by the same authors was presented at the 41st IAF Congress in Dresden, Germany, under the title "Global Impacts of the NASP Program: The NASP-Derived Launch Vehicle," October 6–12, 1990.
6. Allaway, H., *The Space Shuttle at Work*, SP-432, EP-156, (NASA, Washington D.C., 1979), p. 21.
7. The reference to the Shuttle's "relatively gentle acceleration" is an inexplicable misnomer. While it is true that the constant g acceleration loads are lower compared with most expendable launchers, the *vibrational* loading is more severe—and it is this which drives the structural design of satellites much more than constant acceleration loads. Specifically, all payloads must be designed to be stiff enough so that the launcher vibrations do not cause the spacecraft to shake itself to pieces. This is known as vibrational coupling, or resonance. As it turns out, structures which are stiff enough to withstand such vibrations will usually be more than strong enough to withstand constant acceleration loads. It is difficult to understand why this was stated when the Shuttle vibrational loads were well known long before 1979, as otherwise it would have been impossible to have designed the Shuttle.
8. A complete compilation of all these platforms can be found in *Accessing Space*, NP-118, (NASA Office of Commercial Programs, Washington, D.C., 1991).
9. Ibid.
10. Both Solar Max and Landsat 4 and 5 were launched on expendable Delta rockets because of Shuttle delays.
11. Webb, D. C., and R. J. Hannigan, "Global Impacts of the NASP Program: The NASP-Derived Launch Vehicle," *41st Congress of the International Astronautical Federation*, Dresden, Germany, October 6–12, 1990.
12. "Columbus Precursor Flights," published by the European Space Agency, Paris, November 1990., pp. 25–37.
13. Ibid.
14. Wilson, A., ed., *Interavia Space Directory 1991–92*, (Jane's Information Group, Coulsdon, Surrey, 1991), p. 161. The total Eureca program cost in 1983 was estimated at 206 MAU, about $250 million . A further 25.5 MAU or $30 million was spent on the development of the core payload.
15. "Columbus Precursor Flights," p. 3.
16. Ibid., p. 29.
17. Specifically, 4.5 tonnes (Eureca mass) divided by 29.5 tonnes (Shuttle lift capacity when Eureca was designed) gives a mass fraction of 0.15. And, 2.5 meters (Eureca length) divided by 18 meters (Shuttle cargo bay length) gives a volume fraction of 0.14. Hence, mass and volume fraction are about equal.
18. *Interavia Space Directory 1991–92*, p. 99.
19. Ibid.
20. Parkinson, R. C., and R. Longstaff, "The United Kingdom Perspective on the Applications of Aerospace Planes in the 21st Century," AIAA-91-5084, *AIAA Third International Aerospace Planes Conference*, Orlando, Florida, Dec. 3–5, 1991, p. 6. Copyright American Institute of Aeronautics and Astronautics © 1991. Used with permission.
21. Hubble cost about $2,000 million, to which must be added $500 million for the *first* launch. This gives a total mission cost of $2,500 million, one-tenth of which is $250 million. A SPARTAN mission costs on the order of $5 million. Because SPARTANs have a mass between 1 or 2 tonnes, as many as three could be launched and recovered on one aero-space plane mission. For the sake of argument, and to minimize the complexity of orbital retrieval operations, just one mission per aero-space plane will be assumed. Therefore, if an aero-space plane costs $10 million per launch, this would give a single SPARTAN mission cost of $15 million. As a result, $250 million divided by $15 million is about 17. That is, *17 SPARTAN missions could be flown for one-tenth the cost of one Hubble mission*. If two SPARTANs were flown per aero-space plane flight, this would give 25. If three, this would give 30. As another example, one Shuttle mission alone would pay for 30 SPARTAN missions, with dedicated aero-space plane launches for each. Importantly also, with such a high SPARTAN flight rate (e.g., monthly) the volume of activity might be expected to significantly reduce the cost of SPARTAN flights compared with the Shuttle-SPARTAN that must wait for many years before the Shuttle is ready.
22. See "The United Kingdom Perspective on the Applications of Aero-space Planes in the 21st Century," 1991, p. 7.

# Chapter 18

# Unmanned Space Operations in the Aero-Space Plane Era

## Assumptions

This chapter will provide an overview of the impacts aero-space planes might have on a variety of current space activities. To put this into context, it will be assumed that the operational characteristics of aero-space planes will follow current predictions. These characteristics are:

| | |
|---|---|
| • Payload mass to and from orbit | 7–10 tonnes |
| • Dedicated cost per flight | $5–20 million ($1,000/kg) |
| • Annual availability | 50+ missions per fleet |
| • Lead time from order to launch | Months |
| • Rate of safe mission aborts | 1 in 50–100 missions |
| • Rate of catastrophic losses | 1 in 500–1,000 missions |
| • Crew capability | Nominally unmanned<br>Crew module added when necessary |

Aero-space planes hold the potential to achieve these capabilities. If this potential could be achieved, then what would be the impacts on space operations? Would they be so strong that the high cost and risk involved in developing these vehicles would be justified?

The assumption that these capabilities are possible will allow a detailed examination of the ways that aero-space planes could affect and *increase the cost-effectiveness* of how space missions are performed. Technology demonstration, space science, Earth observation, telecommunications, and military systems could be profoundly affected. Comparing today's operations with those done in the aero-space plane era will provide an in-depth understanding of how significant the impacts could be on future space activities.

## Technology Demonstration: The First Mission of Aero-Space Planes

If there is one mission the aero-space plane would profoundly impact—perhaps more than any other—it is to enable *routine technology demonstrations* in space. Specifically, orbital testing provides confidence in the performance of a piece of technology before it is used in later operational spacecraft. This may not grab the imagination in quite the same way as planetary probes or manned space stations, and yet, technology demonstration is absolutely fundamental to these higher profile missions—or at least it should be.

At the present time, few true technology demonstration missions are performed in space. When they do occur, most are classified as either secondary payloads, flown in GAS Cans on the Shuttle or using the Ariane 4 micro-satellite launch structure (ASAP), or they take the form of all-up spacecraft. In the latter case, such spacecraft have to be mass-optimized, highly integrated, and almost always require the development of new spacecraft built to operational standards. The experimental $500 million plus NASA Advanced Communications Technology Satellite (ACTS) [1] is a good example, as are the $800 million ESA Olympus communications technology demonstration satellite [2], and the $300–400 million Japanese Engineering Test Satellite, ETS-6 [3].

For the majority of today's space missions, complete spacecraft must be built on the ground and then launched and used *operationally*, without the intermediary steps of proving whether critical aspects of the spacecraft will work when on orbit. The Hubble Space Telescope is one example, as will be discussed in the next section. This contrasts sharply with terrestrial activities where new technologies can be thoroughly tested and qualified in the environment they are intended to be used, long before operational applications.

An EC Commission's panel report on space has recognized the relative importance of more opportunities to perform technology demonstrations or validations in space with respect to innovation in space science, [4]

> Some concern, however, remains as to whether the innovative developments of the past will still be possible in the age of "big" space science. This concern has been most often voiced as a plea for more frequent launch opportunities for small scientific or technological payloads. In this respect, the validation in space of new technologies has similarities to new scientific developments.

The inability to demonstrate technology in space is truly the bane of the space industry. At one extreme, it inhibits the progress of space development by stalling promising new ideas that simply cannot be proven. At the other, it can lull some organizations into a false sense of security that complex space missions can be achieved, simply because they have no experience to suggest the contrary.

One pertinent example is the concept for using inflatable structures as a means to package large volumes into small spaces. One company, Oerlikon-Contraves of Switzerland, has been working for many years on inflatable reflectors for use on communications satellites providing mobile services [5]. In space, such reflectors could potentially provide the same performance as the current generation of solid reflectors, except that they would consume far less volume and mass at launch. In addition, inflatable reflectors could potentially be larger, simpler to manufacture, and easier to deploy from the side of a spacecraft. They may also be far cheaper (Figure 18.1).

Despite the inherent potential, no commercial communications satellite program has attempted to use inflatable reflectors simply because none has yet flown in space. ESA is in the process of planning a small-scale experiment demonstrating inflatable space rigidizing technologies as part of its in-orbit technology development program, but no firm launch opportunities have been made available [6]. ESA had hoped to include a 4–5 meter diameter inflatable reflector on the $650 million Artemis technology demonstration spacecraft set to be flown in 1995 [7], but it was deleted because of concerns over this unproven piece of technology.

At the other end of the spectrum, the Lawrence Livermore National Laboratories has proposed using inflatable structures to provide large volumes for space stations and planetary habitats [8]. If such an approach were practical, the deployment of large volumes would be very easy. A launch vehicle with a large payload bay wouldn't be needed to orbit the more common rigid metallic modules, such as those proposed for Space Station Freedom. For long-duration manned exploration of the Moon and Mars, inflatable structures would provide much larger living space per kilogram than rigid modules, and they would be far easier to land. Like the inflatable reflectors, there is some uncertainty over the technical feasibility of inflatable habitats, and concerns about puncturing and repairability are often cited. It would only take a small number of test demonstrations to prove the viability of this technology. This is a role the aero-space plane could readily play.

Another example is the use of ion propulsion systems. Such systems hold the potential to significantly reduce the amount of propellant needed by communications satellites to remain in the proper orbital locations. At present, a typical comsat uses 100–200 kg of propellant over 10 years to keep it stationary over a point on Earth by annulling the drift north and south caused by the nonuniformity of the Earth's gravitational field. Potentially, high-performance ion thrusters could reduce this propellant mass by more than half or increase the spacecraft lifetime far beyond 10 years. Given that each kilogram of satellite launched today to geosynchronous transfer orbit costs around $25,000, saving 100 kg

**Figure 18.1** Deployment tests of an inflatable space-rigidizing reflector (*Oerlikon-Contraves/Marco Bernasconi*).

### CURRENT LAUNCH SYSTEMS

***Transportation options to choose from for a technology experiment in space.***

Piggyback launch on ELVs or the Shuttle

- Few opportunities/year.
- Held hostage to primary payload.
- Delays of years likely.
- Minimal experiment interaction in flight.

Dedicated launch on ELVs or the Shuttle

- Very expensive launch costs.
- Few opportunities/year.
- Even fewer options for return.
- Reflights years away.
- Delays of years likely.

***As a direct consequence***

Piggyback experiment

- Restricted to only a handful of very small experiments flown each year.
- Rare reflight opportunities.
- Ensures only relatively simple and conventional experiments are launched.
- Allow low total experiment cost *but* at high cost/kilogram.

Dedicated experiment

- Forces very expensive experiment launched after many years.
- Dictates experiments must work "first time" to justify high cost.
- Demands exhaustive analysis and thorough ground testing.
- Pushes costs still higher and lengthens schedules.
- Restricted to only high priority experiments.

***Space technology developed far more slowly and at very much higher cost compared with terrestrially-developed technology.***

### AERO-SPACE PLANE ERA

***Demonstrated capability to launch regularly, frequently, safely and at a low dedicated cost.***

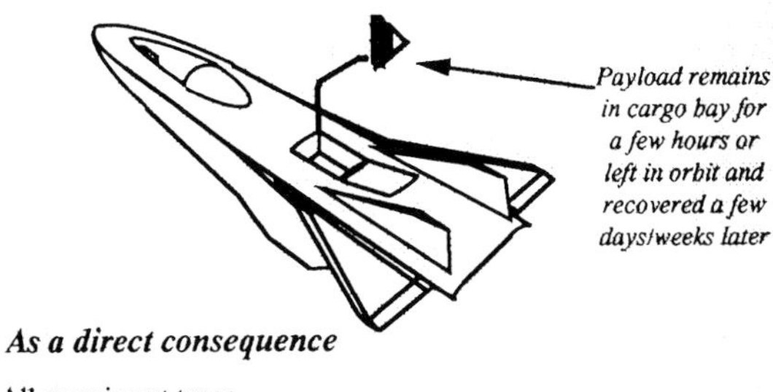


***As a direct consequence***

All experiment types

- Robustly-constructed "breadboard" class experiments can be manufactured quickly and relatively inexpensively.
- Only basic safety-type analysis and ground testing needed.
- Do not have to work the "first time."
- Large number of experiments to be flown - not just high priority experiments.
- Bolted to a standard platform.
- Launched on a "regularly-scheduled" mission.
- Recovered and relaunched weeks or months later.

***Space technology developed in a manner more characteristic of terrestrially-developed technology.***

**Figure 18.2** Technology demonstration today and in the aero-space plane era.

would seem advantageous. Work on ion propulsion systems is being carried out in many parts of the world, but few opportunities to fly such systems are planned. The first Eureca mission carried the German RIT-10 ion thruster [9], and ESA hopes to include the RIT-10 and the British UK-10 thrusters on the Artemis spacecraft [10].

Along with the limitations, the inability to perform technology demonstrations is also responsible for overambitious assertions that some space missions can be readily achieved. Without evidence either way, it is sometimes difficult to find reasons why something *cannot* be done. Stringing together 25 Shuttle flights one after the other over 4 years to assemble Space Station Freedom (1992 version) is one example, as described in Chapter 2. Another example is the assumption that permanently manned lunar bases can be established just because Apollo was successful.

## Regularly Scheduled Access to Space

Technology demonstration is absolutely and unequivocally vital to future space ambitions, as it is for all terrestrial activities. In the aero-space plane era, frequent and low cost access to space will make this sort of activity almost routine. Today, an organization researches a new technology over many years while patiently waiting for someone to pay for a ride to orbit. In the future, as soon as a breadboard version of the particular technology is constructed, it could be attached to a standard platform and launched on an aero-space plane (Figure 18.2).

This standard platform could be manifested on regularly scheduled aero-space plane flights, such as once per month. Each platform could carry as many as 10–20 experiments, with each having a maximum mass of 250 kilograms. With 10 paying passengers, the cost per user would be about $1 million. Such a guaranteed and regular flight opportunity would be analogous to airliner services today, or perhaps a better analogy would be with the transatlantic ships that sailed between Europe and the United States in the early 1900's. Specifically, any organization with an experiment would call up the aero-space plane operators and book a space on the most appropriate flight opportunity. The user would then receive a special adaptor (i.e., a plate or a pressurized canister) to which the experiment would be secured. Then a few days before the launch, the experiment on its adaptor would be placed in a special multipurpose environment chamber and exposed to the extremes of vibrational, acoustic, and thermal conditions that it might encounter during launch and orbital operations. If the experiment failed to survive these tests, the aero-space plane would simply fly without it—just like an airliner which flies regardless of whether it is fully booked or not.

After a few hours or days of experimentation in space, this platform would be returned to Earth for modifications, and the experiment could fly again on the next appropriate or prebooked launch, repeating the process as many times as necessary until the technology is fully characterized. In this example, the process of going from an idea to a proven technology should be achievable in a short period of time. Depending on the program, it should also be considerably less expensive than today—when one-off testing opportunities actually arise.

In addition to shorter development schedules, the incremental ability of technology demonstration also contributes to keeping costs down. Today, every effort (within reason) is expended to ensure spacecraft systems will function properly. This effort comes in the form of detailed engineering and reliability analysis and ground testing, which is an expensive and time-consuming process. In the aero-space plane era, this analysis and ground testing could be largely substituted with actual flight experience—something more valuable than theoretical calculations.

The possibility of low-cost, regularly scheduled flights to space is, of course, impossible today simply because launch dates cannot be guaranteed and, in any case, they are too expensive. When "$1 million" flight opportunities do arise today, they are very rare (e.g., every couple of years or so), subject to lengthy delays, and held hostage to the primary payload. In the aero-space plane era, the demonstrated ability of these vehicles to fly routinely and on time would give users the confidence that will be mandatory for them to take advantage of these frequent opportunities to access space. Further, as aero-space plane launch costs decrease with increasing demand, the cost of each experiment opportunity will fall. For example, a dedicated $1 million per flight cost would translate into a $100,000 user charge for a flight opportunity. Such low costs might lower the threshold of market demand, leading to even more users, and so on. (This subject is also discussed in Chapter 20.) However, such low launch costs may have to wait for the second or third generation aero-space plane.

In the aero-space plane era, the potential exists to provide frequent flight opportunities and avoid the high costs of technology demonstrations. *Low-cost and regularly scheduled flight opportunities would provide a clear and unequivocal demonstration of what can and cannot be done in space.* The result would be an increased pace at which new, more efficient technology systems could be advanced and exploited in operational applications.

## *Space Science: Space-Based Astronomy*

Space science activities would be a major beneficiary of the ability to perform incremental technology demonstrations in space. Consider the development of the Hubble Space Telescope in comparison with ground-based telescopes like the Palomar Observatory in California or the new Keck Observatory in Hawaii. When ground-based telescopes are constructed, they are built in an incremental manner that allows testing throughout the development process. For example, the complex, high-precision guidance system that keeps the telescope pointing precisely on the astronomical object in view can be tested long before the telescope is finally assembled. If it is found not to perform properly, access to the guidance system is, of course, immediate, allowing modifications and adjustments to be made as necessary, until gradually the problems are resolved.

With Hubble, this was definitely not the case. The guidance system *had* to be designed to work properly because once Hubble was in space it would be many years before access would be possible. As it turned out, the guidance system performed well. Nevertheless, three of the six original stabilizing gyros failed. If others had failed, Hubble's fine pointing ability would have been eliminated [11]. It is a testimony to the quality of the engineering that such a complex guidance system works as well as it did.

If a telescope like Hubble were built in the aero-space plane era, not only would it take less time, the type of problems it has encountered would be minimized, and the cost would be far less. If dedicated launch costs could eventually be reduced to well below $10 million with weekly launch options, the cost of the telescope could even be competitive to ground-based alternatives. Aero-space planes would enable critical parts to be tested in space throughout a development program (Figure 18.3). A test article representing the guidance system could be launched several times in order refine the design. Likewise the mirror could be launched and its surface measured in space long before assembly into the final telescope. The value of this type of testing cannot be overemphasized. Because Hubble's mirror was designed to function in zero-gravity, corrections had to be made during ground testing to compensate for the deformation of the mirror's surface due to gravity. Testing in space of this mirror would have made this complex and expensive ground testing unnecessary.

Another example is Hubble's European-built deployable solar arrays that jitter more violently than anticipated [12]. Ground predictions had indicated that these arrays would take about 1 minute to heat up when exposed to the Sun

*Hubble Space Telescope*

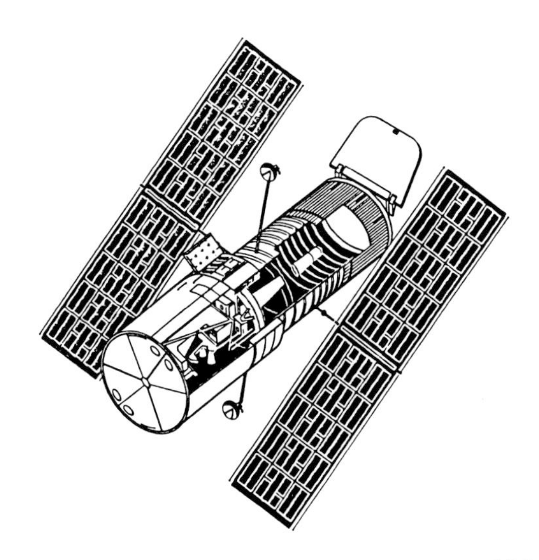

*Aero-space plane-launched space telescopes could be significantly less expensive than Hubble, developed in an incremental manner and serviced when necesssary.*

***Space Shuttle-launched Hubble***

- First "test in space" after 13 years and $2 billion.
- All development work performed on the ground.
- Forced to work the very first time.
- First opportunity to repair telescope in nearly *4 years* after launch.
- Large numbers of repairs needed, but Shuttle limited to just a few days in space.
- First repair is very risky and costly.
- Second repair opportunity several years away, or not at all if Freedom is launched.

***Aero-Space Plane-launched Space Telescope***

- Critical items can be tested in space *long before* the operational telescope is finally assembled (e.g. gyros, optics, solar array, instruments).
- Incremental development is precisely analogous to the construction of Earth-based telescopes.
- Repairs can be undertaken at relatively short notice, as and when necessary.

**Figure 18.3** Approaches for developing Hubble-type space telescopes.

after coming out of the Earth's shadow. Unfortunately, the arrays take just a few seconds to heat up, causing them to deflect more rapidly than anticipated due to thermal expansion. As a result, Hubble's guidance system has difficulty in keeping the target stable in the focal plane every time the telescope passes into sunlight. Replacement of these arrays in December 1993 effectively solved this problem. However, if a development array had been flown first, Hubble's jittery array would have been discovered and the problem eliminated before use on the operational telescope. Such an experiment could be conducted on an aero-space plane in precisely the same way a large deployable solar array was tested on the Shuttle in 1984 [13], i.e., bolted to a standard structure and then launched on a regularly scheduled aero-space plane.

This discussion can be similarly made for the delicate instruments and other sophisticated features of Hubble. In addition, even if an aero-space plane version of Hubble suffered failures in orbit, a repair mission could be launched immediately, not in 3 or 4 years, as is the case with the Shuttle version. It is reasonable to conclude that if aero-space planes were eventually able to meet the capabilities foreseen of them, telescopes like Hubble would not cost $2 billion nor would the designers have to wait 13 years to find out whether it works properly. Perhaps such aero-space plane-launched telescopes could be more competitive with ground-based systems. Further, the aero-space plane would readily allow upgrades and enlargements of the telescope, perhaps beyond what is practical on the ground. As Bob Parkinson notes, [14]

> Major disasters [such as the] the errors in the *Hubble Space Telescope* mirror, are **major** because of the huge investment in the failed systems. Low cost systems would be replaced more readily, and evolve more rapidly due to actual experience, leading to lower costs from accumulated learning. *Hubble* is a major disaster because it cost $1.5 billion and does not work as advertised. Had it cost $150 million, it would have been a major advance in capabilities, and might have been replaced with something even better in a reasonably short time. Indeed, it could actually have had more faults, and needed two or three attempts to get it right, and still cost less . . . overall.

Another space astronomy mission, Astro 1, was conducted in late 1990 on the Shuttle using four telescopes mounted on a Spacelab pallet [15]. While the Astro mission was quite rightly considered a success, it may seem a little irrational to have built these relatively conventional telescopes at a cost in excess of $150 million, for them to be used only for two 10-day missions (the second funded mission is planned for 1994), even though the original plan was to fly them every year or so. An alternative might have been to place these telescopes on their own individual platforms and allow them to free-fly as autonomous spacecraft for as long as necessary. Unfortunately, the cost of this today would have been much larger than the cost of the actual Astro 1 mission because of high launch costs and the cost of building platforms. Two or three individual spacecraft launched on expendable rockets would have been more expensive than Astro 1.

In the era of the aero-space plane, the use of standard space platforms (Chapter 17) might reduce costs substantially, to the point where free-flying spacecraft would be cheaper than the equivalent Shuttle Astro mission. In addition, the ability to test these telescopes in space first (perhaps before being integrated to their independent platforms) would allow many of the problems that plagued Astro throughout its mission to be avoided [16].

## Space Applications: Earth Observation

One of the main criticisms in the early 1990s of the U.S. Earth Observation System, apart from its high projected cost of $11 billion by 1999 [17], is that the first data will not be available much before the end of the decade at the earliest, when the first of a series of large platforms is launched in June 1998 on an Atlas 2AS. The Canadian-led Radarsat is another example. One of its operational roles is to use a synthetic aperture radar to monitor the drift of icebergs in the Hudson Bay and around Canada [18]. Even though such information would be valuable today, Radarsat will not be launched until 1994 or 1995 on a Delta 2. Given that the study of the global environment changes is becoming an increasing international priority, shouldn't every effort be made to obtain critical data as soon as possible?

If available today, the aero-space plane would benefit the Earth observation communities for many of the reasons highlighted in the previous discussions. The development of sophisticated Earth observation instruments is a difficult and complex process, because such instruments must function operationally the first time they are used, and with a lifetime expectancy of many years. For the most part, the instruments function adequately, but only after considerable time and money has been devoted to rigorous ground testing and analysis. If breadboard versions of new instruments could be placed in orbit early in their development cycle, tested in space, and then returned to Earth for modifications and reflights a few weeks or months later, then many of the specific difficulties encountered in a ground development program would be alleviated. This type of activity could also be performed to validate and calibrate the instrument for later operational use (Figure 18.4).

Because mass would not be the overwhelming constraint it is today, the instrument, as with the platform, could actually be constructed in a more robust manner with larger design margins. Although such an option may vary from instrument to instrument, when developing sophisticated instruments the policy should be to try to avoid unnecessary complications and concentrate efforts on meeting the performance specifications. Building a sophisticated instrument while simultaneously being driven to make it as lightweight as possible often raises technical conflicts that would be avoided if mass was not such an overwhelming concern. For example, the performance of some instruments is adversely affected by temperature changes. It is more difficult, though, to obtain a controllable thermal environment with a lightweight structure. Thermal control becomes more manageable with less mass constraints, therefore a heavier option is available. Whether building such instruments in a more robust manner would save significant costs is difficult to quantify, although it would certainly make some activities easier.

Soon after a new instrument has been qualified, it would be secured to its own standard platform and launched into an orbit specifically tailored to maximize the data return, rather than an orbit which is a compromise between its requirements and those of other instruments. If a problem arose in orbit, it would be possible to mount a repair or recovery mission at relatively short notice. This approach would ensure that the time between the start of the program and operational use was minimized, and the complexities of integrating this instrument with other instruments on a lightweight platform would be avoided.

By way of a simplistic example, each of the originally proposed Titan 4-launched EOS platforms was equipped with about 20 instruments. If it were possible to split up this platform and place one or two instruments on separate autonomous platforms, then this could lead to a tenfold increase in the number of payloads compared with the original EOS proposal. Even though the cost of achieving this might be comparable with that of the original Titan 4 and the cur-

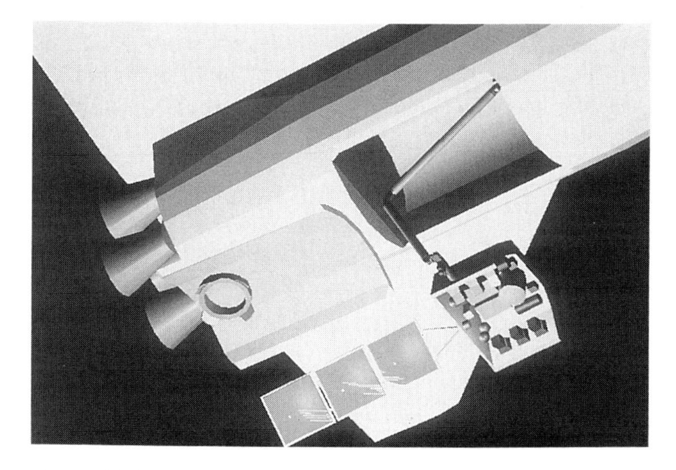

**Figure 18.4** Conceptual standard platform carrying Earth observation instruments (*BAe*).

rent Atlas 2AS-launched version of EOS, elements of the operational system could be deployed far more quickly, and the total program would be far more resilient to launch and spacecraft failures. Ultimately, the Earth observation community benefits, and as a byproduct, the aero-space plane operators have more payloads to launch—and service.

As an alternative to the large spacecraft approach, there is growing interest in the use of sophisticated sensors located on small satellites. Brilliant Eyes, proposed by Edward Teller in 1990, is one concept for launching up to 1,000 small satellites, each equipped with small sensor packages weighing just a few kilograms [19]. It is claimed that such a large constellation of spacecraft would enable constant global monitoring and achieve many of the objectives of the EOS program. More recently, as no firm plans exist to develop the Brilliant Eyes concept, it has been proposed that each of Motorola's 66 Iridium communications satellites be equipped with a modified version of these sensors riding piggyback [20]. Other options are also being studied.

It is, of course, impossible to tell whether aero-space plane capabilities would provide sufficient incentive to change the "large is better" mindset for Earth observation although the "faster, better, cheaper" approach is currently the fashion. However, the ability to readily and cheaply launch development models of these advanced sensor packages and *demonstrate* their performance in space, would go a long way to prove whether such small satellite alternatives hold the promise many organizations claim they do. Further, the ability of the aero-space plane to deploy large numbers of these satellites on each mission, as well as routinely resupply the constellation with new and continuously upgraded satellites, would be a significant benefit to the eventual adoption of the constellation approach to Earth observation.

## Communications Satellites

Communications satellites are well known for being the only truly commercial success of the space industry. The reason is that no tangible goods are manufactured in space. Electromagnetic radiation is simply transferred between the satellite and the Earth. Equally critically, each comsat requires just one launch, and business revenues from their operations are generated over as much as 10 years. If each comsat required one or more launches every year (e.g., for servicing), there wouldn't be a commercial telecommunications industry. Today, about 20 civil satellites are launched worldwide each year for a variety of purposes including telephone calls, television transmission, data transfer, and mobile communications on ships and airliners.

The impact of aero-space planes here may not, initially at least, be as significant as with low Earth orbit platforms for a number of reasons, the most obvious of which is that today's generation of comsats must be positioned in an orbit 36,000 kilometers above the surface of the Earth. The first effect on comsats will be launch cost savings. The simplistic economic analysis in Figure 18.5 shows a large-sized satellite originally designed for launch on an expendable booster, and compares the launch costs of current and aero-space plane launch systems.

| Satellite & Upper Stage Data | |
|---|---|
| Satellite cost | $75 million |
| Satellite mass (minus upper stage) | 2 tonnes |
| Upper stage cost (for aero-space plane) (e.g., PAM-D2 type) | $10 million |
| **Current Launch System** | |
| Launch cost (e.g., Atlas 1) | $60 million |
| Insurance @ 18% | $24 million |
| *Hence, total launch costs* | *$84 million* |
| **Aero-Space Plane** | |
| Launch cost | $10 million |
| Insurance @ 10% | $10 million |
| *Hence, total launch costs* | *$30 million* |
| **Total Savings** | **$54 million** |

**Figure 18.5** Comparison of satellite launch costs between an expendable booster and an aero-space plane.

The transportation cost of an aero-space plane launched comsat might be around one-third to one-half that of current launch systems, with the decrease in the insurance rate reflecting the higher and proven reliability of aero-space planes. (See also the discussion on insurance in Chapter 14.) The $54 million savings in launch costs might only lead to a reduction in the cost to lease a transponder by perhaps 10–15%, because the space segment of a commercial communications organization consumes about one-quarter to one-third the cost of the company's operations [21]. This is not a massive reduction as far as the customer is concerned. However, because the launch cost would be lower and the probability of getting the payload safely to orbit is higher, the satellite owners would probably have little hesitation in choosing an aero-space plane launch.

Even though there might not be a constituency in the telecommunications industry calling for aero-space plane development, if given the choice, they would certainly use them. Indeed, it is reasonable to suggest that the first operational aero-space plane operator will totally dominate the commercial satellite launch industry until a comparable competitor arises.

Several years after the first aero-space plane carries a comsat, savings on launch costs are likely to be contributed to by savings resulting from the potential for simpler satellite designs. The design of a communications satellite is intimately dependent on the launch vehicle it flies on, as discussed in Chapter 2. An Intelsat 7 or Inmarsat 2 is configured the way it is to take maximum advantage of the current generation of very expensive launch vehicles. Therefore, as aero-space planes have the potential to provide a launch capability that is a profound improvement on today's expendable boosters, this will allow a redesign of satellites to best suit the capabilities of an aero-space plane.

Consider the structural design of the comsat. To minimize mass, the structure uses sandwich panels constructed from two thin sheets of aluminium separated by 1–3 cm of aluminium arranged in a honeycomb. Although such panels are stiff and extremely lightweight [22], they are also delicate and expensive to manufacture, making it difficult to securely bolt the different electrical black boxes to these panels. With an aero-space plane, it might be more practical to use a simpler and more robust structural concept. Ribbed panels machined from a solid piece of aluminium are a much simpler and cheaper method of manufacturing satellite sidewalls than using delicate sandwich panels. Such a construction might double or triple the structural mass of the comsat, but it really doesn't matter so long as the total mass is within the aero-space plane's capability [23].

This type of structural design impacts the satellite configuration as a whole, just as with Eureca. (See Chapter 17.) It is easier to bolt equipment onto solid panels compared with honeycomb panels which require the gluing of special inserts into the honeycomb in order to accept a bolt thread. The design process of today's comsats requires considerable effort to ensure that these inserts are in the correct location. If an insert is found to be incorrectly positioned—as occurs many times during satellite integration activities—this leads to retroactive measures, including redesign and some additional structural analysis. As a result, the satellite production schedule is interrupted and a cost penalty incurred. If, on the other hand, a bolt hole is drilled in the wrong location on a solid panel, either the panel can be scrapped or, more likely, the hole can be ignored and a new bolt hole drilled in the proper location. Effectively, the potential for a more robust structure design would allow the satellite to be built in a shorter period of time.

The large, ear-like antennas or reflectors sported by most comsats are another area that can be impacted by redesign. These reflectors, which can be as large as 2 to 3 meters in diameter, are assembled on specially contoured tools using carbon-fiber reinforced plastic for the surfaces and backing structure, and kevlar or aluminium honeycomb for the core. This provides a very stiff, but extremely lightweight, construction that can resist changes in its shape even though such reflectors experience temperature changes of hundreds of degrees every day. Considerable design and quality control effort is expended to ensure that the reflectors will be properly manufactured the first time because the manufacturing process is complex and, therefore, mistakes can be expensive. As a result, reflectors can consume as much as 10–15% of the total satellite cost.

If the mass of the reflectors is allowed to grow by perhaps as much as a factor of five, then alternative concepts could potentially be used. Aluminium might be a possible material if suitable ways of protecting it against extreme changes in temperature could be devised. The manufacturing process would involve machining the surface to the required contour, then testing the surface and, should it be inadequate, it could be re-machined and re-tested. In this respect, the actual reflector also doubles as the tool. This would continue until the desired surface was achieved, after which the excess material could be machined away from the back, and the reflector painted and blanketed for thermal control [24].

The above examples have focussed on the parts of a comsat associated with its mission, specifically the structure on which the communications boxes are mounted and the reflectors. These parts are generally unique to the comsat, and because usually only a handful of satellites are built, it is impossible to get much cost leverage from mass production. In other words, if there are only a small number of unique parts to be built, it is cheaper overall to handcraft these items than to invest in the facilities and techniques needed to make a larger number of them.

In the aero-space plane era, the platform or bus of a comsat could essentially be identical for all missions. Larger mass margins would mean that the propellant tanks could be sized to fit a maximum spacecraft mass, thereby avoiding the need to manufacture several sizes of tank. It would also be cheaper for smaller satellites to use the same bus as a larger satellite because the cost of building a few, perfectly optimized buses for the smaller missions would be greater than the excess costs of having to use an oversized, mass-manufactured bus.

The above discussion isn't new. Recently, the Lockheed Corporation announced that it was developing a new type of large, but low-cost satellite called Frugal Sat or F-Sat. Although this satellite design is only intended for use in low Earth orbit, it embodies a similar philosophy in many areas including a relatively simple aluminium construction for the central structure. In addition, according to a report in *Space News* [25], the Lockheed team

> . . . intends to keep at least 70% of the satellite [bus] design—including electronics and power source—the same for each application. Only 30% of the satellite will have to be altered to fit new payloads for different applications.

Previous studies [26] have indicated that by allowing the

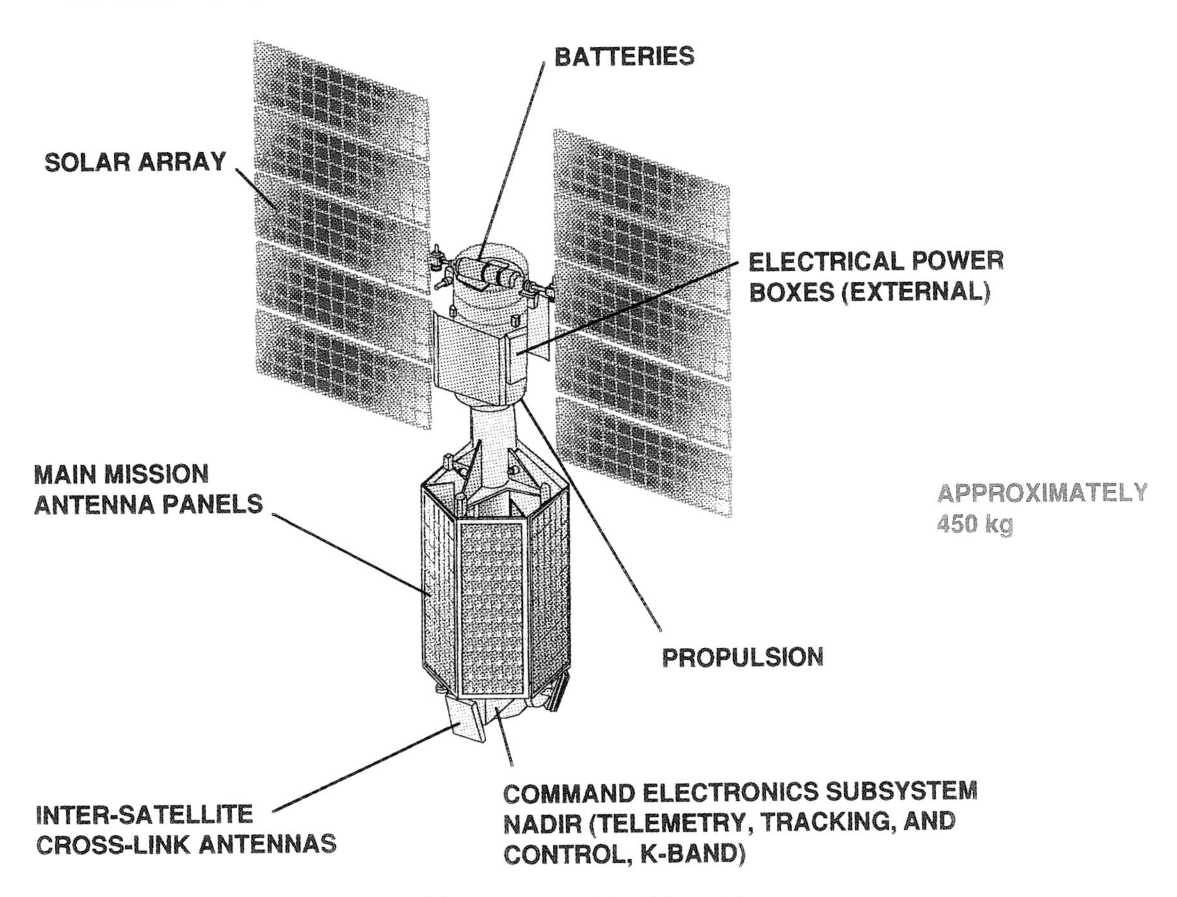


**Figure 18.6** Iridium (*Motorola*).

mass to increase in strategic areas, it should be possible to reduce the cost of constructing typical communications satellites by about 30% (e.g., $50 million versus $75 million in Figure 18.5). This would imply that the total cost to build, launch, and insure a satellite would be reduced by as much as 50% compared with present the situation (e.g., $77 million versus $159 million in Figure 18.5 [27]). Considered from an overall perspective, the net result would be a reduction in the cost the user has to pay to lease the satellite by around 30–35%—a more substantial savings compared with the 10–15% reduction predicted for future comsats using current launch systems.

So far the discussion of communications satellites has dealt with the first and the third impacts of aero-space planes, as defined in Chapter 17. Also, it has been assumed that the aero-space plane launched communications satellites would be identical to current comsats, in terms of their mission performance, life-time, reliability, etc., although spacecraft construction would change to take advantage of relaxed mass constraints and a cheaper launch. Maintaining the same technical objectives as today's comsats, and assuming the number of comsats built worldwide remains unchanged, provides confidence that reductions as large as one-third in satellite user costs are realistic. Given the fierce international competition to build the handful of commercial birds ordered periodically, any opportunity a manufacturer finds to reduce the cost of its bid will be taken to ensure a competitive edge. The emergence of the first aero-space plane clearly has the potential to provide such an opportunity.

If the spacecraft's mission were also designed for an aero-space plane, then a total comsat program restructuring would naturally lead to even greater economic performances gains. Instead of launching a few large spacecraft into the limited slots in GEO, large constellations of smaller satellites could be deployed in low Earth orbits that are within reach of the aero-space plane. Such constellations are being proposed today using very small, low power satellites launched on current expendable boosters. Motorola's Iridium program is perhaps the best known (Figure 18.6) [28]. With an aero-space plane, not only could a number of such satellites be deployed on one mission (as is also possible with expendables), but the constellation would be established much sooner and at lower cost. In addition, such satellites could be much larger and, therefore, more powerful than proposed concepts which must be kept small to maximize the use of a few expensive launch vehicles. Aero-space planes could be used to regularly replace and return failed satellites. In this way the problems of having to use upper stages are negated, reducing transportation and insurance costs.

Overall, it is possible to draw the following basic conclusions for communications satellites as a result of the introduction of aero-space planes:

- Aero-space planes will dominate the satellite launch market.
- Manufacturers of comsats optimized for aero-space planes will also dominate the markets for satellite construction.
- Competition with terrestrial-based systems, such as fiber-optics, will be strengthened.
- Significantly reduced manufacturing and launching costs might make other business opportunities more affordable.

There seems good reason to believe that the impacts of aero-space planes on commercial communications satellites will be positive.

## Military Space Operations

At a first glance, it would appear that large polar orbiting reconnaissance platforms, such as the Key-Hole (optical) or Lacrosse (radar) satellites, could not be launched on an aero-space plane because of their size, reported to be about 15 tonnes [29]. As noted, the size of reconnaissance spacecraft has been defined over the years by the capability of the launch vehicle being used. When the Keyhole satellites were first introduced in the 1970's, they were designed to take maximum advantage of the launch capability of the day. Since then, the drive to increase performance, reliability, and the lifetime of these satellites has led to a growth in size proportional to how easily existing launchers could be stretched.

In the aero-space plane era, other alternatives may be possible. One possibility could involve splitting the satellite into two modules that are launched separately and assembled in orbit. One module would contain the data-gathering payload, while the other would contain the resources and other subsystems needed to support the spacecraft (e.g., communications, power, and propulsion). The advantage of this approach is that either module can be replaced for servicing or repair. Reconnaissance satellites currently have a lifetime of only about 3 years, which is a function of their propellant load. Hence, when all the propellant has been consumed, the satellite is lost. For a satellite launched by an aero-space plane, the empty resource module could be replaced by a new one, thereby increasing the satellite's lifetime and cost-effectiveness. The spectacular success of crew of the Shuttle *Endeavour* to attach a new upper stage to the stranded commercial Intelsat 6 in 1992 demonstrated the basic feasibility of this type of activity (Figure 18.7).

In addition to lifetime, maneuverability is also a function of a reconnaissance satellite's propellant load. During crisis situations it is often necessary to perform plane change and altitude maneuvers to put the satellite in a more effective orbit for image gathering. Such maneuvers can consume large quantities of propellant. Therefore, the degree to which this type of operation is performed is constrained by the available propellant reserves—the faster the propellant is used performing high energy maneuvers, the sooner the satellite will come to the end of its life. For an aero-space plane version, the rate at which propellant is consumed would not be a constraining factor, because the resource module can be readily replaced or refuelled as required. Once the aero-space plane has demonstrated its ability to launch routinely, then the economic and operational advantages of these types of missions become compelling.

Today, such art option is unattractive because rendezvous possibilities are limited by the small number of Shuttle flights. Before 1986, the DOD had begun studying the use of the Shuttle in this role, although no such missions were ever conducted. As an article in *Aviation Week & Space Technology* explained, [30]

> The U.S. Air Force and National Aeronautics and Space Administration are discussing future use of the Space Shuttle as an orbital tanker to retrieve, refuel and repair the service's imaging reconnaissance satellites to extend their ability to photograph the Soviet Union and other strategic intelligence targets . . . The Big Bird and KH-11 spacecraft fly lengthy missions with film and fuel load the main limiting factor on the Big Bird. Fuel for the attitude control and orbit changes is the limiting capability on the KH-11 since it returns its imagery by digital radio transmission and does not have film as a consumable.

Military communications satellites such as the Defense Satellite Communications Systems (DSCS), weather satellites such as the Defense Meteorological Support Program (DMSP), and navigation satellites such as Navstar are all well within the capabilities of the aero-space plane launcher [31]. Because these satellites are much more modest in size and complexity, it might be possible to develop a set of standard modules (e.g., for communications, power, and attitude control) that could be assembled to meet the needs of a particular mission. Although each module might not be optimum for every mission, the advantages are that they could be assembled rapidly, and the total cost of the satellite would

**Figure 18.7** Intelsat-6 reboost mission (*NASA*).

be less. As a pertinent example, the U.S. DOD has proposed a similar philosophy for their Responsive Replacement Vehicle (RESERVES) tactical small satellite program [32]. RESERVES aims at developing standardized spacecraft modules which can be assembled under field conditions into a complete satellite weighing about 250 kg and then orbited by a small launcher. The aim is to go from request for a particular satellite to launch in less than 72 hours. In the aero-space plane era, the mass of the satellite and the payload can be much larger and more capable. A typical small launcher capable of placing 250 kilograms in low Earth orbit costs about $10–15 million, or about the same as an aero-space plane capable of launching and retrieving up to 10 tonnes. The possibility of using aero-space planes to deploy "tacsats" has been studied, as shown in Figure 18.8 for NDVs.

A military space capability that presently does not exist is an on-demand means of accessing space—a launch system constantly in a mode of readiness. An on-demand or rapid response capability is an inherent requirement for all terrestrial military activities. Some military strategists have concluded that a similar space capability would be of value in times of conflict [33]. Because of the limited performance of current launch systems, this has not yet been achieved, although it can be argued that the United States and C.I.S. already have an on-demand capability. Both always have a number of military satellites in orbit at all times, and these can be called upon immediately in any crisis situation. Seen from this perspective, there might seem little immediate need for a responsive launch capability. This is a Catch-22 situation. It is impossible to assess the potential value of on-demand responsiveness because the launch capability does not exist.

In the aero-space plane era, the ability to launch 7–10 tonne payloads on demand for less than a tenth of the cost of current launch systems could provide incentive to evaluate the effectiveness of responsive military space operations. For example, aero-space planes could be used to perform dedicated reconnaissance missions. The aero-space plane would be able to fly directly into an orbit that would pass over the target within 45 minutes of launch. The aero-space plane could either return to Earth after one orbit or make a few more passes before returning. Such a capability would be useful today because there are long periods of several hours every day when reconnaissance information is not available, due to the small number of satellites, their orbital motion, and the weather.

This type of on-demand application for the aero-space plane must be weighed against alternatives, some of which the vehicle itself enables. For the reasons outlined above, the aero-space plane could, potentially, change the way military satellites are built and operated. As a consequence, a

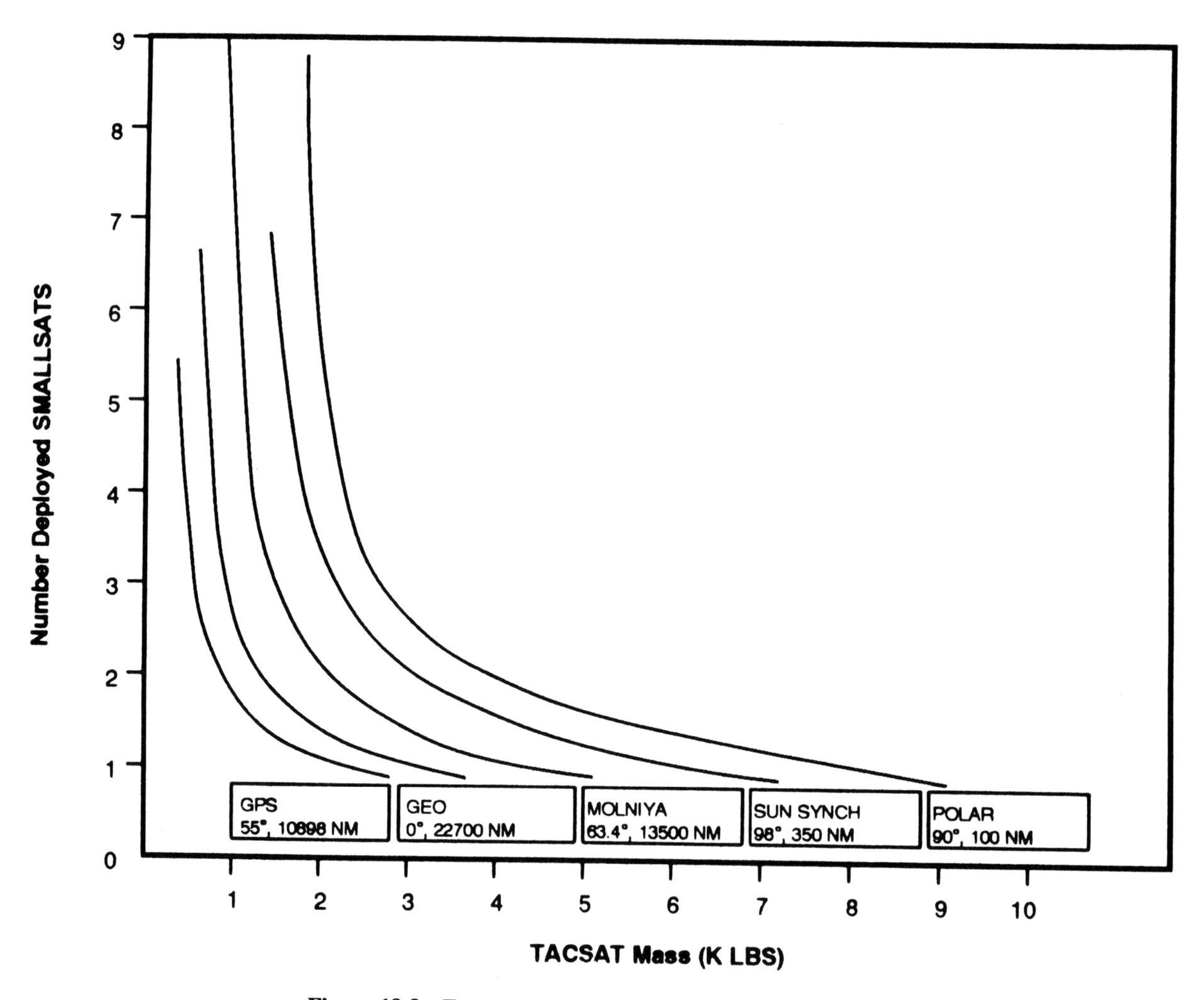


**Figure 18.8** Tacsat deployment from an NDV (*NASP JPO*).

reconnaissance capability structured around a few, very expensive and expendable satellites, could change to one structured around many smaller, less expensive, and serviceable satellites. Aero-space planes have the potential to make technically and economically possible a constant global surveillance capability. According to Bruce Abell, [34]

> As the field of potential conflict spreads to global proportions, we'll depend increasingly on surveillance, especially from space, to monitor changes and to be able to act early to avert or minimize conflict. I can think of a number of ways to take on this task, but they depend on being able to distribute many small, inexpensive satellites in low-earth orbits to provide global, 24-hour, all-weather coverage of the earth. Systems of networked satellites will change the way we think of space surveillance in much the same way that PCs changed the way we think of computing—substituting networks of relatively inexpensive, but highly capable, small satellites we use today. Much of the technology is already available, or can be projected reasonably confidently. But the key technology we lack is flexible, low-cost space access. The kind of system I've just described would require hundreds, maybe thousands of inexpensive satellites. I could foresee their efficient deployment with an NDV [NASP-Derived Vehicle], just as I could see deployment of many new kinds of affordable space technology with an NDV.

Perhaps the most interesting impact the aero-space plane could have on military space activities is that it would reduce the cost to launch and deploy reconnaissance and intelligence gathering type satellites to the point where they become much more affordable to many other countries outside the United States and C.I.S. [36]. If the total cost to build and launch an optical reconnaissance satellite were reduced from the present value of around several hundred million dollars to well under $100 million, then not only would it be attractive to the European nations, but also to Latin and South America, Asia, the Middle East, and other regions. Parallels with other military systems, such as high-performance fighter aircraft and nuclear weapons, can also be made. Once the cost of a system is reduced, the more widespread its proliferation is likely to be.

This scenario clearly poses a number of problems. Would the widespread international use of reconnaissance satellites provide a stabilizing or destabilizing influence on world politics? The current arms reduction agreements are heavily dependent on space-based verification systems. Thus, the ability of many countries to perform independent verification will put more pressure on the disarming nations to obey these agreements. Alternatively, the example of the Gulf War shows the advantages of not allowing the proliferation of such military space systems. Clearly, the ability to minimize an aggressor's information flow is fundamental to winning conflicts. Eventually, many nations will have access to the capabilities provided by aero-space planes, just as they have access to airliners today. As a result, they will have the technical and economic means to establish their own autonomous reconnaissance and other intelligence-gathering space-based capabilities. Evidently, control of the proliferation of such autonomous systems inherently must fall under the auspices of international law.

## Summary—Reassurance of a Positive Future

If aero-space planes eventually are able to meet the type of economic and performance standards described at the start of this chapter, then the only conclusion to be drawn is that the impacts on unmanned space operations of aero-space planes will be positive. The benefits come not only from the opportunity to save money on transportation costs, but also from the ability to demonstrate new technologies over a short period of time, test critical spacecraft components in space long before assembling them into the operational spacecraft, redesign spacecraft to save costs, and undertake new missions currently impractical with existing boosters.

Such new missions might include transitioning from the current philosophy of launching a few, large, and very expensive spacecraft to deploying swarms of many, smaller, and less expensive satellites in low Earth orbit. The constellation approach would have considerable application to Earth observation, telecommunications, and military missions, in particular. Today the constellation approach has only seen limited application because of its reliance on a large number of launches. This situation would change fundamentally in the aero-space plane era.

Suggestions that aero-space planes should be built only to support the limited number of satellites launched presently are too pessimistic. It seems clear that a strong potential exists for a significant *increase* in the number of payloads launched compared with today—a reassuring fact for the would-be developers of aero-space planes.

## References and Footnotes


1. Wilson, A., ed., *Interavia Space Directory 1991–92*, (Jane's Information Group, Coulsdon, Surrey, 1991), p. 405.
2. de Selding, P., "ESA's Olympus Satellite Expected Back in Full Service in November," *Space News*, September 23–29, 1991, p. 40.
3. *Interavia Space Directory 1991–92*, p. 385.
4. Report by an advisory panel on the European Community and Space, *The European Community Crossroads in Space*, (Office for Official Publications of the European Communities, Brussels, 1991), p. 12.
5. Bernasconi, M. C., and G. G. Reibaldi, "Inflatable, Space-Rigidized Structure Overviews of Applications and Their Technology Impact," IAF-85-210, *36th Congress of the International Astronautical Federation*, Stockholm, Sweden, October 7–12, 1985.
6. *ESA Bulletin*, No. 67, August 1991, p. 131.
7. See *Interavia Space Directory 1991–92*, p. 361.
8. Hyde, R., Y. Ishikawa, and L. Wood, "An American-Traditional Space Exploration Program: Quick, Inexpensive, Daring and Tenacious," A

presentation to the senior staff of the NASA Johnson Space Center by the Lawrence Livermore National Laboratories, October 4, 1989.

9. The only other Western ion propulsion systems that have been flown were on SERT 1 and 2 experimental spacecraft in 1964 and 1970, respectively.
10. *Interavia Space Directory 1991–92*, p. 361.
11. "In a Hurry to Fix Hubble," *Space News*, August 5–18, 1991, p. 2.
12. "Hubble: Made in the U.S.A.," *Space News*, August 19–25, 1991, p. 2.
13. Wilson, A., *Space Shuttle Story*, (Crescent Books, Hamlyn Publishing Group Ltd., London, 1986), p. 89.
14. Parkinson, R. C. "Organizational Impediments to the Reduction of Costs of Space Programmes," IAA-91-639, *42nd Congress of the International Astronautical Federation*, Montreal, Canada, October 5–11, 1991, p. 2. Copyright American Institute of Aeronautics and Astronautics © 1991. Used with permission.
15. NASA Kennedy Space Center, "Information Summaries: Space Shuttle Mission Summary 1990," PMS 037, April 1991.
16. Isbell, D., "Makeshift Targeting System Salvages Data From Astro Mission," *Space News*, December 17–23, 1990, p. 6.
17. According to an article in *Space News*, (December 9–15, 1991, p. 16) the first two EOS platforms will cost a total of $7 billion.
18. See *Interavia's Space Directory 1991–92*, pp. 465–466.
19. Lawler, A., "Teller Endorses Brilliant Eyes," *Space News*, April 16–22, 1990, p. 1.
20. Saunders, R., "Iridium May Host Earth-Imaging Sensors," *Space News*, September 16–22, 1991, p. 1.
21. Hannigan, R. J., "The Economic Impacts on Space Operations of Future Launch Systems," IAA-89-697, *40th Congress of the International Astronautical Federation*, October 7–13, 1989, Malaga, Spain, p. 5.
22. Typically, the density of aluminium honeycomb sandwich panels are on the order of 50–100 kilograms per cubic meter compared with solid aluminium which has density of about 2,000 kilograms per cubic meter.
23. See, "The Economic Impacts on Space Operations of Future Launch Systems," p. 8.
24. Ibid., p. 10.
25. Saunders, R., "Lockheed Aims for Cheap Mass Production Satellite," *Space News*, September 9–15, 1991, p. 23. With permission of *Space News*.
26. Webb, D. C., and R. J. Hannigan, for SAIC, "Economic and Socio-Political Impacts of NASP-Derived Vehicles: A Technical Report," *The International Hypersonic Research Institute*, Florida, November 1990, p. 80, and Hannigan, R. J., "The Economic Impacts on Space Operations of Future Launch Systems," IAA-89-697, *40th Congress of the International Astronautical Federation*, October 7–13, 1989, Malaga, Spain.
27. For a current generation comsat plus launch, the total cost is $75 million (satellite) + $60 million (launch) + $24 million (insurance) = $159 million. For the aero-space plane, it is $50 million (satellite) + $10 million (launch) + $10 million (upper stage) + $7 million (insurance) = $77 million. Hence, $77 is about 50% of $159 million.
28. One of the first acknowledgements of Motorola's Iridium plans were reported by Daniel Marcus of *Space News* in an article entitled, "Motorola to Enter Small Satellite Business," June 4–10, 1990, p. 1.
29. See *Interavia Space Directory 1991–92*, pp. 216–217.
30. Covault, C., "USAF, NASA Discuss Shuttle Use for Satellite Maintenance," *Aviation Week & Space Technology*, December 17, 1984, p. 14. Courtesy *Aviation Week & Space Technology*. Copyright 1984, McGraw-Hill, Inc. All rights received.
31. See *Interavia Space Directory 1991–92*, pp. 191–196, 201–206.
32. Marcus, D., "Firms Design Small Satellites for Rapid Launch Capability," *Space News*, June 18–24, 1990, p. 11.
33. General Accounting Office, "National Aero-Space Plane: A Technology Development and Demonstration Program to Build the X-30," GAO/NSIAD-88-122, Washington, D.C., p. 49.
34. Abell, B., "NASP Policy Lessons From Its Perilous Transition From Laboratory to Runway," Remarks at the *AIAA Second International Aerospace Planes Conference*, Orlando, Florida, October 31, 1990.
35. For a discussion of various nations' military space aspirations see an article in *Scientific American* by Jeffrey Richelson entitled, "The Future of Space Reconnaissance," Vol. 264, No. 1, January 1991, pp. 18–24.

# Chapter 19

# Expanding the Human Infrastructure Into the Fourth Domain

## Introduction: A Vision of Populating the Fourth Domain

It is difficult to imagine a growing and dynamic space program without human spaceflight. Today, however, only two people in the Mir Space Station orbit the Earth on a long-term basis. Every now and again they are joined in space by half-a-dozen Shuttle astronauts for week-long sojourns. The largest number of people to have orbited the Earth at any one time is just 12. In this light, does the present situation reflect a growing and dynamic space program?

The presence of a handful of individuals populating space does not mean that there is nothing for people to do there. Space is the fourth domain, after the land, sea, and air, to be explored and exploited. Just as there is plenty for humans to do in the three terrestrial domains, shouldn't there also be much to do in this untapped fourth domain? The reason why we have been able to exploit the land, traverse the oceans, and conquer the skies is because of efficient, reliable, available, safe and, moreover, readily affordable transportation. User-friendly transportation has been the fundamental enabler of all human endeavors on the face of the Earth. Would, therefore, a user-friendly space transportation system enable a similar human expansion into our fourth domain?

The visions—orbital colonies for thousands of people, space factories churning out products, solar power satellites beaming power to the Earth, astronomical bases on the Moon, and outposts on Mars and the outer moons of Jupiter—seem a long way from the reality of today's two-person Mir Space Station—a fragile first settlement in space. Even further away is the International Space Station's planned six-person permanently staffed configuration. Yet, as long as the spacefaring nations have to rely on the current generation of launch systems, Mir and Freedom are probably about the best we can ever hope to do. For the reasons presented in Chapter 2, anything more ambitious is almost certainly unaffordable, or simply impractical or irrational.

What difference would aero-space planes make?

## PART I—Building a Sustainable In-Orbit Infrastructure

### Space Stations Today

User-friendly transportation systems are key to the human exploration and exploitation of space, enabling the construction of a sustainable in-orbit infrastructure, analogous to terrestrial infrastructures such as communications networks, electrical power stations, roads, railways, and other facilities. Infrastructure is, after all, a framework that enables and supports activities.

It is perhaps surprising to note that after three-and-a-half decades in space, the only activity which can be classified as an in-orbit infrastructure is the Mir Space Station. Space Station Freedom was intended to be a quantum improvement over Mir. The original (circa 1984) hope was to build a multipurpose facility whose services would range from a factory for commercial microgravity materials processing to a way-station for manned missions to the Moon and Mars. (See Chapter 1.) Shuttle-related technical and budgetary realities have steadily pruned back the capabilities of Freedom to the point where, at best, it is a facility for life sciences and other research and, at worst, it may be unbuildable within the present timeframe, as discussed at length in Chapter 2. Indeed, the 1993 redesign of Freedom ordered by President Clinton was, in part, a recognition of this very situation.

Assuming that Freedom is successfully built, it is interesting to look at the impacts aero-space planes could have if they were to take over from the Shuttle the role of launching the Station's logistics needs (Figure 19.1). The economics for the current Shuttle-supported Freedom compared to an aero-space plane-supported Freedom are presented in Figure 19.2.

Although the above comparison is very simple, if aero-space planes came anywhere close to meeting their dedicated launch costs, the savings would be on the order of *$2 billion per year*. Even if the aero-space plane launch cost

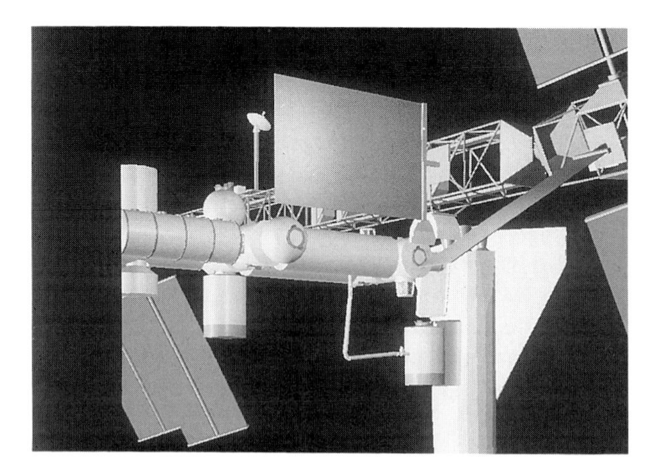

**Figure 19.1** An aero-space plane servicing Freedom (*BAe*).

| | Space Shuttle | Aero-Space Plane |
|---|---|---|
| Missions/year | 5 | 10 unmanned<br>5 manned |
| Payload mass/flight | 18 tonnes | 7 tonnes unmanned<br>3 tonnes manned |
| Cost per flight | $500 m | $10 m |
| **Annual cost** | **$2500 m** | **$150 m** |

**Figure 19.2** Cost comparison of Shuttle-supported Freedom and aero-space plane-supported Freedom.

were five times as much (i.e., $50 million), the savings would still be around $2 billion. It is important to note that realizing these savings would mean that the Shuttle would have to be phased out completely, as the annual Shuttle program costs are effectively independent of flight rates. Similar calculations can be done for expendable launchers, but they would yield roughly the same result.

The staggering difference in costs can make aero-space planes look doubtful, as if this is really "too good to be true." This may indeed be proven to be the case after the first aero-space plane flies. A more credible response is that it *shouldn't* cost $2 or $3 billion a year to haul 80 or 90 tonnes worth of food, water, clothes, spare parts, and experiments to and from orbit.

Another view is that aero-space planes could pay for themselves within a few years purely on the savings of *not* using the Shuttle to support Freedom. For vehicles like the Delta Clipper or the Interim-HOTOL, the 15 flights per year would probably be more than adequate to justify the high cost of their development [1]. Viewed rationally and logically, this would make the case for aero-space planes very compelling. At a minimum, it should be sufficient motivation to find out whether aero-space planes can really be built and achieve the promised operational performance and costs. Again, aero-space planes are *only transportation systems*.

## Aero-Space Plane-Launched Space Stations

While saving money is important, it would not be the only significant benefit of aero-space planes. Space Station Freedom is optimized around the Shuttle, just as much as a communications satellite is optimized around an ELV like Ariane 4. Therefore, it is worth speculating for a moment on what a manned space station would look like in the aero-space plane era, and what type of infrastructure this would enable.

A slowly emerging concern since the inception of the Space Station Freedom program has been the problem of in-orbit construction (Chapter 2). The 1991 Freedom rescaling exercise was a recognition of this fact as was the 1993 redesign. The Freedom (1991) truss structure changed from the 7-year old design that was to be assembled by astronauts in space strut by strut, to one that is split into seven major sections, assembled and integrated on the ground first, and then plugged together in space [2]. Even this may not be good enough. The Mir experience has shown that even if fully integrated and autonomous modules are used, complications can still arise when performing relatively simple plug-in operations.

The ability of aero-space planes to support frequent and relatively inexpensive technology demonstrations, as discussed throughout Section 4, would allow better understanding of in-orbit assembly problems. It would be possible to use space to try out various techniques for docking full-size module test articles, much in the same way neutral buoyancy water tanks are currently used on the ground. More complex orbital assembly demonstrations could eventually involve the construction of pressurized and other complex structures from basic elements, rather like Tinker Toys. The options are many, but the end result is the same: on-orbit demonstrations build experience that is fundamental to the success of future space infrastructure activities.

Once basic technology demonstrations have been performed, one possible scenario for efficiently developing greater experience would be to launch a small space station dedicated to supporting in-space development work. This workshop could then be used as a place to practice on-orbit construction, routinely supported by monthly aero-space plane flights (Figure 19.3). Having such a facility would readily allow experience to be accumulated on what it means to live and work in space. The best way to find out whether something can be done is to go and do it—but keep it simple.

Such a modest space station would have to be developed and deployed rapidly and relatively inexpensively to be worthwhile. To achieve this, the station could be built up from fully integrated modules plugged together on orbit. Most aero-space plane concepts would be able to launch modules about the size of a two-segment Spacelab module.

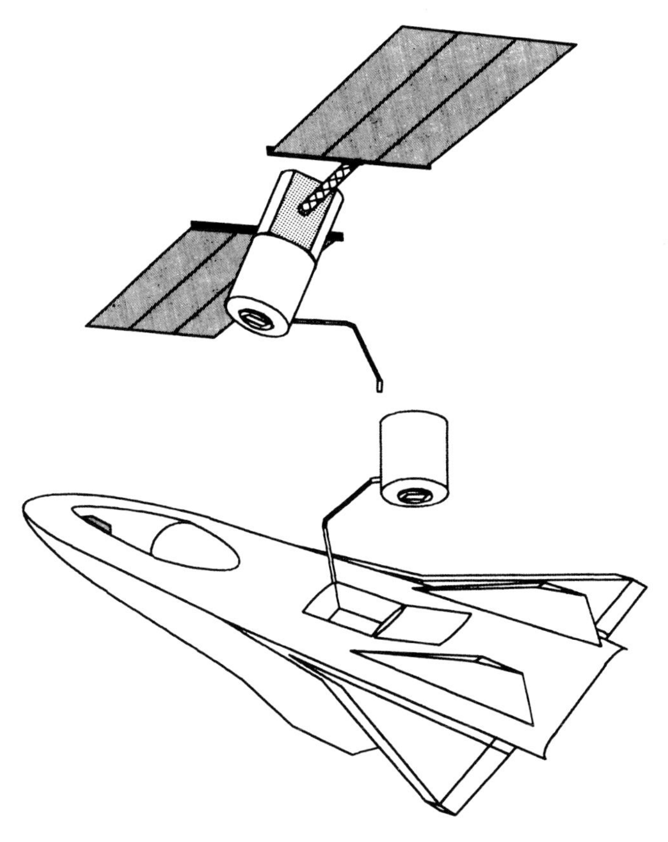

**Figure 19.3** An aero-space plane-launched space station.

After half-a-dozen flights, a small station on the scale of Mir would be assembled. Importantly, the time taken to build this basic station could be short—perhaps months. By contrast, one of the reasons why Freedom (pre-1993 redesign) would take 4 to 5 years to complete, apart from its larger size, is because it is impossible to guarantee a Shuttle flight rate much above five or six per year. This is a problem not encumbering aero-space planes.

A high launch rate facilitates rapid assembly, but does not, by itself, mean a station can be deployed rapidly. Building, integrating, testing, and preparing modules for launch also affect the pace. In the aero-space plane era, because mass is not the overwhelming design driver it is today, individual modules can be made as alike as possible, with only the internal racks being unique. How much the manufacturing process could be accelerated is a complex problem to assess, and additional factors concerning rates of funding and political micromanagement may also have strategic impacts. Nevertheless, it is safe to say that space station elements optimized for an aero-space plane launch would be significantly less expensive and more quickly constructed compared with a Shuttle-launched version. As Maxwell Hunter explains with respect to the SSX concept, [3]

Of course, this large capability [flights per year] is only useful if the payloads can be broken-down to modules small enough to be carried in the SSX and then easily assembled on orbit. In both NASA and the SDIO, the payload designers run unchallenged. Naturally, both organizations feel they fundamentally must have large launch vehicles. Orbital assembly has the social status somewhat lower than a werewolf. There is an easy, and likely entertaining way to change this almost overnight. Simply tell the companies involved (in some seriously credible way) that they have a choice. They can have *serious* deployment money *now* if they can deploy with modules. Otherwise they can *wait ten years* for their *deployment* money while vast sums go to developing the heavy lifter they say they need. Within weeks (probably days) we'll all be inundated with beautiful, simple modular designs, each claiming to be easier than the next. Indeed, until it is presented this way, you'll *never* get a true answer to the question of modularization. This small handful of class A reusable vehicles combined with suitably modularized payloads is awesome.

An aero-space plane-launched space station might take much less time and be significantly less expensive to build than Freedom for many of the reasons outlined above. Other reasons revolve around the way in which such a station would be supported. Again, because of the low Shuttle flight rate and long lead times before launch, the ability to respond to problems as they occur with Freedom will be very limited. With this constraint in mind, Freedom designers must look as far forward as possible and attempt to account for every combination of credible problems that may occur—a time-consuming and expensive activity. Freedom can be repaired after suffering failures, provided preparations for such failures are made several *years* in advance. Unfortunately, some failures cannot offer this kind of luxury.

In the aero-space plane era, short lead times and high flight rates would enable responsive measures (i.e., launch on-demand) to be taken as and when problems with the in-orbit hardware occur. This would mean failure prediction work does not have to be as exhaustive as it is today. In addition, it would be unnecessary to define every detail of the station's future growth years in advance because flexible launch capability allows station modifications as the program grows.

## Evolving Space Stations

Space station evolution is quite obviously a function of its intended use. Ideally, the best solution would be to build more than one type of space station with overlapping capabilities. This would likely become a prerequisite in the future when space stations will be used in an operational sense. One station could be used as a transport node for human exploration of the Moon and Mars (see Part II of this chapter), a second for life science research, and a third for commercial materials processing, should it ever occur. Having more than one station adds essential resilience by ensuring that if one station sustains serious damage (e.g., from collision with orbital debris) all infrastructure activities will not come to a standstill.

Putting the above issues aside, the utility of a space station is in most instances a function of its size. Simply put, the more people a station can hold, the more work they can do. Likewise, the larger the volume and surface area, the greater the number of activities that can be accommodated within and on the external surface of the station. In theory, the Mir-class aero-space plane station described above could be expanded by adding more and more small modules. In reality, this is not efficient and can lead to ungainly configurations with modules sprouting everywhere—plus the safety hazard of a large number of joints.

A better alternative to adding more small modules would be to use the backup expendable heavy-lift launcher to place in orbit the shell of a large space station volume to be attached to an enlarged version of the original Mir-class space station (Figure 19.4). Vehicles like Ariane 5 or Titan 4 could put structural elements in space as large as 4.5 meters in diameter and 20 meters long, while Energia and the originally proposed U.S. National Launch System could launch modules of more than 5.5 meters in diameter and 35 meters long. This large expansion module would only be a basic structure, plus associated plumbing, electrical wiring, and internal and external attachment points—the hooks and scars. The internal layout of this structure could, in theory, also be compartmentalized, like a large ocean-going ship so that, if holed by orbital debris, affected sections could be isolated. This expansion module would not at first contain the expensive internal equipment racks. Once secured to the initial space station, this large module would be gradually outfitted with internal and external equipment.

There are other alternative possibilities to this expansion scenario, including the utilization of inflatable modules, as discussed in Chapter 18. The purpose of the exercise is not to define the most optimum solution. It is intended to demonstrate that if an expendable backup launcher is kept in service, then it makes sense to use it as effectively as possible. Launching just the shell might be an effective method because structures are relatively inexpensive items to manufacture and prepare for launch. Therefore, the cost of a launch failure would be less significant than for a fully outfitted module. Not having to lift the heavy internal equipment means the structure can be manufactured in a more robust fashion, including the incorporation of thicker outer and inner skins to provide greater structural margin and a more resilient orbital debris protection shield. Ideally, such a structure could be built like a tank or a battleship, and not like the highly mass-optimized structures more commonly considered today.

Doesn't this scenario contradict the earlier discussion regarding the use of integrated modules? The purpose of building an initial Mir-class space station with fully integrated modules is because of the absolute lack of experience in

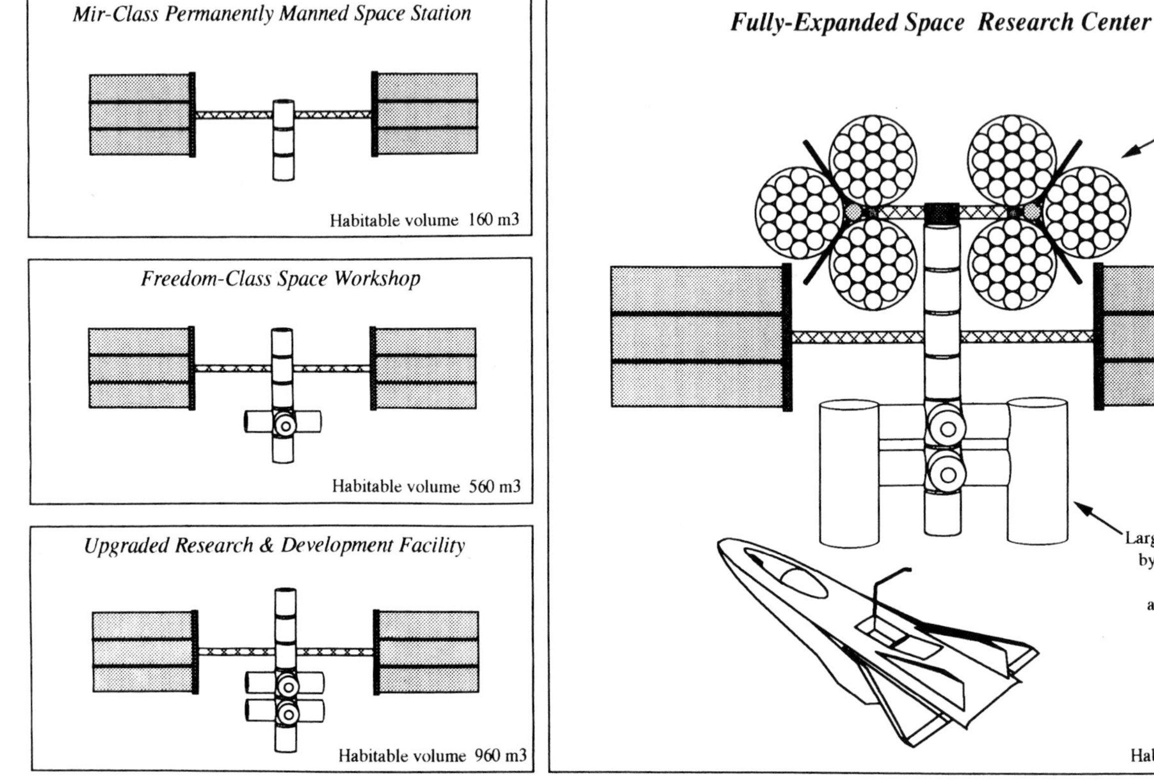


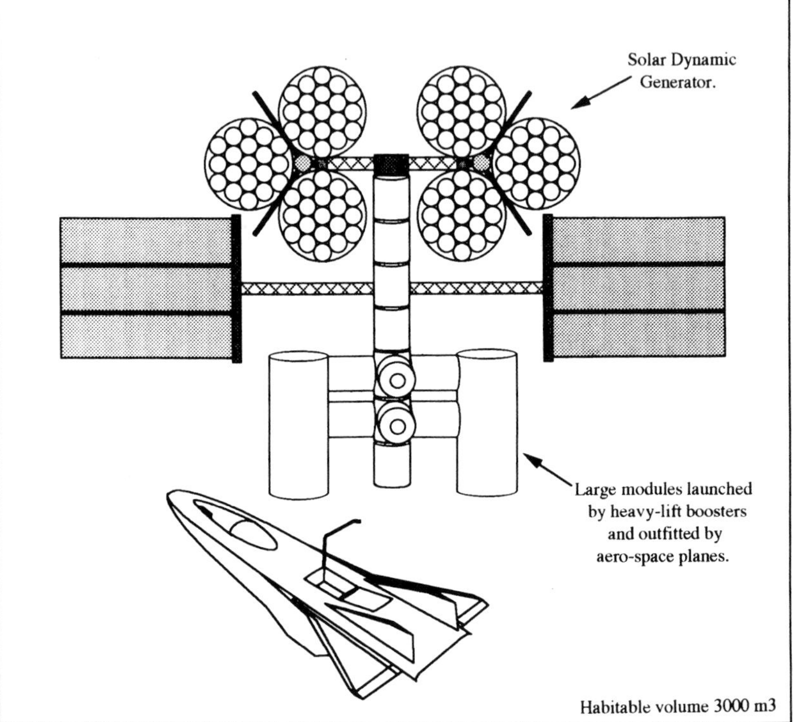


**Figure 19.4** Conceptual evolving space station scenario.

orbital assembly operations. The initial technology demonstration missions that can be performed with aero-space planes, coupled with more thorough demonstrations performed on the initial space station, would pave the way for accumulating the type of experience needed to be confident that large modules can be assembled and outfitted in space. In addition, if problems occurred during outfitting of this expansion module, the availability of a crew already on board the initial and fully functional space station would ensure that such problems could be carefully analyzed and accessed as required. Plus, if a problem occurred that could not be handled by the crew already in space, an aero-space plane could be launched to carry the equipment needed to overcome the problem.

For all of these reasons, current proposals to modify the Shuttle external tanks for habitats in space are unlikely to be viable—some sort of space station is needed first. Toward the transition from the Shuttle to aero-space planes, an interesting situation could develop. Potentially, on the last few Shuttle missions, several specially modified external tanks could be stockpiled in high Earth orbits (400–500 km) [4]. These tanks would be left there for a number of years until the initial aero-space plane-launched station has been constructed and enough experience in orbital assembly is obtained to determine whether these ETs can be used to build what would be a very large (or several large) and relatively inexpensive space station. This is an interesting prospect, and the key is having the demonstrated ability and experience to know how to modify and outfit these tanks in space.

## The Economics of Aero-Space Plane-Launched Space Stations

There is good reason to believe that aero-space plane-launched space stations would be more cost-effective to develop, build, and operate compared with those planned today. Quantifying this is extremely difficult and outside the scope of this discussion [5], but space stations with the following characteristics could be significantly less expensive than the Shuttle-launched Freedom.

- They are constructed from many common and robustly constructed modules that have been preceded by extensive in-orbit technology demonstrations.
- They are assembled over a short period of time.
- They are supported by the responsive ability of aero-space planes to launch as frequently as required for assembly, maintenance, and logistics needs.

To put this into perspective, the average cost of acquiring, building, and operating Space Station Freedom throughout the 1990's will be around $2–2.5 billion per year, while the launch costs will average at least $2.5 billion per year—giving a total of about $5 billion per year. If it is accepted that spacecraft costs tend to gravitate toward the launch costs (Chapter 17), then the yearly development costs for an aero-space plane-launched space station might be expected to be significantly less than that of the present Freedom.

If this is true, the annual cost of an aero-space plane-launched space station could fall to well under $1 billion, perhaps on the order of $500 million per year. Thus, over a 10 year program, the cost of such a station could be in the region of $10 billion, with total annual operational expenditures of around $1 billion. This compares with anywhere from $30–50 billion for Freedom, and a total annual operational cost of about $5 billion. Over a 30 year timeframe (Freedom is now only designed for 10 years), the cost comparison would be $30 billion for an aero-space plane station and, $140 billion for Freedom (Figure 19.5). Again, even though this analysis is not in the least bit rigorous, if the United States and its allies intend to build space stations—for whatever purpose—then the prospect of being able to save such large quantities of money makes the potential of the aero-space plane compelling [6].

Ultimately, aero-space planes could reduce the complexity and bring the cost of establishing and using manned orbiting facilities down to levels more in line with equivalent facilities on Earth, such as oil platforms, submarines, or arctic research stations. Indeed, this may eventually lead to a situation where commercial exploitation can be realistically contemplated, as will be discussed in Chapter 20.

Such costs might eventually be reduced to levels affordable to individual states with access to their own aero-space

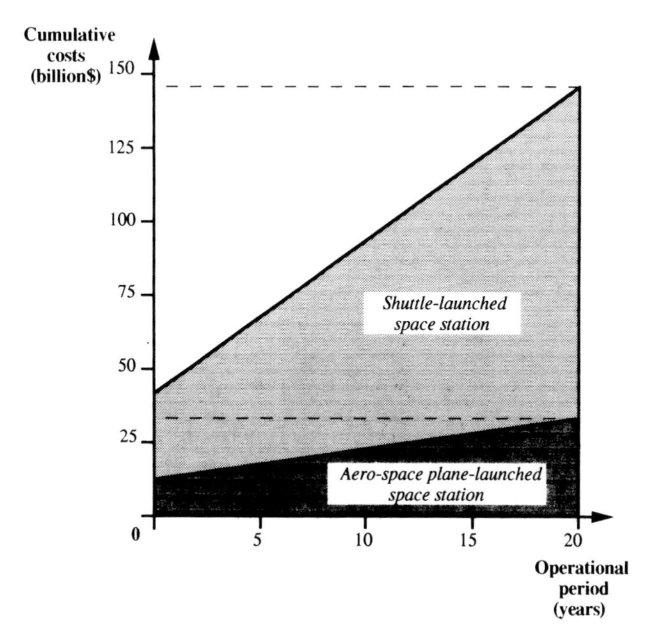


**Figure 19.5** Comparison of aero-space plane and Shuttle-launched space stations.

plane, eliminating the need to always seek cooperation with other nations as a means to secure funds and share costs. Cooperation would still be important, but on the basis of sharing resources for utilization of the infrastructure as opposed to building the infrastructure itself. Building a modest space station that cost a few hundreds of millions of dollars per year (as opposed to the billions today) is well within the reach of nations like Japan, Germany, and France. In this respect, a robust, multielement in-orbit infrastructure might emerge in the natural course of events, and the place of many humans in space will be secure.

## *PART II—Extending the Infrastructure to the Moon and Mars*

### The Impacts of Heavy-Lift Launch Vehicles

The Moon may be 400 thousand kilometers away, and Mars 380 million kilometers at the most, but the first 150 km from the surface of the Earth to low Earth orbit is by far the most difficult to cross. Up to now, the vast majority of manned solar system exploration architecture studies have focussed on using heavy-lift launch vehicles (HLLVs) as the primary means of bridging this narrow gap. The Synthesis Group concludes, [7]

> This investigation [by the Synthesis Group] has led to the very clear conclusion that to achieve these goals, the utilization of heavy lift launch vehicles having a capability to launch 250 metric tons into low Earth orbit is required . . . All of the lunar and Mars architectures have been baselined with such a vehicle. This allows the architectures to be clearly done faster, cheaper, safer and better than with a less capable launch vehicle.

The apparent requirement for such large launchers is a result of the need to propel the crew and their pieces of machinery out of, and sometimes back into, the Earth's deep gravitational well [8]. The equivalent total change in velocity to reach, and then return from, the Moon is about the same as that needed to get into low Earth orbit in the first place. Roughly speaking, a modestly sized manned lunar lander, used to transport people to and from the Moon, requires several times its own mass in propellant—leading to a typical low Earth orbit gross mass of around 100 tonnes or more depending on the strategy implemented [9]. Hence, a role for the HLLV in the first human solar system missions appears assured.

Yet, the use of the HLLV in this role adversely affects the type of architecture chosen and its cost in a profound way. To summarize from Chapter 3:

- Large expendable rockets will be expensive machines to fly.
- For this reason, it will be all but impossible—economically and physically—to demonstrate that such rockets can reliability and regularly launch expensive payloads.
- Therefore, the unproven reliability of the HLLV will tend to force a particular program to maximize use of a minimum number of boosters.
- Hence, not only will solar system exploration payloads need to be highly capable and reliable, but simultaneously, they must also be highly integrated and lightweight in construction to get the most out of a few launches.
- These conflicting requirements increase payload costs and, as a result, constrain launcher and payload production numbers.
- In addition, a small number of very expensive payloads places higher reliability demands on the launcher because of the added assurance needed to protect this high investment.
- This, in turn, further increases the HLLV costs and constrains the rate at which launches occur.

This vicious circle (Figure 19.6) will ensure that large boosters will seldom fly more than a few times per year, and the payloads they carry will be very expensive and only a handful will be built. These factors are considered the primary reasons for the high estimated cost of the earlier proposed U.S. Space Exploration Initiative program, reportedly costing anywhere from $100 billion to $500 billion.

This vicious circle is further exacerbated by the fact that solar system exploration payloads will be critically interdependent on payloads launched by other boosters in a continuous sequence. For example, in the NASA *Report of the 90-Day Study on Human Exploration of the Moon and Mars* [10], three HLLVs would have to be launched one after the other to assemble the hardware for just *one* lunar landing mission. At this rate, after only two or three sorties to the Moon, there would be a better than 50:50 chance of losing one of the next boosters, based on current launch reliability statistics [11]. Yet nowhere within the 90-day report are contingency measures described to absorb such an HLLV loss, and it is certainly unrealistic to expect another HLLV to be risked soon after a failure. This is especially the case if the next payload to be carried was of very high value and only one of a handful remaining. It is even more severe if a crew is to be launched on that next HLLV.

This type of "must-succeed" thinking is typical of many studies and is a throwback to the Apollo era. Statements are often made that "because Apollo was successful, we should also expect a future SEI-type program to be at least as successful." However, launching six or more *interdependent* HLLVs every year over several decades is an entirely different game from launching three *independent* Saturn 5's every year for 5 years. Long-term missions are dependent on the

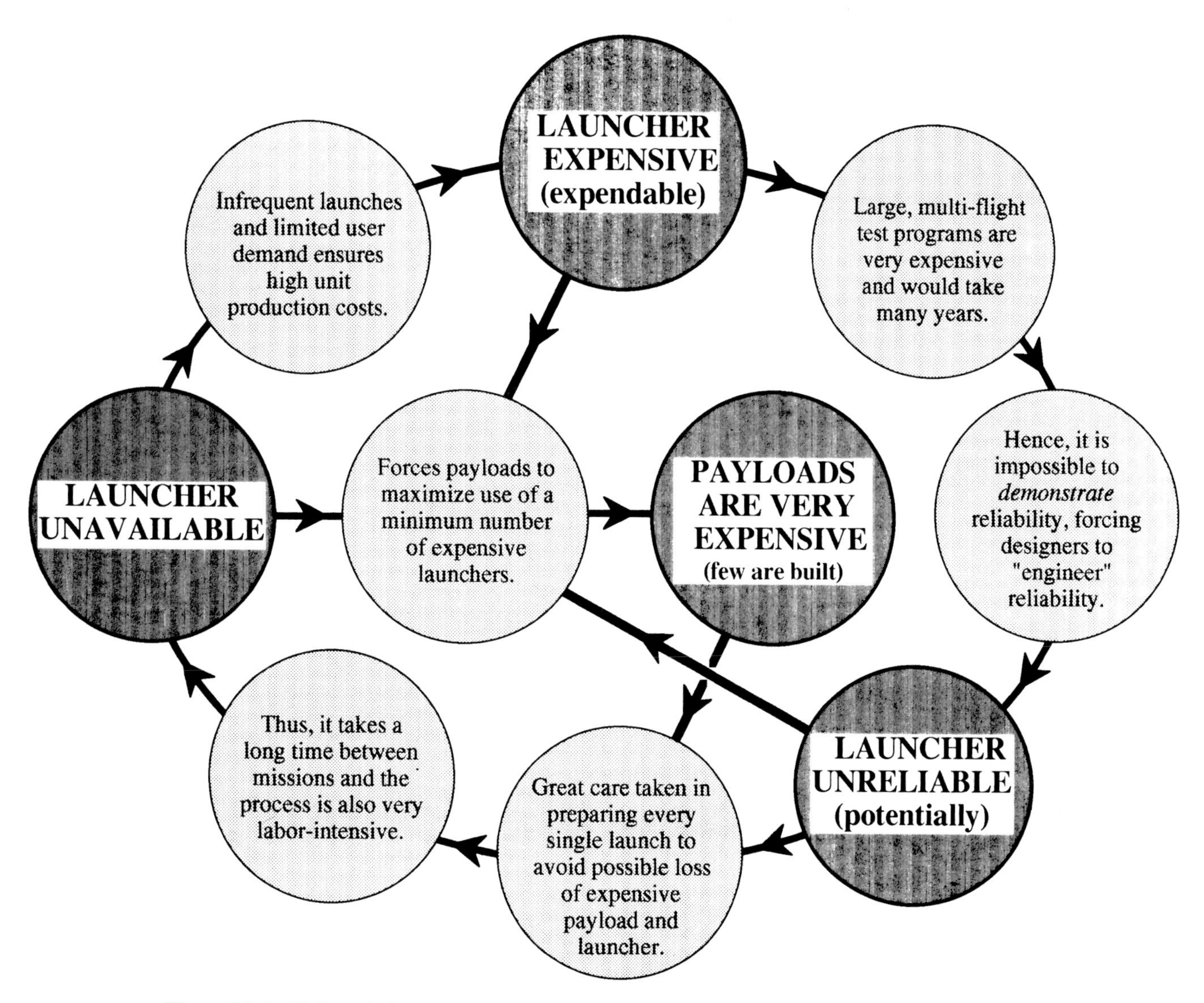


**Figure 19.6** Vicious circle created by using HLLV boosters to launch expensive and unique payloads.

success of other launches, such as for the supply of logistics to a permanent or man-tended lunar base. Each Apollo mission was, by stark contrast, a completely self-contained mission—everything that each Apollo mission needed was launched on a single Saturn 5.

A further problem relates to technology demonstration in space, which is considered an even more urgent requirement for solar system exploration. Life support systems for long duration stays on the Moon or sojourns to Mars must be thoroughly tested in space first, as 1 g ground testing alone will not do. The HLLV or other ELVs can be used for some basic testing. But for the reasons discussed earlier and in Chapters 2 and 18, such tests will be extremely expensive and limited in scope.

Ignoring the need for failure tolerance and routine technology demonstrations in solar system exploration programs, and assuming that such programs will always be successful, might be considered irrational, if not irresponsible, given the likely economic and political impacts of a failure. Programs that rely entirely on a single type of HLLV booster are unlikely to be successful, unless they are extremely lucky. In light of price tags on the order of hundreds of billions of dollars, is it realistic to expect such vast expenditures to be held hostage to the random fortunes of luck?

### A Balanced and Complementary Dual Launch Architecture

The enormous mass needed for a lunar sortie mission nevertheless still requires a capability to lift that mass into low Earth orbit at least for the early missions. Yet, if the HLLV is the sole launch system being used, building and supporting a manned lunar base would be ruinously expensive, it would lack any resiliency, and perhaps it would not even be done.

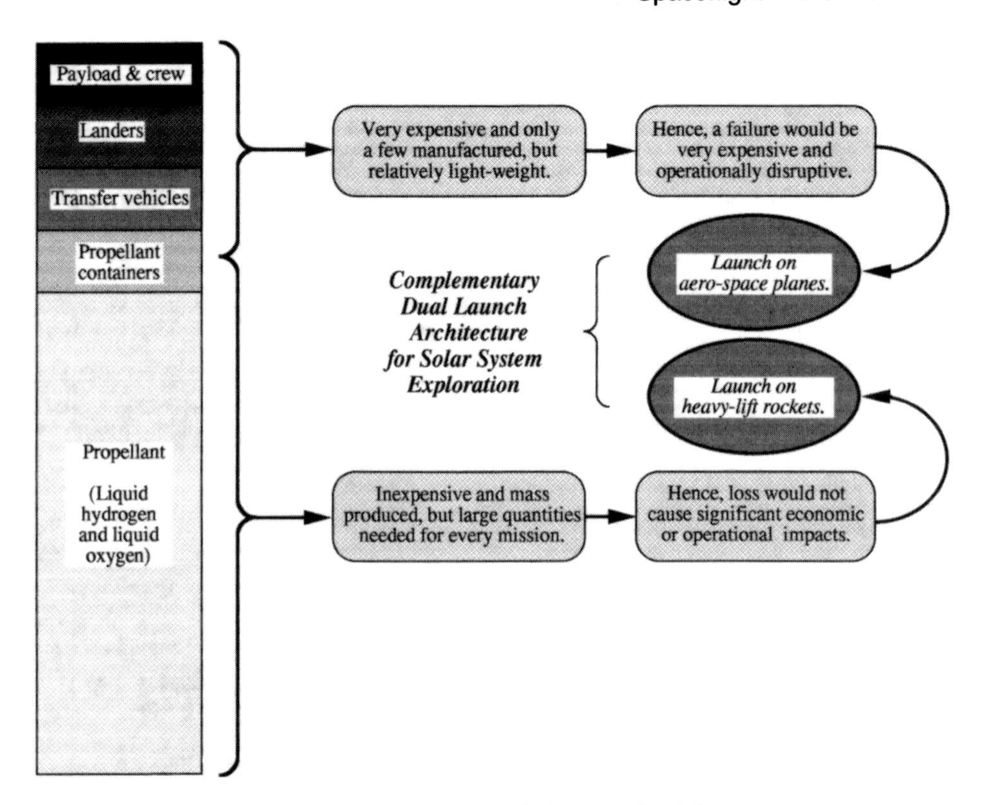


**Figure 19.7** Typical lunar mission mass breakdown.

There is clearly a conflict, one analogous to a worldwide air transportation system relying solely on a handful of Lockheed C-5A Galaxies that are thrown away after each trip.

There may be a way out of this conflict. Some 75% of the LEO mass of a lunar sortie mission is composed of cheap, bulk manufactured, and readily available liquid hydrogen and liquid oxygen. Therefore, it would seem to make sense to limit the HLLV to launching just the propellant. Thus, if it failed, all that would be lost is the rocket—which will be expended anyway—and a payload that is worth a mere $50,000 (1 tonne of propellant costs about $700). Such a loss would obviously be a lot less serious than if an HLLV failed when carrying a manned lunar lander valued in the region of several hundred million dollars (Figure 19.7).

Restricting the HLLV to a tanker role obviously means that another vehicle would be needed to launch the more valuable lunar landers, space transfer vehicles, lunar base modules, crew, and other equipment. The basic characteristics of the aero-space plane would seem to make it ideal for this mission. Its demonstrated reliability and ability to safely abort a mission after an in-flight failure would mean that the expensive elements would stand a much better chance of achieving orbit or not being lost. Further, the responsive launch capability would have other important benefits, not the least of which is routinely scheduling an aero-space plane flight immediately after the HLLV has reached orbit.

Apart from whether aero-space planes can be built, the only real question mark over the use of aero-space planes is whether they can launch these elements. Is their cargo capacity large enough? At first glance it may seem that the answer is no. Solar system exploration architecture studies that have already recognized the potential of aero-space planes usually limit them just to launching people. As the Synthesis Group noted, [12]

> The National Aerospace Plane should be vigorously pursued. The single-stage-to-orbit [Delta Clipper] concept should be carried forward to demonstrate concept feasibility. Both these efforts could hold promise for a cost effective personnel launch system to low Earth orbit.

At present, vehicles like the proposed air-breathing NASP-Derived Vehicle and the Delta Clipper are being designed with a payload capability of about 10 tonnes and 100 cubic meters into LEO. They are clearly more than just personnel transports. The notion that aero-space planes are too small to launch lunar landers and transfer vehicles stems almost entirely from a mindset that has grown accustomed to designing for large heavy-lift launchers. Perhaps this situation would change fundamentally if designers of solar system exploration programs felt that aero-space planes were a realistic proposition.

Even though aero-space planes could conceivably launch propellant, at least 10 aero-space plane launches would be required per lunar sortie. Although the total launch cost could be competitive with an HLLV, other operational reasons might preclude its use in this role. According to Kirkpatrick et al., [13]

For bulk cargo transport, [the NASP-Derived Vehicle] is cost competitive with a heavy lift system, but unless there turns out to be a great disparity in actual costs it may be undesirable to risk a high-value vehicle for a low-value cargo. In addition, the relatively small payload bay implies intrinsic structural inefficiencies for the transport of bulk cargo, relative to a single heavy-lift launch . . . These factors may continue to make it desirable to keep a heavy-lift vehicle in the fleet.

Just as significantly, facilities would be needed in orbit to hold the propellant delivered by aero-space planes. Therefore, propellant storage depots would require support and maintenance, further increasing the quantity, complexities and cost of orbital operations. Launching propellant does not seem to be the best use for an aero-space plane, at least initially.

### To and From the Moon: Small Is Beautiful

Apparently, a modified version of the Delta Clipper could fly from LEO to the lunar surface (Chapter 10), deliver or recover 10 tonnes of payload, and then return to Earth, aerobraking in the upper atmosphere several times before either settling in a low Earth orbit or returning to Earth directly. This is a very clean approach which, if feasible, would clearly make access to the Moon simple and relatively cheap compared with more conventional scenarios. The only problem is that it would require a refuelling stop in LEO of more than 400 tonnes. Launching this 400 tonnes on HLLVs, or building and supporting a large propellant depot in space, represents significant technical problems. Eventually, it may even be possible to scoop most of the oxygen from the upper atmosphere, although the feasibility of this is not proven. Using the Delta Clipper for direct lunar access is clearly a very neat solution, but is probably not viable for the first lunar missions because of the refuelling problems.

A more efficient method would be to adopt a multistaged approach utilizing small, high-performance vehicles designed for use only in space or on the lunar surface. One possible strategy is outlined here. Fundamentally, this strategy takes maximum advantage of the bulk-launch capabilities of the HLLV in order to minimize the size of the aero-space plane-launched lunar landers and transfer vehicles. The smaller the landers and transfer vehicle, the less expensive they will be to develop, and the easier they will be to test at all stages of their development. Critically, this scenario is also tolerant to catastrophic failures of either launcher.

Leaving Earth's deep gravitational well accounts for the largest single increment in energy, in terms of propellant and thrust. Therefore, a sensible approach is not only to use the HLLV to deliver the propellant load into LEO, but also to use it as the upper stage for the first leg of a lunar mission. Most HLLV concepts today are of the two-stage variety, where the second stage is usually a high performance cryogenic system. Thus, rather than the HLLV carrying a payload, the second stage tanks could be stretched to hold more propellants. The most significant problem is developing a reliable engine that can be restarted after 1 or 2 days of exposure to space. However, the Apollo program has already demonstrated the feasibility of reigniting a large cryogenic engine in space after several hours exposure. The single J-2 engine of the Saturn 5's third stage (S-4B) had to be started initially to place the Apollo spacecraft in Earth orbit, then restarted to send these spacecraft toward the Moon.

This strategy eliminates the need to develop a separate upper stage fuelled by the HLLV. It also eliminates the need for the expense and operational complexities of a propellant depot. As HLLVs must be used for the first lunar missions, and each is an expensive piece of machinery that is discarded anyway, it seems reasonable to take maximum use of them in this manner. A further advantage is that the HLLV second stage can also carry all the propellant needed by the lunar landers and transfer vehicles.

For a lunar base, both crew and cargo must be landed on the surface. The approach taken here is to split each into separate crew or cargo missions. Such a strategy simplifies the overall activity and keeps the lunar vehicles small. The vehicles for cargo and manned missions are described below [14].

#### *Cargo Missions*

A cargo mission is relatively straightforward. At the space station, the lunar cargo (such as an inflatable module, power system, or rover) is integrated to a small lander. After the HLLV has been successfully launched, the cargo lander will rendezvous with the HLLV and dock. Propellant is then pumped from the HLLV to the cargo lander's tanks (Figure 19.8). The cargo lander reorientates the composite and commands the HLLV stage to fire. Once the burn is completed and the HLLV stage is discarded, the cargo lander either descends directly to the lunar surface or enters a low Lunar orbit and then lands. As much as 10–15 tonnes could be placed on the Moon using a cargo lander small enough to be accommodated within a typical aero-space plane cargo bay of 4.5 m diameter by 7 m long.

For the first lunar missions, the cargo lander would be expended, because a small expendable lander can land twice as much mass as a reusable lander. Further, a reusable lander would need a space transfer vehicle waiting in lunar orbit to collect and return it to Earth. Intuitively, one expendable cargo lander and HLLV would seem cheaper than developing and servicing a reusable lander, using two HLLVs as well as supporting a space transfer vehicle for every mission. As

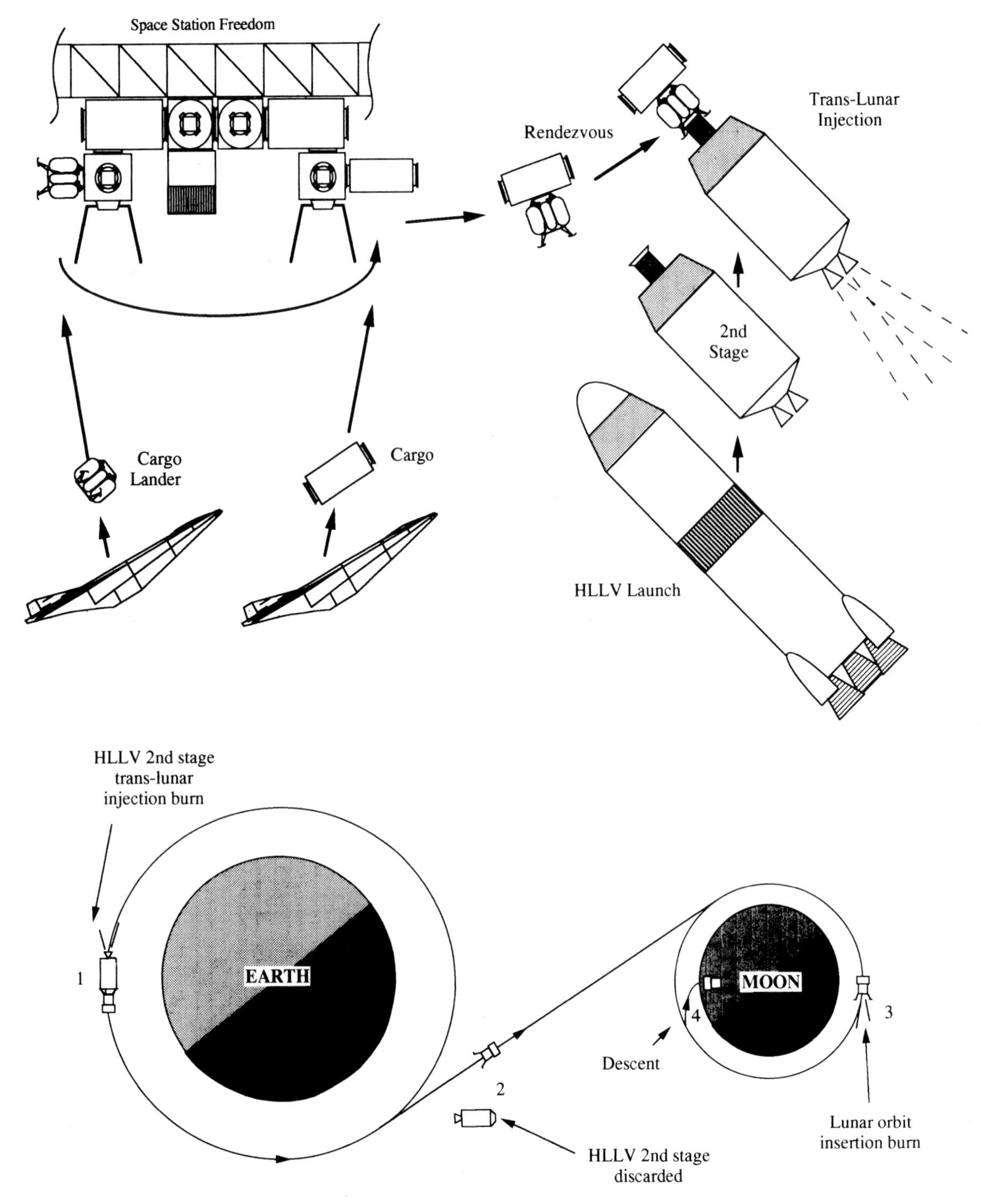


**Figure 19.8** Lunar base cargo missions.

reusable space transfer vehicles are needed for manned missions, it would seem wiser to reserve them for that application and minimize the complex operations associated with their use.

If, in the future, a lunar oxygen production facility [15] is established, the cargo lander on the Moon could be partially refuelled, loaded with excess liquid oxygen, and returned to Earth. The main problem would be that the cargo lander must trade payload for the extra hydrogen fuel needed for the return. Nevertheless, several tonnes of liquid oxygen could potentially be returned to Earth. Whether sufficient oxygen could be returned economically to refuel a fully reusable lunar injection stage has not been estimated. A future scenario that eliminates the role of the HLLV—and, for that matter, *all* expendable hardware—is considered advantageous for economic reasons. This might be feasible when the second aero-space plane fleet enters service to ensure resilient access to space.

### Manned Missions

Landing people on the moon is much more complex than landing cargo, due to the need to safely return the crew to Earth (Figure 19.9). This necessitates the use of a reusable space transfer vehicle and a reusable manned lander. The space transfer vehicle is used after the HLLV second stage has been expended, and its function is to place the manned lander in lunar orbit, and then later collect the lander and return it to LEO using an aero-braking technique. The manned lander's job is to travel from low lunar orbit to the surface of the Moon and then back to lunar orbit. The manned lander would be similar to the cargo lander, except for its cabin and shorter propellant tanks. Typically, each vehicle would weight about 5 tonnes unfuelled. When stowed for launch, the dimensions of each would be about 4.5 meters in diameter and 5 meters in height. These parameters are about the size of a typical aero-space plane cargo bay [16].

The transfer vehicle is the critical element from the point of view of packaging it to fit within a single aero-space plane cargo bay so that it can be returned to Earth for maintenance. Achieving this is not easy, although it is not considered impossible. One concept utilizes an umbrella-like system for deploying a 200 $m^2$ aero-brake covered with a high-temperature-resistant cloth similar to that used by the Shuttle orbiter. A smaller version of such a vehicle has been studied by British Aerospace for launch on HOTOL; it is shown in Figure 19.10 [17]. The aero-brake holds the potential to significantly reduce the amount of propellant needed, compared with pure rocket thrust, by using the atmosphere to slow the returning vehicle down. It is only effective, though, if the amount of propellant that can be saved is greater than the mass penalty of the heat shield.

If the umbrella approach proved unfeasible, an alternative could utilize an inflatable aero-brake or ballute, reminiscent of design studies by Boeing in 1983 [18]. A third possibility is to launch a separate heat shield made from solid structures which would form a rigid aero-brake. This solid aero-brake would be left on orbit or at a space station, while only the core of the vehicle, containing the sensitive electronics and propulsion systems, is launched and returned independently.

It is important to note that the practicality of using an aero-brake is as yet unproven. One significant concern revolves around the controllability of the aero-brake as it skims through the upper atmosphere. If control is off by a small amount, the transfer vehicle could wind up in a different orbital plane from that intended, making a rendezvous with a space station impossible [19]. Repeated technology demonstrations of aero-braking vehicles launched from an aero-space plane should enable these problems to be properly understood long before a choice between an aero-brake or a purely propulsive transfer stage is made. This was one of the goals of NASA's Aero-assist Flight Experiment which was cancelled in the Fiscal 1992 budget in the drive to find money for Freedom.

The scenario outlined above returns the manned lander to Earth orbit after every mission, as opposed to leaving the lander in Lunar orbit and the crew returning on the transfer vehicle. This scenario eliminates the complexity of refuelling the lander or exchanging propellant tanks in lunar orbit, as proposed by the NASA 90-Day Study. More fundamentally, it allows the lander to be fully checked out before the next crew climbs aboard. This is particularly important during the first few years of operations due to safety uncertainties with complex vehicles like lunar landers after being left untended for long periods in space. By virtue of its small size, the manned lander can return to low Earth orbit.

### Is a Space Station Transport Node Required?

The value of on-orbit servicing of the transportation elements is the subject of intense debate. The Synthesis Group report made no specific recommendations for the use of Space Station Freedom—or any other orbital facilities—in its proposed program other than for life science research. The NASA 90-Day Report, by contrast, proposed extensive use of an earlier version of Freedom. The reasoning that each study used to reach its conclusions is not appropriate to an aero-space plane scenario. As Part I of this chapter explains, the unique capabilities of aero-space planes will allow space stations to be more readily supported—*access would no longer be the overwhelming constraint it is today*. The tradeoff may favor on-orbit servicing because access to the facilities is possible and demonstratable in the aero-space plane era. If a manned lander needed a new component, it could be delivered on one of the next scheduled aero-space plane logistics flight to the station. Timely access is no longer an immediate problem.

The ability to service and store the transfer vehicle and manned lander at a space station would simplify manned access to the Moon compared to a scenario without a space station. Initially, these vehicles would each be launched separately on aero-space planes and docked to the station. After that, all that would be required to access the Moon would be the crew and a single HLLV launch. After several re-uses, the lander and transfer vehicle could be returned to Earth for overhaul. Without a space station, three aero-space plane launches would be needed per lunar sortie mission: for the lander, transfer vehicle, and crew.

A conceptual design for a servicing facility is shown as part of Figure 19.11. The design assumes the use of a Freedom-like station purely as a means to demonstrate what might be practical. The concept does not require the use of

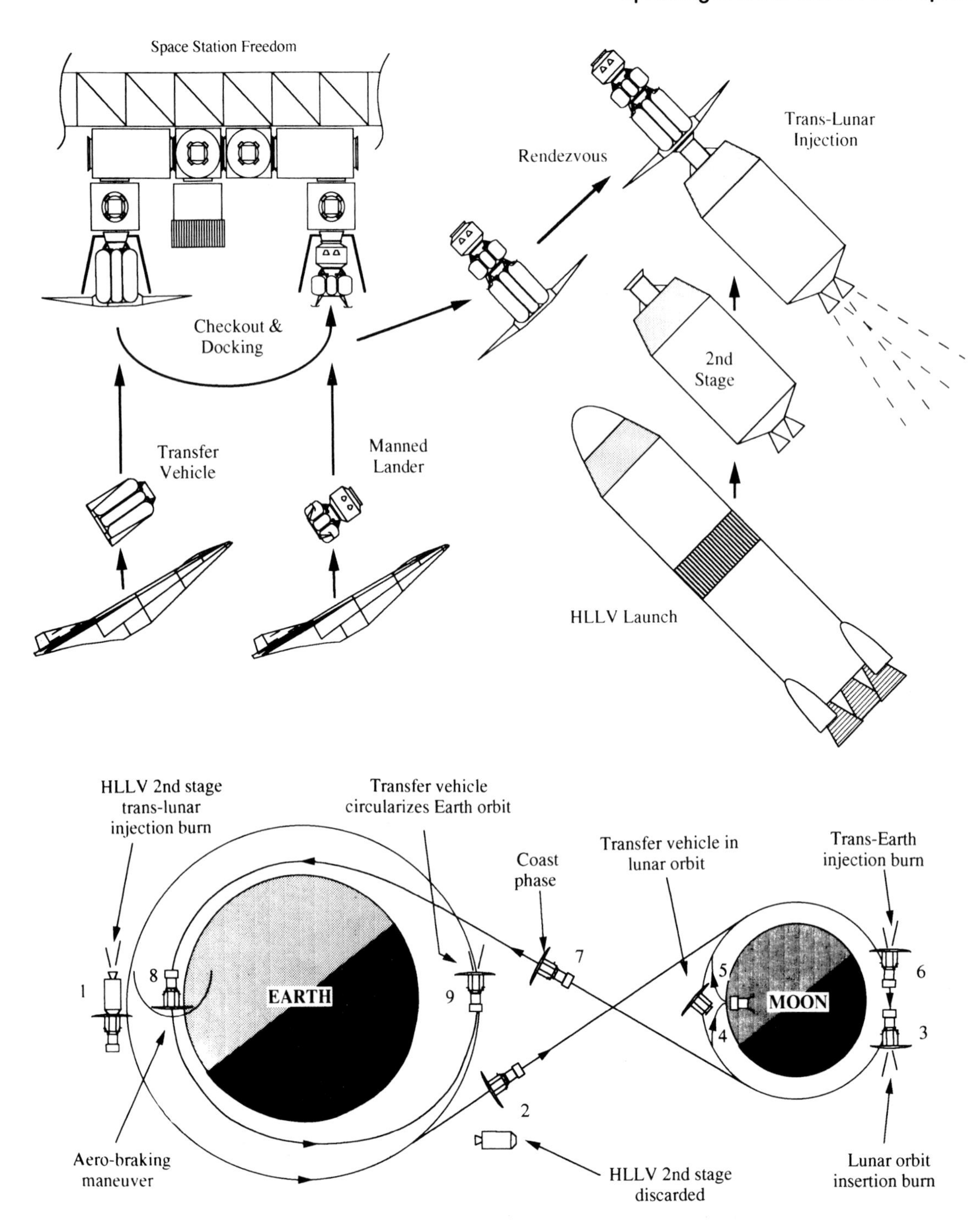


**Figure 19.9** Lunar based crewed missions.

the Freedom truss structure, but instead uses two extra pairs of pressurized modules arranged in an L configuration, with each docked to the station's central nodes. Each of the lower modules would accept either a manned lander or transfer vehicle. In addition, these modules could be equipped with a conical skirt draped around the docked manned lander or transfer vehicle to provide a uniform thermal environment and orbital debris protection [20].

One of the primary reasons for choosing this arrangement is that it allows direct pressurized access to the manned lander and transfer vehicles, limiting any hazardous extra-vehicular activities (EVAs) to unexpected situations, such as the repair of loose thermal insulation. In the case of the manned lander, this arrangement is particularly attractive, as it ensures continuous access to all of the internal systems, allowing systems checks and simulations to be performed before the start of a mission. The transfer vehicle can be designed with all the critical systems located in a pressurized section, perhaps just below the interface between the space station and the transfer vehicle docking ring, to allow some routine pressurized servicing. Other docking ports are provided on these modules for the cargo lander and various lunar base payloads.

To go one step further in the aero-space plane era, lunar landers and transfer vehicles could be serviced at their own dedicated facility. Such a facility would be relatively simple and would not need to accommodate the conflicting requirements of servicing activities and microgravity research. If this type of servicing concept were actually used on Freedom, it would have less disruptive impacts on other station operations compared with concepts using the upper and lower booms, as proposed by the NASA 90-Day Study. Locating heavy lunar equipment on these booms would necessitate EVA activity for servicing activities.

Overall, the reason why the pressurized concept might be feasible is because the transportation components are of manageable size—*they are small*.

### Tolerance to Launch Failures

The critical feature of the lunar base transportation infrastructure is its ability to continue at some reduced level after the catastrophic failure of either the aero-space plane or HLLV. How would this work?

If the aero-space plane fails and the fleet is grounded, access to the Moon would still be possible through use of the HLLV. In a scenario which uses a space station as a servicing node, a manned mission to the Moon would only require one HLLV flight as before, because the lunar lander and transfer vehicle would already be parked at the station. The crew for this mission might already be on the station, or could be launched in a capsule on top of a second cut-down HLLV [21]. This second contingency HLLV would be required, regardless of a Moon mission, due to the need to supply the station with logistics during the aero-space plane downtime, as explained in Chapter 15. If either the manned lander or transfer vehicle were unavailable, they could also be delivered to the station with an HLLV. This is normally an undesirable activity. But if only performed a few times, it is considered an acceptable risk to take, especially in view

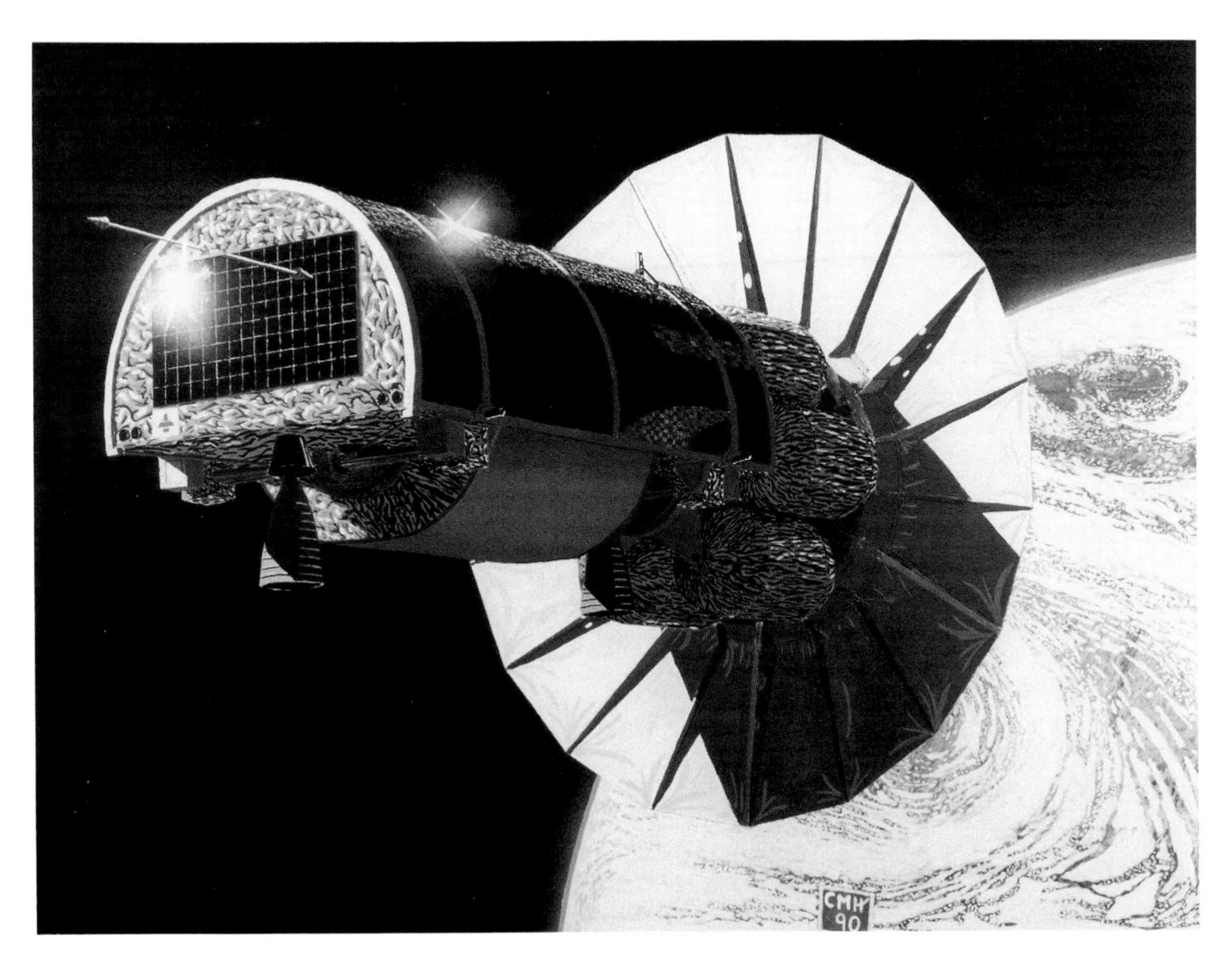


**Figure 19.10** British Aerospace's Orbital Transfer Vehicle (*BAe/Mark Hempsell*).

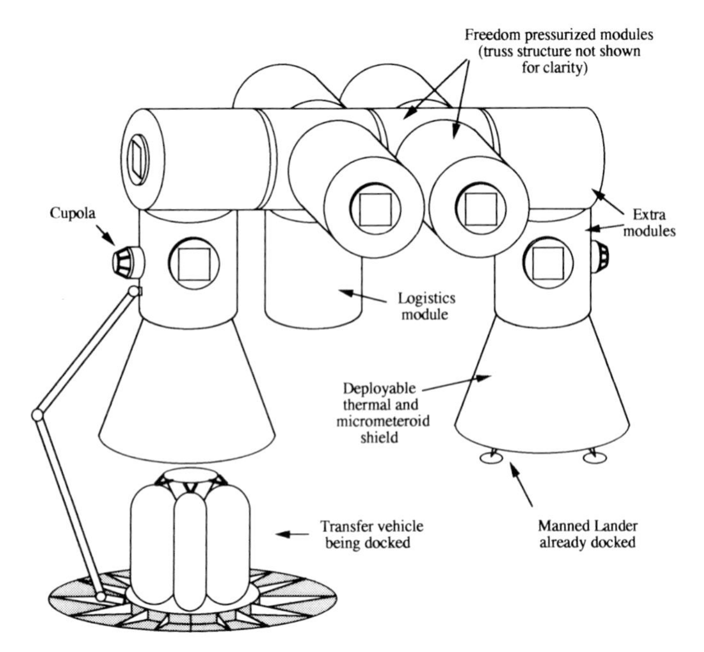


**Figure 19.11** Proposed facilities for servicing lunar transfer vehicles on Freedom.

of the high investment sitting—possibly untended—on the Moon.

Such contingency measures might be avoided altogether if the aero-space plane downtime were only a few months or weeks—a likely situation if it had already accumulated many years of flight experience. Alternatively, if a second type of aero-space plane were available, it could be used in this backup role rather than calling upon the HLLV. Consideration of these issues clearly demonstrates the real operational value of a space station transportation node, because of its ability to reduce the burden on the launch systems. Indeed, a realistic plan for contingency situations might make the need for a space station transportation node essential.

If the heavy-lift launch vehicle fails, access to the Moon would have to be halted for as long as it takes to recertify the HLLV. At first glance this may seem no different than other studies. The crucial difference here is that the HLLV carries only relatively inexpensive payloads, and many identical units are procured every year. As a result, there would be significantly less pressure to keep the HLLV grounded for a long period after a catastrophic failure. For comparison, if the HLLV had to launch one of only a handful of $500 million transfer vehicles and manned landers, as baselined by the Synthesis Group, then there certainly would be considerable pressure to keep the HLLV grounded until the fault had been resolved. If a crew were being launched, then the HLLV would definitely not be launched until the problem was fully solved.

As with the aero-space plane failure above, the risk of re-launching the HLLV after a previous failure has to be traded against the risk to the investment sitting on the Moon. Even though this strategy may immediately endanger a $100 million (for example) worth of HLLV hardware, it might be a small price to avoid jeopardizing a lunar base investment worth many billions of dollars.

### And What About Mars?

Lunar sortie missions can, if necessary, be launched every other week. But because of orbital dynamics, manned voyages to Mars can only be undertaken for a few months every 2 years. For practical purposes, it is physically impossible to regularly support a Mars base with a routine supply of logistics. Therefore, the first Mars missions will have to be undertaken "Apollo-style" as there is no other choice.

The basic difficulty of accessing Mars has profound impacts on way the mission transportation vehicles are launched and assembled, and what type of main propulsion systems are used. In particular, HLLV payloads cannot be classified as cheap or bulk components as for the lunar mission. The propulsion stage is an item that *must* work properly, especially in order to meet the narrow launch window and to safely return the crew from Mars after 1–3 years in space. Therefore, the notion of building a cheap upper stage is meaningless.

As the HLLV must be reliable enough to launch payloads which are just as unique and expensive as those carried by the aero-space plane, then the HLLV might just as well launch the aero-space plane payloads. This is a realistic scenario as assembly of the Mars transportation vehicle can be undertaken at a leisurely pace, ranging from several months to a year or more. This buys the necessary time required to check out and launch the several HLLVs one after another, as well as to provide enough time to recover from a launch failure. (This is as opposed to hundreds of aero-space plane launches.) The aero-space plane would still play a vital role in supporting the construction process. Just as a space station requires maintenance, repairs, and other support needs, so will the Mars vehicle which, if anything, will have even more severe requirements than a space station. Further, the aero-space plane would be used to outfit the Mars vehicle with crew-related consumables, instrumentation, spare parts, and the mission experiments—most of which can be added late into the assembly process.

### Supporting Nuclear Propulsion System Development

Apart from cost, the fundamental objection to using nuclear propulsion is political. This is hardly surprising, given today's relatively high launcher failure rates and instances of nuclear powered satellites reentering the Earth's atmosphere, as occurred with the Soviet Cosmos 954 which split its radioactive core over Canada in 1978. In the United States, nuclear power systems are designed to withstand a catastrophic launch failure. Indeed, it has been reported that such an event occurred during a U.S. military launch, and the nuclear power plant was fished out of the ocean and reused on another spacecraft. Such an approach is only possible for small nuclear systems that produce electricity not in a nuclear reactor, but through the heat given off by the radioactive decay of plutonium. Voyager, Galileo, and Ulysses all used radioisotope thermal-electric generators (RTGs).

By contrast, a Mars nuclear propulsion system will be relatively large and use a nuclear reactor, making protection against failure almost impossible if launched as a single, fully integrated unit on a single HLLV [23]. Alternatively, the availability of an aero-space plane could enable the propulsion system components to be launched on one HLLV, with the radioactive core launched on one or more aero-space plane flights. This might an acceptable arrangement if the aero-space plane has already been in service a number of years and has undertaken hundreds of successful missions, as well as demonstrated various safe aborts—a situation reminiscent of the regular transport of nuclear material by aircraft. It may also be possible to shield the core from a catastrophic aero-space plane failure.

Once in orbit, telerobotic vehicles would be used to install the core, with the procedures for such activities demonstrated long before a live core is ever launched. Further, the aero-space plane could be used to test out a subscale boilerplate version of a nuclear propulsion system. This would avoid the alternative of building a very expensive, fully space-rated nuclear propulsion system just to perform basic demonstration tests, as would be required if only expendable boosters were used. Potentially, the cost and time needed to develop a nuclear propulsion system would be reduced compared with conventionally launched nuclear systems.

### Observations: Solar System Exploration in the Aero-Space Plane Era

Would the use of aero-space planes reduce cost estimates as high as $400 billion over 30 years for solar system exploration? An opinion on this matter is expressed by Gordon Woodcock with respect to lunar bases, [24]

> Given the development of a fully reusable second-generation space transportation system, the cost of operating the transportation system and conducting an aggressive program to permanently occupy and settle the Moon is about what was being spent on operating the space shuttle before the Challenger accident. Manned lunar operations need not be relegated to the "far future", that is, beyond today's planning horizons.

It would seem that aero-space planes would significantly reduce costs compared with conventional programs centering around a single type of expendable heavy-lift rocket. Perhaps a more relevant question to ask is, can solar system exploration be conducted *without* aero-space planes?

Aero-space planes could significantly simplify both the preparations for, and the implementation of, solar system exploration. Undoubtably, HLLVs could be used to return crews to the Moon at some point in the future, but it would appear that without the supportability capabilities of aero-space planes, such endeavors would be all but impossible to do in anything other than self-contained, Apollo-style lunar sorties, except with longer stay times. As such missions are likely to be at least as expensive as Apollo, serious questions must be asked whether it is really worth spending all that money just to "out-Apollo, Apollo."

True exploration, whether on the land, sea, or in space, is an intrinsic aspect of human activity. But it should not take decades to achieve, nor should it be as phenomenally expensive as current estimates seem to suggest. Exploration is important, *but not at any price*, and especially not if that price precludes future exploitation of any discoveries. For example, it has been proposed that He3 be mined on the Moon to fuel fusion reactors on Earth [25]. Such a potential is clearly ludicrous if each lunar mission cost several billion dollars and if only a handful of these risky and complex missions can be performed each year, as seems almost inevitable in an HLLV-only scenario. In the first place, it is doubtful whether the type of systems needed to process the vast quantities of lunar materials to extract He3, and then return it to Earth, could be landed and supported on the Moon. It would be far cheaper to spend a fraction of that money on fusion reactors (or other forms of energy) which can be made to work using fuels available on Earth.

The motivation of using "human exploration as a tool for peace" stands on similarly shaky ground. There are far more practical ways to spend such vast sums of money on Earth for strengthening the friendship between nations. The plight of the people in the former Soviet republics, Africa, South America, and other places in the world will be little enhanced by the knowledge that one of their own has been photographed on Mars. The C.I.S., for example, desperately needs investment in infrastructure (roads, telecommunications, and power generation) if its economy is to grow in the future. Its situation would be little improved by a trip to Mars.

If exploration for the sake of exploration is the true justification, a far cheaper alternative is to use robotic probes, if that exploration has to be restricted to using HLLV boosters. If the introduction of aero-space planes can significantly reduce the cost of solar system exploration, then the time might come when the cost difference between manned and robotic missions converges. This may be a long way off, but it seems clear that if it is to happen at all—as indeed it probably must if solar system resources are to be utilized in the future—an aero-space plane will have to fly first.

## Summary—Building an Infrastructure That Lasts

In the years 1961–62 the raging debate within the U.S. space community was which strategy the United States should adopt to land Americans on the Moon. One particular strategy would have involved the construction of one or more space stations in low Earth orbit. These would have then been used as ports for assembling the ships that would subsequently head off for the Moon. The Earth orbit rendezvous or EOR scenario was pursued vigorously by Wernher von Braun because after the Moon landings were completed there would be something left over *in space* to show for their efforts—the beginnings of an in-orbit infrastructure [26].

EOR was not chosen. Apollo began. It triumphed over the Soviets in July 1969. Apollo finished. And by July 1975 there was nothing solid left to show for the effort; no rockets, no space stations, no infrastructure—just some new technology, spectacular photographs, a few Moon rocks, and, above all, wonderful memories of a time when humans wandered beyond the Earth for the first and only time.

If the nations of the world intend to one day return to the Moon or venture on to Mars, but are forced to base their plans around enormous and expensive one-shot throwaway rockets, it almost certainly won't happen for all the reasons described earlier. HLLV boosters alone are inadequate. If such a program were to happen, then there would be a profound danger of repeating the Apollo experience. An HLLV-based solar system exploration program is incapable of building a useful and easily supportable in-orbit infrastructure. The very fact that the Synthesis Group report suggested that lunar missions should be launched *directly* from the Earth without any intermediate space station steps, is an apt case in point.

Aero-space planes offer a different solution. As Part I shows, their ability to fly regularly, reliably, and safely to and from orbit would have a profound effect on the way space stations are designed, developed, and deployed. Most critically, aero-space planes would also enable regular and timely support of such space stations. Resupply and contingency repair missions would not have to be planned for years in advance, as with the Shuttle-supported Freedom, but could be undertaken as a matter of course. When a station failure occurred, the on-demand capabilities would allow responsive action to be taken, as necessary. Such a capability is fundamental to any future commercial product-making activities, as will be discussed in the last chapter.

Monthly or even weekly flights of aero-space planes would enable experience to be rapidly accumulated in what it really means to live and work in space. The ability to perform technology demonstrations would, in particular, also allow experience to be gained in the servicing of vehicles in space. Performing on-orbit servicing of spacecraft today is clearly a nonstarter. Orbital servicing is not in itself inherently difficult, but the high cost and limited access to space simply preclude it. The Shuttle has only serviced *two* satellites in space in its first 10 years.

Aero-space planes would provide the first real experience in servicing spacecraft on orbit. Such experience must be obtained before any decision can be made on whether or not a space station node is required. From the discussion in Part II, it would appear that one benefit of such a node is its ability to reduce the total launch requirements for every single lunar mission. As the lunar base scenario showed, once the transfer vehicle and manned lander have been delivered to a station, each manned lunar mission would necessitate only one "payload-less" HLLV and a crew delivered by a regularly scheduled aero-space plane flight. Further, aero-space planes working in combination with HLLVs would ensure continuous, although degraded, access to the Moon if either launcher failed catastrophically. Such failure tolerance is considered mandatory for all rational space activities, especially where humans and multibillion dollar investments are concerned.

If aero-space planes can be built, then they seem ideally suited for constructing and supporting space stations, and then helping to expand the human infrastructure to the Moon and later Mars in a cost-effective manner. Aero-space planes would put humans in the fourth domain *to stay*.

## References and Footnotes

1. If the total cost to develop and procure an operational Interim-HOTOL fleet were $10 billion (Chapter 7), then this would be equivalent to just 4 years of Shuttle operational funding for Freedom.
2. Isbell, D., "NASA Trims Costs, Complexity of Station," *Space News*, March 25–31, 1991, p. 4.
3. Hunter, M., "The SSX: SpaceShip Experimental: Draft II," March 11, 1989, p. 25.
4. The major problem is that the ET's spray-on foam insulation lacks the durability needed to survive many years in space. Further, this insulation outgases as it breaks down, leading to a contamination problem around the ET. Solutions to all these and other problems are available, but it seems clear that using a basic ET will probably not be practical. See also Matthew A. Bille's article "External Tanks: Instant Space Stations," in *Ad Astra*, June 1991, p. 12.
5. Hannigan, R. J., "The Economic Impacts on Space Operations of Future Launch Systems," IAA-89-697, *40th Congress of the International Astronautical Federation*, Malaga, Spain, October 7–13, 1989, p. 12–14.
6. An interesting discussion of a possible European space station built and supported by a HOTOL-like vehicle and Ariane 5 is provided by

Parkinson, R. C., and C. M. Hempsell, "Cost Impacts of Supporting a Future European Space Station," IAF-90-602, *41st Congress of the International Astronautical Federation*, Dresden, Germany, October 6–12, 1990.

7. Synthesis Group, *American at the Threshold*, (U.S. Government Printing Office, Washington, D.C., May 1991), p. 31.
8. The term *gravitation well* is sometimes used to describe the degree of difficulty of launching a payload from a planetary body or moon. The deeper the well, the more energy required to get out of it.
9. See, for example, Mendell, W., ed., *Lunar Bases and Space Activities in the 21st Century*, (Lunar and Planetary Institute, Houston, Texas, 1985), pp. 109–186, for an overview of general transportation issues for lunar bases.
10. NASA, *Report on the 90-Day Study on Human Exploration of the Moon and Mars*, November 1989, pp. 3–16.
11. Hannigan, R. J., and D. C. Webb, "The Space Exploration Initiative and the Aero-Space Plane Launcher," presented at the *Twenty-Eighth Space Congress*, Cocoa Beach, Florida, April 23–26, 1991, pp. 2–3.
12. *America at the Threshold*, pp. 65–66.
13. Kirkpatrick, J., R. Simberg, and J. Bryan, "Spaceplane: The Key to a New Era of Space Utilization," AIAA-91-5083, *AIAA Third International Aerospace Planes Conference*, Orlando, Florida, December 3–5, 1991, p. 2.
14. Much of this information is derived from the analysis contained in "The Space Exploration Initiative and the Aero-Space Plane Launcher," as well as in, Webb, D. C., and R. J. Hannigan, "Economic and Socio-Political Impacts of NASP-Derived Vehicles: A Technical Report," *The International Hypersonic Research Institute*, November 1990, pp. 86–93.
15. See, for example, "Lunar Bases and Space Activities in the 21st Century," and Sanders Rosenbery, "A Lunar-Based Propulsion System," p. 169.
16. This discussion emphasizes the importance of having the largest possible payload bay cross section in aero-space planes.
17. Conchie, P. J., C. M. Hempsell, and R. C. Parkinson, "Potential Evolution of an International Moon Base Programme," IAF-90-641, *41st Congress of the International Astronautical Federation*, Dresden, Germany, October 6–12, 1990.
18. "Making Drag Pay," *Flight International*, May 7, 1983, p. 1213.
19. Discussions with Gen. Thomas Stafford, chairman of the Synthesis Group.
20. It is interesting to note that the orbital motion of this debris is such that protection is only needed from the sides as any piece of debris coming from even slightly underneath the station would probably have intercepted the Earth's atmosphere at an earlier point in its orbit.
21. Most HLLV booster concepts use strap-on liquid or solid boosters. Therefore, "cut-down" refers to an HLLV essentially without these boosters.
22. *America at the Threshold*, pp. 22–23.
23. See Henderson, B., "New Thermal Propulsion Gains to Speed Rocket Production," *Aviation Week & Space Technology*, January 20, 1992, p. 20, and Asker, J., "Moon/Mars Prospects May Hinge on Nuclear Propulsion," *AW & ST*, December 2, 1991, p. 38.
24. Woodcock, G., "Logistics Support of Lunar Bases," IAA-86-511 *37th Congress of the International Astronautical Federation*, Innsbruck, Austria, October 4–11, 1986, p. 21.
25. *America at the Threshold*, pp. 52–53.
26. Aldrin, B., and M. McConnell, *Men From Earth*, (Bantam Books, New York, July 1989), p. 89.

# Chapter 20

# The Promise of Commercial Space

## Not for Lack of Trying

Under present circumstances, the full-scale commercialization and industrialization of space cannot be even considered as a potential, *it simply will not happen*. The failure of the commercial space industry to materialize is not due to a lack of imagination or pioneering instinct on the part of entrepreneurs, nor is it due to an unsympathetic regulatory environment. A space industrial revolution will not occur as long as the current generation of unreliable, unavailable, essentially one-directional, and, above all, unaffordable launch systems remains the sole means of accessing space.

It is a remarkable fact that after nearly three and a half decades in space the only truly *self-supporting, profit-making, commercial* activities [1] not dependent on government markets or subsidies are those associated with communications. Beyond communications, attempts have been made to initiate commercial space ventures, but few have come close to succeeding, nor are any others likely to in the foreseeable future except for a few small satellite-based remote sensing projects. But it is not for lack of trying.

Are aero-space planes what the space entrepreneur is looking for?

## Enabling Economic Growth

Are these comments fair? It is possible to argue that even though the vast majority of the present uses of space are beyond the boundaries of commercial enterprise, this does not mean they are uneconomic. Weather satellites are an inexpensive way to provide advanced warning of severe and threatening weather. The formation and path of a hurricane can be precisely monitored over many days, giving accurate and forward predictions of where and when landfall will occur so that coastal residents can take appropriate measures to protect their property and evacuate the endangered area. Without this advanced warning, many more people would perish, properties would be destroyed and businesses would be ruined, and the cost of cleaning up the mess would be far beyond the couple hundred million dollars needed to place a weather satellite in space.

Similar arguments can be made for remote sensing of the Earth's resources or the observance of pollution in the atmosphere. For major industrialized nations and, indeed, the world as a whole, the cost of *not* having this global strategic awareness of the planet would probably be far greater than cost of installing the satellite systems. Realistically, though, it would be all but impractical to try to make money directly from selling this type of global information today while, at the same time, having to recover satellite production, operation, and launch costs.

Weather satellites and resource monitoring systems are, without a doubt, economically beneficial capabilities that came about through a rational process to meet a relevant need. The demand manifested itself, and through non-commercial means, a capability was provided. Therefore, a seemingly logical conclusion is that because there isn't currently a market-driven demand for commercial space products, there won't be a commercial space revolution. No demand, no market. Such logic is basically flawed and, if nothing else, is a premature misrepresentation of the situation. Such a conclusion cannot be drawn before understanding *why* the demand for commercial space is so very weak. If commercial exploitation of the land, sea, and air is so diverse and widespread, why should the fourth domain be any different?

Initially, the fundamental problem is the lack of an efficient infrastructure, where transportation is the first order of concern. Few Earth-bound commercial companies of any description would exist today without the availability of affordable and reliable two-way transportation. Just because space is more difficult to access for technical reasons doesn't mean that commercial space companies won't also be intrinsically dependent on comparable user-friendly transportation infrastructures.

On Earth, commercial companies take advantage of government investment in building and maintaining the infrastructure. Yet, only commercial opportunity and the needs of the marketplace will ultimately lead to the most important benefit: *economic growth*. The growth of the United States, Western Europe, Japan, and other major industrialized nations has been *enabled* essentially due to the multitude and diversity of commercial activities. The collapse (non-growth) of the Soviet Union and the Eastern block nations was due largely to their centralized authority planning structure. The infrastructures that resulted were neither efficient nor robust, and they could not respond to the needs of these countries [2]. Most importantly, the former Soviet and Eastern Block infrastructures were not driven by the positive feedback mechanisms of commercial exploitation. The high price of German reunification is due to the restoration and expansion of the former East German infrastructure so that commercial activities can now be conducted in this former communist state. The C.I.S. and most of the former Eastern block nations recognize the fundamental role infrastructure plays in growth-enabling commercial activities.

The long-term survival of the space industry might well depend on the success of commercial space. The power of commercial space is its potential to enable sustained expansion and growth of the space industry. Not insignificantly, commercial activities reduce the space industry's dependence on the whims of government decisions, enabling more efficient program funding and schedules. As one top government space official said, "Growth will only come about with the emergence of significant non-government dependent markets. Thirty years of government dependence and resulting politicization of the technology selection process has resulted in government dominating 90% of space industry output."

Without the stimulating impacts of a diverse commercial space industry, it is even conceivable that the benefits of today's non-commercial space applications might eventually be superseded by cheaper, non-space alternatives capable of meeting the same ends. Ground-based radar systems are making steadily greater contributions to weather forecasting, for example.

*In the "New World Order," prestige is unlikely to be enough to sustain space efforts.*

## Making Products in Space

The success of commercial communications satellites is due largely to the fact that, once the satellite is launched, the nature of the business does not require the exchange of products between space and the ground, other than electromagnetic radiation. Moreover, all communications satellites require a single *one-way* launch, after which they generate revenues for up to 10 years. If comsats necessitated one or more launches every year, there wouldn't be any commercial comsats today (Figure 20.1).

**Figure 20.1** Communication satellites: The only truly commercial space industry (*Aerospatiale*).

This is precisely why the commercial manufacture of products in space has failed to take off. Opportunities to return products manufactured in space to Earth are few and far between, and when they do arise, either they are restricted to very small payload masses or the cost to launch and return larger payloads is immensely expensive. This became all too clear in the latter half of the 1980's, in no small part because the Space Shuttle didn't, and couldn't, meet its original cost and performance goals (Chapter 11). The early 1980's told a different story. Buoyed by the initial success of the Shuttle and the enthusiasm of President Reagan, some organizations believed that the era of the commercialization of space had finally arrived, and many people believed it would be from here that the next billionaires would emerge [3]. As President Reagan said during his January 25, 1984, State of the Union Address in which he announced the Space Station program,

> A space station will permit quantum leaps in our research in science, communications and in metals and life-saving medicines which can be manufactured only in space.

A Center for Space Policy (CSP) report, which was liberally splashed all over the pages of a special issue of *Aviation Week & Space Technology* in June 1984 [4], claimed by the year 2000, the *annual* gross income of all commercial space activities could be as high as $65 *billion*. Notably, the majority of this revenue was anticipated as coming from products manufactured in space, with the breakdown being:

- Pharmaceuticals $27 billion
- Semiconductors (e.g., gallium arsenide) $ 3.1 billion
- Glasses $11.5 billion

It was estimated that other contributions would include $15 billion for expanded and advanced communications services, $2 billion for remote sensing, $1 billion for space transportation services, and some $3.7 billion for aerospace support services. As Joan Johnson-Freese explained, [5]

> In the heat of the moment and in support of commercialization, some advocates seemingly got carried away with their predictions of potential resources and often forgot to mention the probable time lag between investment and return. These fuelled almost certainly unreachable expectations for individuals and companies.

Today, if the definition of *commercial* is stretched a little, the actual (1990) revenues from commercial space in the United States are only $3.5 billion (about $2.5 billion in 1984 dollars), and perhaps another $1 billion could be added to this by foreign commercial space sales [6]. Of this money, *none* of it comes from materials processing activities, except for a few millions of dollars resulting from universities and companies purchasing services from commercial companies for basic research. Most of the financial resources for these limited activities come directly from government grants and contracts.

Materials processing is not the only commercial "failure." The 1990 annual revenues from the commercial Landsat operations were $30 million, while for Spotimage they were about $32 million—both figures are only sufficient to cover the yearly operations of the two companies, and little goes to the construction and launch of the satellites. According to the president of EOSAT Arturo Silvestrini in a *Space News* interview, [7]

> We have not even started the test of commercialization, which will begin when Landsat 6 is launched. That is the first real misconception: commercialization has failed. It has not failed, because the effort has not yet started. Commercialization does not exist yet in remote sensing.

Commercial communications services can hardly be described as a "failure" with U.S. revenues for transponder leasing remaining steady at about $1 billion per year, with a further $850 million for ground station leasing [8], but it is a far cry from the CSP projections of $15 billion. Interestingly, the only figure which has been more or less met is launch services, which, if Arianespace is included, comes to approximately $1.1 billion.

CSP was not the only organization to make optimistic estimates of the profitability of materials processing. But regardless of "who said what," the failure of materials processing to make the slightest dent in these stupendously overoptimistic expectations has led many to believe that such activities are not inherently commercially viable. Indeed, within some space industry circles, commercial materials processing in space is seen almost as a taboo subject, and an activity reserved either for the lunatic fringe or for people that haven't been awake for a good few years. As George van Reeth notes with regard to the debate over whether ESA should build the Columbus laboratories, "The totally useless thing that we do is Columbus. No man in his right mind would put that sort of investment on the probability that microgravity is going to get somewhere [commercially]."

Such attitudes are understandable under present circumstances. Nevertheless, they tend to cloud opinions on an industry which, apart from a handful of experiments, hasn't had the slightest chance of proving its worth. Given a true space transportation system, and the ability it enables for finding out what it actually means to make things in space, might the predictions of profit-making commercial factories in space eventually turn out to be not all that far away from the truth?

## Research First, Profits Later

Imagine a laboratory on the ground that could shield experiments from the constant tug of gravity, as well as maintain an almost perfect vacuum, and imagine that access to such a laboratory would be taken for granted—like the turning of a door handle. If it were possible to build such a laboratory on the ground, it would clearly see considerable and widespread use.

Of course, the only place where such conditions exist is in a laboratory circling 300 kilometers above the Earth at 28,000 kilometers per hour, and accessing space today is more difficult and expensive than just turning a door handle. Nevertheless, the analogy is still valid—the easier and cheaper space access becomes, the greater will be the potential for microgravity research. This is a first step toward determining whether eventually it will be possible to manufacture commercially viable products in space. *No matter how promising microgravity is, without intensive and extensive research, commercial activities will come to nothing.*

In medicine, for example, the absence of convection mixing, caused by the combination of both gravity and even slight differential heating, is beneficial to the separation of certain types of cells from other similar cells through electrophoresis techniques—a process in which an externally applied electric field is used to separate substances that have minute differences in electrical charge [9]. In a one g environment, convection mixing swamps these small electrostatic differences between like substances. Conversely, microgravity conditions also facilitate uniform mixing of substances for the growth of fragile protein crystals. In one g, a few crystals are all but impossible to grow into the uniform, mono-crystalline structures needed for x-ray crystallography analysis (Figure 20.2). According to Paul Todd, a research professor at the University of Colorado at Boulder, [10]

In the case of liquid-liquid diffusion, larger crystals have always been obtained in space in equivalent amounts of time. However, it is so much easier to create satisfactory growth conditions in space that a valid comparison is very difficult to make. It seems to be impossible to grow crystals by this method as rapidly on earth as in space.

In materials development, the lack of gravity means that metals which cannot be mixed together in a one g environment because of differences in their density (i.e., buoyancy), can be successfully mixed to form new, exotic materials with special properties, rather like oil and water. Also, it might be possible to make extremely lightweight but very strong foam metals out of conventional metals simply by injecting small air bubbles into the molten material. This is something obviously impossible in one g because air bubbles tend to float to the top or take up nonuniform sizes and shapes that can cause material weaknesses.

Other materials activities not only benefit from microgravity but also from the almost total vacuum to facilitate containerless processing, where materials being processed are suspended away from the walls of a container either magnetically or using ultrasonics. In this way contamination can be minimized, something particularly important to the production of thin-film substrates for advanced microelectronics and sensors. In one g, it is difficult (but not impossible) to build up multilayered thin films, but the yield tends to be small due to contaminants. This is why gallium arsenide and other semiconductors are not in wider-scale use and are much more expensive than silicon-based semiconductors despite their high performance properties. These are just a few examples, and because of the depth and breadth of the subject, it is not considered appropriate to go into further detail here.

Research is fundamental to all commercial activities involving the manufacture of a product, whether on the ground or in space. The "failure" of commercial materials processing may seem all the more remarkable given that large amounts of government money are already being spent on microgravity research activities. A number of Spacelab missions—each costing several hundred million dollars not including launch—have been largely dedicated to materials processing work, including the International Microgravity Laboratory and the German Spacelab D missions (Figure

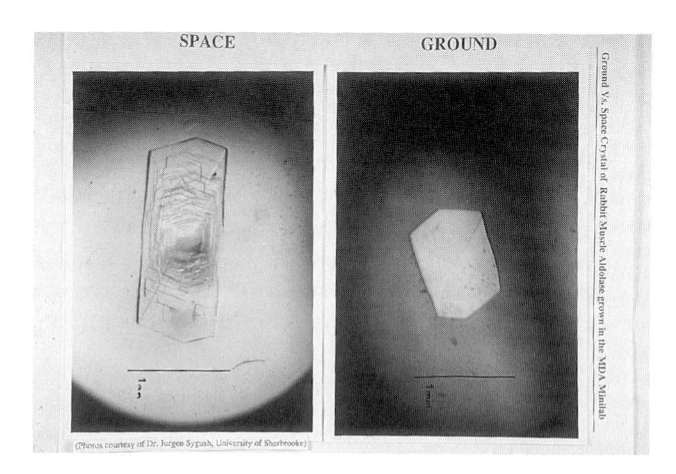


**Figure 20.2** Comparison of protein crystals grown on the ground and in space (*ITA*).

**Figure 20.3** STS-9/Spacelab 1 in 1983 (*NASA*).

20.3). Further Spacelab missions are being planned as precursors to the Space Station. Every time a Space Shuttle is launched, it carries a number of small experiments in the middeck lockers or in Get-Away Special canisters located in the payload bay. A major justification for the $40 billion Space Station Freedom, the $5 billion Columbus, and the $2 billion Japanese JEM programs is materials processing research, all of which were initiated partly by the distant hope of one day undertaking commercial work on these facilities. One of the primary functions of the Soviet Mir is to act as a materials processing laboratory, and it has been reported that the Soviets have actually produced useful semiconductors and other materials in space [11]. From all these examples, it is easy to conclude that materials processing is an extremely well-funded research activity, especially if launch costs are included. Then this begs the question, where are the commercial products to justify this enormous expense?

The problem, initially at least, is not one of money, but

revolves around the way in which research is performed today in space. Generally speaking, to be able to determine the commercial viability of any activity, it is clearly necessary to have the least number of restrictions to get into, as well as return from, space. Current space transportation systems are incapable of responding to the needs of most commercial users. Typically, it is sometimes necessary to wait several *years* before a launch; the date of that launch might unpredictably slip weeks, months, or even years; and opportunities for repeat flights may be several more years and delays away. Even if governments paid for everything, as they more or less already do, this is clearly an inefficient method of conducting research and contrasts sharply with Earth-based laboratories where access to the experiment is taken for granted.

In the case of private sector investment in commercial space, the situation is even easier to understand. First, even though governments must subsidize launch costs, there is certainly no such thing as a free launch. The long lead times and frequent delays invariably mean that funding has to be maintained just to babysit the payload and pay the salaries of the research teams. Many privately financed space companies have been destroyed by this situation. Second, as nearly all of these companies depend on government funding and launch opportunities, their management is compromised by the conflicting operating practices of the commercial and government sectors. While waiting for launch, companies must meet the demanding standards set by government operations, regulations, and safety reviews. Finally, even after all this has been settled, the government can still delay launches or bump payloads, with no guarantee of when the next launch opportunity will occur. Guaranteed and timely access to space will be mandatory for future commercially orientated research.

The experience of McDonnell Douglas and its Electrophoresis Operations in Space (EOS) program is a good example. After an investment of around $20 million, and seven free flights over 3 years of a rack-sized experimental unit in the Shuttle middeck, McDonnell Douglas was preparing to launch its full-scale pharmaceutical production facility to make enough material for clinical trials and thereby pave the way for commercial operations. Then the Challenger accident ensured that this large facility would remain on the ground for at least 3 years, making the program untenable because of the cost of maintaining the hardware and teams together for this uncertain period of time. In addition, McDonnell Douglas's partner in the venture, the pharmaceutical company Johnson & Johnson, pulled out as an alternative and less expensive way was found to separate the erythropoietin on the ground—a drug used in the treatment of anemia [12]. If McDonnell Douglas had had the opportunity to fly its EOS middeck hardware several years sooner, and every 2 or 3 *weeks* for a year, followed *immediately* by several flights of its large payload bay processor, then the story of the first U.S. attempt at making drugs in space might have been different. (Figure 20.4).

Any potential investor looking at a new commercial space endeavor that is intimately dependent on today's government controlled and operated launchers, and also prone to being overtaken by less expensive terrestrial-based technology advances, might be forgiven for believing that the risk of the enterprise was significant and the period before a payback would be many years into the future. The space community has often criticized the unwillingness of commercial investors to finance its programs. But, in this light, is it any wonder? As a survey conducted by *Space News* and KPMG Peat Marwick noted, [13]

> What excites space zealots—advanced technology, futuristic thinking and grandiose market projections—ultimately will hurt their chances of raising money. What matters to investors is whether any new business will provide financial returns that far outweigh the risks . . . investors will be attracted by commercial space ventures that are led by credible managers and take advantage of proven, low-risk technology.

*Proven* and *low-risk* are hardly appropriate descriptions of the current status of space manufacturing.

## Spacehab and Other Examples

A pertinent example in this discussion is the Spacehab program, a privately financed, $100 million venture to build a small pressurized module that sits in the Shuttle cargo bay just behind the flight deck and provides additional locker space for various payloads. Spacehab was originally proposed as a fully commercial program that would pay for itself by selling payload accommodations space within the module. In view of the backlog of small payloads waiting for launch and needing the equivalent of a Shuttle middeck locker, Spacehab was conceived to capitalize on the Shuttle's undercapacity. Unfortunately, even though there are many potential payloads patiently awaiting a launch, few of the owners were able to commit to paying the $2 million for a locker-sized portion of something that did not yet exist. This, in turn, made it difficult to pull together the kind of solid market base needed as a prerequisite to investors releasing capital to fund the project. Instead, to keep Spacehab alive, NASA agreed to be an anchor tenant and pay Spacehab for roughly two-thirds of the space on each of the first six flights [14]. Therefore, in order to recover the cost of each mission and recoup some of the original investment, Spacehab has to sell the remaining capacity to commercial users. This agreement—itself a remarkable triumph for this new company—allowed the private funding to be released,

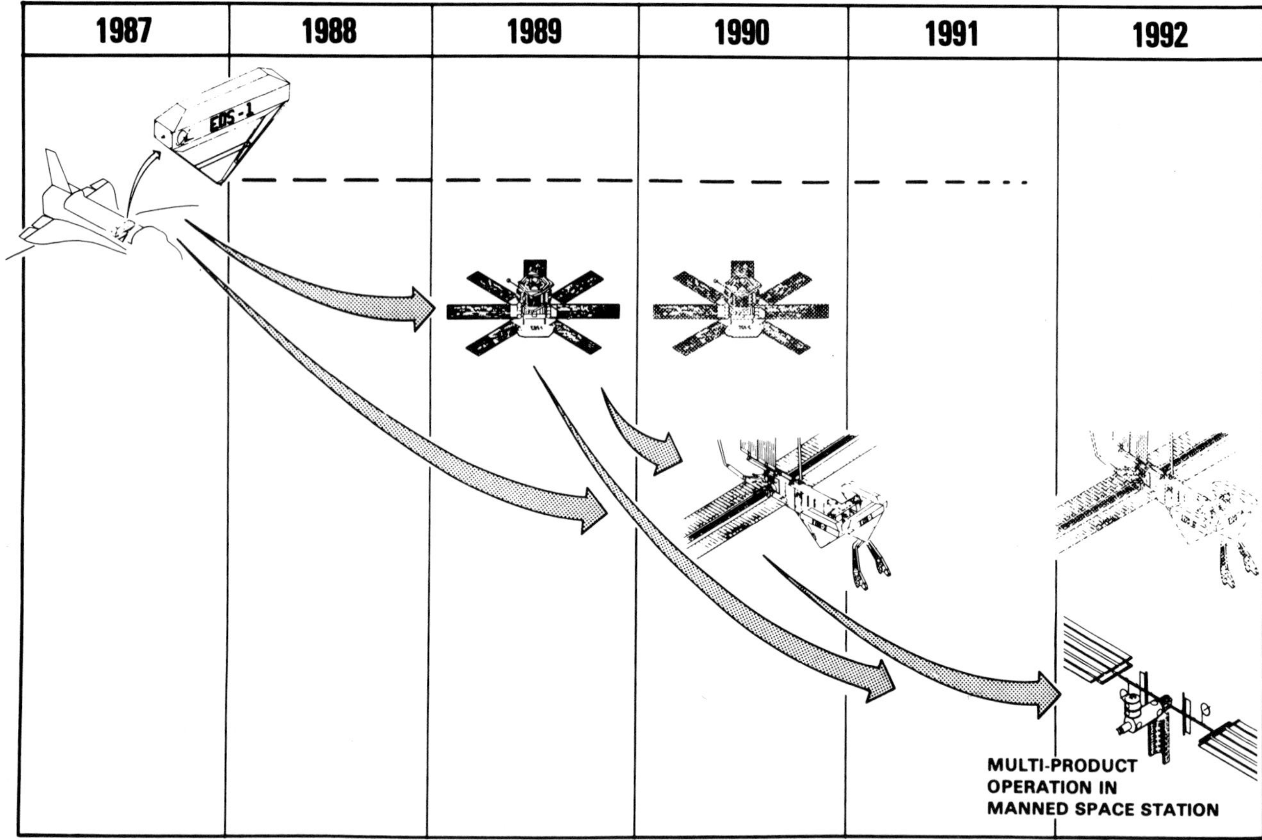


**Figure 20.4** 1983 proposal for pharmaceutical manufacturing in space (*McDonnell Douglas*).

and the first Spacehab module was built and successfully flown in 1993.

Even this agreement and a commitment to fly, Spacehab is still encountering problems to secure enough customers just for the first mission. The reasons are simple. The vast majority of payloads flown in the Shuttle middeck are one-off free flights that take advantage of space as and when it becomes available. These payloads are one-offs because the launch opportunities are so infrequent and so unpredictable that to perform any commercially oriented research would be impractical if not impossible. Building a business based on two annual, and frequently delayed, research flights of a small experiment is inadequate and ponderous for the fast-paced dynamics of the marketplace.

In addition, why should universities or companies pay $2 million and wait a few years for a Spacehab locker when they could wait a few more years and get a free launch for their experiments? The 3M Company was for a while the most prolific commercial company using space for thin films and other materials processing research. It is unlikely to pay for many lockers on Spacehab because its Joint Endeavor Agreement with NASA already provides up to 62 free flights of small payloads in the Shuttle middeck and Get-Away Special canisters [15]. It is also important to point out that 3M is performing space research primarily as a means to help understand its ground processing activities. Even this seasoned R&D company has recognized that much space research has to be performed long before the first products can be manufactured in space and sold at a profit [16].

There are some exceptions, but truly commercial users of Spacehab will be few and far between. These statements are not intended to detract in any way from Spacehab itself, which is considered a valuable facility for maximizing every dollar spent in launching the Shuttle. It has also acted as a precedent-setter for future programs by exposing problems associated with private company involvement with government agencies.

Previous proposals for quasi-commercial microgravity experimentation services include the man-tended Space

Industries Industrial Space Facility (ISF), the Fairchild Leasecraft, and the joint DASA-GE Astro Space Amica platform. Notably, the last two projects were more than just paper studies. Indeed, the Leasecraft was to use the already space-proven Multimission Modular Spacecraft bus, while the Amica is basically a modified version of the ESA Eureca platform (Chapter 17). Yet, the possibility of any of these or similar programs succeeding without governments partially or fully covering their costs seems remote in the extreme.

## The Impacts of Aero-Space Planes

Aero-space planes are not a magic cure for the commercial space industry, nor are the justifications for aero-space planes particularly contingent on new commercial space activities arising. Many other regulatory and institutional problems also stand in the way. However, it is certainly considered that aero-space planes have the potential to remove the *fundamental* inhibitor to commercial materials processing: regular access to and from space. Those policy-related problems might be expected to become less of the hindrance they seem to be today once it becomes clear that there is money to be made in space.

Aero-space plane-type vehicles are certainly a prerequisite to new commercial space activities simply because their capabilities are more in line with what the user needs from a space transportation system. As Bob Parkinson notes, [17]

> Interest in microgravity research and products exists, but successful exploitation will depend on repeated, regular, reliable access to space (with payload recovery), and low transportation costs. Until these features have been demonstrated, the number of committed users will be small. The development of a low cost, reliable, reusable launch vehicle appears to be a necessary prerequisite for such a market.

Instead of the government or private investor being asked to spend a lot of money over many years on a commercial venture that might or might not be proved viable after only one or two launches, a much smaller amount of money would be spent over a shorter period of time on a larger number of launches. In addition, routine access to space also facilitates simpler spacecraft design, reducing costs further. Aero-space planes, therefore, seem to offer the potential for commercial research space programs to be broken down into simpler, cheaper, smaller chunks. This situation is the same as in the discussion of technology demonstrations in Chapter 17.

In order to understand how aero-space planes might initially impact materials processing research in space, both as a technology discipline in its own right and with the incentive of eventually producing saleable products, the following discussion outlines a possible scenario.

### Phase 1: Basic Research

Ideally, initial research activities should be undertaken as inexpensively and as frequently as possible, allowing experiments to be constantly revised and refined. Today, the best that can be done is using aircraft flying parabolic trajectories, providing about 20–30 seconds of low gravity for various sized experiments, and suborbital sounding rockets, providing 6–12 minutes of microgravity for small-sized experiments [18]. Dedicated costs can range from about $0.5 million for aircraft, to $1.5–5 million for a sounding rocket. After that, the costs rise and schedules lengthen dramatically because of the need to use the Space Shuttle, orbital recovery capsules, and Mir.

The aero-space plane could provide a capability first proposed for the original Shuttle and discussed at length in Chapter 17. It is possible to envisage a scenario where first a breadboard experiment is placed inside a pressurized and robust container, thereby avoiding the expense of having to fully space qualify the experiment apparatus. This container would then be bolted to a standard platform, integrated to the aero-space plane's standard payload canister (Chapter 4), and launched. In space, the experiment could remain attached to the aero-space plane for the mission duration of maybe a few hours to a few days, and then returned to Earth.

Alternatively, the platforms could be deployed, and recovered by another aero-space plane a few weeks later. If the mission is short, the platform could be kept simple and use high-energy disposable batteries (e.g., lithium-based) for power. The platform might need to be equipped with a simple telemetry and telecommand capability to determine its health prior to recovery, but it probably wouldn't need much of a propulsion system except for a basic cold gas attitude control system to ensure it was in the proper attitude during recovery operations. Such inexpensive spacecraft designs are not new. Indeed, exactly such a platform exists: the Astro-SPAS built by DASA of Ottobrunn, Germany (Figure 20.5) [19]. Astro-SPAS was successfully launched and recovered by the Shuttle as planned in 1993. Three further flights are planned before the end of the century, subject to Shuttle availability. In the aero-space plane era, Astro-SPAS and other platforms like the U.S. SPARTAN platforms would probably fly more frequently than any other single type of spacecraft.

These are probably the simplest and most effective missions imaginable. The economics of such missions would clearly depend on the cost of the aero-space plane and whether a number of platforms share the same launch. As many as four 2 tonne platforms could be launched on one aero-space plane, giving a total *launch and recovery* cost of perhaps $3–5 million per platform, or about half that if the platforms remain in the payload bay. These costs are com-

**Figure 20.5** Astro-SPAS (*DASA*).

parable to that of a sounding rocket. Multiple manifesting of several independent platforms on one aero-space plane might be considered as a major constraint on this type of activity, as it is for the Shuttle. However, a user community attuned to the possibility of regularly scheduled aero-space plane flights would quickly adapt to sharing space—just as airliners carry many people and various mixed cargoes. Multiple manifesting will become a nonissue in the aero-space plane era, just as it is today with airliners.

## Phase 2: Hands-On Research

Flying autonomous platforms in space for short periods is probably not going to be sufficient for all research or pre-commercial activities. At some point, the ability to adjust or modify a particular research activity in situ will be necessary. Some experiments require periodic, real-time interaction, and others require supervision essential to eliminating the bugs from a commercial production facility, as well as to determine production rates and maintenance needs. A good example of the latter was demonstrated by McDonnell Douglas and its EOS equipment which required almost constant supervision throughout each flight.

It is possible to envisage crewed aero-space plane flights analogous to Spacelab or Spacehab missions, but simpler. Rather than spending in excess of $500 million on one Spacelab mission every few years, it might be possible to redistribute a similar amount of money to fund several cut-down and simplified Spacelab flights every year. Preparation times would be measured in months, as opposed to years with the Shuttle-Spacelab, and like the small platform examples defined for Phase 1, the pressurized crew module could also be placed within a standard canister and loaded into the aero-space plane like just another payload.

Using aero-space planes like small space stations might be an inefficient method of carrying out the required research. Most current aero-space plane concepts actively try to avoid such mission options, first, because of the additional costs associated with providing the systems needed to support long orbital stay times and second, because of the desire to turn around the aero-space plane as quickly as possible. A better option might be to construct small space stations where researchers and their experiments can be left for as long as necessary, allowing uninterrupted research. Is this realistic, given the high cost and problems being encountered with Freedom today? As Chapter 19 attempts to explain, if the aero-space planes come close to meeting the launch cost and performance goals envisaged, then an aero-space plane-launched space station is likely to bear little resemblance to Freedom. Potentially, the aero-space plane would facilitate the development of a space station that not only was readily accessible and, therefore, usable when needed, but also much cheaper and simpler to deploy and assemble than Freedom. The necessity of hands-on development work for future commercial operations might make the construction of such small space stations mandatory.

## Phase 3: The First Commercial Product

The combination of the initial basic research followed by hands-on development activities is probably a prerequisite in determining the viability of commercial materials processing in space. These research activities may reveal that even with all the capabilities of aero-space planes, the cost of accessing space may still be too high to allow truly self-supporting commercial activities. In addition, because these activities are unlikely to begin much before 2010, advancements in ground processing techniques may further reduce the potential for materials processing in space. By the same token, the next 20 years may also reveal many new products that would benefit from in-space processing.

Assuming that there are suitable products, whether semiconductors, pharmaceuticals, or other advanced materials, it is important to appreciate the technical and economic barriers standing in the way of the first commercial product made in space. Intuitively, it might be expected that the most economic means of making products in space is for the processing facility to remain in orbit with only the raw material and finished products transported to and from space. This technique allows the amount of excess mass launched and returned to be minimized, thereby lowering transportation costs. One of the original goals of Freedom was to lease rack space on the station for just such a purpose. Such an opportunity is also foreseeable for the aero-space plane era when space stations are likely to become less expensive and more easily accessible. Using the government-funded infrastructure in support of the commercial sector might be the only way the first commercial product in space can be made, just as has been true on the ground.

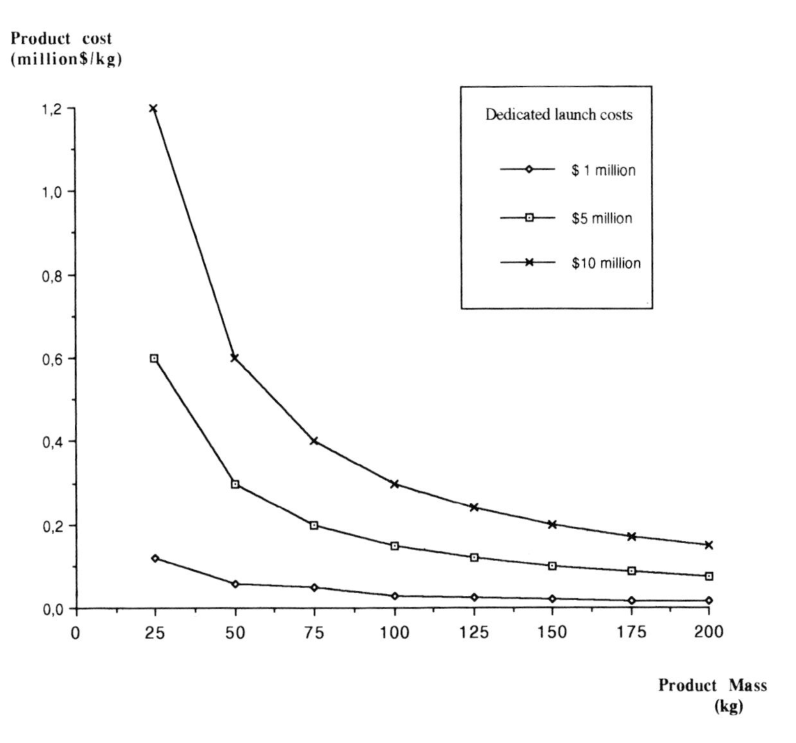


**Figure 20.6** Economics of single-product space platforms.

This scenario is all well and good provided the products can tolerate disturbances induced by the laboratory functions of a space station, as well as the frequent dockings of aero-space planes and other servicing vehicles. If a company has to go to the length—and considerable expense—of manufacturing a product in space, then that product will almost certainly demand the lowest possible microgravity and/or vacuum conditions achievable. If only small quantities of a high value product are being manufactured, ideal conditions are likely to be important for maximum product yield and purity.

Alternatives to space stations may be inevitable. To minimize costs, a dedicated space materials production facility might have to share with many diferent users. Such a facility might consist of one or two pressurized modules stacked together to form a compact man-tended space station. Then, every 1 or 2 months, an aero-space plane would rendezvous with this facility, and replace the old logistics module, containing processed materials and waste products, with a new module full of raw materials, essential resupplies, and spare parts. Any required maintenance would also be undertaken assuming that particular aero-space plane flight was crewed. This describes almost precisely what the Industrial Space Facility (ISF) was intended to be [20].

At this early stage, the business will be searching for that elusive first product to break the ice. Clearly, the commercial funding of an ISF-type facility—still a relatively expensive undertaking in the aero-space plane era—will require solid proof that not only is one product commercially viable, but so are a large number of products. The demand for these products must be sufficiently robust to survive the changes in the marketplace. One product emerging first seems more credible than ten products emerging simultaneously.

An intermediate step for production may require a return to the Phase 1 method of conducting research. Here, a single, small platform would be launched and recovered as necessary, and dedicated to the manufacture of just one product. There are a number of operational advantages to this approach; the most important is that the entire production facility is returned to Earth after every mission for servicing and maintenance. Another advantage is that existing government-developed platforms can be leased, avoiding the need to recoup development costs. Astro-SPAS is perfectly adequate for this type of mission.

At this stage, it is worthwhile to look at the economics of small platform space processing. For argument's sake, it will be assumed that launch costs consume only one-third of the total cost of each mission, and the remaining two-thirds is associated with ground activities, insurance, personnel, development cost amortization, etc. Other overhead costs are obviously dependent on the type of company running the operation. Using these assumptions, the cost per kilogram of processed material is shown in Figure 20.6 in relationship to the amount of products manufactured per flight. This analysis was performed for three user launch costs, irrespective of whether the launch is shared or dedicated, and shows that once the quantity of the material processed exceeds 100 kg, the actual magnitude of the cost differences becomes less significant. This simple analysis shouldn't be taken too literally. However, it indicates that commercial materials processing in space may become attractive if a market can be found for products worth around $200,000 per kilogram, and where the demand for such products is on the order of several hundred kilograms per year. This might be within the range of advanced semiconductors and pharmaceutical products.

By comparison, if an identical mission were flown on the Shuttle today, assuming Shuttle availability and launch costs of around $100 million ($50 million up and down), then the cost per kilogram would be about *$3 million*. There are few products, if any, in this price category. More remarkably, cost isn't the primary inhibitor to using the Shuttle for commercial materials processing, as discussed earlier. This is just one reason of many why commercial materials processing using the Shuttle cannot succeed.

## Phase 4: The First Factories in Space

The first product made in space and sold at a profit may well be a watershed event in the history of spaceflight. This statement is not made lightly, and certainly doesn't mean that future space programs should live or die based entirely on the success of commercial materials processing. But soon after the first space manufactured product makes money, other companies will join this fledging industry, assuming there is more than one viable product. Says John Cassanto, "I wouldn't be in this business if I thought there would only be one or two products. All that it will take is one breakthrough, and then I think the business will snowball from there. Aero-space planes might make all the difference in the world."

When the number of new products reaches five or ten, then the commercial viability of large ISF-like platforms will become clearer. Their deployment would enable the costs of space manufacturing to be significantly reduced compared with using small platforms. How significant this reduction is will obviously depend on a number of factors, including the number of users sharing the same facility, its development cost, the time needed to recoup the investment, and the rate at which it is serviced and resupplied with raw materials. A simple example of the economics of an ISF-type facility is shown in Figure 20.7.

This single-point analysis might be considered as relatively optimistic, especially in terms of the average amount of user material processed each year (1,800 kg of raw materials, giving a yield of 600 kg) and a dedicated launch cost of $5 million per flight. Further, the ground operations costs per user were assumed to be about equal to the launch cost. This compares with the simple platform version where ground operations costs were twice that of the launch costs in order to reflect the need to service the spacecraft on the ground. However, this difference is made up for by the cost the user must pay for orbital servicing operations. Therefore, about 60% of the cost each user is charged to launch raw materials goes toward orbital servicing operations [21]. Given these assumptions, if an average of 600 kg of product is manufactured each year, the cost per kilogram would be around $18,000, approximately an order of magnitude less than that possible with aero-space plane-launched single-product platforms. (For comparison, using the shuttle would cost about $500,000 per kilogram.)

| **Facility Lease Costs** | |
|---|---|
| Life-cycle cost | $500 million |
| Payback period | 10 years |
| No. of users | 10 |
| *Hence, annual lease cost per user* | *$5 million* |
| **Launch & Servicing Costs** | |
| Dedicated launch cost | $5 million |
| Aero-space plane launch capacity | 8 tonnes |
| Raw material per launch (300 kg per user) | 3 tonnes |
| Mass for logistics support of facility | 5 tonnes |
| *Hence, materials/logistics cost per kg of raw material* | *$1,700 per kg* |
| **Annual Costs Per User** | |
| Number of visits per year | 6 |
| Total raw mass launched/year/user | 1,800 kg |
| Ground costs | equal to launch costs |
| *Hence, total annual cost per user* (incl. facility lease cost) (i.e., $1,700 × 1,800 × 2 + $5 million) | *$11 million* |
| **Therefore, cost per kilogram of processed material $18,000** (assuming a 33% yield or 600 kg per year) | |

**Figure 20.7** Economics of an ISF-type facility.

Returning to the aero-space plane example, a more thorough analysis for showing the cost per kilogram of product in relationship to the dedicated launch costs is shown in Figure 20.8. As this figure demonstrates, once the amount of product exceeds around 500 kilograms per year, the cost per kilogram starts to level out below $35,000 per kilogram and declines to as low as $5,000 per kilogram with a sixfold increase in production—about one-half the price of gold. A large number of suitable materials are already within this cost and mass range. This discussion demonstrates the value of investing in aero-space plane configurations that are most likely to achieve the lowest possible dedicated launch costs.

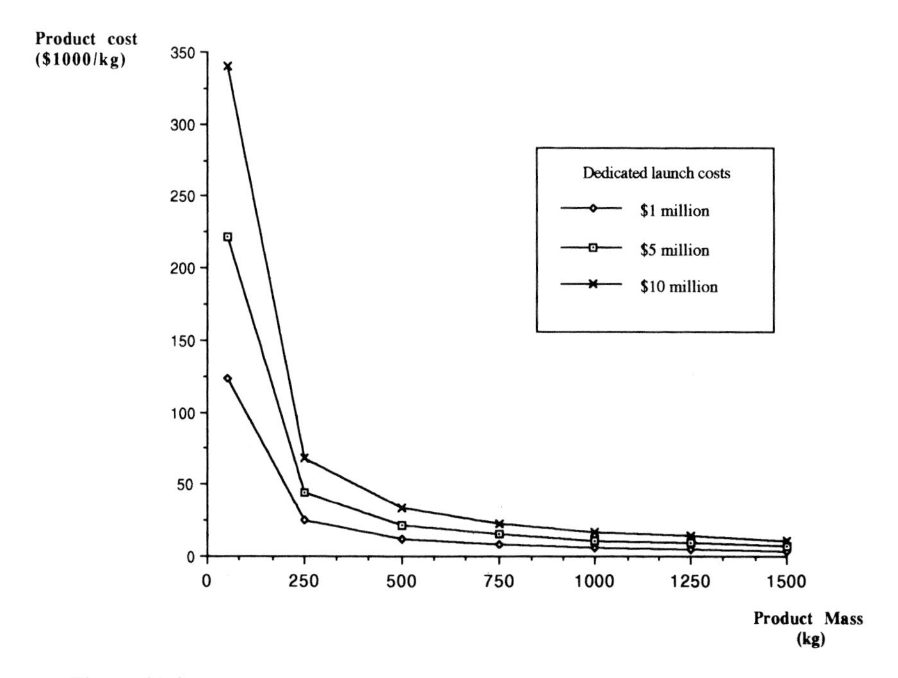


**Figure 20.8** Economics of multi-product commercial manufacturing facilities.

## Expansion, Growth, and a Market-Driven Space Infrastructure

The scenarios outlined above might appear optimistic. Yet, is it really possible to say what is or isn't optimistic when there is nothing sensible to compare it with? At the present, space manufacturing doesn't have the slightest chance of being the boon many people have hoped for because it is impossible to perform the vast amount of research inexpensively, frequently, and over a short time-scale. Plus, as a secondary issue, it costs a fortune just to transport materials to and from space. Clearly, a fundamental improvement in accessing space is required before anyone can gauge whether the above scenarios are optimistic or otherwise.

Having said that, it is worthwhile exploring what might happen if launch costs and ease of accessing space were to improve to the point where the cost to make products falls in the region of a few thousand dollars per kilogram. This would be equivalent to a dedicated aero-space plane launch cost of about $1 million. The following scenario makes only one major assumption: when launch costs are reduced to the point where the *first* fully self-supporting commercial product emerges, demand for space access will be elastic in response to further launch cost reductions. (See Chapter 17.)

### Space Transportation

Increased demand for space access naturally leads to an increase in the launch rate of the first aero-space plane fleet which, in turn, reduces launch costs. As demand continues to rise, a second or third aero-space plane fleet emerges. Competition for commercial launch capacity forces each fleet to strive for lower and lower launch costs by adopting higher-efficiency operational practices. Overall the benefits to the user—whether commercial, civil or military—are reduced launch costs and greater on-demand access to space. This itself leads to further increases in demand, with all the positive impacts that entails. Plus, two or three independent fleets provide the necessary launch redundancy, ensuring practically continuous and uninterrupted access to and from space—an almost mandatory requirement for all commerce, whether in space or on the ground.

### In-Orbit Infrastructure

Increased demand also requires an increase in the number and capability of in-orbit facilities to perform this work. The more these facilities can be shared, the lower the cost to use them becomes. In other words, the initial government-financed space infrastructure is expanded by the growth of a market-driven infrastructure. At this point, within the context of commercial activities, arguments over the value of space stations or debates on whether a mission should be automatic or crewed are irrelevant because the decision to select a specific mission configuration is made on the basis of cost.

Gradually, as the demand for space access grows, the in-orbit infrastructure also grows to the point where there are a number of small, commercially supported platforms in low Earth orbit in the Astro-SPAS and ISF class. It is clearly impractical for a commercial organization to privately finance a permanently staffed space station, even though hands-on research is vital. A more practical measure is for the initial government-financed space station(s) to be expanded with private modules optimized to provide efficient facilities for new commercial research activities. This is something Spacehab has already proposed.

Further significant reductions in launch costs and facility lease costs reduce the cost of producing materials to the region of hundreds of dollars per kilogram. As a result, instead of a company producing relatively small (hundreds of kilograms) amounts of materials every year, a transition is made to the bulk production of large quantities of materials. Foam metal is one intriguing possibility that is only economic if many tens of tonnes can be produced each year. If such a situation materializes, then the future growth of the space industry is secured.

If the initial assumptions are correct, what the above scenario has attempted to describe is the evolution away from the current government-controlled and highly institutionalized space program, toward a space program that gradually becomes self-supporting, leaving the government to do what it does best, namely speculative R&D and science.

## *The Industrialization of Space*

A commercialized, low-cost means of accessing space, coupled with an efficient means to undertake space operations, would probably allow programs like the exploration of the Moon and Mars (Chapter 19) to be conducted at lower global costs because much of the required infrastructure would already be in place. Current solar system exploration plans have called for an Apollo-style straight to the Moon and back approach rather than the alternative of re-using the landers and other vehicles many times. This approach has only been adopted because current launch systems are incapable of supporting the kind of robust servicing facilities needed to handle lunar landers and space transfer vehicles. In an aero-space plane future where there are already a large number of commercial facilities in low Earth orbit, each requiring routine, resilient, and inexpensive support for servicing, maintenance, and resupply, re-using lunar transportation vehicles will not be the unknown it is today. In this light, the first crewed mission back to the Moon might begin with a launch on a commercially developed and operated aero-space plane, followed by use of a privately operated space station as a staging point.

How much the first missions back to the Moon would cost if they could rely on a commercial infrastructure is impossible to estimate, except that there is good reason to believe that the savings would be significant. Opening the technical and economic door to the Moon might also, many decades from now, open up an opportunity for the wide-scale industrialization of space. How might this occur? A realistic proposition is for early lunar sorties to establish a liquid oxygen production plant on the Moon. Dependence on Earth for future lunar sorties would then be cut drastically, as oxygen is by far the heaviest single component of a trip to the Moon. Even the lightweight hydrogen fuel, something apparently very rare on the Moon, could come from the byproducts of many commercial activities (e.g., hydrolyzing wastewater). Once this situation is reached, transportation to and from the Moon can be conducted autonomously of Earth.

Freeing access to the Moon from Earth would further reduce costs of lunar missions considerably, or, alternatively, allow more missions to the Moon for the same amount of money spent prior to the establishment of the first lunar oxygen plant. A continuing reduction in the cost per lunar mission coupled with an accelerating sortie rate to the Moon is vital to the initial establishment and expansion of the first lunar infrastructure—necessary if the Moon is ever to be used as a source of raw materials for large-scale industrialization of space.

## *The Future*

Whether lunar resources can ever be economically supplied to Earth—even with second or third generation aero-space planes—is highly debatable. And whether a near-term use can be foreseen for extraterrestrial materials is also a question without a clear-cut answer. Certainly, the possibilities of using lunar resources for constructing solar power satellites able to beam terrawatts of electricity to Earth, or scraping He3 off the lunar surface to fuel fusion reactors, have received considerable attention [22]. The He3 prospect, in particular, is extremely interesting (Figure 20.9). However, it is asking a lot to hinge hopes for the wide-scale industrialization of space on these two possibilities alone. If technology has advanced to the point where such propositions become economically feasible, technology will probably also have advanced in other areas to the point where fusion reactors can be built and use terrestrial-supplied fuels such as hydrogen. In such a scenario, the potential of solar power satellites or lunar He3 might be considerably weakened.

Yet, many turning points have changed the course of human history—the discovery of fire, iron, bronze. By the time humans are in a position to truly discover the Moon and other solar system bodies, our needs as a species might have changed profoundly. No matter how optimistic the calculations of economic growth, societal restructuring, slowing in population growth, reductions in pollution, conservation of energy, stabilization of the greenhouse effect, and shrinkage of ozone holes, there will be a time when the Earth will just not be capable of sustaining the continued advancement of humanity. And the key word here is *ad-*

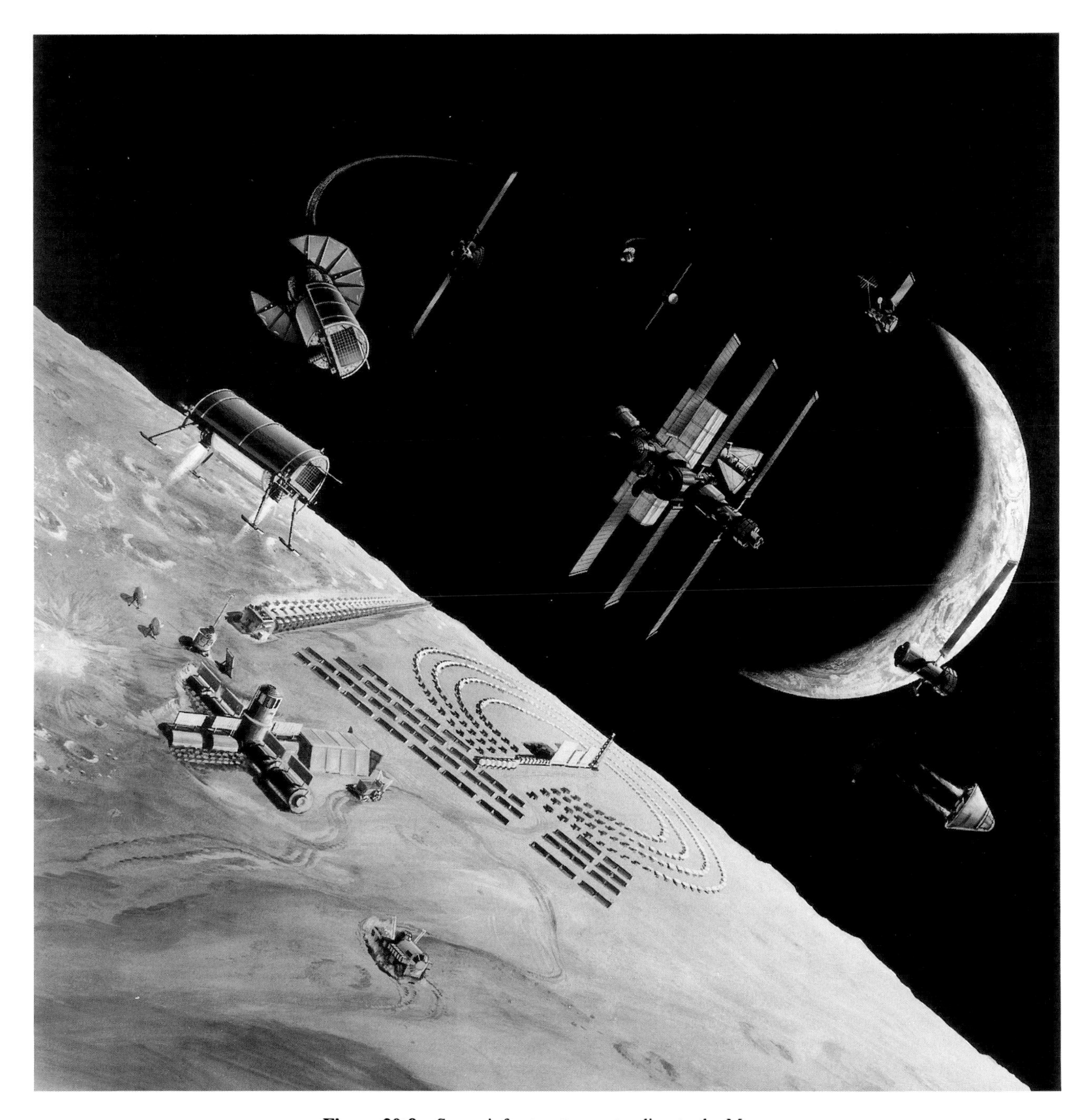

**Figure 20.9** Space infrastructure extending to the Moon.

*vancement*. It can be argued that the Earth is capable of sustaining a select few indefinitely at the expense of everyone else if draconian measures are enforced. Perhaps—but is this stagnating future in character with the history of human achievement, and is it something we really want to look forward to? Moving off the planet and using space as a means of maintaining the continued advancement of humanity might be an alternative worth pursuing.

If we are prepared to look that far into the future, then deciding to build the first aero-space place might well be the single most important action humanity will ever take.

## Summary—Reaching the Profitability Threshold

Many assumptions have been made in this chapter and conclusions have been deduced that may seem, at first glance, to violate the current laws of the space industry. We have gone from the despair inflicted by the failure of commercialization of space to take off in the way many anticipated. The reason for this can be tied directly to the fact that the Space Shuttle that was promised in the early 1970's turned out to be something it really shouldn't have been. It is clear that for long as access to space

- remains as limited and restricted as it is today,
- is frequently subjected to lengthy delays and uncertain launch dates,
- provides only a handful of opportunities to return from space, and
- continues to remain as hopelessly unaffordable as it is,

the commercialization and industrialization of space cannot even be considered to have potential. *It simply will not happen.*

Small decreases in launch costs and small increases in launch opportunities will have little impact on this situation, so far is the threshold of profitability below the best that can be offered with the current generation of launchers today. One or two orders of magnitude reductions in launch costs to and from space, coupled with a similar one or two orders of magnitude increase in launch rates and reliability, might be required just to allow the necessary research to be performed to determine whether the potential of commercial space can be transformed into a self-supporting and profit-making reality.

As the Report of the National Commission on Space states, [23]

The Commission believes that cheaper, more reliable means of transporting both people and cargo to and from orbit must be achieved in the next 20 years. While all space programs would benefit from lower cost orbital transportation, it is especially important that the cost be dramatically reduced for free enterprise to flourish with commercialization of space operations. The Commission is confident that the cost of transportation can and should be reduced below $200 per pound (in 1986 dollars) by the year 2000. If the volume of cargo increases in the early 21st century as it is projected to do, further cost reductions should be achieved.

And in the vision of the Synthesis Group, [24]

Space is a limitless, untapped source of materials and energy, awaiting industrial development for the benefit of humanity. *Commercial products, such as zero gravity derived materials, and service industries, like advanced communications, all become increasingly feasible and profitable once routine, reliable and affordable access to space is available.*

Once the threshold for true commercialization has been crossed, the door may open for new opportunities which would, in turn, further improve access to space in response to increased demand. And continued growth may eventually lead to the wide-scale industrialization of space, where the cost of doing things in space gradually converges with the cost of doing things on the ground. When such a point is reached, debates over whether we really need a space station, or whether money spent in space can be more wisely spent on Earth, will become more or less irrelevant. In the longer term, the possibilities for continuing human socioeconomic advancements may just be as unlimited as space itself.

This is an extraordinary vision of the future, very much in line with everything that has happened on Earth. **Yet none of this will be possible without first having the means to traverse freely the 100 or so kilometers that separate the Earth from space.**

This is the promise of commercial space enabled by aero-space planes.

## References and Footnotes

1. For clarity, it is important to define what is meant by commercialization and its partner privatization. Essentially, *commercialization* is where a company can be sustained essentially by the needs of a non-government market, although the government can still be customers to a certain extent. Such companies are initially privately financed, established with intent to make a profit, and the investors take the risk. *Privatization*, by contrast, is where the government hands over the title, deed, and assets of a company to the private sector. The privatized company then re-sells to government at a lower cost, and only a small fraction, if any, of the business is commercial sales.
2. Despite the proven ability of centralized authorities to develop advanced technology, it is worthless without the opportunity to exploit it.
3. Logsdon, T., "Space Inc.," *Business/Science*, 1984.
4. Covault, C., "Unique Products, New Technology Spawn Space Business," *Aviation Week & Space Technology*, June 25, 1984, pp. 40–201.
5. Johnson-Freese, J., *Changing Patterns of International Cooperation in Space*, (Orbit Book Company, Malabar, Florida, 1990), p. 46.
6. "Space Business Indicators," *Office of Space Commerce*, U.S. Department of Commerce, June 1991, p. 5.
7. Saunders, R., "Newsmaker Forum," *Space News*, January 13–26, 1992, p. 30. EOSAT is contributing $25 million to the construction of Landsat 6. With permission of *Space News*.
8. "Space Business Indicators," p. 15.
9. "EOS: A New Era Dawns in Space," McDonnell Douglas, 1983. See also, "Medicine Sales Forecast At $1 Billion," *Aviation Week & Space Technology*, June 25, 1984, p. 52.
10. From a letter to John Cassanto, January 20, 1992.
11. Discussions with Keith Hindley. See also *Interavia Space Directory 1991–92*, Andrew Wilson, ed., (Jane's Information Group, Coulsdon, Surrey, England, 1991), p. 132.
12. Covault, C., "McDonnell Douglas, 3M Join to Produce Blood Drug in Space," *Aviation Week & Space Technology*, November 18, 1985, p. 16.
13. Marcus, D., "Financiers, Entrepreneurs Far Apart on View of Space," *Space News*, January 13–26, 1992, p. 1. With permission of *Space News*.
14. See, for example, *Interavia Space Directory 1991–92*, Andrew Wilson, ed., (Jane's Information Group, Coulsdon, Surrey, England, 1991), p. 514.
15. Ibid., p. 510
16. "3M Seeks New Materials, Processes," *Aviation Week & Space Technology*, November 18, 1985, p. 65.
17. Parkinson, R. C., and R. Longstaff, "The United Kingdom Perspective

on the Applications of Aerospace Planes in the 21st Century," AIAA-91–5084, AIAA *Third International Aerospace Planes Conference*, Orlando, Florida, December 3–5, 1991, p. 3.

18. NASA, *Accessing Space*, NP-118, NASA Office of Commercial Programs, Washington, D.C., 1991.
19. Brochure supplied by DASA, Ottobrunn, Germany.
20. "Space Industries, Inc., to Begin Marketing Unmanned Facility," *Aviation Week & Space Technology*, June 25, 1984, p. 116.
21. Of the $1,700/kg launch cost, $1,000/kg is dedicated to servicing charges, while the remaining $700/kg is the cost of actually launching the raw material.
22. Synthesis Group, "America at the Threshold," U.S. Government Printing Office, Washington, D.C., May 1991, pp. 52–53.
23. *Pioneering the Space Frontier*, National Commission on Space, (Bantam Books, Inc., New York, May 1986), p. 109.
24. "America at the Threshold," p. 2.

# Conclusions:

# Changing Our Future in Space

Spaceflight of the next century has the potential to be profoundly different from today. It is possible to envisage a diverse range of activities: swarms of spacecraft continuously monitoring the environment of Earth; multiple space factories churning out exotic new materials; massive solar power satellites beaming "clean" energy to Earth; roving vehicles mining He3 on the lunar surface for use in terrestrial fusion power stations; even outposts on Mars. All of these and other activities would be justified on their own merits, many would be self-supporting, and all would contribute directly to the growth of the world's economy and culture. It is even conceivable to envisage a time when space itself becomes just an extension of Earth, where a large percentage of the population will spend most or all of their lives. By this time, humans will have truly become a space-faring species. Star Trek will have finally arrived . . .

To talk in such terms can be hazardous at the best of times, and the reasons are easy to comprehend. It is difficult to imagine mines on the Moon or outposts on Mars while the reality of the 1990s suggests that building even a basic space station is complicated, takes a long time, and is exceedingly expensive. This is compounded by an acute awareness of the budgetary pressures all governments are under. Therefore, to expect governments to drastically increase space spending to accommodate grandiose ventures might be considered naïve. It is difficult enough to justify today's enormously expensive programs like Freedom and Columbus, considering that unambiguous functional requirements cannot even be given.

If space is to become just another domain for human activity, then clearly something fundamental must change. Management reorganizations, wider international cooperation, and stronger political leadership can play a part in bringing about limited change, but they only offer cosmetic improvements. Indeed, support for spaceflight, and especially human spaceflight, may actually wane due to frustration with the perceived low return on investment and as the political justification that gave birth to the space program becomes more and more irrelevant. To expect future large-scale space activities to be justified on the vague promises of technology spinoffs, educational benefits, jobs, and tools for peace—or to expect the public's enthusiasm for space to come to the rescue—is to severely misread reality. Space activities, like all other major terrestrial endeavors, should be fully accountable and achieve rational objectives on a par with the cost.

Until space organizations are prepared to tackle the root of the problem, space activities will never have the chance to grow much beyond current limits. *Transportation* to and from space is at the very core of the problem. Access is the key to space, just as transportation on Earth has been the key to economic and cultural development. Transportation is the one common factor that underpins all space activities. Put simply, if space becomes significantly easier and less expensive to access, then it follows that space activities will similarly become easier, less expensive and more diverse.

*Building aero-space planes is mandatory to realizing at least the first stages of human expansion into space. Aero-space planes will not guarantee the growth of space activities, but significant, self-sustaining growth simply will not occur without them.*

To many observers outside the space industry, the above statements may appear obvious. Yet, why is it that no organization has yet committed funds to building an orbital aero-space plane of any type? If the potential of aero-space planes is as great as many people claim, wouldn't it be reasonable to expect that space agencies and companies would be eagerly pursuing the development of such a capability? Vehicles like the proposed Delta Clipper or Interim-HOTOL, should they be feasible, would be able to make around 50 regularly scheduled flights each year to and from space for about the cost of two Shuttle flights. Low-cost, weekly flights to space would have a dramatic effect on how missions are actually conducted, perhaps reaching the threshold at which new, self-supporting commercial activities could emerge and thrive.

It might be expected that the potential advantages of aero-space planes would be enough reason to encourage their development. Yet the reality is that NASA and the DOD are only investing modest amounts into various aero-space plane technologies, ESA is about to begin a modest technology program for future launch systems, and the programs of other countries are still at basic research levels. Only the SSRT program and McDonnell Douglas seem to offer some reason for optimism. Notably, the DC-X effort has consumed the equivalent cost of launching 3 tonnes on the Space Shuttle.

To supporters of aero-space planes, their potential is compelling. Aero-space planes offer a profound and perhaps even revolutionary improvement in space transportation. Yet, such a radical improvement in space transportation goes against the grain of an industry which has known no alternative other than expensive launch systems with highly restrictive capabilities. It also flies in the face of an industry stung by the false promise of the Space Shuttle. Against this background, the prevailing position taken by most space authorities is to adopt policies that go with what is known—invariably expendable rockets using tried and tested technology. Unfortunately, all that such policies will do is reduce launch costs by small amounts, enough to make them competitive with existing rockets carrying the existing demand for commercial satellites. Beyond this, they will do little else, particularly where human spaceflight is concerned.

Expendable rockets have served the needs of the space industry well up to now. But continuing along a path of modest, incremental improvements to expendable launchers does not seem to be the best route for the future of space now that advances in technology appear to have enabled a real alternative for the first time. Obviously, if space organizations are content with today's level of activity, it might be argued that current launch systems can suffice. That level of activity will focus almost exclusively on launching one-shot, autonomous satellites. Although such missions are important, space has far greater potential. Moreover, to suggest that space activities should never advance much beyond today's modest levels would be out of character with the progress of human development.

A look at terrestrial transportation shows why pursuing reusable aero-space planes is ultimately the proper path to follow. Today, not one single form of transportation on Earth utilizes a large, expendable vehicle to move people and cargo from point A to point B, let alone back. There would obviously be little demand for air or ship services if every trip cost millions of dollars, and if only a handful of frequently delayed voyages were made each year. Further, there would be even loss demand if the aircraft or ship were used just once and then discarded, if it were incapable of returning to its original destination, if it could not be tested at all before carrying cargo or passengers, or if it were incapable of safely aborting the voyage after even a modest failure. Although terrestrial transportation has not been like this since perhaps the time of the Polynesian canoes, from a *functional* perspective current launch systems epitomize this scenario completely.

The technical demands on space transportation are, of course, far more challenging than any form of terrestrial transportation. But as far as the user is concerned, all that matters is how cost-effective and user-friendly the transportation *service* actually is. To the user, *function* matters, not the difficulty in achieving the service.

All reusable transportation systems are inherently capable of being incrementally tested in a manner that demonstrates reliability and safety. This, combined with the fact that such vehicles do not have to be reassembled and checked out before each use, reduces turnaround time, thereby reducing costs and increasing frequency of use. From this perspective, aero-space planes have all the hallmarks of true transportation systems. Almost certainly, the first operational aero-space plane will fly more times in its test program than most current launch systems will fly in a decade.

*The ability to perform thorough shakedown testing of any transportation system is fundamental to achieving a dependable, reliable, and low-cost service. Perhaps the best determinant of the prospects of any future space transportation system is its ability to be incrementally tested.*

The inherent advantages of reusable, incrementally testable, and recoverable aero-space planes seem to stand up to scrutiny. Yet, these reasons alone do not appear sufficient to stir space organizations to build aero-space planes. Perhaps the explanation is the belief that such vehicles cannot be built. Even if aero-space planes could be made to work, some people find it difficult to accept that such complex machines could realize low launch costs. With all the inherent risks and tight margins, some argue strongly that aero-space planes should not be built. Unquestionably, aero-space planes present tough technical challenges, and no one knows for sure whether they are feasible or if they will ever meet the hoped-for performance. This is not an excuse for failing to try. Paper studies alone will never convince anybody.

Few people, of course, are foolish enough to say categorically that aero-space planes can *never* be built. Most have an agnostic viewpoint, hedging their bets by accepting the inevitability of aero-space planes—but only 50 years into the future. This seems to be a widespread attitude, and for as long as it is the case, aero-space planes will always remain 50 years into the future. This is an unfortunate situation, given that aero-space planes would profoundly benefit today's relatively limited space activities. Just because aero-space planes may seem futuristic does not mean they must be reserved for futuristic space missions. Launching a dozen communications satellites along with a few scientific, Earth observation, and technology demonstration payloads would be enough to keep the first aero-space plane fleet fully occupied and economically justified.

Although no aero-space plane is under construction, considerable interest exists in them around the world. Since this interest extends only as far as modest technology research programs, it would seem that the case for aero-space planes has not been properly made. Ironically, it is the aero-space plane community itself that seems responsible for not properly selling the idea. To obtain support, some aero-space plane projects have been touted to the public and politicians as high-speed airliners capable of "flying to Tokyo in 2 hours." The NASP and Sänger concepts in particular have been dogged by this throughout their short history. Another way of obtaining support has been to promote aero-space planes as generators of new technology, leading to spinoffs in other sectors. Clearly, there are better ways to benefit the automobile industry, for one example, than building a $10 billion experimental aero-space plane.

The sellers of aero-space planes seem, in some sense, to have severely underestimated the intelligence of politicians, decision makers, and especially the public. As a consequence, a number of programs have obtained modest startup funds based on peripheral justifications that put to one side the true purpose. This compromises long-term prospects because such nebulous rationale may not, in the eyes of decision makers, be enough to warrant the ultimate development of a multibillion dollar vehicle. Instead of clouding the issue, aero-space plane proponents will benefit from focussing precisely and honestly on the real justifications for building these vehicles. If benefits are realized in side areas, then all the better.

*However complicated the issues may seem, decision makers should be given credit for being able to understand the benefits of aero-space planes. A real need exists for a true space transportation system. This should be enough to sell the basic concept.*

The aero-space plane proponents should also show responsibility in their choice of the technical solution they would like to pursue. There is a tendency for some organizations to believe that they have the right type of vehicle design and that everyone else's solution is inadequate. Discrediting the opposition's efforts can be divisive when support for aero-space planes is still relatively modest. This is especially the case as each organization is ultimately attempting to achieve the same basic objectives. Although it is not necessary to pool resources around a single concept, attempting to destroy the credibility of another organization's efforts will discredit the support base for everyone, delaying the day when the first aero-space plane takes flight. It is better to have a "good enough" aero-space plane flying than none at all.

At this point, no one knows what will be the optimum technical solution because no one has attempted to build an aero-space plane yet. A reasonable approach might be to pursue solutions that are as simple as possible. The first aero-space planes should avoid having to use unproven and complex technology piled on even more unproven and complex technology—all within a highly integrated vehicle. This does not mean simpler solutions will necessarily be feasible, only that if they are, such solutions could be in service sooner and at less cost than their more complex counterparts. That would be good enough to begin with.

Many other concerns must be addressed before the first aero-space plane can hurtle down the runway or lift off from its pedestal. None are considered insurmountable, but nevertheless all must be tackled with diligence and care. Of particular importance is who should pay for building and operating aero-space planes. Ideally, to provide the user with the best possible service, the ultimate goal must be commercial operation of aero-space plane fleets. It is unrealistic to expect the commercial community to take all the technical risk while picking up the tab for the development costs. There needs to be a partnership between government and industry. At the minimum, governments need to spend enough money on an experimental or prototype vehicle to prove conclusively that aero-space planes are feasible and can be turned around rapidly and at demonstratively low cost. Once the risk has been essentially eliminated, the commercial sector can step in and take over.

Assuming all the challenges are met, when the first aero-space plane flies it will change space programs irrevocably. This change may come automatically, without the sweeping cultural transition of the space industry many observers contend must occur first. Initially, the first effect of aero-space planes will be to save money on launch costs. Then, and as soon as a competing fleet enters service and current launchers have been phased out, the ways in which space missions are conceived and payloads are designed will change. As a result, not only will launch cost savings be realized, but similar savings in the cost of payloads and missions will be obtained. This will free up funds for a much broader range of lower-cost missions, flown regularly and at shorter intervals. Overall, the net effect will be greatly increased cost-effectiveness in space operations, and more rapid development of innovative technology—a situation more akin to Earth practices.

After the initial transition to aero-space planes, a more profound change will occur. If dedicated launch costs can be drastically reduced—by at least a factor of 100—this may just create the conditions needed for the emergence of a diverse range of new commercial space activities. Today, the high cost and complexity of accessing space stands in the way of wide-scale commercial space operations. As a result, the space industry as a whole has been essentially devoid of the positive stimulus commercial activities bring. Potentially, the aero-space plane era will see many more space activities being undertaken because the economic benefits can justify the cost. As a result, debates over whether we should build a crewed space station or unmanned space platforms will become irrelevant. Gradually, each piece of the *space infrastructure* will emerge because it brings about the most competitive service for the user. By this point, space will have shrugged off the stigma of being largely a political tool, lacking direction or a real purpose. Space will have become integrated into the daily activities of Earth, contributing positively to economic growth and development.

In the longer term, further reductions in the cost of accessing space may provide the opportunity to clean up Earth. Environment protection can go hand in hand with sustained development provided our closed terrestrial system is given a relief valve—space. To talk in such terms today is not considered fashionable. To suggest that space can become just another domain for human occupancy and exploitation is not considered a serious or plausible option. Yet, the only real reason for this is because today's ruinously expensive "space or bust" launch systems firmly put such a future in the realm of the inconceivable. But what if aero-space planes were able to fly to and from space with all the economy and regularity of airliners today? Would our perception of the future in space change? Would we be able to conceive the many possibilities for human expansion and growth? Indeed, would this happen automatically as entrepreneurs seek out new commercial opportunities?

Although aero-space planes do not need this degree of speculation to justify their development, if this is the type of future we would like to see, there is no other choice than to build aero-space planes. Access is the key to space, and, in the long term, aero-space planes may well be the key to the future survival of planet Earth.

# Index